Springer Proceedings in Mathematics & Statistics

Volume 153

More information about this series at http://www.springer.com/series/10533

Springer Proceedings in Mathematics & Statistics

This book series features volumes composed of select contributions from workshops and conferences in all areas of current research in mathematics and statistics, including OR and optimization. In addition to an overall evaluation of the interest, scientific quality, and timeliness of each proposal at the hands of the publisher, individual contributions are all refereed to the high quality standards of leading journals in the field. Thus, this series provides the research community with well-edited, authoritative reports on developments in the most exciting areas of mathematical and statistical research today.

Slawomir Koziel • Leifur Leifsson
Xin-She Yang
Editors

Simulation-Driven Modeling and Optimization

ASDOM, Reykjavik, August 2014

Springer

Editors
Slawomir Koziel
Engineering Optimization
& Modeling Center
Reykjavik University
Reykjavik, Iceland

Leifur Leifsson
Department of Aerospace Engineering
College of Engineering
Iowa State University
Ames, Iowa, USA

Xin-She Yang
School of Science and Technology
Middlesex University
London, United Kingdom

ISSN 2194-1009 ISSN 2194-1017 (electronic)
Springer Proceedings in Mathematics & Statistics
ISBN 978-3-319-27515-4 ISBN 978-3-319-27517-8 (eBook)
DOI 10.1007/978-3-319-27517-8

Library of Congress Control Number: 2015960792

Springer Cham Heidelberg New York Dordrecht London

Printed on acid-free paper

Springer International Publishing AG Switzerland is part of Springer Science+Business Media (www.springer.com)

Preface

Accurate modeling and simulation of complex systems often necessitates highly sophisticated but very computationally expensive models, and the costs and time of computational simulations can pose challenging issues in many applications. In addition, the search for optimum designs requires multiple simulation-design cycles, often from hundreds to thousands of evaluations of design objectives, which makes modeling and optimization tasks extremely challenging and time-consuming. Among these challenges, a serious bottleneck for realizing an efficient design optimization process is the high cost of computer simulations with simulation times varying from hours to weeks or even months for large complex systems, which means that even a single set of simulations can be very costly. For a typical design process, different design options may require many validations and verifications using computer models in order to test "what-if" scenarios so as to provide decision-makers with robust, realistic design options. Though the speed of the computer power has steadily increased over past decades, however, such speed increase can slightly ease only part of the modeling and simulation problems, and these challenging issues still remain largely unresolved. One of the reasons is the ever-increasing demand of the high-accuracy, high-fidelity models for simulating complex systems.

In addition, the search for more sustainable optimum designs makes such simulation tasks more challenging to solve. Even with a good simulation model, the number of evaluations and validations of design objectives can be very high, varying from several hundreds to thousands or even millions of objective calls. In many cases, such optimization problems can be NP-hard, and thus no efficient algorithms exist. Therefore, some tradeoff is needed in practice to balance the solution accuracy and practicality of simulation and designs.

Furthermore, for many applications in aerospace engineering, microwave engineering, gas transport networks, waste management, and system engineering, alternative methods or approximations methods are often used. A class of approximation techniques and alternative approaches are the surrogate-based modeling, simulation-driven design optimization, and metaheuristic optimization methods. These surrogate-based and simulation-driven approaches provide the possibility of using faster and cheaper surrogate models to represent expensive models with adequate accuracy and sufficiently reduced simulation times, which consequently makes many complex design optimization tasks solvable and achievable in practice.

Among all the key issues and techniques, the main aims are threefold: to increase the accuracy of modeling and simulation, to reduce the simulation and design time, and to come up with more robust and sustainable design options. First, to get more accurate simulation results, sophisticated surrogate models are needed to provide near high-fidelity results; this usually requires many sampling points in the search space in order to produce highly representative surrogates, which will indirectly increase the computational costs. Second, in order to reduce simulation and design time, efficient optimization algorithms are needed in addition to the efficient approximate surrogate models. Traditional algorithms such as gradient-based methods and trust-region methods do not work well. Thus, new methods such as those based on swarm intelligence methods can be promising. In reality, a good combination of new methods with the existing traditional methods can often obtain good results. Finally, researchers and designers have to combine all the techniques and resources so as to produce robust and sustainable design options. Some design options may satisfy all the design requirements, but they may be very sensitive to manufacturing errors, and they may not be sustainable. Thus, designers often have to produce a diverse set of design options and thus allow decision-makers to choose the most suitable design options under given stringent design constraints. Any successful design cycle requires to deal with the above three challenging issues with sufficient accuracy in a practically acceptable time limit.

This edited book provides a timely summary of some of the latest developments in modeling and simulation-driven design optimization. Topics include aerodynamic optimization, gas transport networks, antenna designs, microwave structures, filter designs, waste management, system identification, crystal nanostructures, sparse grids, and other computationally extensive design applications. Therefore, this book can serve as a reference to researchers, lecturers, and engineers in engineering design, modeling, and optimization as well as industry where computationally expensive designs are most relevant. It is our hope that this may help researchers and designs to produce better design tools so as to reduce the costs of the design process aided by computer simulations.

Reykjavik, Iceland
Ames, IA, USA
London, UK
October 2015

Slawomir Koziel
Leifur Leifsson
Xin-She Yang

Contents

Numerical Aspects of Model Order Reduction for Gas Transportation Networks

Sara Grundel, Nils Hornung, and Sarah Roggendorf

Abstract The chapter focuses on the numerical solution of parametrized unsteady Eulerian flow of compressible real gas in pipeline distribution networks. Such problems can lead to large systems of nonlinear equations that are computationally expensive to solve by themselves, more so if parameter studies are conducted and the system has to be solved repeatedly. The stiffness of the problem adds even more complexity to the solution of these systems. Therefore, we discuss the application of model order reduction methods in order to reduce the computational costs. In particular, we apply two-sided projection via proper orthogonal decomposition with the discrete empirical interpolation method to exemplary realistic gas networks of different size. Boundary conditions are represented as inflow and outflow elements, where either pressure or mass flux is given. On the other hand, neither thermal effects nor more involved network components such as valves or regulators are considered. The numerical condition of the reduced system and the accuracy of its solutions are compared to the full-size formulation for a variety of inflow and outflow transients and parameter realizations.

Keywords Gas network simulation • Model order reduction • Proper orthogonal decomposition • Discrete empirical interpolation method • Stiff initial-value problems

MSC code: 65K05 (Numerical Analysis, mathematical programming methods)

S. Grundel
Max Planck Institute Magdeburg, Magdeburg, Germany
e-mail: sara.grundel@mpi-magdeburg.mpg.de

N. Hornung (✉) • S. Roggendorf
Fraunhofer SCAI, Sankt Augustin, Germany
e-mail: nils.hornung@scai.fraunhofer.de; sarah.roggendorf@scai.fraunhofer.de

S. Koziel et al. (eds.), *Simulation-Driven Modeling and Optimization*,
Springer Proceedings in Mathematics & Statistics 153,
DOI 10.1007/978-3-319-27517-8_1

1 Introduction

Gas as an energy source is being transported from producer or supplier to consumer along pipelines via short or long distances. Size and complexity of these transportation networks will thus vary notably. Maybe the most typical question with regard to such gas transportation problems is if the supply can satisfy consumer demands. Often this question is coupled with the goal to operate the network efficiently and to account for variations either in supply, demand or in the network properties itself. The latter problem might, e.g., result from the difficulty to measure pipeline properties such as roughness exactly. Since gas is compressible such that the network may hold strongly varying amounts, transient effects can become important.

All these questions and goals can in principle be answered by numerical simulation provided that the method of simulation is accurate enough while being efficient at the same time. Efficiency and computational costs depend on network size, nonlinearity, and stiffness of the underlying equations. In order to manage size while dealing with nonlinearity we demonstrate how a nonlinear model order reduction method can be applied, which is tailored to the problem in that the general form of the original equations is respected in the reduced order systems.

Certain important assumptions are made in order to simplify the problem, the most prominent of which is the neglect of all network elements except for simple pipelines, junctions, inflow and outflow elements. Moreover, we assume a single phase of an ideal gas flowing through the pipes, disregard most temperature effects, and make further assumptions that allow us to rewrite the problem as a system of ordinary differential equations. Some of these assumptions are made in favor of a simple exposition of the topic, while others originate from numerical considerations.

Transient simulation of gas networks is a very active field of research. Model order reduction for this system is treated on a more basic level by [3]. Many different approaches on how to efficiently compute transient behavior are known in the literature [6, 9, 10]. The main goal of this chapter is to introduce the reader to basic concepts of model order reduction involving its practical application. Thus we present three example networks of different complexity that form the foundation of a comprehensive treatment of a nonlinear reduced order method, including numerical tests with a focus on stiffness, accuracy, and computational cost.

How the laws of continuum mechanics can be applied to model gas flow through a network of pipes is explained in Section 2, including three different example problems to be used in later sections. Basic notions and definitions of stiffness are discussed in Section 3 with the application and example problems in mind. Section 4 introduces basic concepts of model order reduction with a focus on nonlinear methods suitable for gas transport networks, while Section 5 presents and discusses results for the example cases. The final section of this chapter summarizes the main topics and touches further interesting questions.

2 Problems in Network Simulation

We consider the simulation of gas flowing through a system of branching pipelines with influx and outflux defined at determined locations. First the network model itself is described, including continuum mechanics for the flow through a single pipe and mass conservation at pipeline junctions. Subsequently, we present example networks later used to empirically analyze the simulation of resulting full and reduced order systems.

2.1 Pipe Network Modeling

A gas transportation network can be described as a directed graph $\mathscr{G} = (\mathscr{E}, \mathscr{N})$, where $\mathscr{N}$ is the set of nodes and $\mathscr{E}$ is the set of directed edges. Those edges are denoted by tuples of nodes. We distinguish between so-called supply nodes $\mathscr{N}_s$, demand nodes $\mathscr{N}_d$, and interior nodes $\mathscr{N}_0$, where $\mathscr{N} = \mathscr{N}_s \cup \mathscr{N}_0 \cup \mathscr{N}_d$. If we choose not to consider more involved components, each edge represents a pipe and can thus be specified by length, width, and roughness. Nodes, on the other hand, can be imagined as the points where pipes start, end, or meet. The physics of a fluid moving through a single pipe can be modeled by the isothermal Euler equations averaged over the pipe's cross-section $A = \frac{\pi}{4} D^2$ with inner pipe diameter D, see [11].

Now several simplifications are applied. Discarding terms related to kinetic energy, we first get

$$\partial_t \rho + \partial_x q = 0, \tag{1a}$$

$$\partial_t q + \partial_x p + g\rho\partial_x h = -\frac{\lambda(q)}{2D}\rho v|v|, \tag{1b}$$

$$p = \gamma(T)z(p,T)\rho. \tag{1c}$$

Here ρ denotes the fluid's density, p the pressure, and q the volumetric flux. Together they form the set of unknown dynamical variables and depend on space and time (t, x). Velocity is denoted by $v = \frac{q}{\rho}$. The system consists of three equations, two for mass and momentum conservation (1a,1b) and one that specifies material properties (1c). Momentum conservation (1b) includes a term on the left-hand side to incorporate gravity g as a conservative body force depending on height h, as well as a friction term on the right-hand side.

As an approximation to the friction coefficient $\lambda(q)$, we use

$$\lambda = \left(2\log\left(\frac{D}{\kappa}\right) + 1.138\right)^{-2},$$

where D is again the diameter of the pipe and κ a parameter describing its roughness. Notice that this approximation, called the Swamee-Jain equation [12], neglects the dependence of friction on flux q if we assume, for simplicity again, that roughness κ does not depend on q. The field $\gamma = RT$ embodies gas properties by its dependence on a given gas temperature T and on the universal gas constant R. For further simplification we assume that the temperature $T \equiv T_0$ is constant in time and space and that we deal with an ideal gas (i.e., $z \equiv 1$). Thus, $\gamma = RT_0$ is constant and we can rewrite the constitutive assumption as $p = \gamma\rho$.

Without change of notation, volumetric flux is now substituted by mass flux $q \leftarrow Aq$. Along with the relation $v = \frac{q}{\rho}$ the simplified isothermal Euler equations take the form

$$\partial_t \rho + \frac{1}{A}\partial_x q = 0, \tag{2a}$$

$$\frac{1}{A}\partial_t q + \partial_x p + g\rho\partial_x h = -\frac{\lambda}{2DA^2}\frac{q|q|}{\rho}, \tag{2b}$$

$$p = \gamma\rho, \tag{2c}$$

or substituting ρ according to (2c),

$$\begin{aligned} \partial_t p &= -\frac{\gamma}{A}\partial_x q, \\ \partial_t q &= -A\partial_x p - \frac{Ag}{\gamma}p\partial_x h - \frac{\lambda\gamma}{2DA}\frac{q|q|}{p}. \end{aligned} \tag{3}$$

Within the pipe network, (3) is valid for each edge. Any edge stands for a pipe parametrized along its given length L, which means the interval $[0, L]$ establishes the domain of definition of the according partial differential equation. The full system of equations additionally encompasses consistency conditions for each demand node as well as input in terms of supply pressures given at supply nodes.[1] If we assemble all pipes that end in node i into the set I_R^i and, accordingly, all pipes that start in node i into the set I_L^i, then the demand consistency conditions are given by

$$0 = \sum_{l \in I_L^i} q_l(L_l, t) - \sum_{k \in I_R^i} q_k(0, t) + d_i(t) \tag{4}$$

for every node i.[2] If

[1]We always identify flux conditions with demand, i.e., with outflux, and pressure boundary conditions with supply, i.e., with influx. Without further modification, this (somehow arbitrary) identification can be relaxed such that demand can also be modeled as pressure conditions and supply as mass or volumetric fluxes.

[2]The direction given to the edges serves the sole purpose of topology definition and is independent of the direction of the flux within the pipe that results from the laws of continuum mechanics.

$$d_i(t) \equiv 0, \tag{5}$$

this condition reduces to the well-known Kirchhoff law and is valid at any junction of pipes. Strictly speaking, (4) holds for all nodes in $\mathscr{N}_0 \cup \mathscr{N}_d$, whereas for nodes in $\mathscr{N}_0$ we additionally have (5).

Let n_E be the number of edges and assume they are ordered such that every edge has an index in $[1, 2, \ldots, n_E]$. Once we discretize (3) following [4] and add given supply pressures, the resulting overall differential algebraic system of equations is given by

$$\partial_t \frac{p_R^k + p_L^k}{2} = -\frac{\gamma}{A_k}\frac{q_R^k - q_L^k}{L_k} \quad \forall k \in [1, \ldots, n_E], \tag{6a}$$

$$\begin{aligned}\partial_t \frac{q_R^k + q_L^k}{2} = &- A_k \frac{p_R^k - p_L^k}{L_k} - \frac{A_k g}{2\gamma}(p_R^k + p_L^k)\frac{h_R^k - h_L^k}{L_k} \\ &- \frac{\lambda_k \gamma}{4 D_k A_k}\frac{(q_R^k + q_L^k)|q_R^k + q_L^k|}{p_R^k + p_L^k} \quad \forall k \in [1, \ldots, n_E],\end{aligned} \tag{6b}$$

$$0 = \sum_{l \in I_L^i} q_R^\ell - \sum_{k \in I_L^i} q_L^k - d_i(t) \quad \forall i \in \mathscr{N}_0 \cup \mathscr{N}_d, \tag{6c}$$

$$0 = p_i(t) - s_i(t) \quad \forall i \in \mathscr{N}_s. \tag{6d}$$

The vector q_R is the vector of fluxes at the end of the pipes, and the vector q_L is the vector of fluxes at the beginning of the pipes. Except for those at the beginning and end, we do not take any values along the pipes. This means, if the numerical error by this discretization exceeds our needs because the pipes are too long, we have to add artificial nodes (junctions) to the network such that all pipes are short enough to yield accurate enough results.

For a more compact description, we write the system in matrix notation

$$|B_S^T|\partial_t p_s + |B_0^T|\partial_t p_d = -M_L^{-1} q_-, \tag{7a}$$

$$\partial q_+ = M_A(B_S^T p_s + B_0^T p_d) + g(q_+, p_s, p_d), \tag{7b}$$

$$0 = B_0 q_+ + |B_0| q_- - d(t), \tag{7c}$$

$$0 = p_s - s(t), \tag{7d}$$

where

$$M_L = \mathrm{diag}(\ldots \frac{L_k A_k}{4\gamma} \ldots), \tag{8}$$

$$M_A = \mathrm{diag}(\cdots - \frac{A_k}{L_k} \ldots), \tag{9}$$

and the k-th component of the function g is given by

$$g_k(q_+, p_d, p_s) = -\frac{A_k g}{2\gamma}\ell_k(p_d, p_s)\frac{h_R^k - h_L^k}{L_k} - \frac{\lambda_k \gamma}{4D_k A_k}\frac{(q_+^k)|q_+^k|}{\ell_k(p_d, p_s)}, \tag{10}$$

where ℓ_k is the k-th entry of the vector-valued function ℓ

$$\ell(p_d, p_s) = |B_0^T|p_d + |B_S^T|p_s.$$

Both matrices M_L and M_A are diagonal and invertible. The matrix

$$B = \begin{bmatrix} B_0 \\ B_S \end{bmatrix}$$

denotes the incidence matrix of the underlying directed graph, where B_0 corresponds to the demand nodes and junctions and B_S corresponds to the supply nodes. In addition, the notation $q_- = q_R - q_L$ and $q_+ = q_R + q_L$ has been introduced here. To eliminate q_-, we multiply (7a) by $|B_0|M_L$ and then use (7c) to substitute $|B_0|q_-$,

$$\begin{aligned} & |B_S^T|\partial_t p_s + |B_0^T|\partial_t p_d = -M_L^{-1}q_- \\ \Rightarrow\quad & |B_0|M_L|B_S^T|\partial_t p_s + |B_0|M_L|B_0^T|\partial_t p_d = -|B_0|q_- \\ \Rightarrow\quad & |B_0|M_L|B_S^T|\partial_t p_s + |B_0|M_L|B_0^T|\partial_t p_d = B_0 q_+ - d(t). \end{aligned}$$

We also replace p_s according to (7d) to obtain

$$|B_0|\, M_L\, |B_0^T|\partial_t p_d = B_0 q_+ - d(t) - |B_0|\, M_L\, |B_S^T|\partial_t s(t), \tag{11a}$$

$$\partial_t q_+ = M_a B_0^T p_d + g(q_+, s(t), p_d) + M_a B_S^T s(t) \tag{11b}$$

with g as in (10). The structure that (11) implies can be seen more clearly in block matrix notation

$$\begin{bmatrix} |B_0|M_L|B_0^T| & 0 \\ 0 & M_a^{-1} \end{bmatrix}\begin{bmatrix} \partial_t p_d \\ \partial_t q_+ \end{bmatrix} = \begin{bmatrix} 0 & B_0 \\ B_0^T & 0 \end{bmatrix}\begin{bmatrix} p_d \\ q_+ \end{bmatrix} + \begin{bmatrix} -d(t) - |B_0|M_L|B_S^T|\partial_t s \\ M_a^{-1}g + B_S^T s \end{bmatrix}. \tag{12}$$

Compare this derivation of an ordinary differential equation to the very similar approach of [4], where more details are conveyed. From now on we consider (12) of the size $|\mathcal{N}| - |\mathcal{N}_s| + n_E$ (the number of nodes minus the number of supply nodes plus the number of edges). It is an ordinary differential equation in descriptor form where the matrix E,

$$E = \begin{bmatrix} |B_0|M_L|B_0^T| & 0 \\ 0 & M_a^{-1} \end{bmatrix},$$

is positive definite and symmetric. Furthermore, the equation depends on several parameters given by pipe lengths L, diameters D, friction coefficients λ, height differences Δh, and gas properties γ totaling $4 \times n_E + 1$-many parameters.

Remark 1. We are now dealing with an ordinary differential equation of the form

$$E\dot{x} = Tx + f(x, u) + Ku.$$

This means we have been able to decouple the system in such a way that it is no longer written in the form of a differential algebraic equation. During the process of reformulation, the time derivative of the input signal s, which is the pressure at the supply nodes, has been introduced. Since this pressure is usually given explicitly as a slowly changing function of time we can calculate its derivative in most practical applications.

2.2 Example Problems

In this chapter we are going to numerically analyze three example networks. The first consists in a single pipe, the second in a small connected network of 57 pipes, and the third in a larger network with a higher number of elements forming several connected components, the largest of which includes 669 pipes. Network topology is given by a graph which is mainly defined by a list of edges, i.e., ordered tuples of nodes indicating the direction of the edge. For simplicity, the total set of N_N nodes is just described by their numbering, i.e., by a set of integers of the form $\{n \in \mathbb{N} : n \leq N_N\}$. The set of edges given as a list automatically implies a numbering of the edges which we later make use of to set up our equations. Length, height difference, width, and roughness are given as a parameter vector for each pipe. Furthermore, one or several demand fluxes are provided by functions of time. Similarly, one or several time-dependent supply pressures are given.

The topologies of the three networks are visualized in Figures 1, 2 and 3. Supply nodes are marked by triangles, and nodes with nonzero demand are marked by rectangles. Tables 1 and 4 show the corresponding parameter vectors.

At the supply nodes a supply function of the following form is given

$$s(t) = \alpha_s \times s(0) \times \left(0.5 \times \left(\cos\left(\frac{\pi \times (t - T_1^s)}{(T_2^s - T_1^s)}\right) - 1\right)\right) + s(0), \tag{13}$$

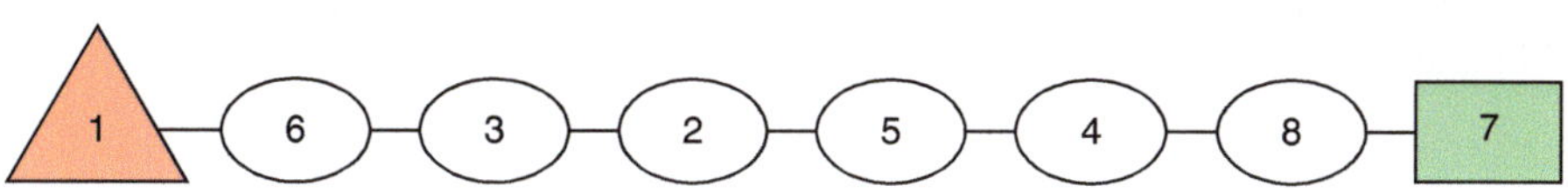

Fig. 1 Network 1, a single pipe divided into subsections of different properties

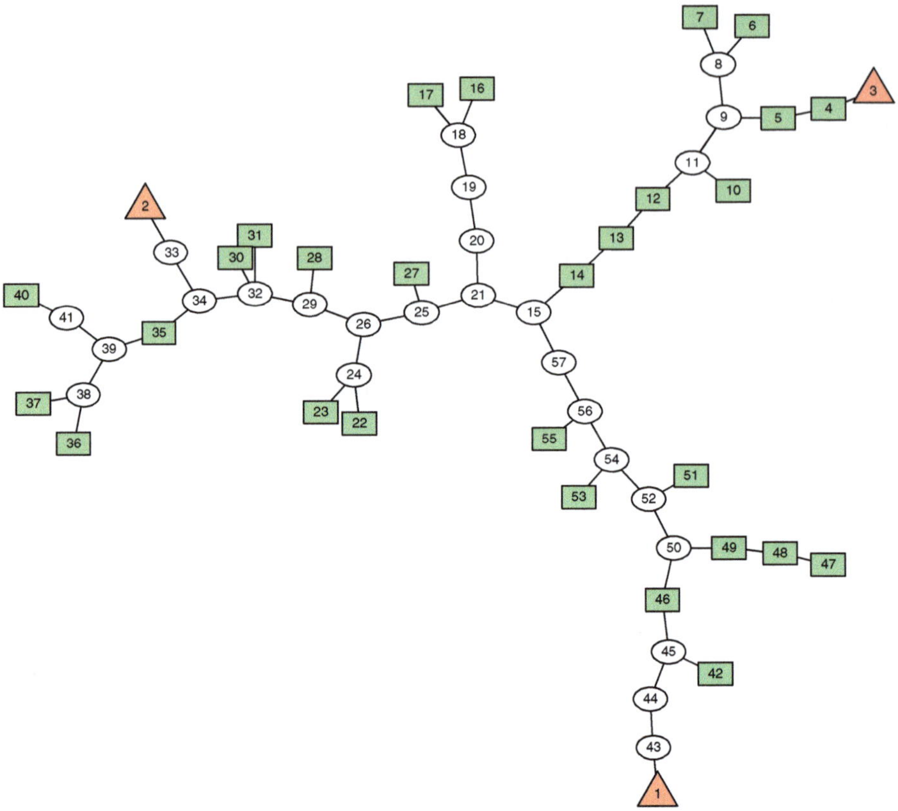

Fig. 2 Network 2, a small network with several supply and demand notes

where α_s is the portion the supply pressure drops during the given time period $T_2^s - T_1^s$. Tables 2 and 3 show the supply pressures at time zero for all supply nodes. Similarly, the demand functions at nodes with nonzero demand are of the form

$$d(t) = d(0) + 0.5 \times \alpha_d \times \left(1 - \cos\left(\frac{\pi \times (t - T_1^d)}{(T_2^d - T_1^d)}\right)\right) \times d(0), \tag{14}$$

where α_d denotes the percentages the demand grows. The corresponding values of $d(0)$, T_1^d, and T_2^d are listed in Tables 2 and 5 for all nodes with nonzero demand. This means that the first example, the single pipe, yields a system of ordinary differential equations of size 14 with 22 parameters, the second network is of size 110 with 29 parameters, and the last network's largest subnet is of size 1341.

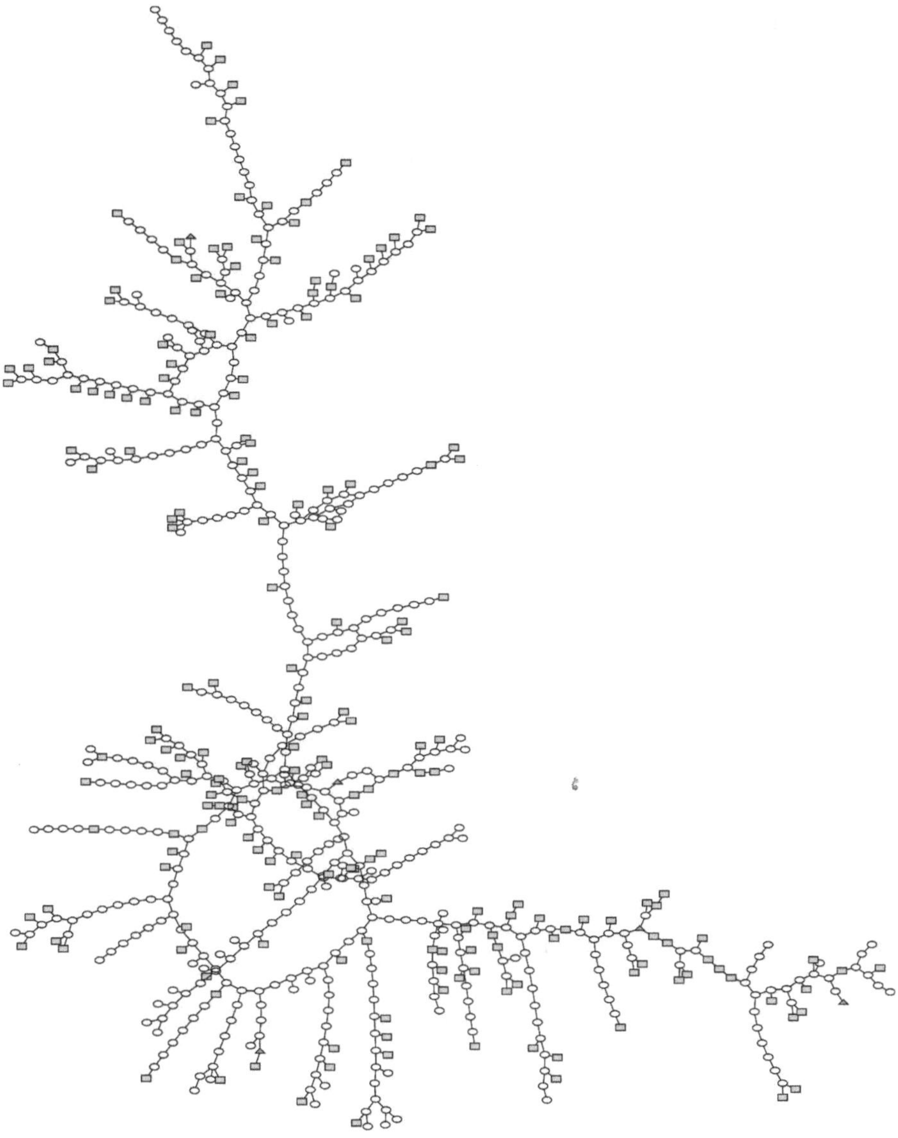

Fig. 3 Connected component of network 3, of larger scale

3 Stiffness in Ordinary Differential Equations

A simple idea in principle, stiffness can be conceived as the intuition that stability is more critical than accuracy or, much simpler, that explicit integration methods might fail. To put this into a mathematical framework can become rather difficult,

Table 1 Parameters for network 1

Pipe element	Length [m]	Diameter [m]	Friction coefficient	Height difference [m]
(8,7)	100.000	1.000	0.012	0.000
(1,6)	100.000	0.900	0.008	0.000
(3,2)	1,000.000	0.500	0.014	0.000
(6,3)	1,000.000	0.500	0.014	0.000
(5,2)	1,000.000	0.500	0.014	0.000
(5,4)	1,000.000	0.500	0.014	0.000
(8,4)	1,000.000	0.500	0.014	0.000

Table 2 Supply and demand for network 1

Node	1	Node	7
$s(0)$ [Pa]	5,000,000	$d(0)$ [kg/s]	219.230
T_1^s [s]	500,000	T_1^d [s]	100,000
T_2^s [s]	1,000,000	T_2^d [s]	500,000

Table 3 Supply for network 2

Node	1	2	3
$s(0)$ [Pa]	5,400,000	2,700,000	2,700,000
T_1^s [s]	700,000	700,000	700,000
T_2^s [s]	1,000,000	1,000,000	1,000,000

though. As a means to analyze stability, we introduce concepts and definitions of stiffness in linear constant-coefficient ordinary differential equations in the following subsection as well as a discussion of practical implications for three example cases in subsequent subsections.

3.1 *Concepts and Definitions*

If a linear constant-coefficient ordinary differential equation is given as

$$\dot{x} = Ax,$$

stiffness is often quantified via a ratio of the minimum and maximum real part of the eigenvalues under the assumption that all eigenvalues of A be negative. Such a ratio is linked with the behavior of the differential equation for $t \to \infty$ (and can be too liberal a condition). The Lipschitz constant of the linear function, which coincides with the largest singular value of A, gives another (too conservative) criterion of stiffness that is tied to the limit $t \to t_0$.

A more realistic measure of stiffness for linear systems is suggested by [7] in terms of a pseudo spectral analysis. We do not repeat details here. It is, however, important to realize that the two notions of stiffness differ more strongly from each

Table 4 Parameters for network 2

Pipe	L [m]	D [m]	λ	$(h_R - h_L)$ [m]
(1,43)	1.0	1.0	0.0120	0.0
(57,56)	1, 661.0	0.8	0.0125	−22.0
(56,55)	1, 550.0	0.8	0.0125	6.0
(56,54)	1, 530.0	0.8	0.0125	6.0
(54,53)	750.0	0.8	0.0125	−3.0
(54,52)	4, 089.0	0.8	0.0125	0.0
(52,51)	229.0	0.8	0.0125	5.0
(52,50)	1, 135.0	0.8	0.0125	3.0
(50,49)	222.0	0.8	0.0125	1.0
(50,46)	1, 948.0	1.0	0.0120	67.0
(49,48)	732.0	0.8	0.0125	−3.0
(48,47)	10.0	0.8	0.0125	0.0
(46,45)	15, 134.0	1.0	0.0120	78.0
(45,44)	1.0	1.0	0.0120	0.0
(45,42)	14, 088.0	1.0	0.0120	−81.0
(44,43)	3.0	1.0	0.0120	0.0
(41,40)	2, 258.0	0.9	0.0122	4.0
(41,39)	2, 010.0	0.9	0.0122	4.0
(39,38)	1, 948.0	0.9	0.0122	2.0
(39,35)	3, 533.0	0.9	0.0122	2.0
(38,37)	35.0	0.9	0.0122	2.0
(38,36)	111.0	0.9	0.0122	−3.0
(35,34)	1, 930.0	0.9	0.0122	5.0
(34,33)	81.0	0.9	0.0122	1.0
(34,32)	1640.0	0.9	0.0122	8.0
(33,2)	34.0	0.9	0.0122	−1.0
(32,31)	1, 666.0	0.9	0.0122	19.0
(32,30)	763.0	0.9	0.0122	16.0

Pipe	L [m]	D [m]	λ	$(h_R - h_L)$ [m]
(32,29)	7, 700.0	0.9	0.0122	22.0
(29,28)	1, 350.0	0.9	0.0122	−12.0
(29,26)	6, 300.0	0.9	0.0122	−7.0
(26,25)	343.0	0.6	0.0132	−10.0
(26,24)	1, 455.0	0.6	0.0132	−8.0
(25,27)	189.0	0.6	0.0132	−1.0
(25,21)	2, 486.0	0.6	0.0132	64.0
(24,23)	392.0	0.6	0.0132	−1.0
(24,22)	814.0	0.6	0.0132	3.0
(21,20)	18.0	0.6	0.0132	1.0
(19,18)	621.0	0.6	0.0132	−24.0
(18,17)	1, 818.0	0.6	0.0132	−47.0
(18,16)	16.0	0.6	0.0132	0.0
(15,14)	1, 040.0	0.6	0.0188	34.0
(14,13)	8, 296.0	0.6	0.0188	−29.0
(13,12)	1, 295.0	0.6	0.0188	2.0
(12,11)	521.0	0.6	0.0188	0.0
(11,10)	470.0	0.7	0.0128	14.0
(11,9)	1, 507.0	0.7	0.0181	−3.0
(9,8)	789.0	0.7	0.0128	4.0
(9,5)	800.0	0.7	0.0161	1.0
(8,7)	275.0	0.7	0.0128	9.0
(8,6)	1, 305.0	0.7	0.0128	2.0
(5,4)	11, 866.0	0.7	0.0161	−7.0
(4,3)	3, 212.0	0.7	0.0161	−6.0
(20,19)	1, 000.0	0.5	0.0137	0.0
(57,15)	1, 000.0	0.8	0.0125	0.0

other the further away A is from a normal matrix, where nonnormality can, e.g., be measured by

$$\|A^*A - AA^*\|_F.$$

Here $\|\cdot\|_F$ denotes the Frobenius norm. As a consequence, both concepts of stiffness are appropriate for normal matrices. The matrices arising from the systems presented in the following are not normal, though, such that we need to expect different stiffness measures.

The whole issue of stiffness becomes even more complicated once we deal with nonlinear equations and source terms, which are present in the systems that we consider. Linear stiffness theory for constant-coefficient ordinary differential

Table 5 Demand for network 2

Node	4	5	6	7	10	12	13	14	16	17
$d(0)$ [kg/s]	0.092	0.028	0.538	0.434	0.807	0.035	0.045	0.452	0.471	1.773
T_1 [s]	100,000	100,000	100,000	100,000	100,000	100,000	100,000	100,000	100,000	100,000
T_2 [s]	500,000	500,000	500,000	500,000	500,000	500,000	500,000	500,000	500,000	500,000
Node	22	23	27	28	30	31	35	36	37	40
$d(0)$ [kg/s]	0.599	0.133	0.040	0.018	0.478	0.678	0.573	0.093	0.710	0.598
T_1 [s]	100,000	100,000	100,000	100,000	100,000	100,000	100,000	100,000	100,000	100,000
T_2 [s]	500,000	500,000	500,000	500,000	500,000	500,000	500,000	500,000	500,000	500,000
Node	42	46	47	48	49	51	53	55		
$d(0)$ [kg/s]	0.318	0.561	0.498	0.493	0.233	0.172	0.278	0.359		
T_1 [s]	100,000	100,000	100,000	100,000	100,000	100,000	100,000	100,000		
T_2 [s]	500,000	500,000	500,000	500,000	500,000	500,000	500,000	500,000		

equations can be applied to a linearization of the nonlinear system: Via the Jacobian matrix we get a constant-coefficient linear equation that can be evaluated at a given time t_0. However, information may get lost in the process of linearization. In the following we empirically evaluate and show the stiffness of the ordinary differential equation (12) for the single pipe and the small network example from Section 2.2.

3.2 *Single Pipe System*

The single pipe introduced in Section 2.2 comprises segments of varying properties and can hence be written as a system of ordinary differential equations of size 14.

In this first excursion in stiffness we are going to compute the common notions of stiffness that are given by the eigenvalues and the singular values of $E^{-1}J_f(0, x_0)$, where x_0 denotes a stationary solution at time $t = 0$. This stationary solution can in principle be chosen arbitrarily and will serve as a starting value later. The Jacobian's singular values and eigenvalues are plotted in Figure 4. Notice that these plots tend to look very similar for other values of t and x. The analysis of singular values and eigenvalues leads to a liberal stiffness estimate of 4 and a conservative estimate of 1.6×10^5. Because the difference between both estimates comprises several orders of magnitude we cannot conclude the extent of stiffness, hence the problem at hand might either become rather stiff or only mildly so.

By a simple numerical test we notice, though, that the step size of a regular solver for non-stiff problems (here MATLAB®'s `ode15`) is chosen much smaller than by the corresponding stiff solver (here MATLAB's `ode23s`). This difference can be interpreted as a sign of more than only mild stiffness. The step sizes chosen by the different MATLAB solvers are shown in Figure 5, where we can

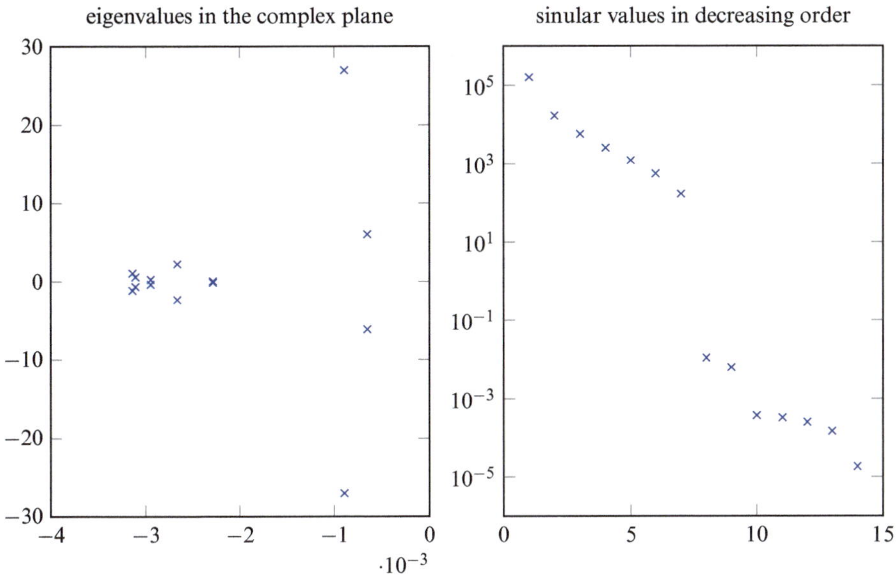

Fig. 4 The eigenvalues and the singular values of the linearized system, network 1

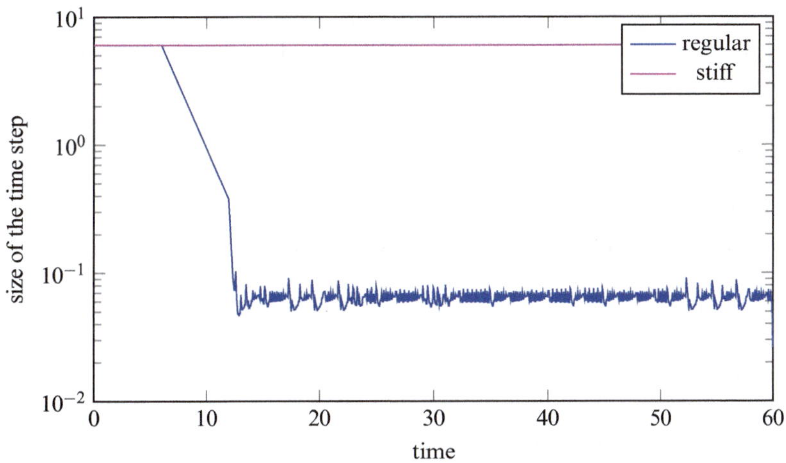

Fig. 5 The time steps chosen by MATLAB's `ode23` and `ode15s` solvers for ordinary differential equations, network 1

observe a difference of two orders of magnitude for this simple example. Even more important, we must expect that we deal with serious stiffness in this class of gas transportation problems due to the fact that, in spite of those small step sizes, the solution found by MATLAB's non-stiff `ode15` blows up in finite time whereas the solution given by `ode23s` does not.

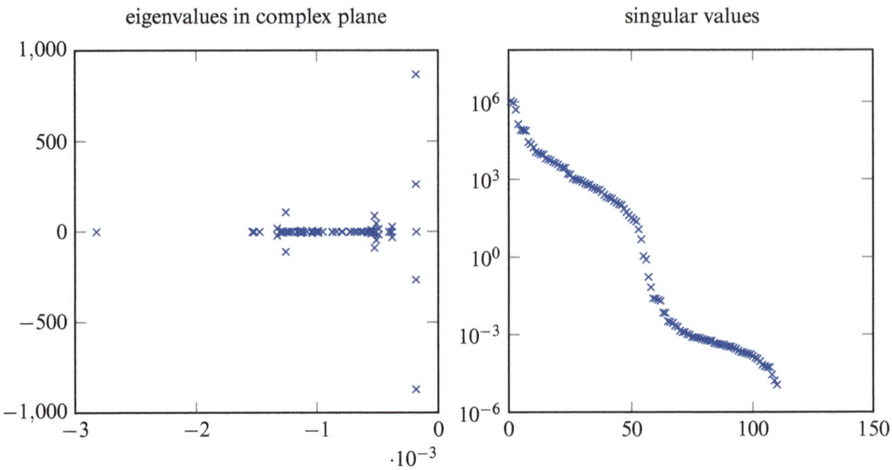

Fig. 6 The eigenvalues and singular values of the linearized system, network 2

3.3 *Small Networks*

For a network of pipes the stiffness of the system becomes clearer yet. Looking at the eigenvalue and singular value plot of the Jacobian matrix, which is shown in Figure 6, we again see that the stiffness estimates vary strongly by several orders of magnitude, this time between 16 and 1.1×10^6. As above this result eludes an exact stiffness analysis of the system of equations. Since both estimates grow higher than in the previous example, stiffness can most probably be assumed to increase with the complexity and size of the network. Conducting the same experiment as before, we compare the resulting MATLAB step sizes as in Section 3.2. We do not include a plot as it shows the same peculiarities as Figure 5, only that the time step of the non-stiff method decreases even more, oscillates around 10^{-3} s and therefore differs by four orders of magnitude from the stiff solver.

Regarding the first six seconds, the solutions computed via the two different solvers are depicted in Figure 7. We can notice again that these systems seem to require a stiff or, possibly even better, a dedicated solver.

4 Model Order Reduction

Seeing that these network problems can become arbitrarily large and stiff, we would be interested in creating a reduced order model which is of small order and hopefully not stiffer than the full system. Since the system is nonlinear, we apply the model order reduction method called proper orthogonal decomposition as this is typically the method of choice for nonlinear systems.

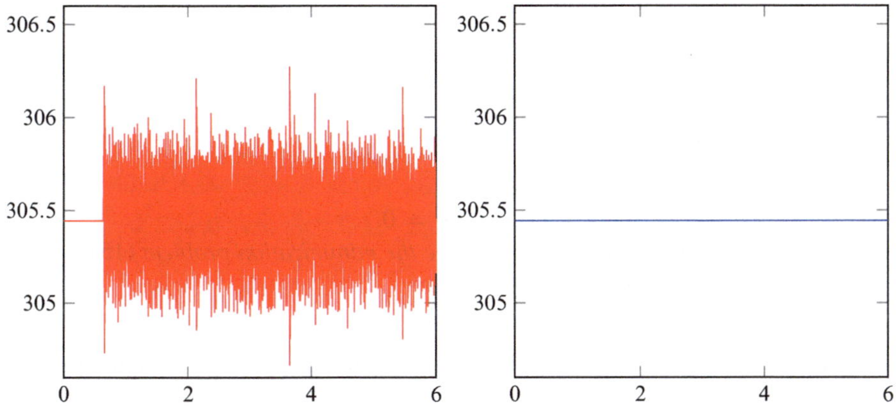

Fig. 7 The time steps chosen by MATLAB's `ode23` and `ode15s` solvers for ordinary differential equations, network 2

4.1 Proper Orthogonal Decomposition

The model order reduction method called proper orthogonal decomposition (POD) starts with multiple snapshots $\{y_j^k\}_{j=1}^m \subset X$, $1 \leq k \leq p$, for a given Hilbert space X (typically $\mathbb{R}^n$ or a function space like $\mathscr{L}^2$). One is interested in finding a subspace spanned by an orthonormal basis $\psi_1, \ldots, \psi_\ell$ within the Hilbert space X that solves the minimization problem

$$\begin{aligned} &\min \sum_{k=1}^{p} \sum_{j=1}^{m} \alpha_j \left\| y_j^k - \sum_{i=1}^{\ell} \langle y_j^k, \psi_i \rangle_X \psi_i \right\|_X^2 \\ &\text{s. t. } \{\psi_i\}_{i=1}^{\ell} \subset X \text{ and } \langle \psi_i, \psi_j \rangle_X = \delta_{ij},\ 1 \leq i,j \leq \ell. \end{aligned} \tag{15}$$

This minimization can also be written as a maximization problem due to the orthogonal nature of the elements ψ_i

$$\begin{aligned} &\max \sum_{k=1}^{p} \sum_{j=1}^{m} \alpha_j \sum_{i=1}^{\ell} \langle y_j^k, \psi_i \rangle_X^2 \\ &\text{s. t. } \{\psi_i\}_{i=1}^{\ell} \subset X \text{ and } \langle \psi_i, \psi_j \rangle_X = \delta_{ij},\ 1 \leq i,j \leq \ell. \end{aligned} \tag{16}$$

Theorem 1. *Let X be a separable Hilbert space, $\mathscr{R} : X \to X$ a summation operator on X defined by*

$$\mathscr{R} : X \to X, \quad \mathscr{R}\psi \mapsto \sum_{k=1}^{p} \sum_{j=1}^{m} \alpha_j \langle \psi, y_j^k \rangle_X y_j^k.$$

The following assertions are all true:

(a) $\mathscr{R}$ *is linear compact self-adjoint and nonnegative.*
(b) *A basis of* X *of eigenfunctions* ψ_j *exists such that*

$$\mathscr{R}\psi_j = \lambda_j \psi_j$$

and $\lambda_1 \geq \lambda_2 \geq \ldots \geq \lambda_d > \lambda_{d+1} = \cdots = 0.$
(c) *The first* ℓ *of these eigenfunctions solve the minimization problem* (15) *and the maximization problem* (16).

For a proof of Theorem 1 see [13].

In the remainder of this chapter, let us assume that $X = \mathbb{R}^n$ and the inner product is defined by $\langle x, y\rangle_X = x^T E y$ for a given positive definite symmetric matrix E. We will denote by Y_k the matrix of snapshots $\{y_j^k\}_{j=1}^m$. This means Y is an $n \times m$ matrix.

Corollary 1. *The eigenvectors of the operator* $\mathscr{R}$ *equate to the eigenvectors of the matrix*

$$\left(Y_1 D Y_1^T + \cdots + Y_p D Y_P^T\right) E, \tag{17}$$

where $D = \mathrm{diag}(\alpha_1, \cdots, \alpha_n)$.

So in order to find the best approximation space of size ℓ for a given set of snapshots, all we have to do is compute the first ℓ eigenvectors of (17). Notice that the matrix (17) is an $n \times n$ matrix, the size of the state space X. This matrix is not symmetric. If we, however, knew the eigenvalues and eigenvectors of the following symmetric system

$$E^{1/2}\left(Y_1 D Y_1^T + \cdots + Y_p D Y_P^T\right) E^{1/2}, \tag{18}$$

we could compute the eigenvectors of (17). In fact, these eigenvectors are obtained if we multiply the eigenvectors of (18) by $E^{-1/2}$. Since E is usually a large matrix and its square root is hard to compute, we are going to describe a faster alternative. The matrix (18) can be written as $\hat{Y}\hat{Y}^T$ for

$$\hat{Y} = \left[\hat{Y}_1\ \hat{Y}_2\ \ldots\ \hat{Y}_p\right],$$

where $\hat{Y}_i = E^{1/2} Y_i D^{1/2}$. Since this matrix is symmetric and positive definite its eigenvalue and singular value decompositions are equivalent. Assuming we know the singular value decomposition of $\hat{Y}$,

$$\hat{Y} = USV^T, \tag{19}$$

we also know the singular value decomposition of $\hat{Y}\hat{Y}^T = US^2U^T$ and the singular value decomposition of $\hat{Y}^T\hat{Y} = VS^2V^T$. This means, if we compute V we can recover

U by (19), with $U = \hat{Y}S^{-1}V$. The matrix U is the matrix of singular vectors or eigenvectors of $\hat{Y}\hat{Y}^T$ (recall that $\hat{Y}\hat{Y}^T$ equals (18)). Hence, we get the eigenvector matrix of (17) by $E^{-1/2}U = E^{-1/2}\hat{Y}S^{-1}V$. If $\phi_1, \ldots, \phi_\ell$ are the singular vectors of $\hat{Y}\hat{Y}^T$ we can, consequently, compute the eigenvectors ψ_i,

$$\psi_i = \frac{1}{\sqrt{\sigma_i}} \left[Y_1 D^{1/2},\ Y_2 D^{1/2},\ \ldots\ Y_p D^{1/2} \right] \phi_i,$$

where σ_i and ϕ_i are the singular values and singular vectors of $\hat{Y}^T\hat{Y}$. The objective function is then given by

$$\begin{aligned}
&\sum_{k=1}^{p}\sum_{j=1}^{m} \alpha_j \left\| y_j^k - \sum_{i=1}^{\ell} \langle y_j^k, \psi_i \rangle_X \psi_i \right\|_X^2 \\
&= \sum_{k=1}^{p}\sum_{j=1}^{m} \alpha_j \left(\|y_j^k\|_X^2 - 2\langle y_j^k, \sum_{i=1}^{\ell} \langle y_j^k, \psi_i \rangle_X \psi_i \rangle + \| \sum_{i=1}^{\ell} \langle y_j^k, \psi_i \rangle_X \psi_i \|_X^2 \right) \\
&= \sum_{k=1}^{p}\sum_{j=1}^{m} \alpha_j \left(\|y_j^k\|_X^2 - \sum_{i=1}^{\ell} \langle y_j^k, \psi_i \rangle_X^2 \right) \\
&= \sum_{i=1}^{n} \sigma_i - \sum_{i=1}^{\ell} \sigma_i = \sum_{i>\ell} \sigma_i.
\end{aligned}$$

which follows from the fact that ψ_i are eigenfunctions of the operator $\mathscr{R}$ [13]. Not to mention, this result can be utilized to determine ℓ, i.e., to decide where to cut off eigenvalue computations. In practical applications, a heuristic choice for ℓ can, e.g., be implied by the requirement that $\mathscr{E} \geq 99\,\%$ where

$$\mathscr{E} = \frac{\sum_{i=1}^{\ell} \sigma_i}{\sum_{i=1}^{n} \sigma_i} = \frac{\sum_{i=1}^{\ell} \sigma_i}{\sum_{k=1}^{p} \sum_{j=1}^{m} \alpha_j \|y_j^k\|_X^2}. \tag{20}$$

We now need to understand how to use the described subspace of X in order to create a reduced order model of a dynamical system given in the form

$$E\dot{x} = Tx + f(x, u) + Ku,$$

where the function f depends on the dynamic variable x and the time-dependent input function u. We are now going to briefly address the matter that the system matrices E, T, K, and the function f additionally depend on parameters. The basic idea is to generate snapshots for certain sampled parameter values. This parameter sampling can be picked as a uniform grid. An alternative way to set up a sampling in a semi-optimized way by using a Greedy algorithm is, e.g., described by [5].

Algorithm 1 POD

Require: $Y_1, \ldots, Y_p \in \mathbb{R}^{N \times n}$, inner product matrix E, diagonal weight matrix D, reduction order ℓ
Ensure: metamodel $\psi_1, \ldots, \psi_\ell$

1: $B_{jk} \leftarrow D^{1/2} Y_j^T E Y_k D^{1/2}, j, k \in \{1, \ldots p\}$
2: B $\leftarrow$ matrix with B_{ij} blocks
3: $\sigma_i, \phi_i \leftarrow$ singular values and singular vectors of B
4: **for** $i \leftarrow 1 : \ell$ **do**
5: $\quad \psi_i \leftarrow \frac{1}{\sqrt{\sigma_i}} [Y_1 D^{1/2}, \ldots, Y_p D^{1/2}] \phi_i$

In fact, (15) can be seen as the discrete version of a continuous minimization problem. Seen in this light, the objective function would read

$$\sum_{k=1}^{p} \int_0^T \left\| y^k(t) - \sum_{i=1}^{\ell} \langle y^k(t), \psi_i \rangle_X \psi_i \right\|_X^2 ,$$

which, in turn, *implies* the minimization problem (15) by using a quadrature rule to compute the integral. A classical choice of α_i are trapezoidal weights and the inner product is typically given by $\langle x, y \rangle = x^T E y$ with the inner product matrix denoted by E. This allows us to compute $\psi_1, \ldots, \psi_\ell$ by the methods discussed above and summarized in Algorithm 1. We then determine the matrix W as $W = [\psi_1, \ldots, \psi_\ell]$ and project the large scale system by Galerkin projection onto

$$W^T E W \dot{\hat{x}} = W^T T W x + W^T f(W\hat{x}, u) + W^T K u, \tag{21}$$

where $W^T E W = I$ by the mutual orthogonality of ψ_i with respect to said inner product.

4.2 *Nonlinearity and Problem Structure*

In the following, we are going to describe the rest of the algorithm that is used in the numerical experiments of Section 5. There are mainly two questions left to be answered carefully which were not mentioned in the theoretical derivation of POD.

1. Do we need to consider the special structure of the problem when we set up the reduced order model? Pressures and fluxes are not necessarily of a similar magnitude such that a direct application of the POD matrix as in (21) might neglect important effects [8]. We therefore reduce the pressure and the flux vector separately.
2. How do we handle the nonlinearity of the function? The term $W^T f(W\hat{x})$ contributes ℓ-dimensional function values with an ℓ-dimensional vector-valued

argument. However, a higher-dimensional vector must be evaluated to compute $W^T f$. To obtain a fast simulation, this evaluation needs to be avoided.

Let us first discuss implications of the special structure of our problem. For the convenience of the reader, the ordinary differential equation (12) is repeated here:

$$\begin{bmatrix} |B_0|M_L|B_0^T| & 0 \\ 0 & M_a^{-1} \end{bmatrix} \begin{bmatrix} \partial_t p_d \\ \partial_t q_+ \end{bmatrix} = \begin{bmatrix} 0 & B_0 \\ B_0^T & 0 \end{bmatrix} \begin{bmatrix} p_d \\ q_+ \end{bmatrix} + \begin{bmatrix} -d(t) - |B_0|M_L|B_S^T|\partial_t s \\ g(q, p, s) + B_S^T s \end{bmatrix}.$$

For given input functions $s(t)$, $d(t)$, and given parameters, we solve this differential equation within the time interval $[0, T]$ in time steps of size $\Delta t > 0$. This means we obtain pressure values $p(0), p(\Delta t), \dots, p(T)$ and flux values $q(0), q(\Delta t), \dots, q(T)$, representing snapshots that we assemble into matrices

$$Y_i^p \text{ (pressures) and } Y_i^q \text{ (fluxes)}, \tag{22}$$

where i indicates the snapshots of different given parameter values. They are used to calculate projection matrices W_p and W_q separately for pressures and fluxes by Algorithm 1, such that the reduced order system obtained via Galerkin projection of the full order ordinary differential equation onto the matrix

$$\begin{bmatrix} W_p & 0 \\ 0 & W_q \end{bmatrix}$$

reads

$$\begin{bmatrix} W_p^T|B_0|M_L|B_0^T|W_p & 0 \\ 0 & W_q^T M_a^{-1} W_q \end{bmatrix} \begin{bmatrix} \partial_t p_d \\ \partial_t q_+ \end{bmatrix} =$$

$$\begin{bmatrix} 0 & W_p^T B_0 W_q \\ W_a^T B_0^T W_p & 0 \end{bmatrix} \begin{bmatrix} p_d \\ q_+ \end{bmatrix} + \begin{bmatrix} -W_p^T d(t) - W_p^T|B_0|M_L|B_S^T|\partial_t s \\ W_q^T f(W_p p, W_q q, s) + W_p^T B_S^T s \end{bmatrix}. \tag{23}$$

Since all involved matrices can be precomputed, the only issue left is the computation of the nonlinear term $W_q f(W_p p, W_q q, s)$. So as to approximate this nonlinear term by a function which no longer requires any higher order operation, the discrete empirical interpolation method (DEIM in short, see [2]) is employed. If we precompute a snapshot matrix F that consists of function evaluations of f at a number of given values of p and q, matrices P and U can be computed via Algorithm 2, such that the resulting reduced (i.e., approximate) nonlinear function evaluation amounts to

$$W_q^T f(W_p p, W_q q, s) = W_q^T U(P^T U)^{-1} P^T f(W_p p, W_q q, s).$$

Algorithm 2 DEIM

Require: $F \in \mathbb{R}^{N \times n}$, reduction order m
Ensure: metamodel $\psi_1, \ldots, \psi_\ell$

1: $[U, S, V] \leftarrow \text{svd}(F)$
2: ind $\leftarrow$ index of the maximal entry of $U(:, 1)$
3: $P \leftarrow e_{\text{ind}}$ coordinate basis vector
4: **for** $i \leftarrow 1 : m$ **do**
5: ind $\leftarrow$ index of maximal entry of $U(:, i+1) - U(P^T U)P^T U(:, i+1)$
6: $P \leftarrow [P, e_{\text{ind}}]$

We immediately see that the matrix $W_q^T U(P^T U)^{-1}$ is of size $r_1 \times m$, where m is the reduction order for DEIM and r_1 the reduction order for the flux component in POD. The function $P^T f(W_p p, W_q q, s)$ is truly a function of order m since the matrix P is a matrix of zeros and ones, where exactly m entries of f are picked. Owing to the structure of the function f, namely the k-th entry only depending on the k-th entry of $W_q q$ and $B_0^T W_p p + B_S^T s$, this computation is straightforward.

5 Numerical Examples

The purpose of the following section is to experimentally show the extent of reduction of the system we can achieve for our three network models (Section 2.2), while still being able to reproduce the general dynamical behavior of the gas within the network.

5.1 The Single Pipe

Our first example is one of the most trivial networks possible and serves as proof of concept. To begin with, a single solution trajectory, which we again call a snapshot, is calculated for the parameters given in Section 2.2. Its singular values are shown in the top panel of Figure 8. By the methods explained in Section 4.2, the snapshot is used to set up a reduced order model. In short, this means we can determine projection matrices W_1 and W_2 via POD as well as projection and picking matrices U and P via DEIM. The resulting reduced order model parametrically depends on length, diameter, and friction coefficient of each pipe segment as well as on the field γ that characterizes properties of the gas flowing through it. In order to better understand the behavior of the system for different choices of these parameters, we make use of a Latin hypercube sampling of size 10, where we vary all 22 parameters in a cube of $\pm 5\,\%$ of the original parameter set. In this particular test, the order of the reduced model is 4, which results from the choice $r_1 = r_2 = m = 2$. For the

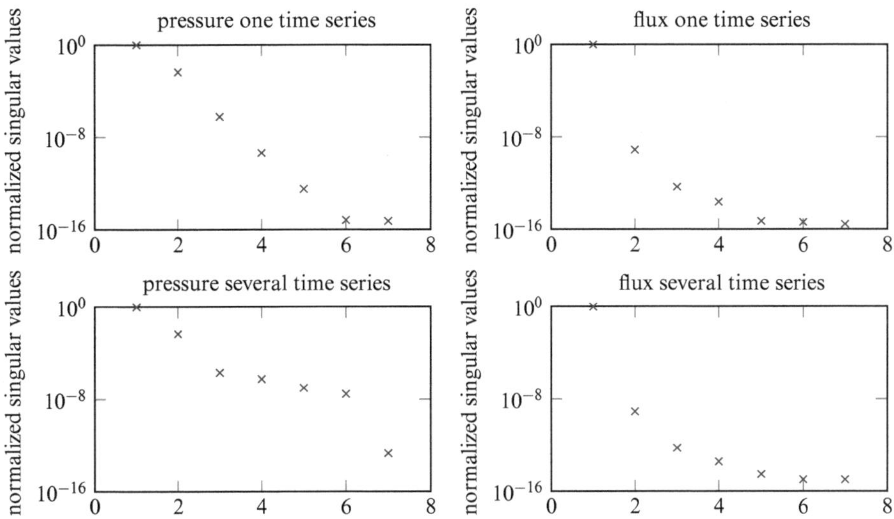

Fig. 8 The singular values for one time series as well as for 10 time series, each the result of a different parameter set. The values that are shown along the y-axis are ordered decreasing in size, along the x-axis

different test networks, which only differ by the said parameter values, we compute the reduced order model as well as, for reference, the full model and compute the relative error in the pressure and the absolute error in the flux, displaying the respective maximal error in Table 6. Furthermore, the relative error at a distance from $t = 0$ is also given as the maximal difference over all pressure and flux components. We also display the error of the reduced order system compared to the full system of the original parameter distribution. This shows that, when we vary the parameters, we still stay within the same order of magnitude of the error.

Figure 8 emphasizes this result. Here we see that the singular value decay in the flux looks similar, whether we have one time series or several time series for several parameters. The difference becomes noticeable in the left column that refers to pressure. However, in both cases the third singular value, after which DEIM cuts off the series of singular values, is of order 10^{-8}.

Of course, the input and output functions of this problem are not varied. And due to the nature of the problem we cannot expect to find a reduced system of small order representing all possible input and output functions. The difficulty consists in that the partial differential equation describing the physics in the pipe, the isothermal Euler equation, is a transport dominated partial differential equation. Thus shocks might travel through the system and we can, therefore, not expect the solution to lie in a low-dimensional subspace.

The following numerical experiment, however, shows that, if we assume the supply pressures and demand nodes only vary within a certain range of given values, we are able to still use the reduced system. We are now going to repeat

Table 6 Numerical error, network 1

Sample	1	2	3	4	5	6	7	8	9	10	original parameters
Absolute error (flux) [kg/s]	0.0393	0.4779	0.3269	0.2307	0.5856	0.1450	0.4893	0.2796	0.2631	0.2666	0.2479
Relative error (pressure)	0.0711	0.0214	0.0783	0.0301	0.0737	0.0263	0.0777	0.0571	0.0660	0.2401	0.021
Relative error (both, $t \in [10^5, 10^6]$)	0.0711	0.0214	0.0783	0.0301	0.0737	0.0263	0.0777	0.0480	0.0660	0.2401	

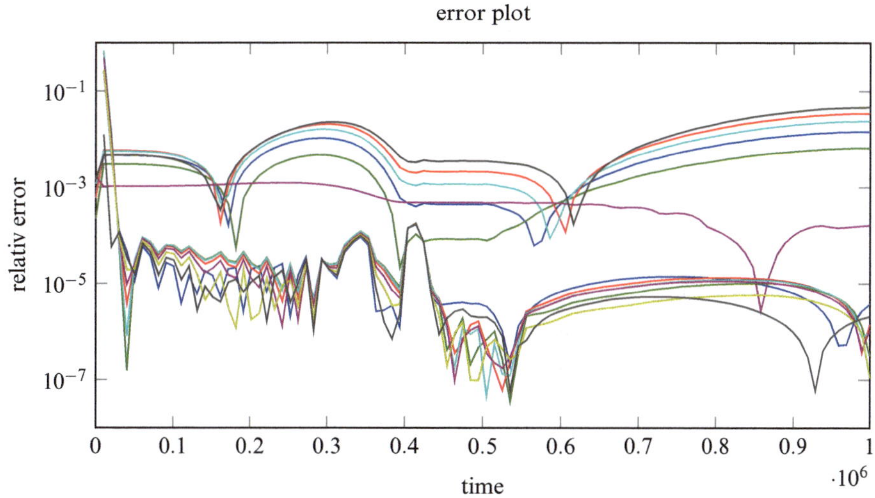

Fig. 9 The maximal relative error over the 10 sampled scenarios

the comparison of the full and reduced simulation. Again, we pick 10 simulations, varying the percentage the supply pressure and the demand flux change over time within $\pm 10\,\%$. Remember the reduced order was created with these values all varying by $2\,\%$ only. The relative error plot, where the maximum is taken over all 10 scenarios, is displayed in Figure 9.

5.2 Network of 57 Pipes

Having a manageable size but realistic structure, the following example supports the idea that model order reduction already helps to reduce simulation time and stability of the simulation for networks of smaller scale. We test parameter variations as well as changes of the input function receiving robust and accurate results. The system is first solved for the parameters and input function given in Section 2.2, $T = 1 \times 10^6\,\text{s} \approx 280\,\text{h}$. We store 100 time steps for a system size of 110, of which 54 dimensions refer to pressure and 56 to flux. By and large, we gather a snapshot matrix of dimension 100×110 in this fashion. The singular value decay of YEY^T is illustrated in the top panel of Figure 10, where the singular values are displayed for the fluxes and the pressures independently again. On the subject of the criteria given in (20), we can conclude that a reduced order of size 1 for pressures and of size 1 for fluxes is sufficient to obtain a "good" approximation in that case with $\mathscr{E} \geq 0.99999$ for the pressures as well as for the fluxes. We, however, select $r_1 = r_2 = 3$. Typically, the reduction order for DEIM has to equal at least this

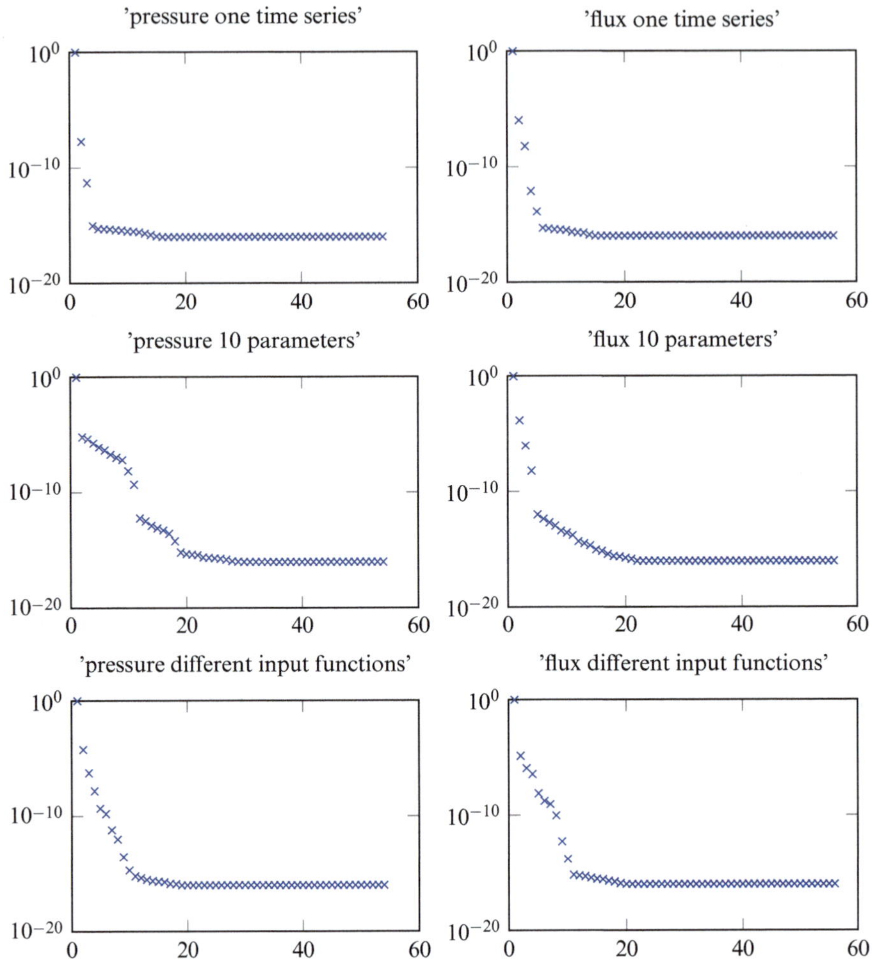

Fig. 10 Singular values of the snapshot matrices

reduced order (3 in our case). It turns out that this minimum choice is enough for the example at hand, see the error values listed in Table 7. In fact, $r_1 = 3, r_2 = 3, m = 3$ yield a reasonably accurate reduced order system.

The projection matrices are applied to create a parametric reduced order model. The parameter space of this parametrized model possesses a dimension of size 29. We allow this parameter vector to vary in a box around $\pm 5\,\%$ of its given values again. We pick 10 values in that box by Latin hypercube sampling to obtain 10 scenarios. For each scenario we compute the reduced solution and the full solution. We realize here that in order to run MATLAB's ode solver `ode15s` we need to have a very accurate starting value for the full model. Therefore, we take the original initial value and compute a truly stationary solution by Newton's method to be used as initial value. For the reduced order system this step is not necessary,

Table 7 Numerical error, network 2

Sample	1	2	3	4	5	6	7	8	9	10
Absolute error (flux) [kg/s]	10.6854	3.4017	2.3185	10.6690	4.3558	4.0095	9.1439	14.2883	12.9632	5.0359
Relative error (pressure)	0.0296	0.0426	0.0421	0.0340	0.0300	0.0224	0.0388	0.0312	0.0429	0.0171
Relative error (pressure, $t \in [10^3\,\text{s}, 10^6\,\text{s}]$)	0.0296	0.0348	0.0221	0.0916	0.0346	0.0096	0.0388	0.0831	0.0573	0.0216
Time (full model) [*s*]	1,388	904	1,029	1,320	1,060	878	941	940	811	914
Time (reduced model) [*s*]	2	1.7	1.6	1.8	1.7	1.7	1.8	1.4	1.1	1

which in turn means that the solution of the full order system becomes even slower in comparison. However, this procedure also implies that the solutions near the initial time are not as close to each other as later because they do not start from the same initial value. We thus compare the maximal relative error only after the first 1000 s have already passed. This error is shown in Table 7. We, furthermore, compare the timings, for the solution of the reduced system and for the solution of the full system, where the identification of a starting value is included as well as the simulation time itself.

With regard to the full solution of the 10 scenarios from above, creating a snapshot matrix as in Section 4.1, we can compute the singular values that are plotted in the middle panel of Figure 10. The singular values show the size of a linear subspace to be used in order to create a good approximation of the space in which the solutions lie.

Furthermore, we have conducted some test with varying input functions. As in the case of the single pipe, we vary the percentages the demand and supply values change during the simulated time in a box of $\pm 10\,\%$. We can match the general behavior even though the errors between the full and the reduced system can be large. However, measurements often differ from the computed solution even stronger than our reduced model, see, e.g., [1]. We, furthermore, display the singular value decay for the snapshot matrix created by 10 such scenarios (last panel of Figure 10).

5.3 *Network of 669 Pipes*

A system of 669 elements is rather large such that computations of a single time series can already take from several hours up to weeks with standard solvers (it is possible to accelerate the implementation, though). The goal here is to show that we can predict the behavior of the system with the model order reduction method as described above up to a certain extent. We only compute the solution of this large system for the first 100 s. From the results of this computation we calculate the projection matrix W as well as the DEIM matrices P and U with reduction order $r_1 = 4, r_2 = 2, m = 4$. These matrices are needed to set up the reduced system. We then run the reduced system for different values of $d(0)$, which are a random perturbation of the original $d(0)$. We cannot compare the reduced time series to the time series of the full system, as the latter is too expensive to compute. Since we run the reduced system in time until it should arrive at a new stationary solution, we can assess the extent of "stationarity" of our result from the right-hand side of the system. In Table 8 we compare the value of the right-hand side of the reduced system to the value of the right-hand side of the full system at the state we arrived at following the trajectory of the reduced system. Even though these values are not numerically zero, they are reputably closer to zero than if we evaluated the right-hand side at the initial configuration. Compared to the initial configuration, these values are hence better starting values to use a Newton-type method to find a truly stationary solution.

Table 8 Value of the right-hand side, network 3

Sample	1	2	3	4	5	6	7	8	9	10
$f_r(r_E, x_E)$	0.4	0.3	2.7	0.9	1.3	0.4	0.6	1.4	0.6	1.5
$f(t_E, Wx_E)$	18	22	29	19	20	17	26	22	18	22
$f(t_E, x_0)$	10^6	10^6	10^6	10^6	10^6	10^6	10^6	10^6	10^6	10^6

The simulation time for simulating those reduced order systems amounts to approximately 25 s. If we use the end value obtained from the reduced simulation and apply a Newton method to find a stationary solution such that $|f(x_S)| \leq 10^{-3}$, the Newton method takes around 10 to 100 s. Whereas, if we use x_0 to start our Newton method, it takes about 10,000 s. So in the presented case, the reduced method can certainly help to speed up the computation of a feasible stationary solution.

6 Conclusions

For gas transportation problems through networks of pipelines, from supplier to consumer, the underlying continuum mechanics, practical simplifications, and a numerical model have been compiled and explained to the reader. Notably, a state-of-the-art transformation of the resulting discretized differential algebraic system of equations to a system of ordinary differential equations has been presented. Among the properties of this system, stiffness has been spotlighted since it affects the choice of suitable numerical solvers strongly. Empirical evidence is shown that stiffness plays an important role and that it even grows with the complexity of the systems.

With the goal of exhaustive studies of varying input and output in mind, we indicate how to apply model order reduction methods for nonlinear systems efficiently, making use of POD with DEIM. Special attention is given to the preservation of the general form of the underlying ordinary differential equation lest pressure and flux values may be mixed. A parameter domain is defined by varying network properties within a limited range to indicate uncertainty. The domain is used to obtain empirical results that illustrate possible gain in simulation speed by order reduction, while the induced loss in accuracy is discussed.

The numerical behavior of three networks of different complexity is examined in this light. Results indicate that a speed-up of 10 to 1000 can be achieved, where we have to anticipate certain inaccuracy within the obtained solutions. In some cases inaccuracy can be of an order of magnitude such as 10 % of the typical solution values. The stiffness of the reduced system has been monitored and not found to be increased by the applied model order reduction technique.

One of the most obvious extensions to the case presented here can be found in the treatment of more types of network elements, as well as in the inclusion of more involved temperature effects. From a numerical point of view, the development of dedicated solvers would probably be interesting.

References

1. Alamian, R., Behbahani-Nejad, M., Ghanbarzadeh, A.: A state space model for transient flow simulation in natural gas pipelines. J. Nat. Gas Sci. Eng. **9**, 51–59 (2012)
2. Chaturantabut, S., Sorensen, D.C.: Nonlinear model reduction via discrete empirical interpolation. SIAM J. Sci. Comput. **32**(5), 2737–2764 (2010). doi:10.1137/090766498
3. Grundel, S., Hornung, N., Klaassen, B., Benner, P., Clees, T.: Computing surrogates for gas network simulation using model order reduction. In: Surrogate-Based Modeling and Optimization, pp. 189–212. Springer, New York (2013)
4. Grundel, S., Jansen, L., Hornung, N., Clees, T., Tischendorf, C., Benner, P.: Model order reduction of differential algebraic equations arising from the simulation of gas transport networks. In: Schöps, S., Bartel, A., Günther, M., ter Maten, E.J.W., Müller, P.C. (eds.) Progress in Differential-Algebraic Equations, Differential-Algebraic Equations Forum, pp. 183–205. Springer, Berlin/Heidelberg (2014). doi:10.1007/978-3-662-44926-4_9
5. Haasdonk, B.: Convergence rates of the POD–greedy method. ESAIM: M2AN **47**(3), 859–873 (2013). doi:10.1051/m2an/2012045. http://dx.doi.org/10.1051/m2an/2012045
6. Herty, M., Seaïd, M.: Simulation of transient gas flow at pipe-to-pipe intersections. Int. J. Numer. Methods Fluids **56**(5), 485–506 (2008). doi:10.1002/fld.1531. http://dx.doi.org/10.1002/fld.1531
7. Higham, D., Trefethen, L.: Stiffness of ODEs. BIT Numer. Math. **33**(2), 285–303 (1993). doi:10.1007/BF01989751
8. Hinze, M., Kunkel, M.: Discrete empirical interpolation in POD model order reduction of drift-diffusion equations in electrical networks. In: Scientific Computing in Electrical Engineering SCEE 2010, pp. 423–431. Springer (2012)
9. Kiuchi, T.: An implicit method for transient gas flows in pipe networks. Int. J. Heat Fluid Flow **15**(5), 378–383 (1994)
10. Osiadacz, A.: Simulation of transient gas flows in networks. Int. J. Numer. Methods Fluids **4**(1), 13–24 (1984). doi:10.1002/fld.1650040103. http://dx.doi.org/10.1002/fld.1650040103
11. Schmidt, M., Steinbach, M.C., Willert, B.M.: High detail stationary optimization models for gas networks. Optimization and Engineering, **16**(1), 131–164 (2015). doi:10.1007/s11081-014-9246-x
12. Swamee, P.K., Jain, A.K.: Explicit equations for pipe-flow problems. J. Hydraul. Div. **102**(5), 657–664 (1976)
13. Volkwein, S.: Optimal control of a phase-field model using proper orthogonal decomposition. ZAMM – J. Appl. Math. Mech./ Z. Angew. Math. Mech. **81**(2), 83–97 (2001). doi:10.1002/1521-4001(200102)81:2<83::AID-ZAMM83>3.0.CO;2-R

Parameter Studies for Energy Networks with Examples from Gas Transport

Tanja Clees

Abstract The focus of this chapter is on methods for the analysis of parameter variations of energy networks and, in particular, long-distance gas transport networks including compressor stations. Gas transport is modeled by unsteady Eulerian flow of compressible, natural gas in pipeline distribution networks together with a gas law and equations describing temperature effects. Such problems can lead to large systems of nonlinear equations with constraints that are computationally expensive to solve by themselves, more so if parameter studies are conducted and the system has to be solved repeatedly. Metamodels will thus play a decisive role in the general workflows and practical examples discussed here.

Keywords Metamodeling • Response surfaces • Design of experiments • Parameter analysis • Robustness • Gas networks

MSC codes: 93B51, 65D17, 62K20, 62K25

1 Introduction

Networks rule the world. The well-known social networks are just one example. Infrastructure for transport of gas, electricity, or water, but also electrical circuits inside technical devices are other important instances. Due to the ongoing transformation of our energy production and incorporation of increasingly larger amounts of renewable energy sources, energy networks of different types (electrical grid, gas, heat, etc.) have to form an integrated system allowing for balancing of supplies and demands in the future. Conversions between different energy media (power-to-gas, power-to-heat, etc.) and storages provided by, for instance, pipeline systems and caverns will play a decisive role. This increases the demand for enhanced cross-energy simulation, analysis and optimization tools.

T. Clees (✉)
Fraunhofer SCAI, Sankt Augustin, Germany
e-mail: tanja.clees@scai.fraunhofer.de

S. Koziel et al. (eds.), *Simulation-Driven Modeling and Optimization*,
Springer Proceedings in Mathematics & Statistics 153,
DOI 10.1007/978-3-319-27517-8_2

Transport networks for energy or water as well as circuits can be mathematically modeled in a very similar fashion, based on systems of differential-algebraic equations. Their numerical simulation can be performed based on the same or at least similar numerical kernels.

In Section 2, the physical model for flow of compressible, natural gas in pipeline distribution networks with several technical elements is sketched. In a similar fashion, one can model electrical grids, pipeline systems and other energy transport networks as well.

Afterwards, in Section 3, we describe analysis tasks which are common for many energy networks and define some terms important there. Section 4 gives a short overview on a selection of methods frequently used. A flow chart asking some key questions and several workflows are outlined in 5. Section 6 summarizes several versatile visualization techniques. Based on these workflows, several examples from gas transport analysis are studied in some detail in Section 7. Finally, Section 8 concludes this chapter.

2 Gas Transport Network Modeling

In the following, the physical model considered here is sketched. More details can be found in [24]. Several ongoing research aspects such as model order reduction, ensemble analysis, coupled network and device simulation are discussed in [5, 13, 18, 20, 21, 28], for instance. Throughout this chapter, we use MYNTS (MultiphYsical NeTwork Simulation framework), see [1, 4]. In MYNTS, the physical model described in the following is implemented.

2.1 *Isothermal Euler Equations*

A gas transport network can be described as a directed graph $\mathcal{G} = (\mathcal{E}, \mathcal{N})$ where $\mathcal{N}$ is the set of nodes and $\mathcal{E}$ is the set of directed edges denoted by tuples of nodes. If we choose to not consider more involved components as a start, each edge constitutes a pipe and can thus be specified by length and width. Nodes, on the other hand, can be imagined as the points where pipes start, end or meet. The physics of a fluid moving through a single pipe can be modeled by the isothermal Euler equations. Discarding terms related to kinetic energy for simplification we get:

$$\partial_t \rho + \partial_x q = 0 \tag{1}$$

$$\partial_t q + \partial_x p + \partial_x \left(\rho v^2\right) + g\rho\partial_x h + F = 0 \tag{2}$$

$$p = R_s T z \rho. \tag{3}$$

The hydraulic resistance is given by the Darcy-Weisbach equation:

$$F = \frac{\lambda(q)}{2D}\rho v|v| \tag{4}$$

Here, $\rho = \rho(x,t)$ denotes the density, $p = p(x,t)$ the pressure, $q = \rho * v$ the flux, $v = v(x,t)$ the velocity, $T = T(x,t)$ the temperature, $h = h(x)$ the geodesic height, $D = D(x)$ the pipe diameter, z the compressibility, R_s denotes the specific gas constant (see also Section 2.3), and $\lambda = \lambda(q)$ being the friction coefficient.

Together with Kirchhoff's equations, the nonlinear system to be solved is defined. The meaning of the equations is as follows:

- the continuity (1) and Kirchhoff's equations modeling the mass (or molar) flux
- a pipe law (2) and Darcy-Weisbach modeling pressure-flux (see Section 2.2)
- a gas law (3) modeling pressure–density–temperature (see Section 2.3)

ρ, p, q, v, T form the set of unknown dynamical variables.

2.2 Pipe Laws

The friction coefficient in the Darcy-Weisbach equation can be modeled by, e.g.:

$$\lambda = \left(2\log\left(\frac{4.518}{\mathrm{Re}}\log\left(\frac{\mathrm{Re}}{7}\right) + \frac{k}{3.71D}\right)\right)^{-2} \quad \text{Hofer} \tag{5}$$

$$\lambda = \left(2\log\left(\frac{D}{k}\right) + 1.138\right)^{-2} \quad \text{Nikuradze} \tag{6}$$

where Re is the Reynolds number and κ a parameter describing the roughness of the pipe currently considered. For high Reynolds numbers, Hofer approaches the easier Nikuradze equation.

2.3 Gas Laws

For ideal gases, $pV = nRT$ holds where p denotes pressure, V volume, n amount (in moles), R ideal gas constant, and T temperature of the gas. Defining the specific gas constant R_s as the ratio R/m with m being the mass, we get $p = \rho R_s T$. For the non-ideal case, compressibility z is introduced, and Eq. 3 holds. Several gas laws are frequently used:

$$z = 1 + 0.257p_r - 0.533\frac{p_r}{T_r} \quad \text{AGA(upto70bars)}$$

$$z = 1 - 3.52\exp(-2.260T_r)p_r + 0.247\exp(-1.878T_r)p_r^2 \quad \text{Papay(upto150bars)}$$

$$z = 1 + B\tilde{\rho} - p_r\sum_n C_n + \sum_n C_n\left(b_n - c_n k_n \rho_r^{k_n}\right)\rho_r^{b_n}\exp(-c_n\rho_r^{k_n}) \quad \text{AGA8} - \text{DC92}$$

Here, p_r, T_r, ρ_r denote the reduced pressure, temperature, and density, respectively. A reduced property is obtained by dividing the property by its (pseudo-)critical value. If AGA8-DC92 shall be applied, the fractions of all components in the gas mix have to be computed as well, cf. 2.5. AGA8-DC92 contains several further coefficients and constants which are not further explained here. Papay is quite popular. However, for an accurate simulation, AGA8-DC92 should be used though.

2.4 Network Elements

Several types of nodes should be distinguished:

- Standard supplies: Input pressure, temperature and gas composition are defined.
- Standard demands: Either output mass flow or volume flow or power is defined.
- Special supplies: Either input mass flow or volume flow or power is defined.
- Interior nodes: The remaining nodes, where nothing is defined (besides (h).

For all nodes, their geodesic height h has to be given.

Besides nodes and pipes, several elements are present in many gas transportation networks. The following elements are considered here, a typical selection:

- Compressors, described by a characteristic diagram (engine operating map; see, for instance, Figure 4).
- Regulators: Input pressure, output pressure, output volume flow are typical regulation conditions; regulators might be described by a characteristic curve.
- Coolers, heaters.
- Valves, resistors, flaptraps.
- Shortcuts: A special element which can be seen as an extremely short pipe.

2.5 Gas Mixing and Thermodynamics

Natural gas consists of 21 components, and the by far largest fraction is methane. For modeling the molar mix of gas properties, such as combustion value, heat capacity, fractional composition, etc., the system has to be enlarged and reformulated to take 21 gas components into account.

Also for modeling thermodynamical effects, the composition is incorporated. Several important effects have to be considered:

- Heat exchange (pipe-soil): A (local) heat transfer coefficient is used.
- Joule-Thomson effect (inside the pipe): Temperature change due to pressure loss during an isenthalpic relaxation process.

The system is then enlarged by an equation describing the molar mix of enthalpy (a form of internal energy). Arguments include, in particular, the heat capacity and the critical temperature and pressure, as modeled by the gas law chosen. For modeling the gas heating inside compressors, the isentropic coefficient has to be considered as well. Several models for approximating this effect exist, cf. [23], for instance.

2.6 Outputs, Parameters and Criteria

The following terms are used here:

- Input: Settings/values which have to be defined before starting a simulation.
- Output: Results stemming from a simulation.
- Parameter: An input which shall be varied for a specific analysis task.
- Criterion: A single value or distribution computed from one or more outputs; a criterion might be a global value or one with only a local meaning for the network given.

Several general questions arise for each parameter:

- Of which type is the parameter: discrete (e.g. state of a valve) or continuous (e.g. input pressure)?
- Which (range of) values shall be considered for the parameter for the specific analysis task? Examples are:
  - "on" and "off" for the state of a valve
  - the interval [52.0; 60.0] for an input pressure
- Can all parameters be varied independently? If not, is the dependency be known in advance or result of another process?
- Which type of distribution shall be considered for the parameter for the specific analysis task?
- How is this distribution be defined?
  - Function: Analytic (physical model) or fitted to data stemming from measurements or simulations (attention: the method for and assumptions behind the fitting might have a large impact).
  - Histogram (raw data) resulting from another process.

See Table 1 for a selection of input parameters, varied in the examples (see Section 7), as well as output functions, analysed in more detail.

Table 1 Input and output functions (selection).

shortcut	function	node/edge	remark
pset	input: pressure at supply	node	values in *bar* here
qset	input: volume flow at demand	node	values in $1000Nm^3/h$ here
m	mass flow	edge	values in kg/s here
p	pressure	node	values in *bar* here
pslope	pressure difference	edge	$p_2 - p_1$ on the edge
t	temperature	node	values in Kelvin here
tslope	temperature difference	edge	$t_2 - t_1$ on the edge
had	head of compressor	edge	change of isentropic enthalpy $[kJ/kg]$
qvol	volume flow through compressor	edge	volume per second $[m^3/s]$

A General Remark. Depending on the concrete physical scenario to be solved and, in particular, conditions for compressors and regulators, either a set of equations or a set of equations and constraints (and possibly an objective function) has to be solved. We call both cases "simulations" in the following, though.

3 Analysis Tasks and Ensembles

Parameters of the model can vary, depending on their meaning and physical and/or numerical scenarios considered. A parameter variation might cover a tiny up to a huge range of values, in one or more intervals, and different types of distributions might be considered. Typical ones are

- uniform
- (skewed) Gaussian (see, for example, Figure 9)
- based on a histogram harvested from measurements

Different tasks can be solved. Important ones include

- comparison of scenarios and visualization of differences on the net (Section 6.3)
- stability analysis (Section 3.1)
- parameter sensitivity and correlation analysis (Section 3.2)
- robustness analysis and analysis of critical situations (Section 3.3)
- robust design-parameter optimization (RDO, Section 3.4)
- calibration of simulation models (history matching, Section 3.5)
- analysis of the security of energy supplies: this has to be carried out, in particular, for electrical grids; the so-called $(N-1)$-study is usually performed. If N is the number of (at least) all (decisive) components, N simulations are performed in each of which another one of these components is assumed to fall out.

If the data basis for such an analysis task is a collection of results from several simulation runs or measurements, we call this data basis *ensemble* here.

In the following, we describe the tasks listed above in more detail. Afterwards, we will present and discuss methods for creating ensembles and analysing them.

3.1 *Stability Analysis*

We call (the design of) a scenario *stable* if tiny changes of initial conditions and physical properties have a tiny impact on the results only. We call it *instable*, if tiny changes in the initial conditions lead to substantially different results with a large portion of purely random scatter. We call it *chaotic*, if tiny changes in the initial conditions lead to substantially different and even unpredictable results. Instability might stem from physical and/or numerical issues, and it is difficult to separate these effects in many cases.

A pragmatic way to perform a stability analysis is given by workflow STAT, see Section 5.1, where many parameters of the model as well as numerical settings are randomly varied in a tiny range each.

3.2 Parameter Sensitivity and Correlation Analysis

We call (the design of) a scenario *sensitive*, if small changes of the initial conditions lead to substantially different, still predictable results.

There are several ways to measure sensitivity and to perform sensitivity analysis. A simple method and quick check for nonlinear behaviour is based on a star-shaped experimental design (design-of-experiment (DoE), see Section 4.1). Per parameter, 3 values (left, center, right) for checking dependencies are available then. This simple analysis might be the first part of workflow ADAPTIVE (Section 5.3).

If a deeper analysis shall be performed directly, or if only one set of simulation runs is possible, either workflow STAT (Section 5.1) or workflow UNIFORM (Section 5.2) can be performed. Workflow UNIFORM has the advantage that a metamodel is constructed as well. Global impacts are reflected by correlation measures (see Section 4.2). Local sensitivities, 2D histograms and (approximations of) cumulative density functions give a deeper insight, see also Sections 6.1 and 6.1.

3.3 Robustness Analysis and Analysis of Critical Situations

We call (the design of) a scenario *robust* if small changes of the initial conditions will only have small and w.r.t. to the "quality" affordable impacts on the results. In particular, robustness analysis is a typical way to examine critical situations based on simulations.

One has to carefully distinguish between robustness and reliability. Roughly speaking, robustness aims at the behaviour for the majority of the cases (between the 5- and 95-percent-quantile or the 1- and 99-percent-quantile, say), whereas reliability asks for the seldom cases (outside the area considered for robustness analysis). In practice, reliability might be very difficult to compute accurately, whereas one can at least characterize robustness. Some robustness measures are (cf. [25, 27]):

- If a certain objective should be optimal in average, an expected value can be minimized (Attention: The distribution of the target function itself is allowed to have big outliers).
- If a certain objective should vary as less as possible, dispersion can be minimized. It is mandatory to combine this measure with one for controlling quality itself.

- If a certain objective must not fall below a given threshold, worst-case analysis and a respective measure can be used.
- If a given percentage of values of a target isn't allowed to fall below a threshold, a quantile measure can be applied.

Each measure is reasonable, there are more alternatives, and several measures can even be used simultaneously. Note that the decision for one or more measures depends on the intention of the designer.

3.4 Robust Design-Parameter Optimization (RDO)

RDO means parameter optimization with robustness aspects. One or more robustness criteria can be added to the optimization process. However, minimization of the value of a target function can lead to a higher dispersion and vice versa: usually, you cannot achieve both. Compromises might be found by a weighted objective function or the computation of Pareto fronts or a more substantial change of design.

3.5 History Matching

The adjustment of parameters of a model with respect to (historical) data stemming from physical measurements is quite often called *calibration* or *history matching*.

The goal is to ensure that predictions of future performance are consistent with historical measurements. History matching typically requires solving an ill-posed inverse problem, and thus, it is inherently non-unique. One can obtain a set of matching candidates by means of solving a multi-objective parameter-optimization problem.

Besides the parameters and their ranges, one or more optimization criteria have to be set up measuring the quality of the match. Often, differences of decisive properties such as pressures, fluxes, temperatures, etc., measured in, e.g., the L1- or L2-norm, are used.

4 Methods

In order to solve one of the analysis tasks discussed above, methods have to be selected and a workflow set up. Here, methods are outlined. In Section 5, several workflows as well as a flow chart supporting the selection of a workflow are discussed.

4.1 Experimental Designs

Experimental designs (design-of-experiment, DoE) considered here are based on some standard sampling schemes. Among them can be Monte Carlo (MC), Quasi Monte Carlo (QMC), Latin hypercube sampling (LHS), stratified sampling (SS), Centered stratified sampling (CSS). A detailed description of sampling schemes can be found in [19, 25], for instance.

A special DoE useful for a rough sensitivity analysis is the star-shaped DoE. It consists of $2n_p + 1$ experiments where n_p is the number of parameters. The central design plus, per parameter, a variation to a smaller as well as a larger value (typically with the same distance to the central point) is performed.

Note that the choice of the DoE is depending on the analysis task and the concrete step performed.

4.2 Correlation Measures

The Pearson correlation measure is an often-used, yet easily misleading one, because only monotonous correlations are captured (a typical example is depicted in Figure 1 (on the left)), and particularly for the case depicted in Figure 1 (on the right), it completely fails.

A measure reflecting nonlinearities is necessary as, for instance, the DesParO correlation measure, developed for RBF metamodels (see next section). In order

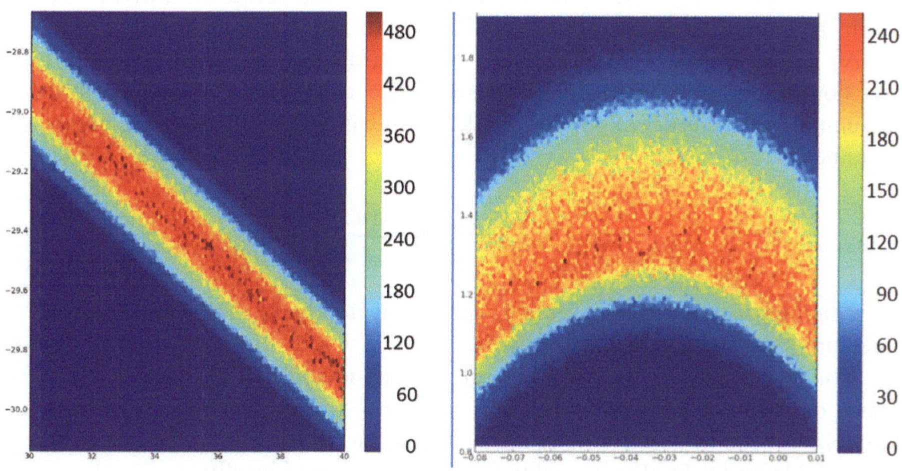

Fig. 1 Exemplary correlation plots. For the situation on the right, the Pearson measure will be 0 which is quite misleading. The color represents the number of samples in the respective hexagonal bin. Note that the actual range of values on the x- and y-axis are not relevant here

to roughly check which parameter-criteria dependencies are still linear or already nonlinear, both measures can be compared. This has been done exemplarily in Figs. 5, 8 and 15.

4.3 *Metamodeling (Response Surfaces) and Adaptive Refinement*

Classically, ensemble evaluations and Pareto optimizations rely on many experiments (simulation runs) - usually a costly procedure, even if the number of parameters involved is reduced beforehand.

For drastically reducing the number of simulation runs, one can set up fast-to-evaluate metamodels (response surfaces). This way, dependencies of objective functions on parameters are interpolated or approximated. Metamodels are quite often a good compromise for balancing the number of simulation runs (or physical measurements) to set up the model and a sufficient accuracy of approximation.

In the DesParO software [6], we use radial basis functions (RBF; e.g. multi-quadrics, ANOVA), see [3], with polynomial detrending and an optional adjustment of smoothing and width.

We developed a measure for the *local tolerance* of the model w.r.t. leave-one-out cross-validation. It has some similarities to what can be done when working with Kriging. By means of this measure, interactive robustness analysis can be supported, in addition to quantile estimators. Analogously, we developed a nonlinear global correlation measure as well. Figs. 5, 8, 15 show examples of DesParO's metamodel explorer. Current tolerances are visualized by red bars below the current value of each criterion. Visualization of correlations is explained in Section 6.1.

As an orientation for the number of experiments n_{exp} which shall be used for constructing a basic RBF metamodel, one can use the following formula, assuming that the order of polynomial detrending is 2 and n_p denotes the number of parameters:

$$n_{exp} \geq C\left(2 + n_p + \frac{n_p(n_p + 1)}{2}\right) \tag{7}$$

C is an integer which can be set to 3, 4, or 5, say, to obtain a rough, small-sized, or medium-sized metamodel, respectively.

A standard measure for the quality of a metamodel is PRESS (predicted residual sums of squares). Originally, it stems from the statistical analysis of regression models, but can be used for other metamodels as well. If a local tolerance estimator is available, as for DesParO, quality can also be assessed locally.

More details and applications are discussed in [2, 8–12, 31], for instance.

A metamodel can be adaptively refined, in particular, if a local tolerance measure is available, as is the case in DesParO. We developed an extension and modification of the expected-improvement method (cf. [30] and references given therein to

Keane's method) for determining points (i.e. sets of parameters), the addition of which to the DoE is expected to improve interpolation. De facto, a hierarchical model results. See [7, 15], for instance.

4.4 *Quantiles and Robust Multi-Objective Optimization*

Several methods for computing quantiles and their applications are discussed in, e.g., [14, 17, 25–27, 29]. Methods for robust multi-objective optimization are presented and their practical applicability discussed in, e.g., [7, 15, 16, 22, 30].

5 Workflows

In the following, several workflows for tackling the analysis tasks summarized in Section 3. The specific task to be considered is denoted with **Analysis Task**.

The flow chart sketched in Figure 2 can be used as an orientation. In order to balance computational effort and accuracy while working with simulations, one should know, in addition, how fast a single simulation run is.

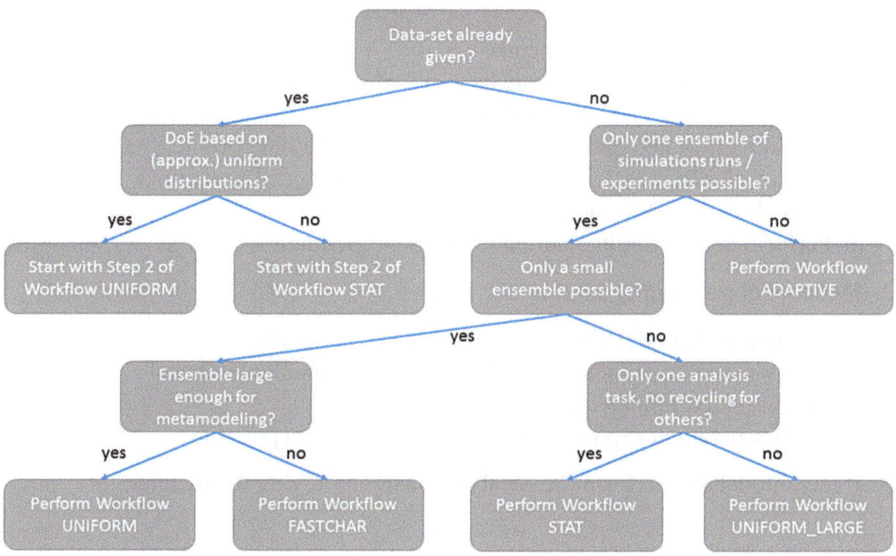

Fig. 2 Flow chart

5.1 Workflow STAT

A standard workflow for performing the **Analysis Task** directly based on simulation runs or experimental data is sketched below:

1. Ensemble set-up
   a. Determine the set of parameters
   b. For each parameter, determine its range of values and distribution according to the **Analysis Task**
   c. Set up a DoE according to the parameter ranges and distributions determined above; for determining the size of the DoE, find a balance between effort and quality
   d. Perform corresponding simulation runs / experiments
2. Perform the **Analysis Task** based on the ensemble

5.2 Workflows UNIFORM and UNIFORM-LARGE

A standard workflow for performing the **Analysis Task** employing a metamodel is sketched below:

1. Ensemble setup
   a. Determine the set of parameters
   b. For each parameter, determine its range of values
   c. Set up uniform DoE; for determining the size of the DoE, find a balance between effort and quality; in case of UNIFORM, find orientation in Eq. 7; in case of UNIFORM-LARGE, estimate the number of experiments necessary for a classical QMC method, say
   d. Perform corresponding simulation runs / experiments
2. Metamodel and quality assessment
   a. Set up a metamodel using the ensemble created above
   b. Check model tolerances (*global PRESS value*, local tolerances)
   c. Check *correlation measures*
   d. Reduce parameter space for analysis task, as far as possible
3. If metamodel ok: Perform the **Analysis Task** employing the metamodel

5.3 Workflow ADAPTIVE

An iterative workflow for adaptive hierarchical metamodeling and optimization of decisive metamodeling parameters is sketched now:

1. (Optional:) Set up a star-shaped DoE for performing a basic sensitivity analysis
2. (Optional:) Based on its results, reduce the parameter space
3. Perform workflow UNIFORM
4. This includes the first run of the **Analysis Task**. In case of RDO, a rough Pareto optimization for finding candidate regions should be performed
5. If necessary, perform *model refinement* (cf. Section 4.3), then go to step 4
6. Perform the final run of the **Analysis Task** employing the metamodel

6 Visualizations

Besides the methods for approximating dependencies and statistical measures, visualization techniques play a decisive role. Without appropriate representation of results, the sometimes immense output cannot be digested and efficiently interpreted. Visualization can efficiently support, for instance, pointing to interesting features of a problem and interactive exploration of parameter-criteria dependencies. Some important techniques are summarized in the following.

6.1 Correlations

Global correlation measures can be visualized by means of tables with boxes. The magnitude of the box represents the magnitude of the absolute correlation value, its color the direction of correlation, e.g., blue for monotonously decreasing, red for monotonously increasing, black for nonmonotonous behaviour.

Examples can be found in Figure 5 (on the right), for instance.

Correlations can directly be visualized by means of two-dimensional scatter plots of all pairs of values involved. However, especially if larger areas of the two-dimensional space are filled this way, a good alternative are two-dimensional histograms (see the next section).

6.2 Histograms and Alternatives

Classical techniques to visualize distributions are

- (one-dimensional) histograms: an example can be found in Figure 14 (on the left)
- approximate CDF curves (CDF: cumulative density function) an example can be found in Figure 10 (on the bottom)
- boxplots

In addition to a histogram, a *plot of sorted values* is often of help. To create one, all values of interest have to be sorted first, decreasing or increasing by (absolute)

value. All values (or selected ranges only) of the resulting vector v are plotted then, i.e., all data points $(i,)v(i)$ are plotted. An example can be found in Figure 14 (on the right).
2D histograms (hexbins) are a good alternative to scatter plots if enough data points are available. Examples can be found in Figs. 1, 7 and 12, for instance.

6.3 2D Network Representations

Colors and thicknesses of nodes and lines can be chosen differently representing values of different inputs or outputs. A classical version is shown in Figure 13. Pressure values are used for coloring the nodes, averaged pressure values for coloring the edges, and pipe widths for determining the thickness of the edges. A 2D network representation is also a good choice for showing differences of values for two scenarios of an ensemble. In order to find areas with large differences, node and/or edge sizes should dependent on the local difference of the output function considered. Typical applications are the comparison of different physical laws, parameter variations (extreme cases), or local differences between the maximal and minimal value in the ensemble for the output function considered.

Manipulating the coordinates is another possibility. Quite often, the coordinates do not reflect the geometrical situation but is a compromise of real locations and a schematic view. Important areas can be given more space this way. Alternatively, algorithms might be used which perform mappings of coordinates in order to highlight areas with, for instance, large mass flows.

6.4 3D Network Representations

A classical setting is to use (x, y, h) for 3D plots. This setting allows for debugging of height data - drastic drop downs to zero, for instance, might indicate missing or wrongly specified height values.

For analysing function profiles, one might use, for instance, pressure or temperature as z-coordinate. Figure 3 provides an example for a realistic pressure profile. Analogously, means, medians, quantiles or differences between minimum or maximum values or two selected quantiles can be visualized.

7 Examples from Gas Transport

Applications of the methods discussed above are illustrated by means of two examples from long-distance gas transport, namely a compressor station and a mid-sized pipeline network with several supplies and regulators, for instance.

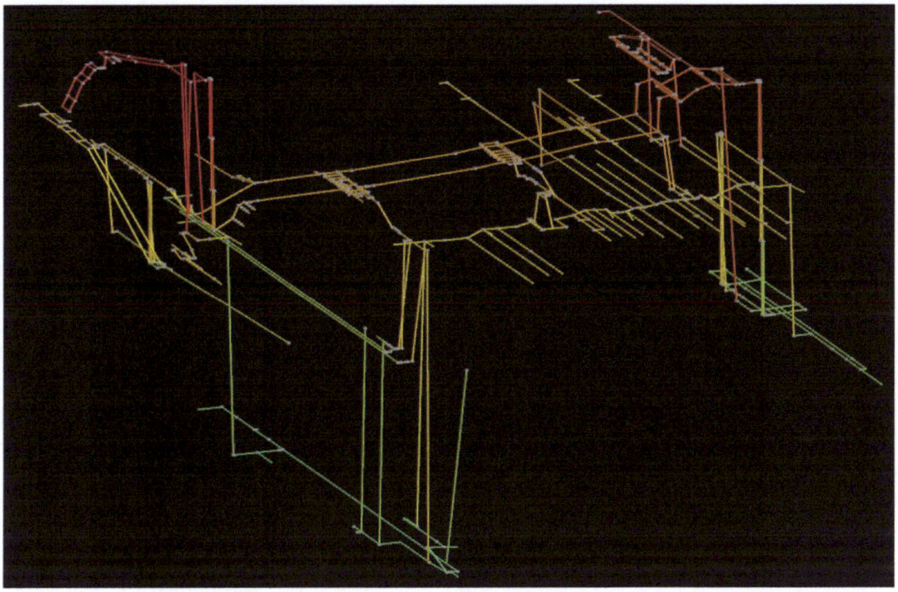

Fig. 3 3D plot (MYNTS' interactive OpenGL-based 3D viewer) for a decisive part of a large realistic network - pressure profile: color and z-coordinate are based on pressure values, clearly showing the different pressure levels and transition areas (due to compressor stations, for example)

7.1 Example 1 - Compressor Station

The first example, see Figure 4, is a simple compressor station consisting of two machine units with a compressor, a drive, a cooler, a regulator, two resistors and a master each, as well as a master and several valves and pipes for controlling the station and switching tracks. One supply and one demand node are attached.

The parameters and criteria investigated overall are listed in Table 2. Two scenarios are analysed. Their settings and results are described and discussed in the following sections.

7.1.1 Scenario 1

In Scenario 1, only the first compressor is up and running, and only QSET is varied.

Workflow UNIFORM is used. The DesParO metamodel resulting from a standard uniform DoE with 50 experiments (after automatically excluding 9 parameter values which are very close to others) is shown in Figure 5. 50 experiments are not really necessary here, given that only one parameter is varied. The metamodel would react more or less identically if only 10 experiments are used, say. However, the same ensemble can be used to create a model, evaluate it randomly and plot 1D and 2D histograms, see Figure 6, as well as to plot finely resolved curves for parameter-criteria dependencies directly, see Figure 7.

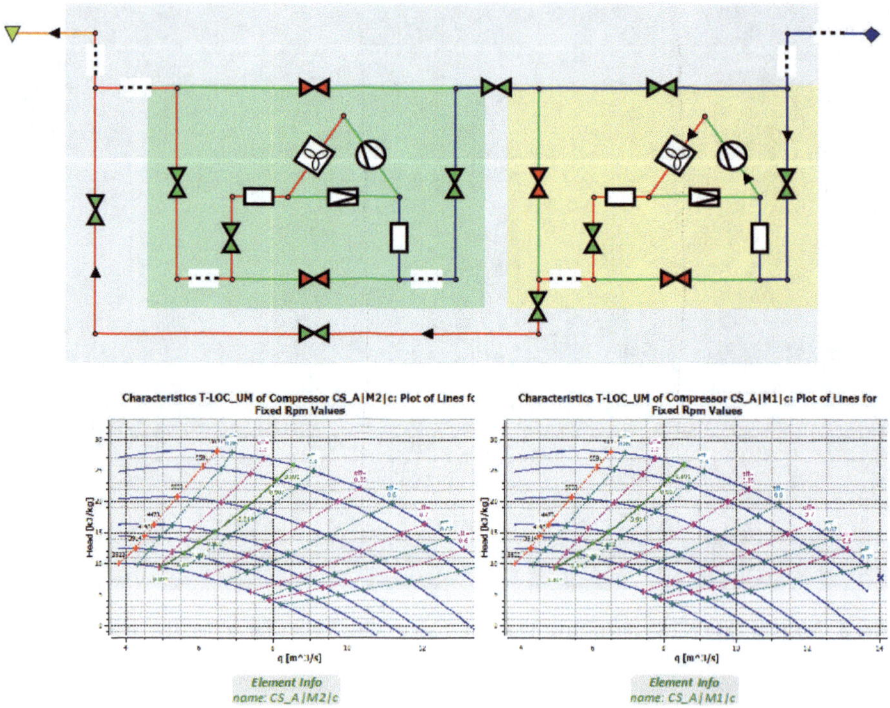

Fig. 4 Example 1: (*top*) schematics of the network (*diamond* on the right marks the supply, *triangle* on the left the demand node, other elements as explained in the text); (*bottom*) characteristics maps of the two compressors

Table 2 Example 1: parameters and criteria

property	element	type	description
PSET	junc0	parameter	input pressure at supply
QSET	junc1	parameter	output volume flow at demand
Had	CS_A|M1|c	criterion	head of compressor 1
QVOL	CS_A|M1|c	criterion	volume flow through compressor 1
tslope	CS_A|M1|c	criterion	tslope of compressor 1
pslope	CS_A|M1|c	criterion	pslope of compressor 1
m	CS_A|M1|c	criterion	mass flow through compressor 1
pslope	junc0^locE	criterion	pslope at pipe beginning at the supply
tslope	junc0^locE	criterion	tslope at pipe beginning at the supply
m	junc0^locE	criterion	mass flow at pipe beginning at the supply
T	junc1	criterion	temperature at demand
P	junc1	criterion	pressure at demand
m	locA^junc1	criterion	mass flow at pipe ending at the demand
tslope	locA^junc1	criterion	tslope at pipe ending at the demand
PSLOPE	locA^junc1	criterion	pslope at pipe ending at the demand

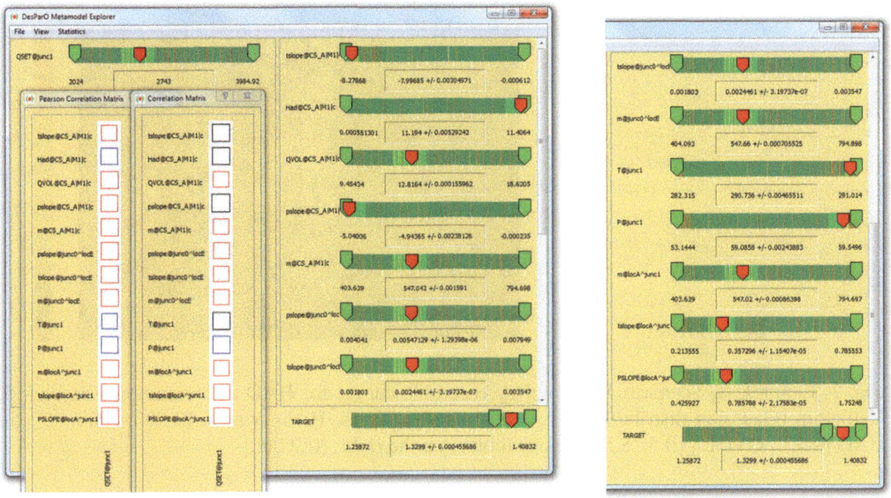

Fig. 5 Example 1, Scenario 1: metamodel and correlations for the parameters and criteria listed in Table 2. Criteria from top to bottom: tslope, Had, QVOL, pslope, m of the compressor, pslope, tslope and m of junc0^locE, T and P of junc1, m, tslope and pslope of locA^junc1

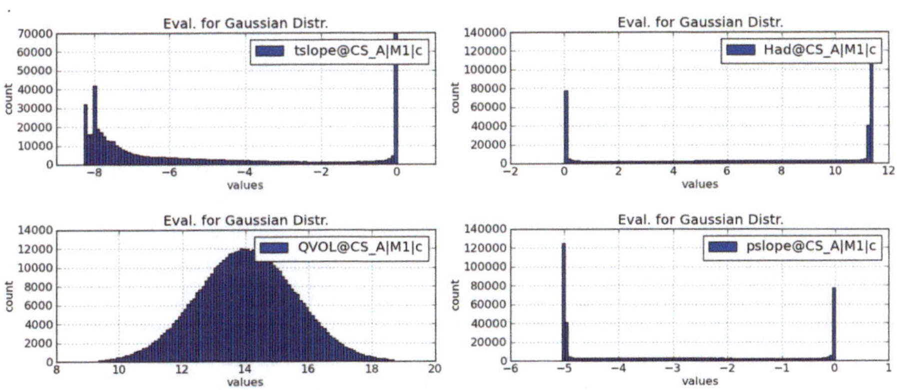

Fig. 6 Example 1, Scenario 1: histograms (Gaussian distribution for QSET, one million evaluations) of the first 4 criteria. x-axis: values (clockwise: approx. [-8.5;0.5], [-2;12], [-6;1], [8;20]), y-axis: counts (clockwise: up to 70,000, 140,000, 140,000, 14,000)

Figure 5 shows both Pearson and DesParO correlation results. The magnitude of the correlation values is very similar among the parameters. However, several correlations are nonlinear (black box in the DesParO correlation plot), and Pearson indicates a large monotonous correlation instead. Especially, Had@CS_A|M1|c reacts in a strongly nonlinear fashion to changes of QSET, see also Figs. 6 and Figure 7. This criterion is decisive for describing the behaviour of the compressor. Hence, the nonlinearity cannot be neglected, and a linear model and Pearson correlation are not sufficient in this small and quite simple test case involving one compressor only.

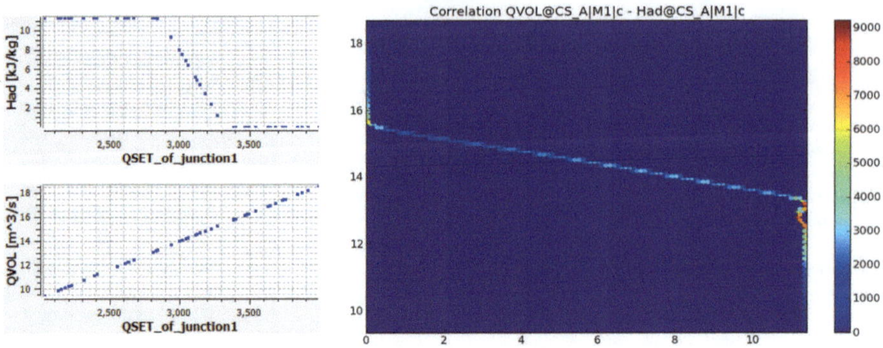

Fig. 7 Example 1, Scenario 1: exemplary ensemble curves for HAD of the compressor, and 2D histogram (Gaussian distribution for QSET) for QVOL vs. HAD. x-axis: [2,000;4,000], y-axis: HAD [0;12], QVOL [10.5;18.5]. Counts (color) on the right: [0;9,000]

7.1.2 Scenario 2

In Scenario 2 both compressors are up and running in parallel, and both parameters (PSET and QSET) are varied: $[50; 60]$ for PSET, and $[2,000; 10,000]$ for QSET. We already learned from analysing Scenario 1 that several parameter-criteria dependencies are expected to be strongly nonlinear, depending on the concrete range of variations.

Again, workflow UNIFORM is used, and, since we have only two parameters here, a standard full-factorial DoE with $7^2 = 49$ experiments is chosen as a basis for direct analysis as well as creation of a DesParO metamodel. Ensemble curves, i.e., raw-data plots of parameter-criteria dependencies, are shown in Figure 11.

Indeed, Scenario 2 shows several interesting effects. The ranges are de facto chosen here so that the compressors cannot completely fulfill their task of creating an output pressure of 60 bars. Fig. 11 (top-right plot) clearly shows that 60 bars cannot be reached for most combinations. Analogously, Figure 11 (bottom-left plot) shows that the compressor goes to de facto bypass mode (zero head) for QSETs above approximately 7,000.

The metamodel for the ensemble is shown in Figure 8. The model has a reasonable quality (see PRESS values), however, DesParO's tolerance measure cannot be neglected here. Based on the parameter distributions depicted in Figure 9, the metamodel is evaluated and 1D and 2D histograms for several exemplary interesting dependencies shown in Figs. 10 and 12. Note that skewed Gaussian distributions are used, in case of QSET only for a part, namely $[2,000; 7,000]$, of the range covered by the metamodel (Figure 8).

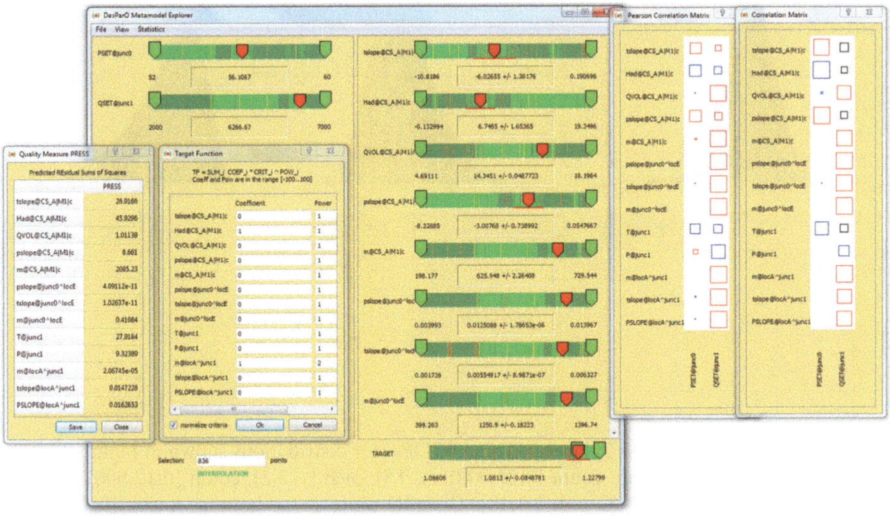

Fig. 8 Example 1, Scenario 2: metamodel, correlations, PRESS values, target function

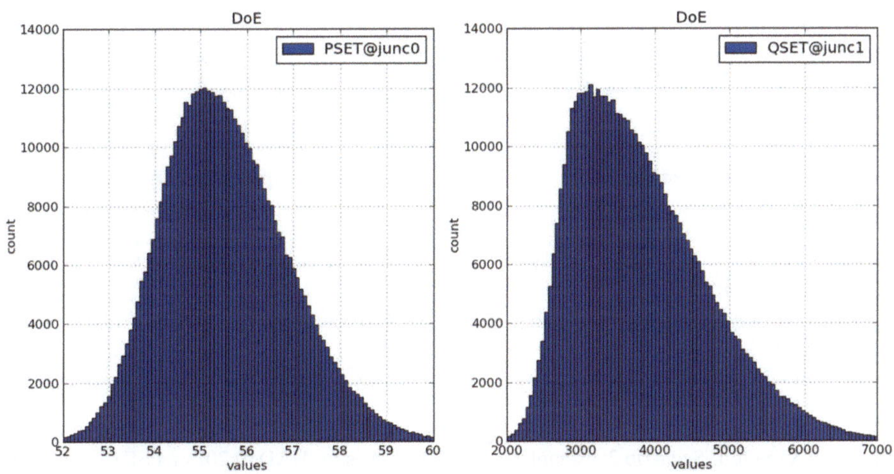

Fig. 9 Example 1, Scenario 2: histograms of the skewed Gaussian distributions of parameters PSET [52;60] and QSET [2,000;7,000]. Counts (y-axis): [0;14,000]

Looking at Had@CS_A|M1|c and QVOL@CS_A|M1|c, one can study effects of nonlinear, linear or very weak dependencies on variations of PSET and QSET here. QVOL has to react linearly on variations of QSET, as the 1D histogram in Figure 10 as well as the 2D histogram in Figure 12 show. As long as the compressors are able to completely fulfill their common task (PSET large enough), QVOL does weakly react on PSET. For smaller PSETs and large QSETs, the compressors go to their limit, and the distribution of QVOL is wide.

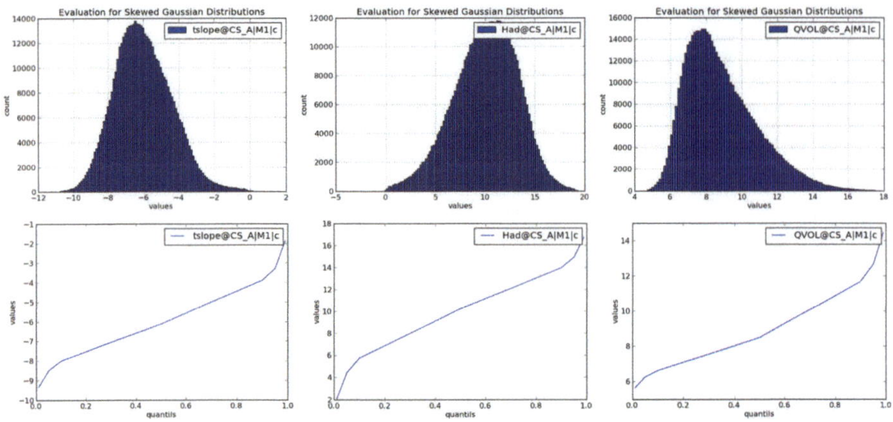

Fig. 10 Example 1, Scenario 2: histograms and CDFs (for distributions shown in Figure 9) of the first 3 criteria. Figures on top: count [0;14,000] vs. TSLOPE [-12;2], count [0;12,000] vs. HAD [-5;20], count [0;16,000] vs. QVOL [-4;18]. Quantile plots at the bottom: x-axis [0;1], y-axis: TSLOPE [-10;1], HAD [2;18], QVOL [5.5;14.5]

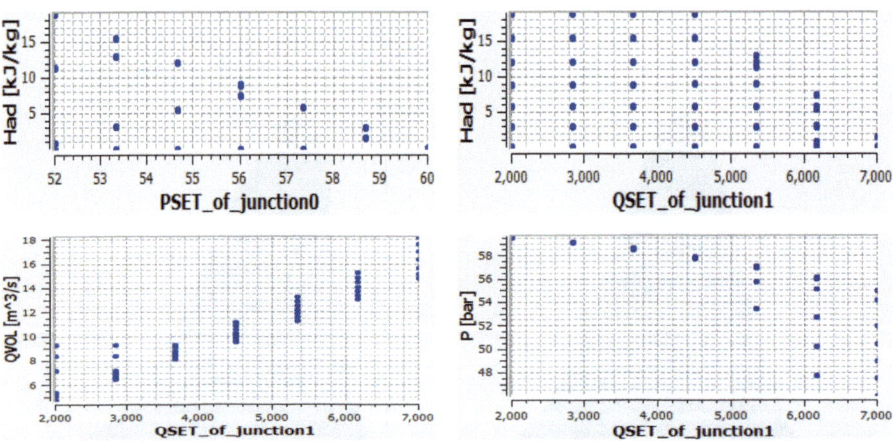

Fig. 11 Example 1, Scenario 2: exemplary ensemble curves for HAD [0;20] vs. PSET [52;60] and QSET [2,000;7,000] as well as QVOL [5;18] and P@junction1 [46;59.5] vs. QSET [2,000;7,000]

Based on the metamodel, RDO tasks can be set up and solved. Figure 8 shows such a task and its visual exploration. Here, the following target is set:

$$\max \text{Had@CS\_A}|\text{M1}|\text{c} + (\text{m@locA\^{}junc1})^2 \tag{8}$$

One could also use, for instance, QVOL@CS_A|M1|c instead of m@locA^junc1. By visual inspection, one can see that the parameter space separates into two parts: one is with mid-sized PSET and QSET, one with small PSET and large QSET. DesParO's tolerance measure gives a first indication of robustness of results.

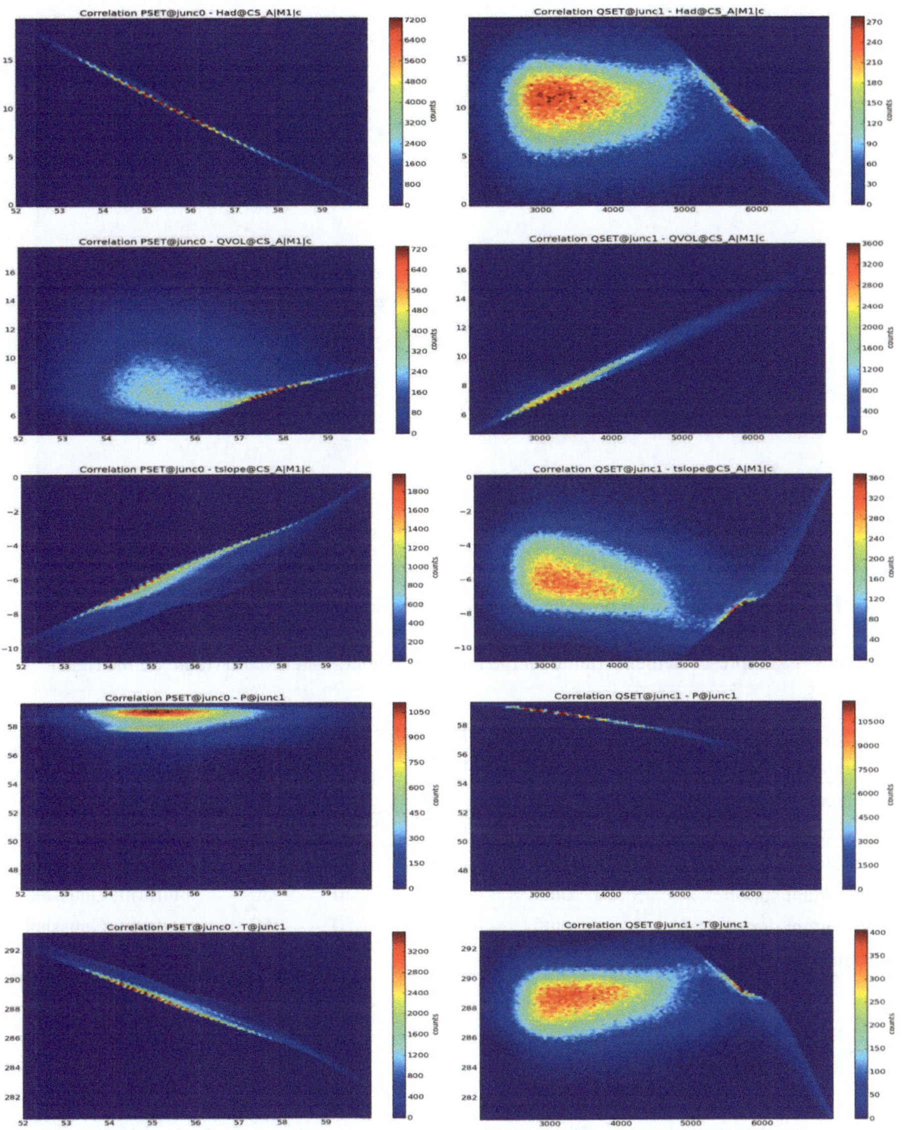

Fig. 12 Example 1, Scenario 2: exemplary 2D correlation plots for exemplary criteria vs. PSET (on the left) or QSET (on the right). Ranges for values analogously as before, counts up to (left) 7200, 720, 2000, 1100, 3600, (right) 270, 3600, 360, 12000, 400

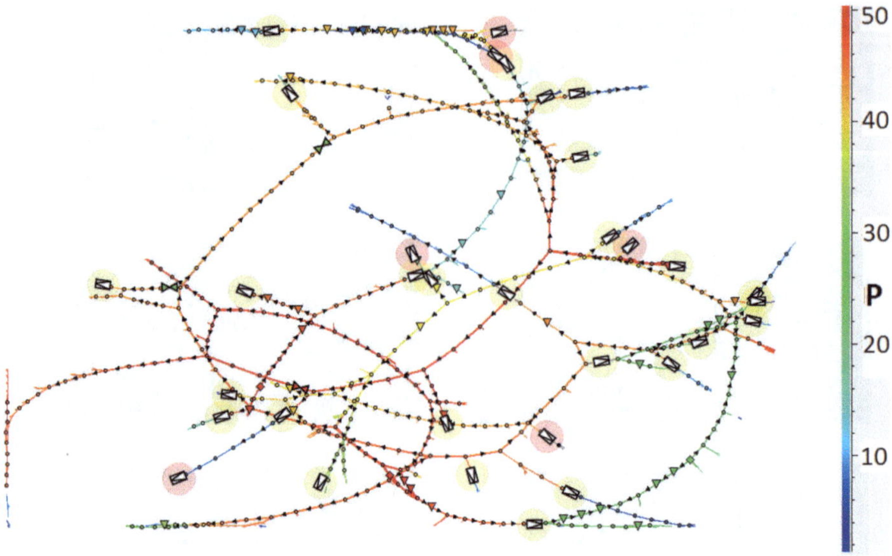

Fig. 13 Example 2: net with pressure results of the basic scenario considered

Table 3 Example 2: elements (table on the left), parameters and criteria (table on the right)

element type	number
pipes	907
regulators	45
heaters	45
valves	52
resistors	89
important supplies	5
important demands	5
remaining nodes	1069

element	parameter	min	max	distributions DoE; analysis
(all)	tsoil	267.15	287.15	uniform; Gaussian
out1	QSET	20	30	uniform; Gaussian
out2	QSET	50	60	uniform; Gaussian
out3	QSET	30	40	uniform; Gaussian
out4	QSET	20	30	uniform; Gaussian
out5	QSET	50	60	uniform; Gaussian
in1...in5	m	–	–	(criteria)

7.2 *Example 2*

The second example, see Figure 13, is a mid-sized network consisting of the elements mentioned in Table 3 (on the left). A typical distribution of pressures resulting from the simulation of a typical scenario is shown in Figure 14. In contrast to Example 1, this network does not contain compressors. The task is here to determine the influence of variations of the 5 largest demands as well as the soil temperature on the network, see Table 3 (on the right).

Again, workflow UNIFORM is used, and a standard uniform DoE with 50 experiments is chosen as a basis for direct analysis as well as creation of a DesParO metamodel. A thin full-factorial DoE for checking nonlinearities would already have $6^3 = 216$ experiments. As can be seen from Figure 15, the criteria depend

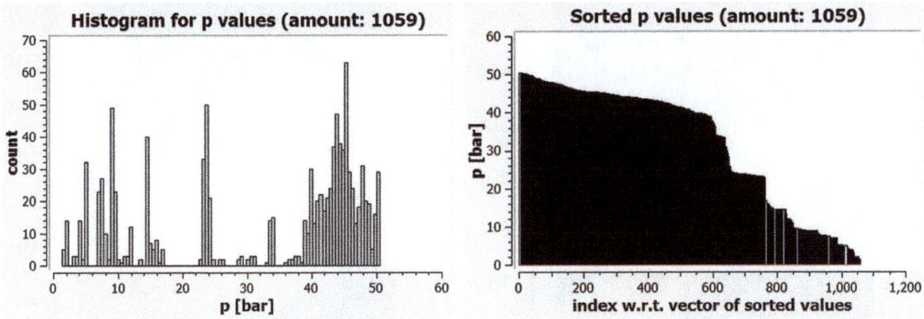

Fig. 14 Example 2: exemplary histogram (left: count [0;70] vs. pressure [0;60]) and plot of sorted values (right) for pressure (P) (y-axis: pressure [0;60], x-axis: index [0;1,200])

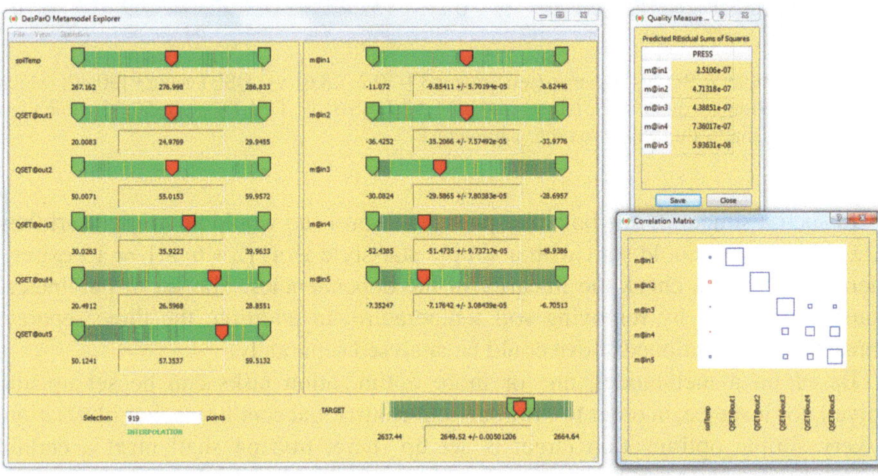

Fig. 15 Example 2: metamodel exploration, PRESS quality measure and correlation plot for the parameters and criteria listed in Table 3

monotonously from the parameters. They are de facto quite linear: the Pearson and DesParO correlation measures are more or less identical, the PRESS values for the quality of the metamodel tiny, of course. As can be seen from the correlation plots, the soil temperature does not play a decisive role. The interplay of the different supply demand combinations reveals a partition into three parts: two one-to-one constellations (out1-in1 and out2-in2), one three-to-three constellation. Figure 16 shows exemplary combinations of strongly dependent parameter-criteria combinations.

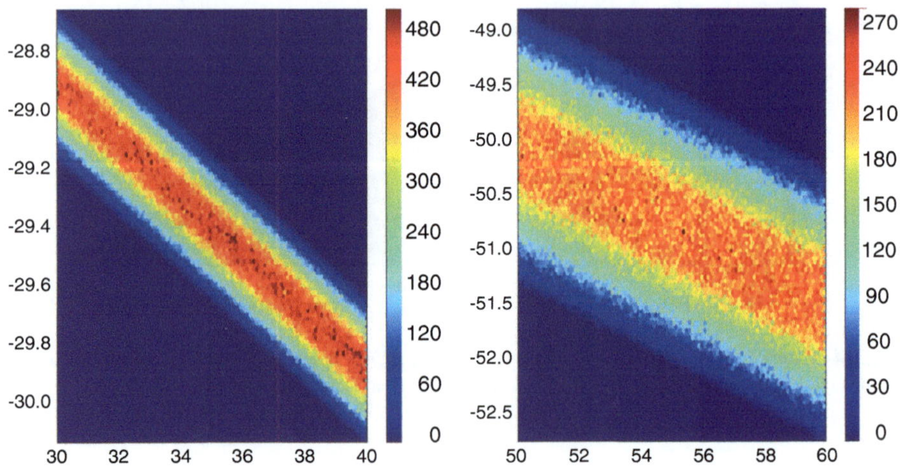

Fig. 16 Example 2: 2D histograms: (left) m@in3 [-30.2;-28.6] vs. QSET@out3 [30;40] which has a decisive impact on m@in3, (right) m@in4 [-52.7;-48.8] vs. QSET@out5 [50;60] which has an impact but shares this with two other parameters

In such a situation, maybe a cheaper way to proceed would be to use workflow ADAPTIVE. We could start with a simple star-style DoE ($2 * 6 + 1 = 13$ experiments) in order to check the linearity of the dependencies. The set of parameters could be reduced by removing soil temperature. In addition, the three separate constellations mentioned above could be analysed separately.

Based on a metamodel, one or more optimization tasks can be set up and solved, for instance, in order to fulfill certain contractual conditions. For illustration, a very simple optimization target is set up (here: m@in4 shall meet a certain value). Figure 15 shows how a visual exploration already reveals influences on the parameter combinations.

8 Conclusions and Outlook

Several methods for approximating parameter-criteria dependencies and determining statistical quantities, corresponding workflows and versatile visualization methods for analysing parameter variations in energy networks have been described and discussed. A common physical model for flow in gas transport networks has been described. Two exemplary gas networks with some typical effects have been studied in some detail using the methods and workflows discussed.

From examining the examples, one can see that nonlinearities might play a decisive role, especially in networks with compressor stations. Numerical methods for measuring correlations and approximating parameter-criteria dependencies have to be chosen which are able to reflect nonlinearities. Regression-based linear methods can sometimes support the analysis though - a comparison of the values provided by Pearson's correlation measure with the ones provided by DesParO's measure indicates nonlinearities in an intuitive fashion.

The challenge of transforming our energy production by means of incorporating increasingly larger amounts of renewable energy sources can only succeed if also our energy networks are transformed to build up an integrated system allowing for balancing of supplies and demands with the use of energy conversion and storages provided by pipeline systems and caverns, to give just some examples. This motivates the author and her colleagues a lot. They will continue their work on physical modeling of energy networks, their efficient simulation, statistical analysis and optimization.

References

1. Baumanns, S., Cassirer, K., Clees, T., Klaassen, B., Nikitin, I., Nikitina, L., Tischendorf, C.: MYNTS User's Manual, Release 1.3. Fraunhofer SCAI, Sankt Augustin, Germany (2012). www.scai.fraunhofer.de/mynts
2. Borsotto, D., Clees, T., Nikitin, I., Nikitina, L., Steffes-lai, D., Thole, C.A.: Sensitivity and robustness aspects in focused ultrasonic therapy simulation. In: EngOpt 2012 – 3rd International Conference on Engineering Optimization. Rio de Janeiro, Brazil (2012)
3. Buhmann, M.: Radial Basis Functions: Theory and Implementations. Cambridge University Press, Cambridge (2003)
4. Cassirer, K., Clees, T., Klaassen, B., Nikitin, I., Nikitina, L.: MYNTS User's Manual, Release 2.9. Fraunhofer SCAI, Sankt Augustin (2015). www.scai.fraunhofer.de/mynts
5. Clees, T.: MYNTS – Ein neuer multiphysikalischer Simulator für Gas, Wasser und elektrische Netze. Energie — Wasser-Praxis **09**, 174–175 (2012)
6. Clees, T., Hornung, N., Nikitin, I., Nikitina, L., Pott, S., Steffes-lai, D.: DesParO User's Manual, Release 2.2. Fraunhofer SCAI, Sankt Augustin, Germany (2012). www.scai.fraunhofer.de/desparo
7. Clees, T., Hornung, N., Oyerinde, A., Stern, D.: An adaptive hierarchical metamodeling approach for history matching of reservoir simulation models. In: SPE/SIAM Conference on Mathematical Methods in Fluid Dynamics and Simulation of Giant Oil and Gas Reservoirs (LSRS). Istanbul, Turkey (2012). Invited presentation (T. Clees)
8. Clees, T., Nikitin, I., Nikitina, L.: Nonlinear metamodeling of bulky data and applications in automotive design. In: Günther, M., et al. (eds.) Progress in industrial mathematics at ECMI 2010. Mathematics in Industry, vol. 17, pp. 295–301. Springer, Berlin (2012)
9. Clees, T., Nikitin, I., Nikitina, L., Kopmann, R.: Reliability analysis of river bed simulation models. In: Herskovits, J. (ed.) CDROM Proceedings of the EngOpt 2012, 3rd International Conference on Engineering Optimization, no. 267. Rio de Janeiro, Brazil (2012)
10. Clees, T., Nikitin, I., Nikitina, L., Thole, C.A.: Nonlinear metamodeling and robust optimization in automotive design. In: Proceedings of the 1st International Conference on Simulation and Modeling Methodologies, Technologies and Applications SIMULTECH 2011, pp. 483–491. SciTePress, Noordwijkerhout, The Netherlands (2011)
11. Clees, T., Nikitin, I., Nikitina, L., Thole, C.A.: Analysis of bulky crash simulation results: deterministic and stochastic aspects. In: Pina, N., et al. (eds.) Simulation and Modeling Methodologies, Technologies and Applications, AISC 197. Lecture Notes in Advances in Intelligent and Soft Computing, pp. 225–237. Springer, Berlin, Heidelberg (2012)

12. Clees, T., Steffes-lai, D., Helbig, M., Sun, D.Z.: Statistical analysis and robust optimization of forming processes and forming-to-crash process chains. Int. J. Mater. Form. **3**, 45–48 (2010). Supplement 1; 13th ESAFORM Conference on Material Forming. Brescia, Italy (2010)
13. Grundel, S., Hornung, N., Klaassen, B., Benner, P., Clees, T.: Computing surrogates for gas network simulation using model order reduction. In: Koziel, S., Leifsson, L. (eds.) Surrogate-Based Modeling and Optimization, pp. 189–212. Springer, New York (2013)
14. Harrell, F.E., Davis, C.E.: A new distribution-free quantile estimator. Biometrika **69**, 635–640 (1982)
15. Hornung, N., Nikitina, L., Clees, T.: Multi-objective optimization using surrogate functions. In: Proceedings of the 2nd International Conference on Engineering Optimization (EngOpt). Lisbon, Portugal (2010)
16. Jones, D., Schonlau, M., Welch, W.: Efficient global optimization of expensive black-box functions. J. Glob. Optim. **13**(4), 455–492 (1998)
17. Jones, M.C.: The performance of kernel density functions in kernel distribution function estimation. Stat. Probab. Lett. **9**(2), 129–132 (1990)
18. Klaassen, B., Clees, T., Tischendorf, C., Soto, M.S., Baumanns, S.: Fully coupled circuit and device simulation with exploitation of algebraic multigrid linear solvers. In: Proceedings of the Equipment Data Acquisition Workshop. Dresden (2011)
19. Kleijnen, J.: Design and Analysis of Simulation Experiments. Springer, New York (2008)
20. Lorenz, J., Bär, E., Clees, T., Evanschitzky, P., Jancke, R., Kampen, C., Paschen, U., Salzig, C., Selberherr, S.: Hierarchical simulation of process variations and their impact on circuits and systems: results. IEEE Trans. Electron Devices **58**(8), 2227–2234 (2011)
21. Lorenz, J., Clees, T., Jancke, R., Paschen, U., Salzig, C., Selberherr, S.: Hierarchical simulation of process variations and their impact on circuits and systems: methodology. IEEE Trans. Electron Devices **58**(8), 2218–2226 (2011)
22. Maass, A., Clees, T., Nikitina, L., Kirschner, K., Reith, D.: Multi-objective optimization on basis of random models for ethylene oxide. Mol. Simul. Special Issue: FOMMS 2009 Conference Proceedings, vol. **36**(15), pp. 1208–1218(11) (December 2010)
23. Maric, I., Ivek, I.: Natural gas properties and flow computation. In: Potocnik, P. (ed.) Natural Gas. InTech (2010). ISBN: 978-953-307-112-1. doi: 10.5772/9871. Available from: http://www.intechopen.com/books/natural-gas/natural-gas-properties-and-flow-computation
24. Mischner, J., Fasold, H.G., Kadner, K.: gas2energy.net - Systemplanerische Grundlagen der Gasversorgung. Div Deutscher Industrieverlag München (2011). ISBN 978-3835632059
25. Rhein, B., Clees, T., Ruschitzka, M.: Robustness measures and numerical approximation of the cumulative density function of response surfaces. Commun. Stat. Simul. Comput. **43**(1), 1–17 (2014)
26. Rhein, B., Clees, T., Ruschitzka, M.: Uncertainty quantification using nonparametric quantile estimation and metamodeling. In: Eberhardsteiner, J., et.al. (eds.) European Congress on Computational Methods in Applied Sciences and Engineering (ECCOMAS 2012). Vienna, Austria (2012)
27. Rhein, B., Ruschitzka, M., Clees, T.: A simulation framework for robust optimization based on metamodels. In: Proceedings of NAFEMS World Congress 2013, International Conference on Simulation Process and Data Management, Salzburg, 9–12 June 2013
28. Schöps, S., Bartel, A., Günther, M., ter Maten, E.J.W., Müller, P.C. (eds.): Progress in Differential-Algebraic Equations, Differential-Algebraic Equations Forum. Proceedings of Descriptor 2013, pp. 183–205. Springer, Berlin, Heidelberg (2014)
29. Sfakianakis, M.E., Verginis, D.G.: A new family of nonparametric quantile estimators. Commun. Stat. Simul. Comput. **37**, 337–345 (2008)
30. Sobester, A., Leary, S., Keane, A.: On the design of optimization strategies based on global response surface approximation models. J. Glob. Optim. **33**(1), 31–59 (2005)
31. Steffes-lai, D., Clees, T.: Statistical analysis of forming processes as a first step in a process-chain analysis: novel PRO-CHAIN components. Key Engineering Materials (KEM) **504–506**, 631–636 (2012). Special Issue Proceedings of the 15th ESAFORM Conference on Material Forming. Erlangen, Germany (2012)

Fast Multi-Objective Aerodynamic Optimization Using Space-Mapping-Corrected Multi-Fidelity Models and Kriging Interpolation

Leifur Leifsson, Slawomir Koziel, Yonatan Tesfahunegn, and Adrian Bekasiewicz

Abstract The chapter describes a computationally efficient procedure for multi-objective aerodynamic design optimization with multi-fidelity models, corrected using space mapping, and kriging interpolation. The optimization procedure utilizes a multi-objective evolutionary algorithm to generate an initial Pareto front which is subsequently refined iteratively using local enhancements of the kriging-based surrogate model. The refinements are realized with space mapping response correction based on a limited number of high-fidelity training points allocated along the initial Pareto front. The method yields—at a low computational cost—a set of designs representing trade-offs between the conflicting objectives. We demonstrate the approach using examples of airfoil design, one in transonic flow and another one in low-speed flow, in low-dimensional design spaces.

Keywords Multi-objective optimization • Aerodynamic design • Multi-fidelity CFD models • Space mapping • Kriging interpolation

MSC code: 65K10

1 Introduction

Aerodynamic design problems are mostly solved using single-objective optimization. However, aerodynamic design is by nature a multi-objective task. Handling the problem as a multi-objective one is often impractical due to the cost of

L. Leifsson (✉)
Department of Aerospace Engineering, Iowa State University, Ames, IA 50011, USA
e-mail: leifur@iastate.edu

S. Koziel • Y. Tesfahunegn • A. Bekasiewicz
Engineering Optimization & Modeling Center, School of Science and Engineering, Reykjavik University, Menntavegur 1, 101 Reykjavik, Iceland
e-mail: koziel@ru.is; yonatant@ru.is; bekasiewicz@ru.is

S. Koziel et al. (eds.), *Simulation-Driven Modeling and Optimization*, Springer Proceedings in Mathematics & Statistics 153,
DOI 10.1007/978-3-319-27517-8_3

running high-fidelity computational fluid dynamics (CFD) simulations, which are ubiquitous in aerodynamic design [1, 2]. Although CFD-based parameter sweeps and engineering experience are common practice when searching for better designs in a single-objective sense, automation using numerical optimization techniques is becoming increasingly popular and necessary [3–6]. A multitude of methods for single-objective optimization problems are available such as conventional gradient-based algorithms [7] (including those utilizing inexpensive adjoint sensitivities [8, 9]), and surrogate-based optimization (SBO) techniques [10–15] that offer efficient global optimization as well as a substantial reduction of the design cost when compared to the traditional methods [12]. In this chapter, we describe an efficient approach to solve multi-objective aerodynamic design problems using high-fidelity simulations and surrogate-based methods.

A Pareto front is a common way of representing solutions to multi-objective problems. It contains a set of the best possible designs which are non-commensurable in the conventional (single-objective sense) [16]. The most widely used multi-objective optimization methods include multi-objective evolutionary algorithms (MOEAs) [17–20]. The computational complexity of MOEAs is high due to the fact that a (usually large) set of designs is being processed in each iteration of the algorithm. SBO [10–15] can be used to reduce the computational cost of multi-objective aerodynamic design by using inexpensive/less costly surrogate functions in lieu of the CPU-intensive high-fidelity models.

In general, the surrogate models can be created by either approximating the sampled high-fidelity model data using regression (so-called data-driven surrogates [10, 11, 15]), or by correcting physics-based low-fidelity models (so-called multi-fidelity surrogates [12–14, 21–23]) which are less accurate but computationally inexpensive/less costly representations of the high-fidelity models.

Data-driven surrogate models can be constructed using polynomial regression, radial basis function interpolation, kriging, and support vector regression [10, 11]. Typically, a substantial amount of data samples, selected using Design of Experiments [11], are required.

Multi-fidelity surrogates [12] are constructed using low-fidelity models that are manipulated to become a reliable representation of the high-fidelity models. The two main parts in constructing a multi-fidelity surrogate are (1) developing the low-fidelity models, and (2) modifying them to achieve a better representation of the high-fidelity models. Techniques to modify (also called correct) the low-fidelity models include bridge functions [24–26], calibration [21, 22], space mapping [27–29, 41], shape-preserving response prediction [30, 31], adaptive response correction [32], and adaptive response prediction [33]. The multi-fidelity models are usually more expensive to evaluate than the data-driven surrogates, but fewer high-fidelity model data are required to obtain a given accuracy level as compared to when using the high-fidelity data in the data-driven surrogates directly.

This chapter describes a multi-objective procedure for aerodynamic design exploiting low-fidelity CFD simulations, space mapping, kriging interpolation, and MOEAs. The procedure is illustrated using examples involving the design of transonic and low-speed airfoil shapes.

2 Methodology

In this section, we provide a formulation of the multi-objective design problem, a procedure for constructing multi-fidelity surrogate model, as well as outline the multi-objective optimization algorithm.

2.1 Problem Formulation

Let $\boldsymbol{x}$ be an $n \times 1$ vector of the design variables, and $\boldsymbol{f}(\boldsymbol{x}) = [f_1(\boldsymbol{x})\, f_2(\boldsymbol{x})\ \ldots\ f_q(\boldsymbol{x})]^T$ be a $q \times 1$ vector of the high-fidelity model responses. Examples of responses include the airfoil section drag coefficient $f_1(\boldsymbol{x}) = C_{d.f}$ and the section lift coefficient $f_2(\boldsymbol{x}) = C_{l.f}$. Let $F_{obj,i}(\boldsymbol{x})$, $i = 1, \ldots, N_{obj}$, be the ith design objective. A typical performance objective would be to minimize the drag coefficient, in which case $F_{obj,i}(\boldsymbol{x}) = C_{d.f}$. Another objective would be to maximize lift, in which case $F_{obj,i}(\boldsymbol{x}) = 1/C_{l.f}$ or $F_{obj,i}(\boldsymbol{x}) = -C_{l.f}$ (normally, the objectives are supposed to be minimized so the maximization problem has to be transformed into a minimization one before carrying out the design process). Yet another objective could be to minimize a noise metric NM_f, in which case $F_{obj,i}(\boldsymbol{x}) = NM_f$.

If $N_{obj} > 1$, then any two designs $\boldsymbol{x}^{(1)}$ and $\boldsymbol{x}^{(2)}$ for which $F_{obj,i}(\boldsymbol{x}^{(1)}) < F_{obj,i}(\boldsymbol{x}^{(2)})$ and $F_{obj,l}(\boldsymbol{x}^{(2)}) < F_{obj,l}(\boldsymbol{x}^{(1)})$ for at least one pair $i \neq l$, are not commensurable, i.e., none is better than the other in the multi-objective sense. We define Pareto-dominance relation Υ (see, e.g., Fonseca [16]), saying that for the two designs $\boldsymbol{x}$ and $\boldsymbol{y}$, we have $\boldsymbol{x}\ \Upsilon\ \boldsymbol{y}$ ($\boldsymbol{x}$ dominates over $\boldsymbol{y}$) if $F_{obj,i}(\boldsymbol{x}) \leq F_{obj,i}(\boldsymbol{y})$ for all $i = 1, \ldots, N_{obj}$, and $F_{obj,i}(\boldsymbol{x}) < F_{obj,i}(\boldsymbol{y})$ for at least one i. In other words, the point $\boldsymbol{x}$ dominates over $\boldsymbol{y}$ if it is not worse than $\boldsymbol{y}$ with respect to all objectives, and it is better than $\boldsymbol{y}$ with respect to at least one objective. The goal of the multi-objective optimization is to find a representation of a Pareto front X_P of the design space X, such that for any $\boldsymbol{x} \in X_P$, there is no $\boldsymbol{y} \in X$ for which $\boldsymbol{y}\ \Upsilon\ \boldsymbol{x}$ (Fonseca [16]).

2.2 Multi-Fidelity Surrogate Model Construction

The process of finding the Pareto front is realized using evolutionary algorithms (EAs) [17]. EAs iteratively process the entire set of potential solutions to the problem at hand. Therefore, they typically require numerous evaluations of the objective function. Consequently, using the expensive high-fidelity model, $\boldsymbol{f}(\boldsymbol{x})$, directly in the multi-objective optimization is normally prohibitive. This difficulty is alleviated using a surrogate model, $\boldsymbol{s}(\boldsymbol{x})$, constructed using a corrected low-fidelity model, $\boldsymbol{c}(\boldsymbol{x})$.

A low-fidelity model can be developed based on, for example, (1) simplified physics, (2) reduced grid discretization, and (3) reduced solver convergence criteria, or any combination of the aforementioned approaches [13]. Here, we use a combination of approaches (2) and (3).

Given the low-fidelity model $\boldsymbol{c}(\boldsymbol{x})$ the process of constructing the surrogate is as follows. Let $\boldsymbol{c}(\boldsymbol{x}) = [c_1(\boldsymbol{x})\ c_2(\boldsymbol{x})\ \ldots\ c_q(\boldsymbol{x})]^T$ denote a $q \times 1$ vector of responses from the low-fidelity model. We will use a response with $q = 3$, and $c_1(\boldsymbol{x}) = C_{l.c}(\boldsymbol{x})$, $c_2(\boldsymbol{x}) = C_{d.c}(\boldsymbol{x})$, and $c_3(\boldsymbol{x}) = NM_c(\boldsymbol{x})$ to demonstrate the construction process (here, $C_{l.c}$ and $C_{d.c}$ are the lift and drag coefficients, respectively, and NM_c is the noise metric). The surrogate model is constructed in two steps.

In Step 1, the multi-point space mapping correction is applied to the low-fidelity model. The initial surrogate model $\boldsymbol{s}_0(\boldsymbol{x})$, a vector of the same dimension as $\boldsymbol{c}(\boldsymbol{x})$, is obtained by applying a parameterized output space mapping [27, 28, 34]. The mapping uses the correction terms that are directly applied to the response components $C_{l.c}(\boldsymbol{x})$, $C_{d.c}(\boldsymbol{x})$, and $NM_c(\boldsymbol{x})$ of the low-fidelity model. The initial surrogate model is defined as [27]

$$\boldsymbol{s}_0(\boldsymbol{x}) = \boldsymbol{A}(\boldsymbol{x}) \circ \boldsymbol{c}(\boldsymbol{x}) + \boldsymbol{D}(\boldsymbol{x}) = \Big[a_l(\boldsymbol{x})\, C_{l.c}(\boldsymbol{x}) + d_l(\boldsymbol{x}) \quad a_d(\boldsymbol{x})\, C_{d.c}(\boldsymbol{x}) + \\ + d_d(\boldsymbol{x}) \quad a_N(\boldsymbol{x})\, NM_c(\boldsymbol{x}) + d_N(\boldsymbol{x})\Big]^T, \tag{1}$$

where ° denotes component-wise multiplication. Both the multiplicative and additive correction terms are design-variable-dependent and take the form of

$$\boldsymbol{A}(\boldsymbol{x}) = \Big[a_{l.0} + [a_{l.1}\ a_{l.2}\ \ldots\ a_{l.n}]\cdot(\boldsymbol{x} - \boldsymbol{x}^0) \quad a_{d.0} + [a_{d.1}\ a_{d.2}\ \ldots\ a_{d.n}]\cdot(\boldsymbol{x} - \boldsymbol{x}^0) \\ a_{N.0} + [a_{N.1}\ a_{N.2}\ \ldots\ a_{N.n}]\cdot(\boldsymbol{x} - \boldsymbol{x}^0)\Big]^T, \tag{2}$$

$$\boldsymbol{D}(\boldsymbol{x}) = \Big[d_{l.0} + [d_{l.1}\ d_{l.2}\ \ldots\ d_{l.n}]\cdot(\boldsymbol{x} - \boldsymbol{x}^0) \quad d_{d.0} + [d_{d.1}\ d_{d.2}\ \ldots\ d_{d.n}]\cdot(\boldsymbol{x} - \boldsymbol{x}^0) \\ d_{N.0} + [d_{N.1}\ d_{N.2}\ \ldots\ d_{N.n}]\cdot(\boldsymbol{x} - \boldsymbol{x}^0)\Big]^T, \tag{3}$$

where $\boldsymbol{x}^0$ is the center of the design space. The response correction parameters $\boldsymbol{A}$ and $\boldsymbol{D}$ are obtained as

$$[\boldsymbol{A}, \boldsymbol{D}] = \arg\min_{[\overline{\boldsymbol{A}}, \overline{\boldsymbol{D}}]} \sum_{k=1}^{N} ||\boldsymbol{f}(\boldsymbol{x}^k) - (\overline{\boldsymbol{A}}(\boldsymbol{x}^k) \circ \boldsymbol{c}(\boldsymbol{x}^k) + \overline{\boldsymbol{D}}(\boldsymbol{x}^k))||^2, \tag{4}$$

i.e., the response scaling is supposed to (globally) improve the matching for all training points $\boldsymbol{x}^k$, $k = 1, \ldots, N$, where N is the number of training points.

A training set combining the following subsets is used: (1) a star-distribution design of experiments with $N = 2n + 1$ training points (n being the number of design variables) allocated at the center of the design space $\boldsymbol{x}^0 = (\boldsymbol{l} + \boldsymbol{u})/2$ ($\boldsymbol{l}$ and $\boldsymbol{u}$ being the lower and upper bound for the design variables, respectively), and the centers of its faces, i.e., points with all coordinates but one equal to those of $\boldsymbol{x}^0$, and the remaining one equal to the corresponding component of $\boldsymbol{l}$ or $\boldsymbol{u}$; this sampling scheme is also referred to as the star distribution [12], (2) design space corners, and (3) additional points allocated using the Latin Hypercube Sampling (LHS) [40]. In the application examples given in Sections 3 and 4 of this chapter, we use all three subsets (with $N = 10$ LHS points) for transonic airfoil design, and only (1) and (3) for low-speed airfoil design.

The problem (4) is equivalent to the linear regression problems $\boldsymbol{C}_l[a_{l.0}\ a_{l.1}\ \ldots\ a_{l.n}\ d_{l.0}\ d_{l.1}\ \ldots\ d_{l.n}]^T = \boldsymbol{F}_l$, $\boldsymbol{C}_d[a_{d.0}\ a_{d.1}\ \ldots\ a_{d.n}\ d_{d.0}\ d_{d.1}\ \ldots\ d_{d.n}\ ]^T = \boldsymbol{F}_d$, and $\boldsymbol{C}_N[a_{N.0}\ a_{N.1}\ \ldots\ a_{N.n}\ d_{N.0}\ d_{N.1}\ \ldots\ d_{N.n}\ ]^T = \boldsymbol{F}_N$, where the matrices $\boldsymbol{C}_l$, $\boldsymbol{C}_d$, $\boldsymbol{C}_N$, $\boldsymbol{F}_l$, $\boldsymbol{F}_d$, and $\boldsymbol{F}_N$ are defined as

$$C_l = \begin{bmatrix} C_{l.c}(\boldsymbol{x}^1) & C_{l.c}(\boldsymbol{x}^1)\cdot(x_1^1 - x_1^0) & \cdots & C_{l.c}(\boldsymbol{x}^1)\cdot(x_n^1 - x_n^0) & & 1 & (x_1^1 - x_1^0) & \cdots & (x_n^1 - x_n^0) \\ C_{l.c}(\boldsymbol{x}^2) & C_{l.c}(\boldsymbol{x}^2)\cdot(x_1^2 - x_1^0) & \cdots & C_{l.c}(\boldsymbol{x}^2)\cdot(x_n^2 - x_n^0) & & 1 & (x_1^1 - x_1^0) & \cdots & (x_n^1 - x_n^0) \\ \vdots & \vdots & \ddots & \vdots & \vdots & \vdots & & \vdots & \vdots \\ C_{l.c}(\boldsymbol{x}^N) & C_{l.c}(\boldsymbol{x}^N)\cdot(x_1^N - x_1^0) & \cdots & C_{l.c}(\boldsymbol{x}^N)\cdot(x_n^N - x_n^0) & & 1 & (x_1^1 - x_1^0) & \cdots & (x_n^1 - x_n^0) \end{bmatrix} \tag{5}$$

$$\boldsymbol{F}_l = \left[C_{l.f}(\boldsymbol{x}^1) \quad C_{l.f}(\boldsymbol{x}^2) \quad \ldots \quad C_{l.f}(\boldsymbol{x}^N)\right]^T \tag{6}$$

$$C_d = \begin{bmatrix} C_{d.c}(\boldsymbol{x}^1) & C_{d.c}(\boldsymbol{x}^1)\cdot(x_1^1 - x_1^0) & \cdots & C_{d.c}(\boldsymbol{x}^1)\cdot(x_n^1 - x_n^0) & & 1 & (x_1^1 - x_1^0) & \cdots & (x_n^1 - x_n^0) \\ C_{d.c}(\boldsymbol{x}^2) & C_{d.c}(\boldsymbol{x}^2)\cdot(x_1^2 - x_1^0) & \cdots & C_{d.c}(\boldsymbol{x}^2)\cdot(x_n^2 - x_n^0) & & 1 & (x_1^1 - x_1^0) & \cdots & (x_n^1 - x_n^0) \\ \vdots & \vdots & \ddots & \vdots & \vdots & \vdots & & \vdots & \vdots \\ C_{d.c}(\boldsymbol{x}^N) & C_{d.c}(\boldsymbol{x}^N)\cdot(x_1^N - x_1^0) & \cdots & C_{d.c}(\boldsymbol{x}^N)\cdot(x_n^N - x_n^0) & & 1 & (x_1^1 - x_1^0) & \cdots & (x_n^1 - x_n^0) \end{bmatrix} \tag{7}$$

$$\boldsymbol{F}_d = \left[C_{d.f}(\boldsymbol{x}^1) \quad C_{d.f}(\boldsymbol{x}^2) \quad \ldots \quad C_{d.f}(\boldsymbol{x}^N)\right]^T \tag{8}$$

$$C_N = \begin{bmatrix} NM_c(\boldsymbol{x}^1) & NM_c(\boldsymbol{x}^1)\cdot(x_1^1 - x_1^0) & \cdots & NM_c(\boldsymbol{x}^1)\cdot(x_n^1 - x_n^0) & & 1 & (x_1^1 - x_1^0) & \cdots & (x_n^1 - x_n^0) \\ NM_c(\boldsymbol{x}^2) & NM_c(\boldsymbol{x}^2)\cdot(x_1^2 - x_1^0) & \cdots & NM_c(\boldsymbol{x}^2)\cdot(x_n^2 - x_n^0) & & 1 & (x_1^1 - x_1^0) & \cdots & (x_n^1 - x_n^0) \\ \vdots & \vdots & \ddots & \vdots & \vdots & \vdots & & \vdots & \vdots \\ NM_c(\boldsymbol{x}^N) & NM_c(\boldsymbol{x}^N)\cdot(x_1^N - x_1^0) & \cdots & NM_c(\boldsymbol{x}^N)\cdot(x_n^N - x_n^0) & & 1 & (x_1^1 - x_1^0) & \cdots & (x_n^1 - x_n^0) \end{bmatrix} \tag{9}$$

$$\boldsymbol{F}_N = \left[NM_f(\boldsymbol{x}^1) \quad NM_f(\boldsymbol{x}^2) \quad \ldots \quad NM_f(\boldsymbol{x}^N)\right]^T \tag{10}$$

The dimension of the vectors $\boldsymbol{F}_l$, $\boldsymbol{F}_d$, and $\boldsymbol{F}_N$ is $N \times 1$, and the dimension of the matrices $\boldsymbol{C}_l$, $\boldsymbol{C}_d$, and $\boldsymbol{C}_N$ is $N \times 2(n+1)$. The correction matrices $\boldsymbol{A}$ and $\boldsymbol{D}$ can now be found analytically as a least-square optimal solution to the aforementioned regression problems

$$\begin{bmatrix} a_{l.0} \\ a_{l.1} \\ \vdots \\ a_{l.n} \\ d_{l.0} \\ \vdots \\ d_{l.n} \end{bmatrix} = \left(\boldsymbol{C}_l^T \boldsymbol{C}_l\right)^{-1} \boldsymbol{C}_l^T \boldsymbol{F}_l, \quad \begin{bmatrix} a_{d.0} \\ a_{d.1} \\ \vdots \\ a_{d.n} \\ d_{d.0} \\ \vdots \\ d_{d.n} \end{bmatrix} = \left(\boldsymbol{C}_d^T \boldsymbol{C}_d\right)^{-1} \boldsymbol{C}_d^T \boldsymbol{F}_d, \quad \begin{bmatrix} a_{N.0} \\ a_{N.1} \\ \vdots \\ a_{N.n} \\ d_{N.0} \\ \vdots \\ d_{N.n} \end{bmatrix} = \left(\boldsymbol{C}_N^T \boldsymbol{C}_N\right)^{-1} \boldsymbol{C}_N^T \boldsymbol{F}_N \tag{11}$$

Note that the matrices $\boldsymbol{C}_l{}^T\boldsymbol{C}_l$, $\boldsymbol{C}_d{}^T\boldsymbol{C}_d$, and $\boldsymbol{C}_N{}^T\boldsymbol{C}_N$ are non-singular for $N > n + 1$, which is the case for our choice of the training set.

In Step 2, we construct the kriging surrogate model. Having the space-mapping-corrected low-fidelity model from Step 1, we sample the initial surrogate and create the kriging interpolation model of it [10, 11]. The kriging surrogate model, $\boldsymbol{s}$, is very fast, smooth, and easy to optimize. In particular, a large number of model evaluations can be done at low cost in the context of multi-objective optimization using evolutionary methods. Step 1 of the surrogate modeling process allows us to reduce misalignment between the low- and high-fidelity models. The surrogate model created in Step 2 is a data-driven model so that it is fast and yet accurate because relatively dense sampling of the design space is utilized. Nevertheless, the computational cost of the surrogate is reasonably low because it is based on low-fidelity model data.

2.3 Optimization Method

The multi-objective design optimization flow can be summarized as follows:

1. Correct the low-fidelity model $\boldsymbol{c}$ using parameterized output space mapping as in Eqn. (1), in particular, identify the correction matrices $\boldsymbol{A}(\boldsymbol{x})$ and $\boldsymbol{D}(\boldsymbol{x})$;
2. Sample the design space and acquire the $\boldsymbol{s}_0$ data (i.e., evaluate the low-fidelity model $\boldsymbol{c}$ at the selected locations and apply the correction in Eqn. (1) using $\boldsymbol{A}(\boldsymbol{x})$ and $\boldsymbol{D}(\boldsymbol{x})$ from Step 1);
3. Construct the kriging interpolation surrogate model $\boldsymbol{s}$ using the sample values obtained in Step 2;
4. Obtain the Pareto front by optimizing $\boldsymbol{s}$ using MOEA;
5. Evaluate high-fidelity model at selected geometries from the Pareto front.
6. If the termination condition is not satisfied, add the new high-fidelity data set to the existing one, and go to Step 1.
7. END

The first four steps of the method lead to obtaining an initial representation of the Pareto front by optimizing the surrogate model $\boldsymbol{s}$ in a multi-objective sense using a multi-objective evolutionary algorithm (MOEA). In the illustration examples in Sections 3 and 4, we use a standard multi-objective evolutionary algorithm with fitness sharing, Pareto-dominance tournament selection, and mating restrictions [16, 17].

The surrogate model is updated upon conclusion of the verification stage (Step 5) by executing the output space mapping procedure using the entire high-fidelity model data set (the original one and the Pareto front representation). The geometries in Step 5 are selected uniformly along the Pareto front. The number of selected geometries is not critical, in the illustration examples, we use around 10 samples per iteration. The improved surrogate model is then re-optimized in a multi-objective sense.

The computational cost of each iteration of the above procedure is only due to the evaluation of the high-fidelity model at the geometries picked up from the Pareto front (in practice, a few points are sufficient). Moreover, the design space in the refinement iterations can be restricted to only the part that contains the Pareto set (the remaining part of the space is irrelevant) to reduce the number of required evaluations. The termination condition is based on comparison between the Pareto front produced by optimizing the current surrogate model and the high-fidelity verification samples.

3 Example 1: Transonic Airfoil Design

The first test case considers the design of airfoils in transonic flow involving lift and drag as design objectives.

3.1 Problem Formulation

A specific case of transonic airfoil shape design with the aim at maximizing the section lift coefficient and minimizing the section drag coefficient at the same time is considered. In other words, we have two objectives, $F_{obj,1}(\boldsymbol{x}) = 1/C_{l.f}$ and $F_{obj,2}(\boldsymbol{x}) = C_{d.f}$. We fix the operating conditions at a free-stream Mach number of $M_\infty = 0.75$ and an angle of attack of $\alpha = 1$ deg. The airfoil shapes are parameterized by the NACA four-digit parameterization [35], where the airfoil shape design variables are m (the maximum ordinate of the mean camber line as a fraction of chord), p (the chordwise position of the maximum ordinate), and t/c (the maximum thickness-to-chord ratio). The design variable vector is $\boldsymbol{x} = [m\ p\ t/c]^T$.

The airfoils are constructed by combining a thickness function $z_t(x/c)$ with a mean camber line function $z_c(x/c)$. The x/c- and z/c-coordinates are [35]

$$(x/c)_{u,l} = (x/c) \mp (z_t/c)\ \sin\theta,\ \ (z/c)_{u,l} = (z_c/c) \pm (z_t/c)\ \cos\theta, \qquad (12)$$

where u and l refer to the upper and lower surfaces, respectively, and

$$\theta = \tan^{-1}\left(\frac{d\,(z_c/c)}{d\,(x/c)}\right), \tag{13}$$

is the mean camber line slope. The NACA four-digit thickness distribution is given by

$$(z_t/c) = t\left(a_0(x/c)^{1/2} - a_1\,(x/c) - a_2(x/c)^2 + a_3(x/c)^3 - a_4(x/c)^4\right), \tag{14}$$

where $a_0 = 1.4845$, $a_1 = 0.6300$, $a_2 = 1.7580$, $a_3 = 1.4215$, $a_4 = 0.5075$, and t is the maximum thickness. We use the following parameter bounds: $0.0 \le m \le 0.03$, $0.2 \le p \le 0.8$, and $0.08 \le t \le 0.14$. There is one nonlinear constraint regarding the cross-section area A (non-dimensionalized by the chord squared), i.e., $A \ge 0.075$.

3.2 High-Fidelity Model

The flow is assumed to be steady, inviscid, and adiabatic with no body forces. The compressible Euler equations are taken to be the governing fluid flow equations. The solution domain boundaries are placed at 25 chord lengths in front of the airfoil, 50 chord lengths behind it, and 25 chord lengths above and below it, see Fig. 1a. The computational meshes are of structured curvilinear body-fitted C-topology with elements clustering around the airfoil and growing in size with distance from the airfoil surface. The computer code ICEM CFD [29] is used for the mesh generation. An example mesh is shown in Fig. 1b.

The free-stream Mach number, static pressure, and angle of attack are prescribed at the farfield boundary. The flow solver is of implicit density-based formulation and the inviscid fluxes are calculated by an upwind-biased second-order spatially

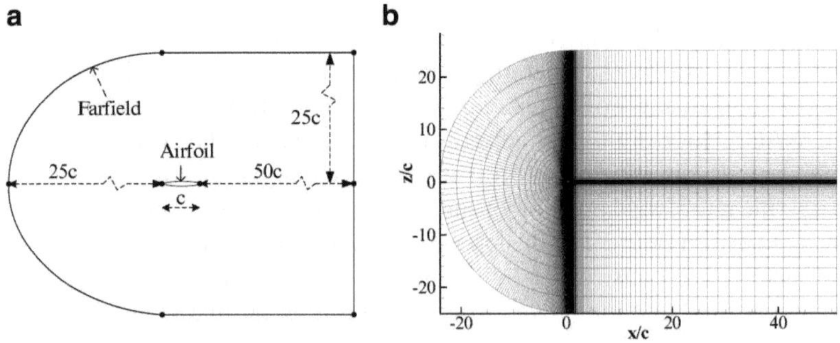

Fig. 1 Elements of the computational mesh: (**a**) sketch of the computational domain, (**b**) an example mesh

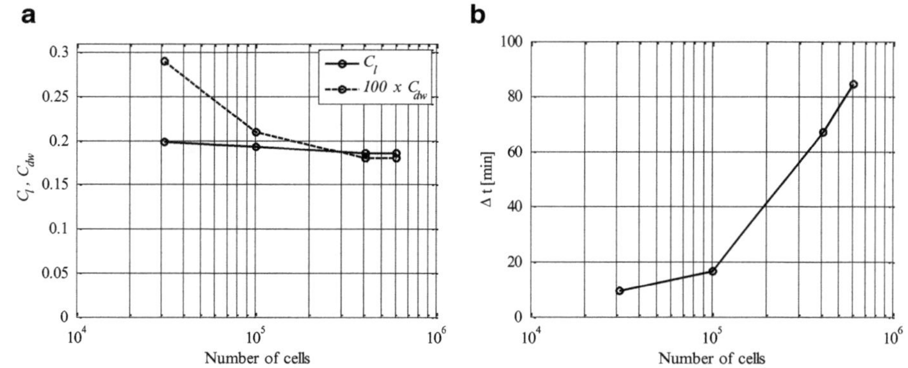

Fig. 2 Grid convergence study using the NACA 0012 airfoil at a Mach number of $M_\infty = 0.75$ and an angle of attack of $\alpha = 1°$: (**a**) the lift and drag coefficients versus the number mesh cells, and (**b**) the simulation time versus the number of mesh cells

accurate Roe flux scheme. Asymptotic convergence to a steady-state solution is obtained for each case. The solution convergence criterion for the high-fidelity model is the one that occurs first of the following: a reduction of all the residuals by six orders, or a maximum number of iterations of 1000. Numerical fluid flow simulations are performed using the computer code FLUENT [29].

A grid convergence study was performed using the NACA 0012 airfoil at a Mach number of $M_\infty = 0.75$ and an angle of attack of $\alpha = 1°$. The study, shown in Fig. 2a, revealed that roughly 400,000 mesh cells are needed for mesh convergence, and that particular mesh was used for the high-fidelity model. The overall simulation time for the case considered is around 67 min (Fig. 2b) using four processors on an Intel (R) Xeon CPU E5-2620@2.00 GHz machine. The flow solver reached a converged solution after 352 iterations. The other meshes required around 350–500 iterations to converge, except the coarsest mesh, which terminated after 1000 iterations, with the overall simulation time around 9.5 min on the same four processors as the high-fidelity model.

3.3 Low-Fidelity Model

The low-fidelity CFD model is constructed in the same way as the high-fidelity model, but with a coarser computational mesh and relaxed convergence criteria. For the low-fidelity model, we use the coarse mesh in the grid study presented in Fig. 3a, with roughly 30,000 mesh cells. The flow solution history for the low-fidelity model, shown in Fig. 3a, indicates that the lift and drag coefficients are nearly converged after 80–100 iterations. The maximum number of iterations is set to 100 for the low-fidelity model. This reduced the overall simulation time to 1.5 min. A comparison of the pressure distributions, shown in Fig. 3b, indicates that the low-fidelity model,

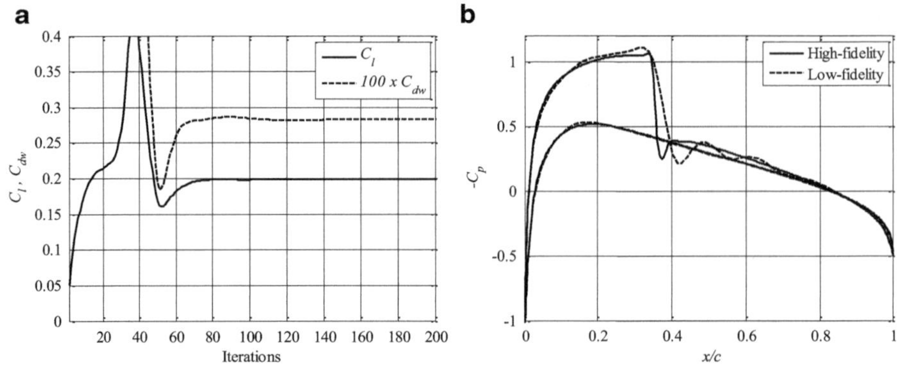

Fig. 3 Simulation results for NACA 0012 at a Mach number of $M_\infty = 0.75$ and an angle of attack of $\alpha = 1°$: (**a**) the evolution of the lift and drag coefficients obtained by the low-fidelity model; (**b**) a comparison of the pressure distributions obtained by the high- and low-fidelity models

in spite of being based on much coarser mesh and reduced flow solver iterations, captures the main features of the high-fidelity model pressure distribution quite well. The biggest discrepancy in the distributions is around the shock on the upper surface, leading to an over estimation of both lift and drag (Fig. 3a).

The ratio of the simulation times of the high- and low-fidelity model in this case study is 43.8. In many cases the solver does not fully converge with respect to the residuals and goes on up to 1000 iterations. Then the overall evaluation time of the high-fidelity model goes up to 170 min. In those cases, the ratio of the simulation times of the high- and low-fidelity models is around 110. For the sake of simplicity, we will use a fixed value of 80 in the numerical computations presented in the results section.

3.4 Results

The Pareto set obtained in the first iteration is shown in Fig. 4a. It is generated by optimizing the initial kriging model, i.e., the one constructed from the space-mapping-corrected low-fidelity model (Eqns. (1)–(11), but without the noise metric values) with the training samples allocated as described below Eqn. (4). The new set of ten high-fidelity samples allocated along that initial Pareto front representation (only nine of them are shown in Fig. 4a) is used for verification purposes but also to enhance the surrogate model by re-running the space mapping corrections (Eqns. (1)–(11)). It can be observed that there is some discrepancy between the Pareto-optimized surrogate model and the sampled high-fidelity model, which means that the optimization process has to be continued. The final Pareto set shown in Fig. 4b is obtained after four iterations of the algorithm. Its verification carried out using an additional set of high-fidelity model samples indicates that the surrogate model

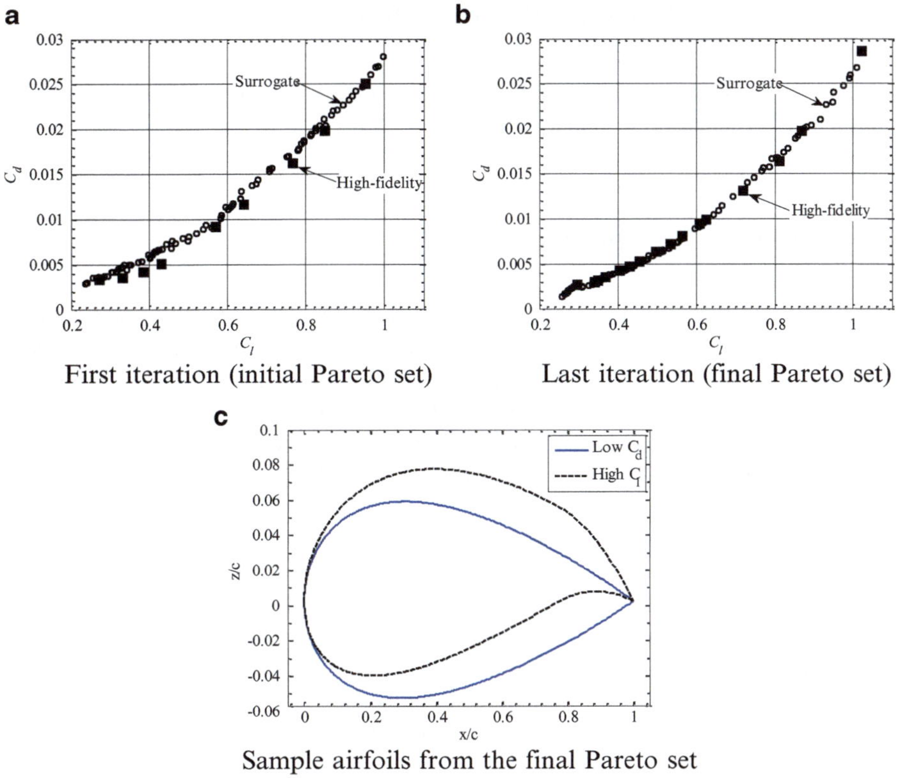

Fig. 4 Pareto front representation obtained by optimizing the surrogate model: (**a**) and (**b**) show the surrogate model points (*circles*), and selected high-fidelity model points (*squares*), and (**c**) shows airfoil shapes from the final Pareto set (**b**) for low drag coefficient and high lift coefficient

is a sufficiently good representation of the high-fidelity model (both data set are practically aligned). The total optimization cost is about 60 high-fidelity model evaluations: 30 to obtain the initial set as mentioned above, and ten per each additional iteration of the algorithm.

Figure 4c shows the airfoil shapes from the final Pareto set (Fig. 4b) with a low drag coefficient ($\boldsymbol{x} = [0.0013\ 0.5326\ 0.1117]^T$), i.e., the left-most high-fidelity sample on Fig. 4b, and a high lift coefficient ($\boldsymbol{x} = [0.0267\ 0.7725\ 0.1134]^T$), i.e., the right-most high-fidelity sample on Fig. 4b. Both designs fulfill the cross-sectional area constraint, i.e., $A_{low\ Cd} \geq 0.075$ and $A_{high\ Cl} \geq 0.075$. The two airfoil shapes have similar thickness ($t_{low\ Cd} \approx t_{high\ Cl} \approx 0.11$), but the design with lower drag has a much lower camber than the one with higher lift ($m_{low\ Cd} = 0.0013$ vs. $m_{high\ Cl} = 0.0267$).

4 Example 2: Low-Speed Airfoil Design

The second test case involves low-speed airfoil design with trade-offs of aerodynamic performance versus noise signature.

4.1 *Problem Formulation*

This example involves the trade-off between the aerodynamic and aeroacoustic performances of low-speed airfoils. We consider only the clean wing, trailing-edge (TE) noise in this example, and we aim at minimizing the section drag coefficient ($C_{d.f}$) for a given section lift coefficient ($C_{l.f}$), and at the same time minimize the TE noise (given by a noise metric NM_f explained in the next section). Therefore, we have two objectives, $F_{obj,1}(\boldsymbol{x}) = C_{d.f}$ and $F_{obj,2}(\boldsymbol{x}) = NM_f$, both obtained by a high-fidelity simulation.

The specific design case considered is optimization for a target lift coefficient of $C_l = 1.5$. The operating conditions are a free-stream Mach number of $M_\infty = 0.208$ and a Reynolds number of $Re_c = 0.665$ million. The angle of attack α is a dependent variable utilized to obtain, for any given airfoil geometry, the target lift coefficient. We use again the NACA four-digit parameterization [35] (as described in Section III.A.1) with the following parameter bounds: $0.0 \leq m \leq 0.1$, $0.3 \leq p \leq 0.6$, and $0.08 \leq t \leq 0.14$. There are no other constraints considered in the optimization process.

The kriging surrogate model constructed for the purpose of evolutionary-based multi-objective optimization was obtained using 343 low-fidelity model samples allocated on a uniform rectangular $7 \times 7 \times 7$ grid. The surrogate was further corrected using the multi-point output space mapping (1), and utilizing nine high-fidelity model samples: seven samples allocated according to the star-distribution factorial design of experiments, and two additional random samples necessary to ensure that the regression problem has a unique solution.

4.2 *Noise Metric Model*

The noise metric model was developed by Hosder et al. [36] to give an accurate relative noise measure suitable for design studies. The noise metric therefore does not give an accurate prediction of the absolute noise level. However, it does give an accurate measure of the change in noise due to changes in the wing shape. The noise metric model is recalled here for convenience. A rigorous derivation of the noise metric can be found in Hosder et al. [36].

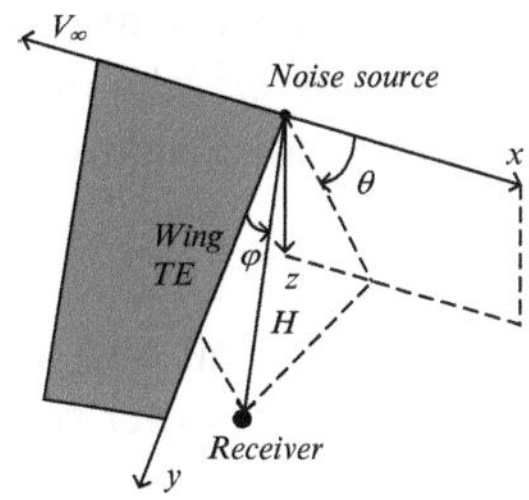

Fig. 5 Geometrical definition of the noise source, receiver, and directivity angles (note that the trailing-edge sweep angle β is 0° here) [36]

The objective is to estimate the acoustic noise perceived by an observer at a distance H from a clean wing (Fig. 5). An intensity indicator for the clean wing turbulent trailing-edge noise is defined as [36]

$$I_{NM} = \frac{\rho_\infty}{2\pi^3 a_\infty^2} \int_0^{b/2} u_0^5 l_0 \cos^3\beta \frac{D(\theta, \phi)}{H^2} dy, \tag{15}$$

where b is the wing span, ρ_∞ is the free-stream density, a_∞ is the free-stream speed of sound, u_0 is the characteristic velocity scale for turbulence, l_0 is the characteristic length scale for turbulence, and β is the trailing-edge sweep angle. $D(\theta,\varphi)$ is the directivity term and is defined as

$$D(\theta, \phi) = 2\sin^2\left(\frac{\theta}{2}\right)\sin\phi, \tag{16}$$

where θ is the polar directivity angle, and φ is the azimuthal directivity angle (Fig. 5).

The characteristic turbulent velocity at a spanwise location of the wing trailing edge is chosen as the maximum value of the turbulent kinetic energy (*TKE*) profile at that particular spanwise station, that is

$$u_0(y) = \max\left[\sqrt{TKE(z)}\right], \tag{17}$$

where z is the direction normal to the wing surface. The characteristic turbulence length scale at each spanwise station is modeled as

$$l_0(y) = \frac{u_0(y)}{\omega}, \tag{18}$$

where ω is the turbulent frequency (dissipation rate per unit kinetic energy) observed at the maximum *TKE* location.

The noise metric (*NM*) for the trailing-edge noise (in dB) is written as

$$NM = 120 + 10\log(I_{NM}), \tag{19}$$

where the noise intensity indicator has been scaled with the reference noise intensity of $10^{-12}\ W/m^2$ (the minimum sound intensity level that the human ear can detect). The total noise metric is

$$NM = 10\log\left(10^{\frac{NM_u}{10}} + 10^{\frac{NM_l}{10}}\right), \quad (20)$$

where NM_u and NM_l are noise metric values obtained by Eqn. (19), evaluated for the upper and lower surfaces, respectively.

4.3 High-Fidelity CFD Model

The flow is assumed to be steady, compressible, and viscous. The steady Reynolds-averaged Navier–Stokes equations are taken as the governing fluid flow equations with the k-ω SST turbulence model by Menter [37]. The solution domain boundaries are placed at 25 chord lengths in front of the airfoil, 50 chord lengths behind it, and 25 chord lengths above and below it. The computational meshes are of structured curvilinear body-fitted C-topology with elements clustering around the airfoil and growing in size with distance from the airfoil surface. The grids are generated using the hyperbolic C-mesh of Kinsey and Barth [38]. The high-fidelity models grid has around 400,000 mesh cells.

Numerical fluid flow simulations are performed using the computer code FLUENT [39]. The flow solver is of implicit density-based formulation and the fluxes are calculated by an upwind-biased second-order spatially accurate Roe flux scheme. Asymptotic convergence to a steady-state solution is obtained for each case. The solution convergence criterion for the high-fidelity model is the one that occurs first of the following: a reduction of the residuals by six orders of magnitude, or a maximum number of iterations of 4000.

4.4 Low-Fidelity CFD Model

The low-fidelity CFD model is constructed in the same way as the high-fidelity model, but with a coarser computational mesh and relaxed convergence criteria. The low-fidelity mesh has around 30,000 mesh cells. Although the flow equation residuals are not converged, the lift and drag coefficients and the noise metric typically converge within 1200 iterations. Therefore, the maximum number of iterations is set to 1200.

4.5 Results

Figure 6 shows the distribution of the solutions in the feature (output) space at the first iteration of the evolutionary algorithm. The population size used was $N = 500$.

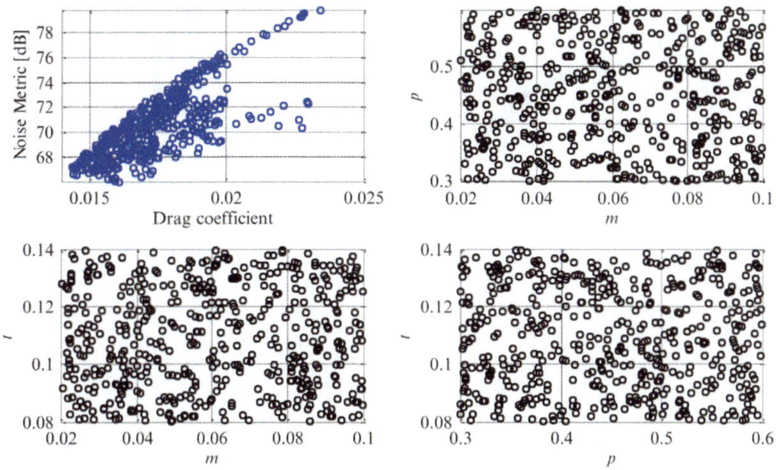

Fig. 6 Multi-objective optimization of the surrogate model: distribution of the initial population in the feature (output) space and the design space

Random initialization with uniform probability distribution is utilized. One can observe a strong correlation between the drag coefficient and the noise metric in the majority of the feature space. Moreover, one can also observe that in a small region of the feature space the two objectives are weakly conflicting, i.e., in the region with low values of both objectives.

The Pareto set obtained after optimizing the surrogate model is shown in Fig. 7, together with the allocation of the solution in the design space. Note that all the Pareto-optimal solutions correspond to the thinnest possible airfoil shapes (here, $t = 0.08$). As there are no thickness constraints, the optimizer reaches the lower bound on the thickness parameter to reduce drag, in particular skin friction drag, while still maintaining the prescribed lift.

Figure 8a shows the high-fidelity model verification samples, indicating certain discrepancies between the drag/noise figures predicted by the surrogate model and actual values. The Pareto front refinement has been subsequently executed in the refined design space of $0.045 \leq m \leq 0.075$, $0.3 \leq p \leq 0.6$, and $0.08 \leq t \leq 0.14$. The verification samples obtained in the previous step have been utilized to update the surrogate model.

The results of the refinement iteration are shown in Fig. 8b. The overall optimization procedure is terminated at this point because the assumed accuracy of <1 drag count (where a drag count is defined to be $\Delta C_d = 0.0001$) and <0.1 dB with respect to the noise metric is met. The final Pareto front shows that the range of the drag coefficients for the trade-off solutions is from 148 to 156 drag counts with the corresponding noise metric from 66.4 dB to 65.6 dB. Thus, improvement of the noise performance by 0.8 dB can be obtained by increasing the drag by eight counts.

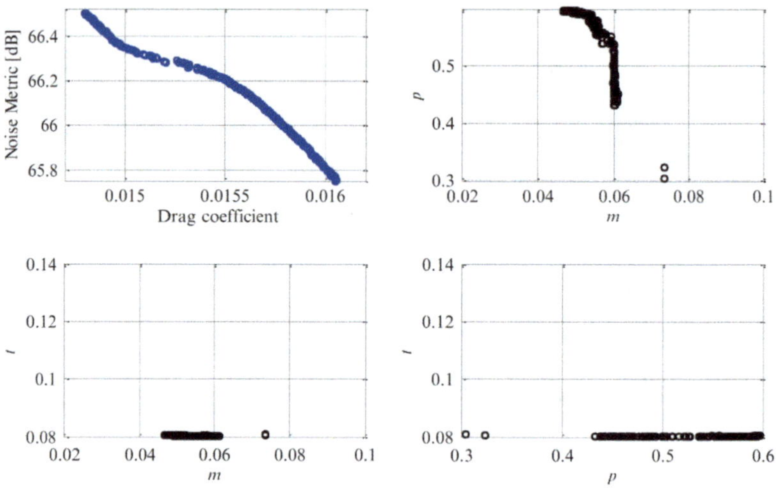

Fig. 7 Multi-objective optimization of the surrogate model: Pareto set found by optimizing the surrogate model and the corresponding allocation of Pareto-optimal solutions in the design space

Figure 8c shows the airfoil shapes from the final Pareto set (Fig. 8b) with a low drag coefficient ($\boldsymbol{x} = [0.0628\ 0.4789\ 0.080]^T$), i.e., the left-most high-fidelity sample in Fig. 8b, and with a low noise metric ($\boldsymbol{x} = [0.0510\ 0.5978\ 0.080]^T$), i.e., the right-most high-fidelity sample in Fig. 8b.

5 Conclusion

A computationally efficient procedure for multi-objective optimization of aerodynamic surfaces has been described. The approach exploits a fast surrogate constructed using kriging and space mapping corrected low-fidelity CFD simulation data, as well as the multi-objective evolutionary algorithm that finds a set of designs representing the best trade-offs between design objectives, here, the lift and drag coefficients. A refinement procedure allows for improving the initial Pareto front representation at a low cost depending on the number of high-fidelity verification samples used in the process. The design examples demonstrate a consistent performance of the described method. Future work will extend the approach for higher-dimensional cases, where the initial computational effort related to construction of the response surface approximation model may become a serious issue.

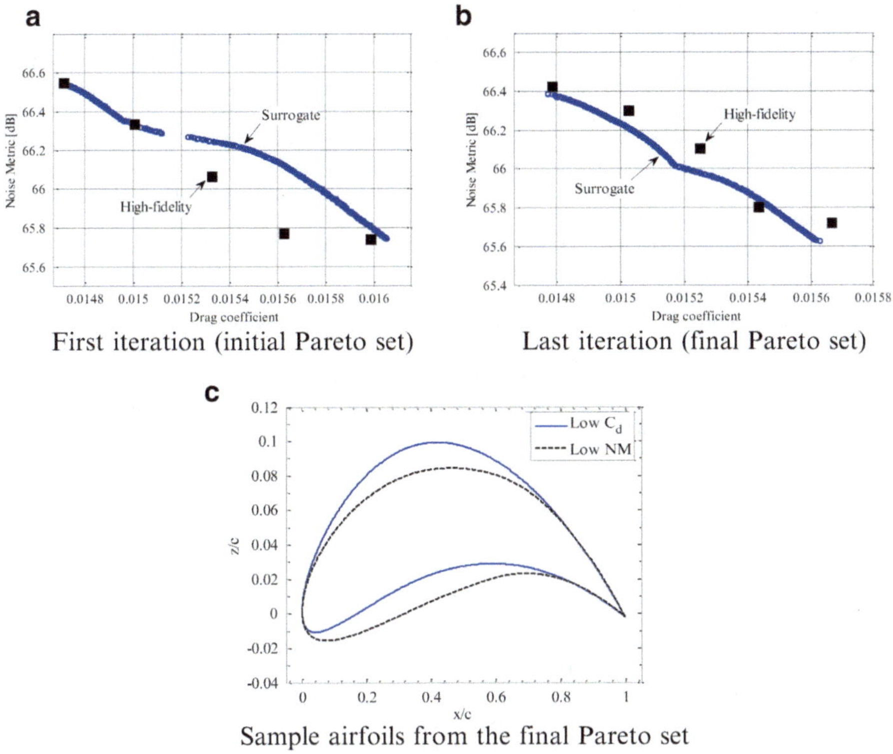

Fig. 8 Pareto front representation obtained by optimizing the surrogate model: (**a**) and (**b**) show surrogate model points (*circles*), and selected high-fidelity model points (*squares*), and (**c**) shows airfoil shapes from the final Pareto set (**b**) for a low drag coefficient and for a low noise metric

References

1. Leoviriyakit, K., Kim, S., Jameson, A.: Viscous aerodynamic shape optimization of wings including planform variables. In: 21st Applied Aerodynamics Conference, Orlando, Florida, 23–26 June 2003
2. Braembussche, R.A.: Numerical optimization for advanced turbomachinery design. In: Thevenin, D., Janiga, G. (eds.) Optimization and Computational Fluid Dynamics, pp. 147–189. Springer, Berlin (2008)
3. Mader, C.A., Martins, J.R.R.A.: Derivatives for time-spectral computational fluid dynamics using an automatic differentiation adjoint. AIAA J. **50**(12), 2809–2819 (2012)
4. Mousavi, A., Nadarajah, S.: Heat transfer optimization of gas turbine blades using an adjoint approach. In: 13th AIAA/ISSMO Multidisciplinary Analysis Optimization Conference, AIAA Paper 2010-9048, Fort Worth, Texas, 13–15 September 2010
5. Leung, T.M., Zingg, D.W.: Aerodynamic shape optimization of wings using a parallel Newton-Krylov approach. AIAA J. **50**(3), 540–550 (2012)
6. Epstein, B., Peigin, S.: Constrained aerodynamic optimization of three-dimensional wings driven by Navier-Stokes computations. AIAA J. **43**(9), 1946–1957 (2005)
7. Nocedal, J., Wright, S.J.: Numerical Optimization. Springer, New York (2006)

8. Kim, S., Hosseini, K., Leoviriyakit, K., Jameson, A.: Enhancement of class of adjoint design methods via optimization of parameters. AIAA J. **48**(6), 1072–1076 (2010)
9. Schmidt, S., Gauger, N., Ilic, C., Schulz, V.: Three dimensional large scale aerodynamic shape optimization based on shape calculus. In: 41st AIAA Fluid Dynamics Conference and Exhibit, AIAA Paper 2011-3718, Honolulu, Hawaii, 27–30 June 2011
10. Queipo, N.V., Haftka, R.T., Shyy, W., Goel, T., Vaidyanathan, R., Tucker, P.K.: Surrogate-based analysis and optimization. Prog. Aerosp. Sci. **41**(1), 1–28 (2005)
11. Forrester, A.I.J., Keane, A.J.: Recent advances in surrogate-based optimization. Prog. Aerosp. Sci. **45**(1–3), 50–79 (2009)
12. Koziel, S., Echeverría-Ciaurri, D., Leifsson, L.: Surrogate-based methods. In: Koziel, S., Yang, X.S. (eds.) Computational Optimization, Methods and Algorithms. Studies in Computational Intelligence, pp. 33–60. Springer, Berlin (2011)
13. Alexandrov, N.M., Lewis, R.M., Gumbert, C.R., Green, L.L., Newman, P.A.: Optimization with variable-fidelity models applied to wing design. In: 38th Aerospace Sciences Meeting & Exhibit, Reno, NV, AIAA Paper 2000-0841, January 2000
14. Robinson, T.D., Eldred, M.S., Willcox, K.E., Haimes, R.: Surrogate-based optimization using multifidelity models with variable parameterization and corrected space mapping. AIAA J. **46**(11), 2814–2822 (2008)
15. Booker, A.J., Dennis Jr., J.E., Frank, P.D., Serafini, D.B., Torczon, V., Trosset, M.W.: A rigorous framework for optimization of expensive functions by surrogates. Struct. Optim. **17**(1), 1–13 (1999)
16. Fonseca, C.M.: Multiobjective genetic algorithms with applications to control engineering problems. Ph.D. thesis, Department of Automatic Control and Systems Engineering, University of Sheffield, Sheffield, UK (1995)
17. Coello Coello, C.A., Lamont, G.B.: Applications of Multi-Objective Evolutionary Algorithms. World Scientific, Singapore (2004)
18. Epstein, B., Peigin, S.: Robust hybrid approach to multiobjective constrained optimization in aerodynamics. AIAA J. **42**(8), 1572–1581 (2004)
19. Nemec, M., Zingg, D.W., Pulliam, T.H.: Multipoint and multi-objective aerodynamic shape optimization. AIAA J. **42**(6), 1057–1065 (2004)
20. Zerbinati, A., Minelli, A., Ghazlane, I., Desideri, J.-A.: Meta-model-assisted MGDA for multi-objective functional optimization. Comput. Fluids, Elsevier **102**(10), 116–130 (2014)
21. March, A., Willcox, W.: Provably convergent multifidelity optimization algorithm not requiring high-fidelity derivatives. AIAA J. **50**(5), 1079–1089 (2012)
22. March, A., Willcox, W.: Constrained multifidelity optimization using model calibration. Struct. Multidiscip. Optim. **46**, 93–109 (2012)
23. Koziel, S., Yang, X.S., Zhang, Q.J. (eds.): Simulation-Driven Design Optimization and Modeling for Microwave Engineering. Imperial College Press, London (2013)
24. Han, Z.-H., Gortz, S., Hain, R.: A variable-fidelity modeling method for aero-loads prediction. Notes Numer. Fluid Mech. Multidiscip. Des. **112**, 17–25 (2010)
25. Han, Z.-H., Gortz, S., Zimmermann, R.: On improving efficiency and accuracy of variable-fidelity surrogate modeling in aero-data for loads context. In: Proceedings of CEAS 2009 European Air and Space Conference, Manchester, UK, 26–29 October 2009, London, UK: Royal Aeronautical Society
26. Han, Z.-H., Zimmermann, R., Gortz, S.: A new cokriging method for variable-fidelity surrogate modeling of aerodynamic data. In: 48th AIAA Aerospace Sciences Meeting Including the New Horizons Forum and Aerospace Exposition, Orlando, FL. AIAA 2010-1225, 4–7 January 2010
27. Koziel, S., Leifsson, L.: Knowledge-based airfoil shape optimization using space mapping. In: 30th AIAA Applied Aerodynamics Conference, New Orleans, Louisiana, 25–28 June 2012
28. Koziel, S., Cheng, Q.S., Bandler, J.W.: Space mapping. IEEE Microw. Mag. **9**(6), 105–122 (2008)
29. ICEM CFD, ver. 14.0, ANSYS Inc., Southpointe, 275 Technology Drive, Canonsburg, PA 15317, 2012.

30. Koziel, S., Leifsson, L.: Surrogate-based aerodynamic shape optimization by variable-resolution models. AIAA J. **51**(1), 94–106 (2013)
31. Leifsson, L., Koziel, S.: Surrogate modeling and optimization using shape-preserving response prediction: a review. Eng. Optim. 1–21 (2015). doi: 10.1080/0305215X.2015.1016509
32. Koziel, S., Leifsson, L: Adaptive response correction for surrogate-based airfoil shape optimization. In: 30th AIAA Applied Aerodynamics Conference, New Orleans, Louisiana, 25–28 June 2012
33. Koziel, S., Leifsson, L: Multi-fidelity airfoil optimization with adaptive response prediction. In: 14th AIAA/ISSMO Multidisciplinary Analysis and Optimization Conference, Indianapolis, Indiana, 17–19 September 2012
34. Eldred, M.S., Giunta, A.A., Collis, S.S.: Second-order corrections for surrogate-based optimization with model hierarchies. In: 10th AIAA/ISSMO Multidisciplinary Analysis and Optimization Conference, Albany, NY, August 30–September 1 2004
35. Abbott, I.H., Von Doenhoff, A.E.: Theory of Wing Sections. Dover Publications, New York (1959)
36. Hosder, S., Schetz, J.A., Mason, W.H., Grossman, B., Haftka, R.T.: Computational-fluid-dynamics-based clean-wing aerodynamic noise model for design. J. Aircr. **47**(3), 754–762 (2010)
37. Menter, F.: Two-equation eddy-viscosity turbulence models for engineering applications. AIAA J. **32**, 1598–1605 (1994)
38. Kinsey, D.W., Barth, T.J.: Description of a hyperbolic grid generation procedure for arbitrary two-dimensional bodies, AFWAL TM 84-191-FIMM, 1984
39. FLUENT, ver. 14.0, ANSYS Inc., Southpointe, 275 Technology Drive, Canonsburg, PA 15317, 2012.
40. Beachkofski, B., Grandhi, R.: Improved distributed hypercube sampling, AIAA Paper 2002-1274. In: 43rd AIAA/ASME/ASCE/AHS/ASC Structures, Structural Dynamics, and Materials Conference, Denver, CO, 22–25 April 2002
41. Bandler, J.W., Cheng, Q.S., Dakroury, S., Mohamed, A.S., Bakr, M.H., Madsen, K., Søndergaard, J.: Space mapping: the state of the art. IEEE Trans. Microwave Theory and Tech. **52**(1), 337–361 (2004)
42. Couckuyt, I.: Forward and inverse surrogate modeling of computationally expensive problems. Ph.D. thesis, Faculty of Engineering and Architecture, Ghent University (2013)

Assessment of Inverse and Direct Methods for Airfoil and Wing Design

Mengmeng Zhang and Arthur William Rizzi

Abstract The goal of aerodynamic design for airfoils and wings is to improve the performance of the lifting surfaces, e.g., by minimizing the drag. We consider here two approaches, the classical inverse design approach that finds the surface which produces desired pressure distributions, and the direct mathematical optimization based on local parameter searches, that is usually enabled by fast gradient computation, for example, by the adjoint method. The hybrid approach is to combine both of them. Each approach has its own pros and cons. In this chapter the approaches are assessed by application to the design of transonic RAE2822 airfoil and ONERA M6 wing.

Keywords Inverse design • Gradient-based optimization • Parametrization • Drag reduction • Wing surface curvature • Adjoint solver

MSC code: 76N25 (Flow control and optimization)

1 Introduction

Aircraft design activities are concerned with the determination of designs that meet a priori specified performance features of the vehicle. The specified design objectives are traditionally met through an iterative process of analyses, evaluations, and modifications of the design. In this sequential trial-and-error procedure, the designer must rely on experience, intuition, and ingenuity for every re-design, and this makes aircraft design an exciting creative discipline. In practice, however, designers are often forced to depend on tried concepts to cut a path through an incomprehensible number of feasible designs, historically characterizing the process as a slow gradual improvement of existing types of concepts.

The task of designing an aircraft is among the most complex in engineering. The complexity can be simplified by sequential decision that divides the design

M. Zhang (✉) • A.W. Rizzi
Royal Institute of Technology (KTH), 10044 Stockholm, Sweden
e-mail: mzha@kth.se; rizzi@kth.se

S. Koziel et al. (eds.), *Simulation-Driven Modeling and Optimization*,
Springer Proceedings in Mathematics & Statistics 153,
DOI 10.1007/978-3-319-27517-8_4

into the conceptual design, preliminary design, and the detailed design. While the greatest freedom to exploit potential trade-offs between aircraft subsystems for the optimization of the design occurs earliest in the design stage, since the decisions taken during the earlier stage commit up to 80 % of the life-cycle costs, although, of course, the actual costs incurred appear on the books much later. Mistakes here must be avoided because they are very costly to remedy later and delay acceptance. Matters involving the interaction of aerodynamics with structures and controls are particularly prone to errors due to the low fidelity of the analysis methods traditionally used. If these complicated problems are not resolved in an integrated sense, the sub-optimal design might be led to in a global sense and become "myoptic."

1.1 MDO and Aerodynamic Design Approaches

Multidisciplinary Design Optimization, or MDO [1, 2], combines analysis and optimizations in several individual disciplines with those of the entire system concurrently through formal mathematical processes. It puts into place a formal integrated system design process for better product quality by effectively exploiting the synergism of interdisciplinary couplings. MDO, as a discipline, itself comprises of many areas of research. It is "a methodology for design of complex engineering systems that are governed by mutually interacting physical phenomena and made up of distinct interacting subsystems (suitable for systems for which) in their design, everything influences everything else" [1].

One of the significant factors holding back the widespread adoption of MDO is its computational cost when the number of design variables becomes very large (*the curse of dimensionality*). The use of high-fidelity models can raise the cost from merely expensive to unbearable. Parallel computing helps, but cannot overcome computational inefficiency. The revolution in computing speed and memory capacity of digital computers together with persistent systematization of design methodology has led to tools for computational aircraft design (e.g., MDO) that aim at automation of the conventional design process through integration of numerical methods for analysis, sensitivity analysis and mathematical programming so that the best design in terms of a pre-defined criterion can be determined. Traditionally the process of selecting design variations has been carried out by trial-and-error, relying on the intuition and experience of the designer, the engineer in the loop. Increasing the level of automation by computational means has reduced, but not eliminated, the engineer-in-the-loop activities. The overall success of the design process depends heavily not only on reliability and accuracy of the computational methods but also on how well the designer has set his goals.

1.1.1 Aerodynamic Design: Three Approaches from a User Perspective

Optimization of aircraft wings is not new. The thing that makes wings so hard to design is that their aerodynamics and structure are not just interdependent, they are variable. Computational aerodynamic design, one of the disciplinary subsets of the aircraft design process, aims directly at determining the geometrical shape of the aircraft hull that produces certain specified aerodynamic properties, with or without constraints on the geometry. Usually termed aerodynamic shape optimization (ASO) [3–5], this is the subject of this chapter. ASO is a very attractive technology because it replaces workable designs with optimal ones, and cuts down design times, thus enabling faster responses to the economic pressure of the marketplace.

For wing design there are generally three approaches: the direct optimization design (mathematics-skill dependent) [6], the inverse design (engineering-skill dependent) [5, 7–9], and the hybrid approach which combines both of them. The direct approach requires the user to define the cost function (usually the drag) along with the constraints, and then seeks the solution to the constrained optimization problem by mathematical algorithms for non-linear optimization. When the algorithm involves gradient searches, the sensitivities that indicate how to change the geometry in order to reduce the cost function can be computed also for very many design parameters by solution of an adjoint to the flow problem. This approach is the most popular way of doing optimization nowadays as the computer capacities are continuously increasing. However, it is always trapped in a local optimum due to the limitation of the algorithms, and it may overexploit the flow localities. The second approach works by first finding a well-posited pressure distribution that fulfills the design requirements and then determining a geometry that yields this target pressure. One big issue of this approach is that it needs to formulate a good "target" pressure distribution first, which requires engineer in the loop. The last method can employ both approaches under user control, but very manual. One example is `LINDOP` optimizer [10, 11] in the `MSES` [12] package.

2 Introduction to Direct Optimization, Inverse Design, and Hybrid Approach

2.1 Direct Mathematical Optimization

A straightforward way to search for an optimal design is to construct a non-linear constrained optimization problem,

$$
\begin{aligned}
\text{min}: \quad & I = I(w, X) \\
\text{subject to}: \quad & \\
& C_L(w, X) \geq C_L^0, \\
& C_m(w, X) = C_m^0, \\
& g_j(X_\Gamma) \leq 0, \quad 1 \leq j \leq m,
\end{aligned}
\tag{1}
$$

where X is the mesh, X_Γ is the surface of the geometry, w is the flow-field variables, and g_j are the geometric constrains. The cost function I is selected by the designer, which might be the drag coefficient $I = C_D$, the drag to lift ratio $I = \frac{C_D}{C_L}$, or the pressure difference $I = \int (p - p_d)^2 d\Omega$ if an inverse design problem is being posed. Numerous optimization algorithms [6] are available for attempted solution of the mathematical problem. We have the mesh generation algorithms

$$\mathsf{M}(X, X_\Gamma) = 0 \tag{2}$$

and the surface parametrization algorithm S

$$\mathsf{S}(X_\Gamma, \ell) = 0 \tag{3}$$

where ℓ are the design variables which determine the surface X_Γ.

The change in I can be estimated by a small variation $\delta\ell$ to the parameter vector and recalculating the flow to obtain the change in I, thus approximating the directional derivative,

$$I(\ell + \delta\ell) = I(\ell) + \frac{dI}{d\ell} \cdot \delta\ell + O(||\delta\ell||^2). \tag{4}$$

Most optimization algorithms employ a line search along search direction d,

$$\ell^{n+1} = \ell^n - \lambda d \tag{5}$$

with λ a step size parameter. The search direction d is composed of gradients $\frac{dI}{d\ell}$; quasi-Newton methods also compute approximations to the Hessian matrix of second derivatives, $H_{ij} = \frac{d^2 I}{d\ell_i d\ell_j}$, by differences of gradients in an updating scheme. When the parameter space is high-dimensional, this approach using the gradient itself entails high computational cost.

2.1.1 Gradients by Adjoint Equations

The adjoint method was originally applied to aerodynamics by Jameson [13] adapting ideas originally formulated by Lions [14] on optimal control of systems governed by partial differential equations. The adjoint equations can be conveniently formulated in a framework to calculate the sensitivity of a given objective function I to parameters ℓ which control the geometry. The derivation is easy when R, etc. below are interpreted as the finite dimensional discretization of the flow equations, objective functions, etc. The residual R of the governing equations for a given flight state(s) which expresses the dependence of flow variables w on the mesh X is:

$$R(w, X) = 0 \tag{6}$$

Thus a small change in X produces a small change δI to the cost function,

$$\delta I = \frac{\partial I}{\partial w}\delta w + \frac{\partial I}{\partial X}\delta X \tag{7}$$

and a small change δw to the flow w,

$$\delta R = \frac{\partial R}{\partial w}\delta w + \frac{\partial R}{\partial X}\delta R = 0. \tag{8}$$

The mesh deformation δX is calculated from the corresponding displacements of the nodes that define the surface of the geometry X_Γ by parametrization S.

Equation (8) is multiplied by a Lagrange multiplier vector Ψ, subtracted from Eq. (7) and the result re-arranged,

$$\delta I = (\frac{\partial I}{\partial w} - \Psi^T \frac{\partial R}{\partial w})\delta w + (\frac{\delta I^T}{\delta X} - \Psi^T \frac{\partial R}{\partial X})\delta X. \tag{9}$$

Choosing Ψ so that the first term on the right vanishes gives

$$[\frac{\partial R}{\partial w}]^T \Psi = (\frac{\partial I}{\partial w})^T. \tag{10}$$

This is a linear PDE known as the adjoint equation. There are two main ways to characterize the adjoint approach, as a discrete method, in which the discretized governing equations are used to derive the adjoint equations, and as a continuous method, in which the adjoint equations are derived from the analytical PDEs [15]. The discrete and continuous approaches are found to have relative advantages and disadvantages over each other [16]. The discrete adjoint equations derived directly from the discrete flow equations become very complicated when the flow equations are discretized with higher order schemes using flux limiters. On the other hand it can provide an exact gradient of the inexact cost function which results from the discretization of the flow equations. In theory a discrete method can handle PDEs of arbitrary complexity without significant mathematical development and can treat arbitrary functionals I. In comparison, the continuous adjoint requires significant theoretical development but is better connected to the underlying physics and can be solved by a method independent of the flow solution scheme. However, it is more limited in the types of functionals and governing equations that can be treated, and the gradient calculated will differ more from that found by finite differencing. But as the mesh is refined, all three gradients, discrete, continuous, and finite difference, converge to the same limit.

A few words are needed here to explain why we can consider this (the discretization of) a linear PDE known as the adjoint to the flow equations: we see no derivatives operating on Ψ. The key here is the scalar product: for the continuous PDE formulation, an integral over the domain. The trick is to perform suitable integrations by part in the integral to move derivatives from the primary variables (the flow variables) to the dual—the Lagrange parameters. Notice also that it is the linearized flow equation that appears. This implies that the adjoint of the Euler flow equations is very similar to the linearized Euler equations—they are almost self-adjoint, which in turn implies that the adjoint equation can be solved by much the same procedures as the primal (flow equations).

The total perturbation δI now depends only on the change of the mesh δX, but is independent of the flow solution perturbation δw. Unlike the gradient calculation by finite difference, for each optimization step, the gradient of I with respect to an arbitrary number of design variables (usually a large amount) can be determined without the need for additional flow-field evaluations. To solve the adjoint equation (10), it costs approximately as much as a flow solution. Note, however, that the boundary conditions in the adjoint PDE are usually chosen to eliminate boundary integral contributions rather than efficient expulsion of waves through the boundaries and this may hamper convergence of the numerical solution. Finite difference methods can also be used to find these sensitivities but are in general significantly more expensive, requiring at least one additional flow solution per parameter.

Examples shown here of direct optimization design are computed by the `SU2` [15] software suite from Stanford University: an open-source, integrated analysis and design tool for solving complex, multi-disciplinary problems on unstructured computational grids. The built-in optimizer is a Sequential Least SQuares Programming (SLSQP) algorithm [6] from the *SciPy Python* scientific library. The gradient is calculated by continuous adjoint equations of the flow governing equations [15, 17]. `SU2` is in continued development. Most examples pertain to inviscid flow but also RANS flow models with the Spalart–Allmaras and the Menter SST k-ω turbulence models can be treated.

2.2 *Inverse Design*

Inverse design is a classical way of designing airfoils and wings, which was popular several decades ago before the advent of high performance computing as a tool in aircraft design.[1] The method consists of *predictor* and *corrector* processes which require engineering know-how at the very beginning of the design stage. The predictor/corrector design approach systematically modifies a given geometry based on direct solutions for the flow around the airfoils or wings. The calculated pressure distribution is compared with a prescribed target distribution and the resulting differences are used by a geometry "corrector" module to modify the current geometry to a shape more likely to generate the desired pressure. The corrector module may be an optimization procedure such as `LINDOP` [10–12] described

[1]This section is adapted from [5, 18].

below, or a design algorithm that directly relates pressure changes to geometry changes. Examples of the latter type include Barger and Brooks [9] methods for designing super-critical airfoils. The method was developed and coupled with several two- and three-dimensional transonic codes by Campbell [7].

Dulikravich [19] solved a similar 3D problem using a Fourier series method. Later Campbell [20] and Obayashi [21] raised some ideas on setting up the target pressure distribution and reasonable constraints for inverse design problem.

Inverse design approach has a long history but it is not out of date. Dealing with the surface curvatures is robust, and the aerodynamicist sees more physical properties of the wing. The approach was recently re-visited and improved by German Aerospace Center (DLR) [22] with good results on laminar wing design. Zhang developed the `SCID` toolbox with the resulting surface curvature [7, 20] inverse design method. The flow chart of `SCID` inverse design is shown in Fig. 1. It connects streamline curvatures on the wing surface with pressure changes to iteratively modify an initial shape. It is combined with under-relaxation chosen

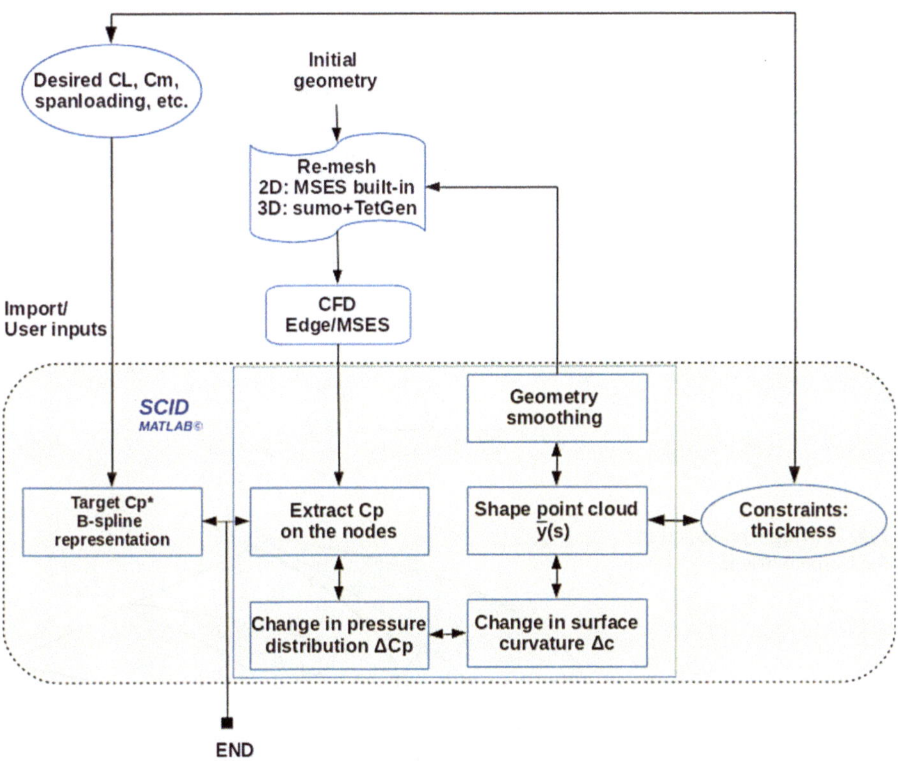

Fig. 1 Flow chart for wing inverse design using `SCID` algorithm, retrieved from [18]

to help convergence, and smoothing procedures to ensure a smooth surface and curvature. Examples shown here for inverse design are computed by SCID [18, 23]. In SCID the CFD code MSES [12] is used for airfoils and EDGE[2] is used for wings.

2.2.1 Pressure–Curvature Relations

The relation is derived from the normal component of the momentum equation for inviscid flow along the streamline on the wing surface as long as the flow is attached. The steady Euler equations with $\mathbf{e}_s$ the unit vector along a streamline, so that the velocity vector is $\mathbf{u} = U\mathbf{e}_s$, $U = |\mathbf{u}|$,

$$UU_s\mathbf{e_s} + U^2\frac{d\mathbf{e}_s}{ds} + \nabla p/\rho = 0 \tag{11}$$

where s is the arc-length along the streamline. In the Darboux frame in Fig. 2, $\mathbf{e_n}$ and $\mathbf{e_t}$ are the surface unit normal, and the second unit normal $\mathbf{e_s} \times \mathbf{e_n}$ to the streamline, there holds

$$\frac{d^2\boldsymbol{\gamma}}{ds^2} = \frac{d\mathbf{e}_s}{ds} = c_n\mathbf{e}_n + c_g\mathbf{e}_t \tag{12}$$

where c_n and c_g are the *normal* and *geodesic* curvatures. The normal component of the streamline Euler equation is:

$$0 + \rho U^2 c_n + \frac{\partial p}{\partial n} = 0 \tag{13}$$

from which we can derive a relation between the curvature and the pressure coefficient,

$$c_n + \frac{C_p}{2L(1 - C_p)} = 0$$

Fig. 2 Wing represented in the Darboux frame

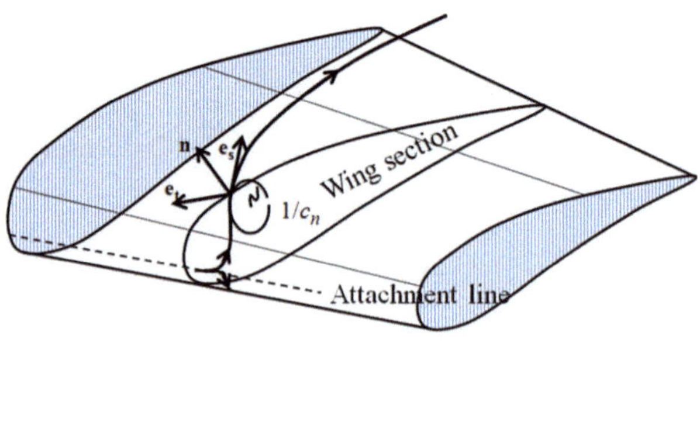

---

[2]www.foi.se/en/Customer--Partners/.../Edge1/Edge/.

where L is a length scale for the pressure gradient $L \approx \frac{p-p_\infty}{\partial p/\partial n}$. Since L is unknown and varies along the streamline, we introduce an inverse length scale coefficient F. The relation used in the shape modification step is the proportionality between changes in normal curvature and pressure coefficient, for unit chord-length, so curvature becomes non-dimensional,

$$c_n = F \cdot C_p = \frac{d^2\boldsymbol{\gamma}}{ds^2} \cdot \mathbf{e}_n - F \cdot C_p. \tag{14}$$

The F-coefficient was proposed by Barger and Brooks [9]. Campbell [7] suggests the c_n-dependence $F = A(1 + {c_n}^2)^B$ for perturbations ΔC_n, ΔC_p, to produce

$$\Delta c_n = A(1 + {c_n}^2)^B \Delta C_p \tag{15}$$

where A and B are adjustable constants.

There remains to relate the surface normal curvature change to geometry change itself. The surface analogue of the Frenet–Serret formulas for the surface coordinates $\boldsymbol{\gamma}(s)$ along the streamline is

$$\boldsymbol{\gamma}_{ss} = c_n \mathbf{e_n} + c_g \mathbf{e_t}. \tag{16}$$

For small change on the surface, it gives

$$\Delta\boldsymbol{\gamma}_{ss} = \Delta c_n \mathbf{e_n} + c_n \Delta\mathbf{e_n} + \Delta c_g \mathbf{e_t} + c_g \Delta\mathbf{e_t} \tag{17}$$

where only the first term on the right-hand side is kept. Note that the last two terms vanish for airfoils. Since only displacement normal to the surface will change the surface, it makes sense to so restrict the geometry change, say

$$\Delta\boldsymbol{\gamma} = h(s)\mathbf{e_n}$$

and then the normal projection of Eq. (17) gives precisely

$$h_{ss} = \Delta c_n.$$

In the shape modification step A is chosen as large as possible without creating divergence in the iteration. Reported values range from 0 to 0.5. Smaller values give slow convergence, larger values may cause divergence. The correct coefficients must be chosen as compromise between speed of convergence and robustness. An adjustment algorithm is applied to select A and B according to the status of convergence.

2.2.2 Shape Modification of Streamline Sections

The following applies to airfoils but is also used in the wing design, with the assumption of surface streamlines not deviating much from the surface traces of the sections used to build the wing. With Δc_n from Eq. (15), the new shape $\mathbf{y}(s)$ is computed from the two-point boundary value problem

$$\frac{d^2\Delta\boldsymbol{\gamma}}{ds^2} = \mathrm{coeff}\Delta C_p \mathbf{e_n}, \quad \Delta\boldsymbol{\gamma}(s_{0,\,\mathrm{lower}}) = \Delta\mathbf{y}(s_{\mathrm{max,\,upper}}) = 0. \tag{18}$$

The arc-length s starts from the trailing edge on the lower surface. The boundary conditions are applied to ensure a sharp and closed trailing edge. The section geometry is represented by point clouds $\boldsymbol{\gamma}^i = (x, z)$, $i = 1, 2, \ldots, N$, N is the number of total points of the airfoil.

2.3 *Hybrid Design*

This chapter describes a hybrid scheme which combines inverse design with optimization of the aerodynamic shape based on the above as shown in Fig. 3. The "hybrid" means we use mathematical optimization with gradients produced by the adjoint technique or by finite differences in a loop together with inverse design that integrates the streamline curvature to produce the shape associated with the target pressure distribution to find the shape (right). One key point is that as the iteration proceeds, the engineer, with some insight from the direct optimization, can modify the current target pressure to guide the design process [7, 21, 23].

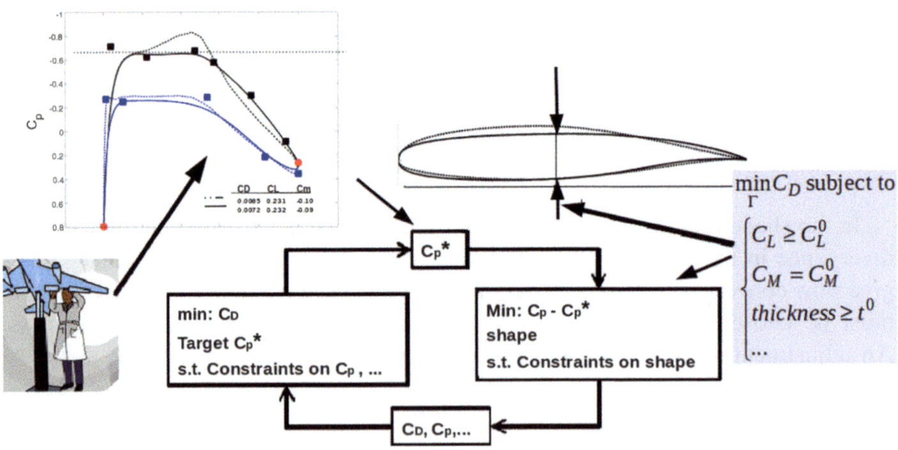

Fig. 3 The feedback design loop

The target pressure distribution on the wing planform is constructed by considering the span loads, isobar patterns, etc. [24], as well as other best practice guidelines provided by experience. Keeping the engineer in the loop emphasizes wing design rather than accurate solution of a (possibly not-so-well formulated) mathematical optimization task:

> Engineer in loop to minimize drag by finding the best feasible target pressure distribution, for which a feasible shape can be found by inverse design.

`LINDOP` [10, 11] is an external code apart from `MSES` [12] which is used for airfoil optimization, it intelligently breaks down the evolution of design into reasonably small number of optimization cycles and allows the engineer in the design process during each cycle. `MSES` estimates the flow field by solving the Euler equations in the internal flow field coupled to thin boundary layer equations in the boundary layer [12, 25]. The overall equation system

$$R(\mathbf{w}; \alpha, AoA, M) = 0 \tag{19}$$

consists of the interior steady Euler equations, the boundary layer equations, and the necessary coupling and boundary conditions. The flow field is solved by Newton-based methods.

The optimization method used in `LINDOP` is also based on Newton iteration, with gradients easily available from the (exact) Jacobians employed in the flow solution. The designer can use the gradients to interactively try out various objective functions I (e.g., aerodynamic forces C_L, C_D, $\frac{C_L}{C_D}$) with respect to small perturbations in design parameters α, and flow parameters AoA and M with almost no additional cost. The Hessian matrix necessary for quasi-Newton optimization is approximated by the BFGS updating scheme [6].

There are two types of optimization problems defined in `LINDOP`:

(I) Least-square problem (modal-inverse design): e.g., $I = \frac{1}{2} \int (f(s) - f_{\text{spec}}(s))^2 ds$, where $f(s)$ is usually the pressure distribution.
(II) General optimization problem (direct design): e.g., $I = C_D$, always with some constraints, on, for example, lift, pitching moment, wing volume, or thickness, etc.

Indeed, those two problems are the most common ones among many design cases. The former one (2.3) is always solved by inverse design, if specified pressure distributions ($f_{\text{spec}}(s)$) are given. The latter one (2.3) is usually solved directly by optimization. The following chapter shows how to solve those two problems using hybrid design with `MSES-LINDOP` an exemplary tool. The designer is allowed to generate design-parameter changes in many ways regarding to different problem to be solved, it can be from direct keyboard inputs, or indirect posing & solving

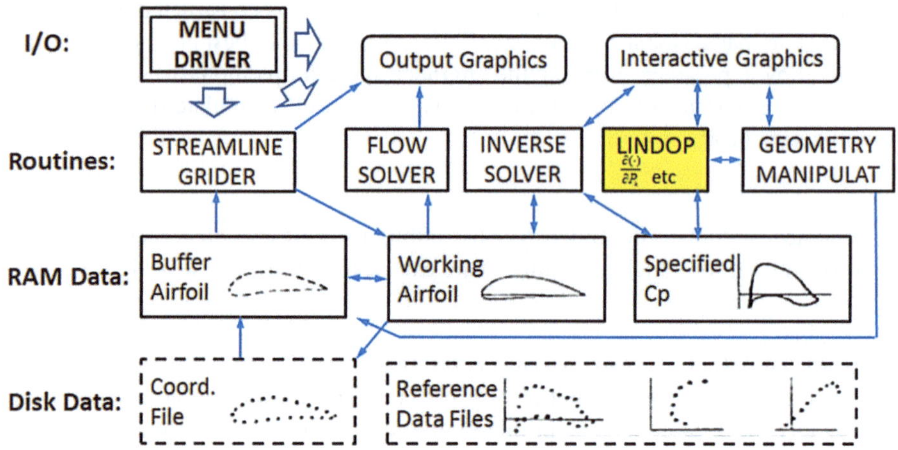

Fig. 4 `LINDOP` work flow, retrieved from Drela [11]

optimization problems [10]. The hybrid approach can be complicated but it is particularly useful for *complex* design problems, for example, multi-element airfoils with multi-point design [26].

Figure 4 spells out how `LINDOP` works. By specifying the target pressure distribution the user can interactively work through the feedback graphics where the flow solver and inverse solver are applied. The procedures require substantial user intervention. The engineer is in the loop to lead the design towards the correct direction in every small step, however it is also very manual and tedious.

3 Parametrization

There are many ways to parametrize a wing, to produce either the lofted wing surface, or the set of surface mesh points. For example the wing surface can be lofted through airfoil stacks (Fig. 5), or the geometry can be represented by modeling the perturbations of the "baseline" shape [27]. The latter technique can also perturb mesh points, or so-called mesh deformation [28]. This section shows several popular parametrization methods used in the test cases, and discusses the mesh update methods, namely, re-meshing and mesh deformation.

For the overall shape definition, mapping from surface mesh to volume mesh is usually done by "re-meshing," i.e. re-creation of the complete grid for each shape to analyze. The generated grids will have well-formed cells, and usually mesh generation takes only a small fraction of the flow solution time. This allows loose coupling but also means that each flow solution must be done essentially from scratch since the number of flow variables is different from the previous calculation. However, if the CFD package supports interpolation between arbitrary grids it is possible to obtain a good initial guess for the flow which can speed up the solution.

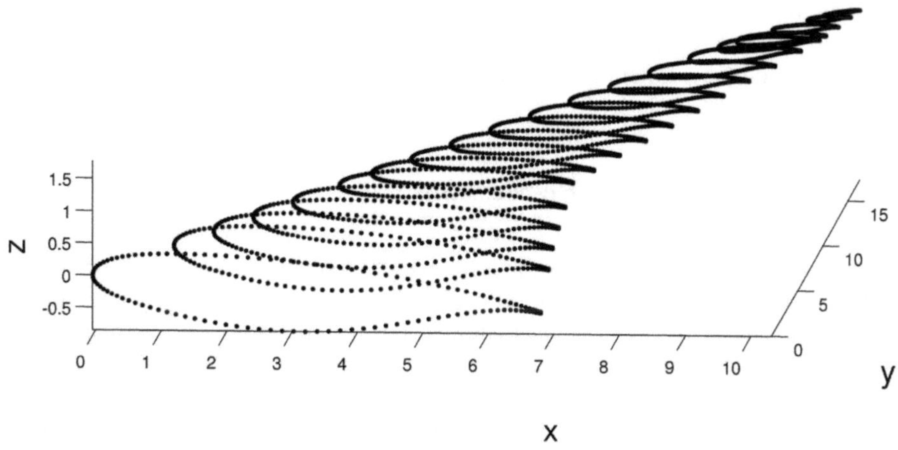

Fig. 5 Airfoil stacks

In the deformation approach, only the coordinates of grid points change, so no interpolation is necessary for initial guesses. A mesh deformation algorithm for propagating the deformation of surfaces through the whole grid is needed. The cheapest alternative is interpolation methods for arbitrary points, based on, e.g., radial basis functions or Kriging. Also PDE-based methods employing the equations of elasticity or the Laplace equations are in use. However, these techniques are most easily implemented closely coupled to the flow solver. The PDE methods provide some guard against creation of bad computational cells. But although they deal with deformations which are very small compared to wing dimensions, they can be large compared to mesh cell sizes, e.g., at a sharp trailing edge when the twist is changed.

3.1 Shape Definition

There are many ways to parameterize a wing, to produce either the lofted wing surface, or the set of surface mesh points. For example, the wing surface can be lofted through airfoil stacks, or the geometry can be represented by modeling the perturbations of a "baseline" shape [27]. The latter technique can also perturb off-surface mesh points, by so-called mesh deformation [28]. This section shows popular parametrization methods used in the test cases, and discusses the mesh update methods, namely, re-meshing and mesh deformation. Although the CAD-free parametrization techniques have been proposed [29, 30], we believe that the re-meshing technique has some advantages. Re-meshing is easy if a smooth geometry is provided. A reliable and fast meshing tool is a key. `SCID` uses `sumo` [31], a tool for rapid automatic Euler and RANS meshing. If re-meshing rather than mesh deformation is applied in finite difference approximation of derivatives, changes to design variables cannot be too small, lest the unavoidable, random-looking, mesh changes resulting from the detailed workings of the generation algorithm hide the gradient information.

3.2 *Airfoil Stacks and Re-Meshing*

Airfoil sections are the most important building block of aerodynamic geometry. Vassberg and Jameson [3] state that: "Airfoils are used to define wings, pylons, nacelles, struts, winglets, features, horizontal stabilizers, verticals, propellers, turbomachinery blades and stators, cowls, blimps, sailboat sails, keels and ballast-bulbs, cascades, helicopter rotors, fins, chines, strakes, vertical/horizontal-axis wind turbines, flaps, frisbees, and boomerangs." In most software systems for aircraft shape definition, the defining stations are chordwise cuts. The wing surface parametrization is decomposed into parameterization of n stations of airfoils. It is customary for the first defining station to be at the symmetry plane (wing root), and the last defining station to be at the wing's theoretical tip. Each airfoil (defined as scaled to leading edge at the origin to trailing edge at [1,0]) is rotated by an incidence, translated to the defining station leading edge, then scaled to match the projected planform chord. The wing surface is usually lofted by Bézier/Bspline surfaces [32].

In `SCID` as well as many software systems for aircraft shape definition, the defining stations are spanwise cuts. The wing surface parametrization is decomposed into parametrization of n stations of airfoils. The geometry is updated (and smoothed) in every design cycle, then a *re-meshing* is carried out, as indicated in Fig. 1.

3.3 *Airfoil Shape Definition*

Some airfoil families are defined by a number of parameters with geometric interpretation, such as the NACA four, five, and six-digit families. But those families are of limited interest for the super-critical airfoils for transonic speeds, so more general schemes must be devised.

3.3.1 Bézier/Bspline Curves

Using Bézier/Bspline polynomials to parametrize the airfoil shape is simple and robust [32–34], it ensures geometrical properties including leading edge radius, trailing edge shape by solving a least-square fitting problem. It usually gives good representation of an airfoil (and smoothing) chosen by the number of control points. Melin et al. [35] developed a technique that uses four pieces of cubic Bézier

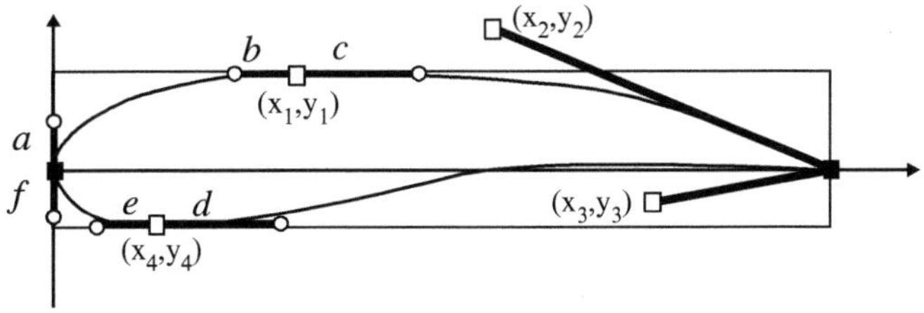

Fig. 6 Airfoil is parametrized by four pieces of cubic Bézier curves

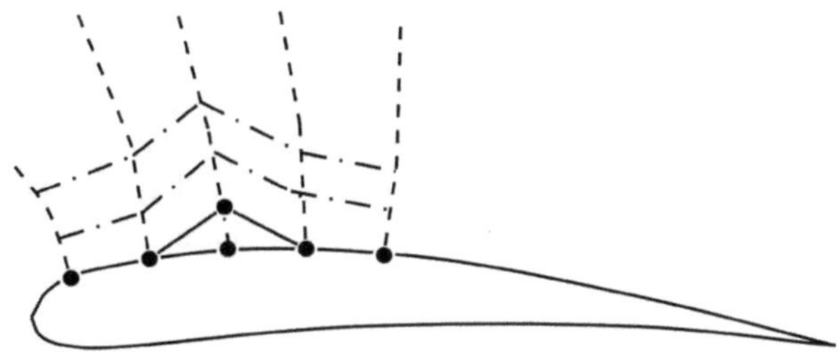

Fig. 7 The surface mesh point at i, j is moved by $\delta X_{\Gamma_{i,j}}$

curves [33] to parametrize an airfoil within a reasonable error level. A similar parametrization is used in SCID [18, 23] for geometry update and smoothing purpose with wing represented by airfoil stacks (Fig. 6).

3.4 Parameterization of Shape Perturbations

3.4.1 Mesh Points and Mesh Deformation

When the mesh points are used to represent the surface, the design variables are the coordinates of the mesh points. The main advantage of parameterizing a shape with mesh points is that there is no restriction on the attainable geometry. Also, this parametrization technique can be easily implemented in any design problem. However, the use of mesh points does present some difficulties. First, the independent displacement of points may create non-smooth surfaces which are unsuitable as lifting surfaces and give the flow solver a hard time. Second, if all surface mesh points are used, the method is very costly for 3D problem since we will deal with a large number of design variables (i.e., surface mesh points). Both difficulties can be easily resolved by using a set of smooth functions to perturb the initial mesh so that the surface mesh points are *mapped* from a limited number of design variables, such as in MSES-LINDOP [10–12], SU2 [15]. Figure 7 shows the surface mesh point at i, j which is moved by $\delta X_{\Gamma_{i,j}}$, Fig. 8 shows the surface mesh points moved by Hicks–Henne bumps.

Deforming the computational mesh is an efficient alternative to re-meshing and it enables a smooth *mapping* from the design parameters to the cost function. One issue for mesh deformation is that by deforming the surface boundary of the mesh

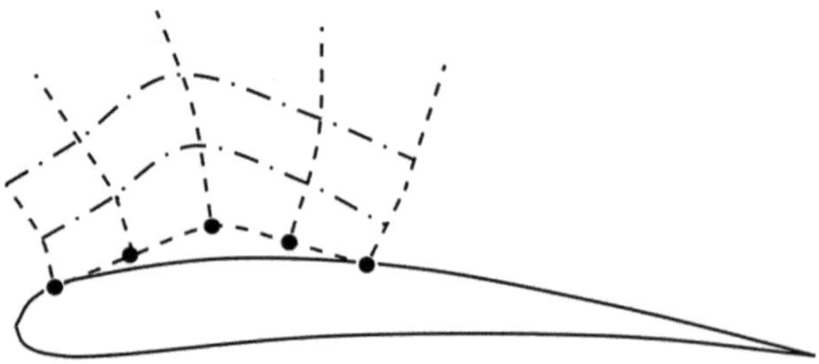

Fig. 8 The surface mesh points are moved by Hicks–Henne bumps [37]

points, the rest of the grids must be deformed accordingly. SU2 uses the linear elasticity equations [15, 36] to compute the volume mesh displacement from the displacement of the perturbed surface. If the computational cells are small, this prevents creation of negative volume cells by deformation. In certain circumstances, further mesh smoothing [28] will be required.

3.4.2 Hicks–Henne Bumps

A single Hicks–Henne (HH) bump function [37] perturbs the airfoil shape (y coordinates) by a "bumps," so that with a sequence of HH bumps there obtains a perturbation of airfoil shape,

$$\Delta y(x) = \sum_{k=1}^{N} \alpha_k \sin\left(\pi x^{\frac{\log 0.5}{\log x_k}}\right)^t \tag{20}$$

with the x-locations of max-points are $x_k, k = 1, 2, \ldots, N$, and the coefficients α_k are design variables. Figure 9 shows an example of the fourth order bumps ($t = 4$) with $N = 10$, x_k is equally distributed over $[0.5/N, 1 - 0.5/N]$.

3.4.3 Free-Form Deformation

Free-form deformations (FFD) provide a method of deforming an object by adjusting the control points of a lattice. The technique was first described by Sederberg and Parry in 1986 [38] and its effect is used in computer animation. In 2D the shape perturbations are simply modeled by Bézier/Bspline/NURB control points [33, 34, 38, 39]

$$d\boldsymbol{\gamma}(x, y) = \sum_{i,j=0}^{nx-1,ny-1} d\mathrm{CP}_{i,j} B_i^{nx}(u) B_j^{ny}(v) \tag{21}$$

$$x = x_{\min} + u(x_{\max} - x_{\min})$$

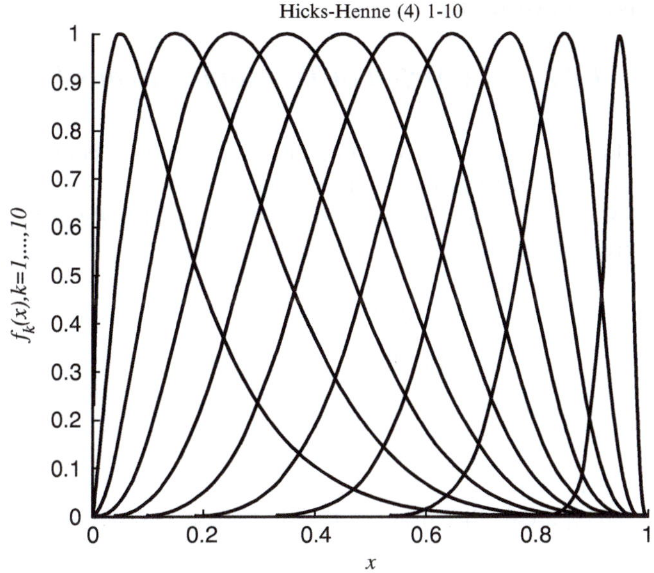

Fig. 9 The Hicks–Henne bump functions, $t = 4$, with $N = 10$, x_k is equally distributed over $[0.5/N, 1 - 0.5/N]$

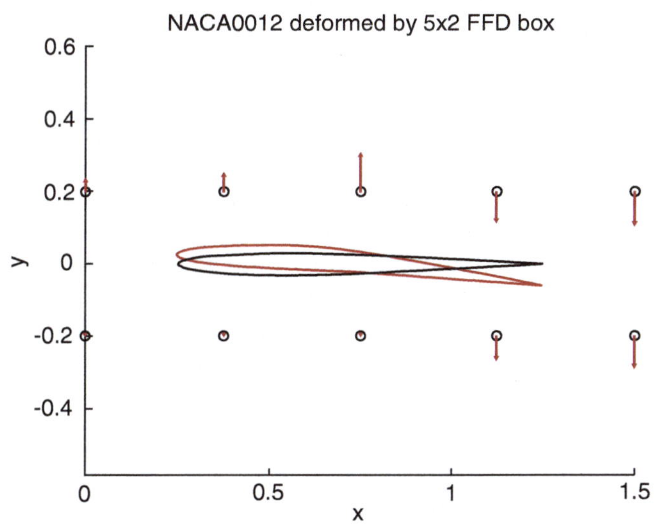

Fig. 10 NACA-0012 is deformed by a 5×2 FFD Bézier box

where nx, ny are the degrees of the FFD functions, $u, v \in [0, 1]$ are the parametric coordinates, $\mathrm{CP}_{i,j}$ is the $nx \times ny$ array of control points, Bs are the Berstein polynomials [32]. Fig. 10 shows an example that deforms an NACA-0012 airfoil by a 5×2 FFD Bézier box. In SU2 deformation of the baseline wing is done by a 3D FFD Bézier box [15, 38] in a similar way.

4 Test Cases Statement

4.1 Case I: RAE-2822 Airfoil in Transonic Viscous Flow

The drag should be minimized at Mach number 0.734 and lift coefficient of 0.824, and the cross section area must exceed or equal that of the baseline. Initial angle of attack is 2.79°. The flow is viscous with Reynolds number $Re = 6.5 \times 10^6$. The optimization problem is

$$\begin{aligned} \text{min}: &\quad c_d \\ \text{subjectto}: &\quad c_\ell = 0.824 \\ &\quad c_m \geq -0.092 \\ &\quad \mathrm{A} \geq A_{baseline} \end{aligned} \tag{22}$$

where c_l, c_d, and c_m are the lift, drag, and pitch moment coefficients, A is the airfoil cross section area.

4.2 Case II: ONERA M6 Wing Optimization in Transonic Inviscid Flow

The ONERA M6 wing is a classic computational fluid dynamics (CFD) validation case for external flows because of its simple geometry combined with complexities of transonic flow (i.e., local supersonic flow, shocks, etc.). It has almost become a standard for CFD codes because of its inclusion as a validation case in numerous CFD papers over the years [40]. In the proceedings of a single conference, the 14th AIAA CFD Conference[3] (1999), the ONERA M6 wing was included in 10 of the approximately 130 papers. This wing configuration is used here as a baseline for drag minimization. The drag should be minimized at Mach number 0.8395 and the flow is assumed to be inviscid. The maximum thickness t of each section should be preserved to a specified value. Initial angle of attack is 3.06°.

$$\begin{aligned} \text{min}: &\quad c_d \\ \text{subjectto}: &\quad c_\ell = 0.2864 \\ &\quad \mathrm{t}_{i,max} = \mathrm{t}_{i,specified}. \end{aligned} \tag{23}$$

[3]http://www.aiaa.org/.

5 Results from Direct Optimization SU2

5.1 *Test Case I*

The Spalart–Allmaras turbulent model [41] is used in this test case. Figure 11 shows the airfoil grids with 140,573 nodes. The mesh is perturbed by Hicks–Henne bump functions [37] with 19 design variables. Table 1 shows the optimization solution table for RAE 2822 airfoil (Fig. 12). The KKT condition [6] is met after 35 design cycles. The shock at around 55 % chord is weakened, with a drag benefit of 70 drag counts.[4] Figure 13 shows the pressure distribution and airfoil shape for both baseline and optimized airfoils. The C_p of the optimized shape has wiggles in between 0.5 and 0.6 chord, that the shock starts to re-build a little. Figure 14 shows the Mach contours of both RAE 2822 and its optimized shape, with weakened shock on the optimized shape. Note that the Mach is re-developed between 0.5 and 0.6 chord.

5.2 *Test Case II*

It is an unstructured mesh with 36,454 tetrahedral cells, half geometry with a symmetric plane at $y = 0$. The wing tip is capped. The wing is parametrized by FFD Bézier box [38] as discussed in previous section by 176 design variables, with root section unchanged. The optimized wing is obtained after 14 design cycles, the drag coefficient C_D is reduced by around 17.9 *counts*, while the lift coefficient

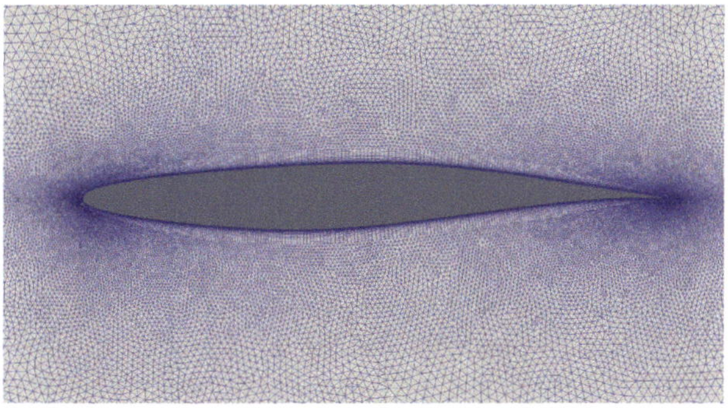

Fig. 11 RAE-2822 airfoil mesh with 140,573 nodes

Table 1 RAE 2822 optimization results table

Airfoil	c_ℓ	c_d	c_m
Baseline RAE2822	0.8092	0.01949	−0.09679
Optimized	0.8431	0.01263	−0.08631

[4] 1 drag count is defined as 10^4 drag coefficient; 1 lift count is defined as 10^3 lift coefficient.

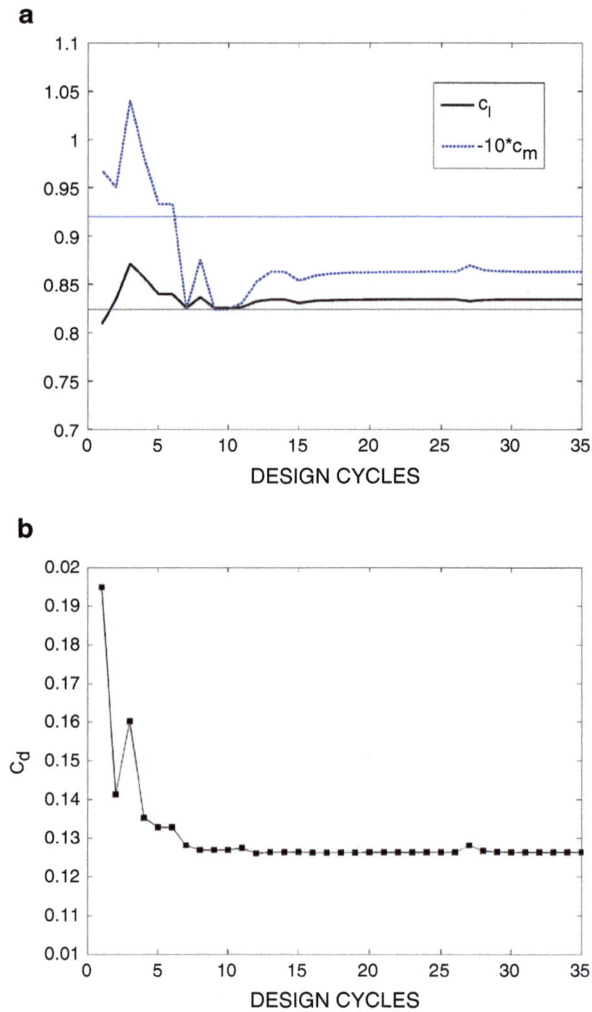

Fig. 12 The design cycles history for RAE 2822 with 19 design variables on the 140,573 nodes mesh. (**a**) Convergence of constraints on c_ℓ and $-10 * c_m$ (**b**) Convergence of cost function c_d

C_L is increased by 4 % as it can be seen in Table 2. Figure 15 shows the pressure coefficient comparisons for M6 baseline and its optimized wing shape from SU2. The shock is reduced such that 17 drag counts benefit is obtained.

The wing section profiles are studied at five stations from root (0 m) to tip (1.1963 m) on both baseline and optimized geometries. The maximum thickness varies from section to section, and its maximum locations even shifted a bit forward on inboard sections, see Figures 16a, b. The optimizer changes twist by less than 1 degree, and maximum chamber by less than 1 % to arrive at a point believed to be a local optimum, using 176 design variables. This indicates that the wing is hard to improve on.

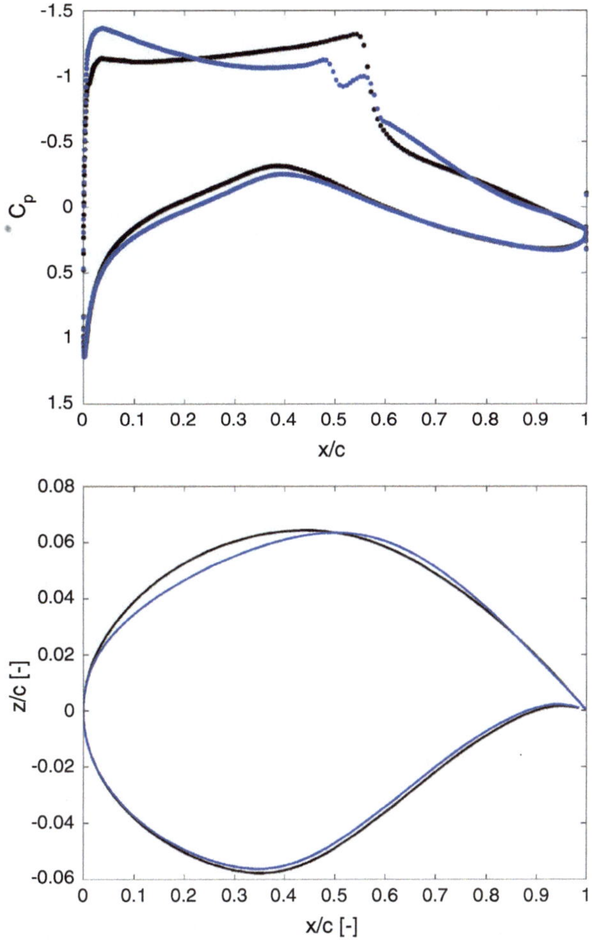

Fig. 13 Pressure distributions and airfoil shapes for RAE 2822, *black*: baseline; *blue*: optimized

5.3 *Assessment: Direction Optimization*

The direct optimization gives a rapid indication of possible directions for improvement when traditional inverse design and/or geometric cut-and-try are impractical. It also provides the possible design improvement paths when unusual non-aerodynamic design variables are present, for example, the r.m.s. strain constraints. Due to the fact that it minimizes the cost function and the cost function can be defined in multi-points, it is suitable to handle multi-point design problems [42], whereas single-point design is better handled with traditional inverse design. For the 2D problem, the optimized result in the test case was obtained after only 35 iterations. This approach is clear to extend to 3D (as the ONERA M6 wing), but needs even longer computation time.

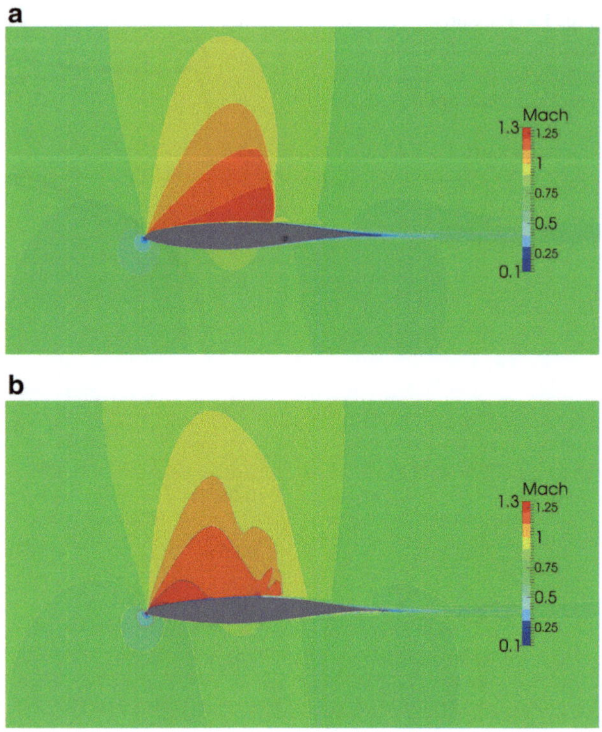

Fig. 14 Mach contour for RAE 2822 and its optimized shape at Mach 0.734, $c_\ell = 0.824$. (**a**) Baseline RAE 2822 Airfoil (**b**) Optimized shape

Table 2 M6 optimization results table

Shape	C_L	C_D
Baseline (ONERA M6)	0.28641	0.0117909
Optimized	0.298087	0.010016

The cons are also considerable. First of all, it requires tough learning curve, and one cell with high surface gradient can influence the overall search direction. For example, SU2 has an option that removes the sharp edge sensitivities from the gradient calculation to guarantee a descent direction in optimization [17]. This treatment to sharp trailing edge is easier to find the mean gradient. However, the drawback is that it removes the gradient from trailing edge, resulting little geometric changes around trailing edge, if we recalled the ONERA M6 wing case, the optimized wing has little twists and cambers compared with the baseline configuration, see Figure 16.

Setting up the optimization problem requires engineering skill as well. The cost function and the constraints should be well defined to ensure convergence. Palacios et al. [17] claimed a "sequential way" to apply constraints when designing a simple wing-body configuration in transonic viscous flow using SU2 otherwise the optimizer would fail. The gradient is sensitive to the mesh deformation method/strategy, if the gradient is not on the order of a meaningful dimensional perturbation of the design variables (control points), the first step of the optimizer will cause the mesh deformation to fail due to too large of a step being taken.

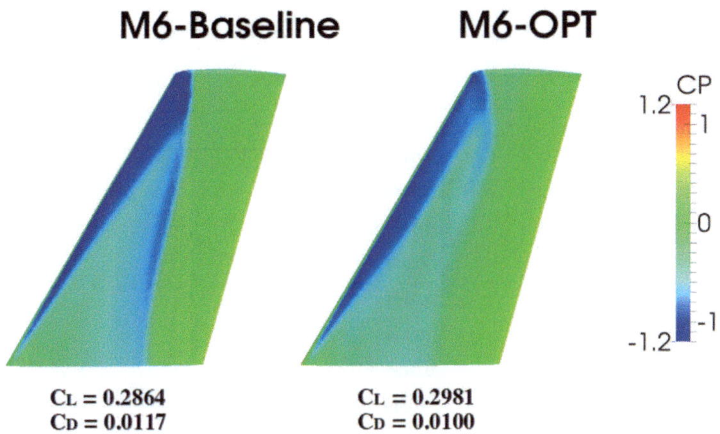

Fig. 15 Upper surface C_p for ONERA M6 baseline and optimized wing from SU2 optimization

This approach is always too "myoptic," it is trapped in the local optimum rather than finding the global optimum, exploiting the smallest significant physical scales. For 3D problem, it requires long computing time on big computers, although the adjoint approach reduces the gradient calculation time a lot.

6 Results from Inverse Design SCID

6.1 *Test Case I*

The target pressure is defined as a pressure distribution with weakened shocks. The target is found perfectly with 50 iterations by SCID-inviscid mode to get quick convergence, see Figure 17. Due to zero pressure gradient through boundary layer we would rather use SCID-inviscid mode to compute once the target pressure is given. However there are form drag and skin friction drag that SU2 can give while SCID-inviscid mode cannot. A compromise is made by re-running the solution from inviscid SCID in MSES [12] viscous mode, see Figure 18. The drag is reduced from 170 counts to 115 counts, with c_m constraint perfectly held (Table 3).

6.2 *Case II Variation*

A variation design of test case II is carried out, which is a similar exercise as Jameson did [43] by the adjoint code for inverse design. The wing planform is ONERA-M6 and the initial geometry was made up of NACA 0012 sections and

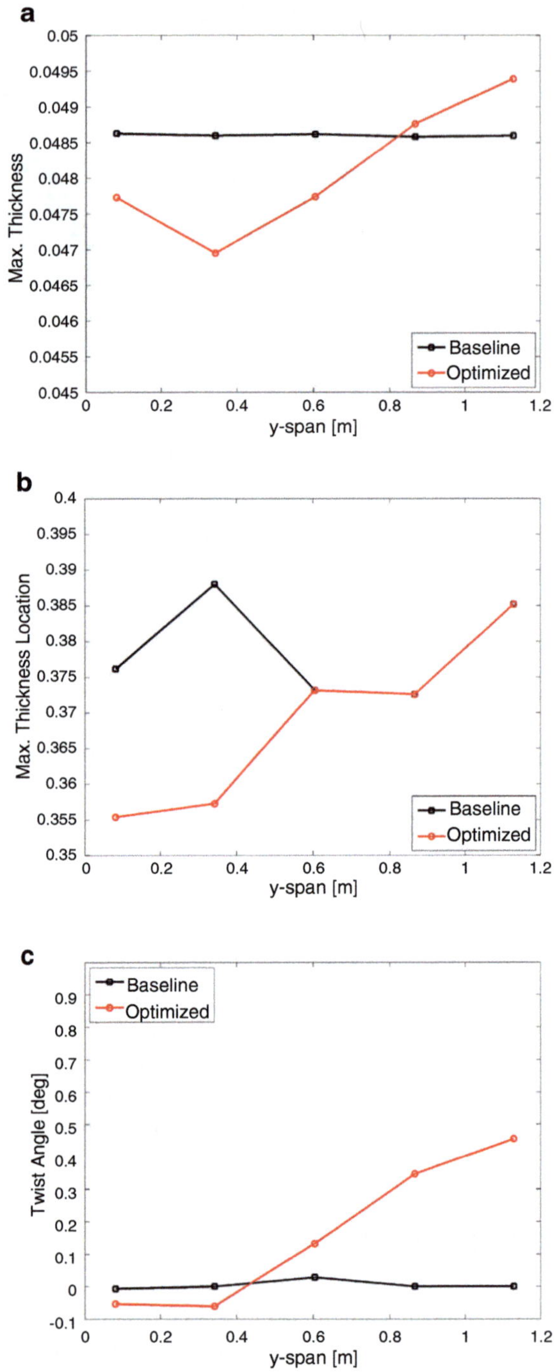

Fig. 16 Geometric comparison for M6 baseline vs. optimized results on five spanwise stations. (**a**) Maximum thickness (**b**) Maximum thickness locations (**c**) Local twist distributions

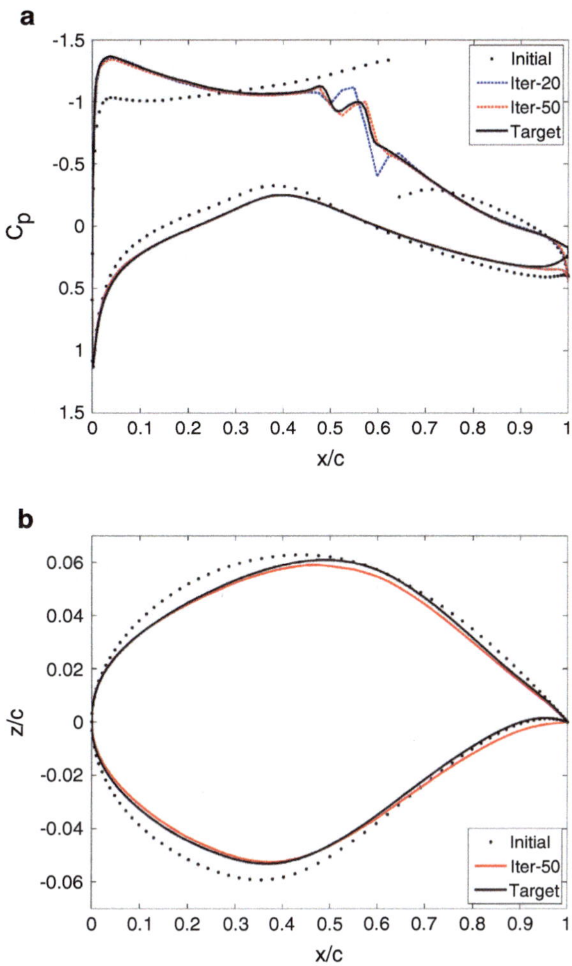

Fig. 17 The resulting pressure distributions and the airfoil shape from SCID-inviscid; baseline/initial shape is RAE-2822 airfoil. (**a**) Pressure distributions computed from SCID-inviscid (**b**) Resulting airfoil shapes computed from SCID-inviscid

the target pressure distribution was the pressure distribution over the ONERA-M6 wing. The target pressure distribution was computed by SU2 in inviscid flow with the same mesh and under the same conditions as computed the ONERA M6 baseline in the previous section, namely, unstructured grids with 36,454 cells (rather coarse), Mach number 0.8395, angle of attack is fixed at 3.06°. Eight equally spaced sections are designed for half wing, from 0 % of the wing semi-span (root) to 94.32 % of the wing semi-span. Figures 19 and 20 show the pressure distribution and the section geometries over the initial NACA 0012 airfoil wing and the final design by SCID. The final design was achieved by 110 designs. Note that after 40 designs the target pressure distribution was already almost found, with only slightly deviations. More design cycles would not make significant difference. Smoothing plays an important

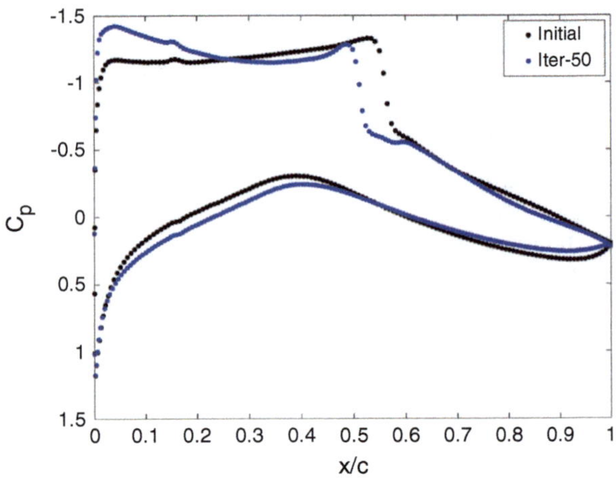

Fig. 18 Re-run the solution Iter-50 in viscous mode

Table 3 RAE 2822 design by re-running the baseline and the optimized solution from inviscid SCID in MSES viscous mode

Airfoil	c_ℓ	c_d	c_m
Baseline RAE2822 -vis	0.824	0.016987	−0.09394
Iter-50 -vis	0.824	0.011512	−0.05896

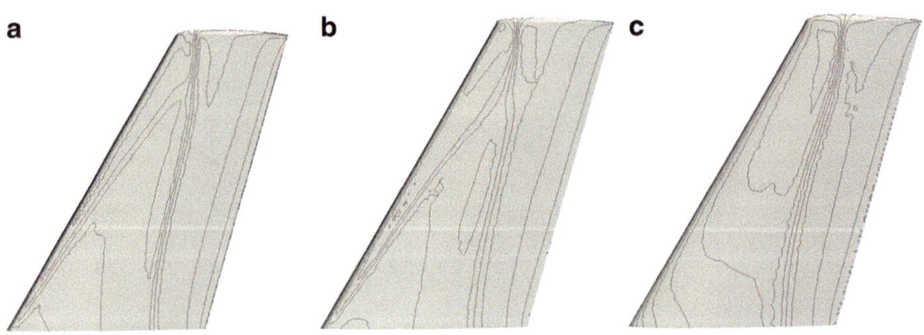

Fig. 19 Initial and Final C_p contours comparisons of M6 wing planform. (**a**) Target pressure distribution contours over an M6 wing computed by SU2; (**b**) Final pressure distribution contours obtained by SCID-inviscid mode; (**c**) Initial pressure distribution contours over an M6 wing with NACA 0012 profile

role, the different smoothing technique would lead different pressure distributions especially for the last design cycles. As Jameson [43] claimed, this is a particularly challenging test, because it calls for the recovery of a smooth symmetric profile from an asymmetric pressure distribution containing a triangular pattern of shock waves, (Table 4, Fig. 21).

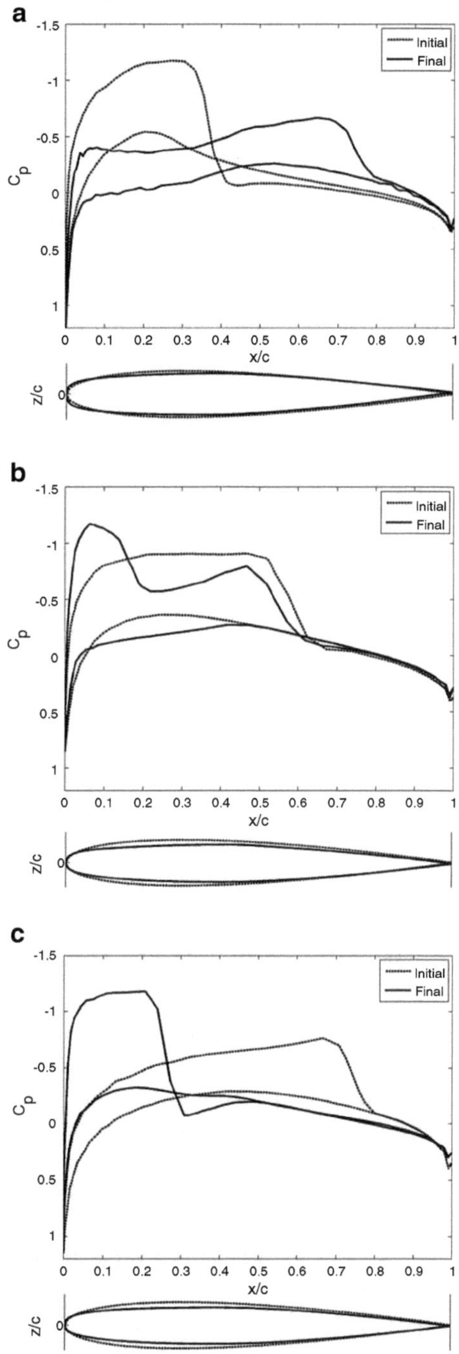

Fig. 20 Initial and Final pressure distribution and modified section geometries along the wing, computed by SCID-inviscid mode, compared with target pressure distribution computed by SU2 Root. (**a**) section, 0 % semi-span (**b**) Mid section: 54.26 % semi-span (**c**) Tip section: 94.32 % semi-span

Table 4 Wing design for ONERA M6 planform by SCID

ONERA M6 Planform	C_L	C_D	C_m
Initial SCID	0.29828	0.020906	−0.133737
Final SCID	0.28634	0.01540	−0.11941
Target SU2	0.28641	0.0117909	−0.120047

6.3 Assessment: Inverse Design

The significant advantages of inverse design are that the engineer applies her knowledge and (thus) a realistic airfoil is obtained at every iteration (e.g., MSES inverse design mode [12]). The pressure gradient is zero through boundary layer [44] so that it can be chosen to maintain laminar flow. The changes in streamline curvature–pressure relationship are robust that it ensures fast convergence to target pressure with good quality. The engineer in the loop is seeking favorable properties of the pressure distributions such as:

1. the flattened leading edge pressure peak on the upper surface which avoids leading edge flow separation;
2. weakened or eliminated shock waves which reduce the wave drag;
3. monotonic trailing edge pressure recovery that avoids boundary layer separation.

However, using the inverse method, the engineer must be well experienced and knowledgeable to know how to set the target pressure. Because this method works on pressure distributions rather than the lift or drag coefficients, it may not reach a true optimum, at best only the target pressure. Moreover, the streamline curvature–pressure relation used in SCID is more tricky and complex for transonic flow in 3D since the shocks and cross-flow are introduced [18].

7 Results from Hybrid Design: Case I

Figure 22 shows the pressure distributions on the baseline airfoil calculated in MSES inviscid mode, and the target pressure distributions (solid). The target pressure is obtained from the optimized solution of RAE 2822 airfoil in inviscid flow using EDGE adjoint solver, which is out of scope here. Directly driving the C_p towards the "target" pressure distributions causes divergence. What we need to do in LINDOP is to make the design *iteratively*. There are many options that user can interact, for example, the search direction, the search step, and quasi-Newton toggle. Figures 23 and 24 show the gradient-inverse design procedures in LINDOP inviscid mode. In the former figure only the upper surface is modified, and the C_p value at the trailing edge (right endpoint of freewall segment) is fixed. In the latter figure only the lower surface is modified, and the stagnation C_p value (left endpoint of freewall segment) is fixed.

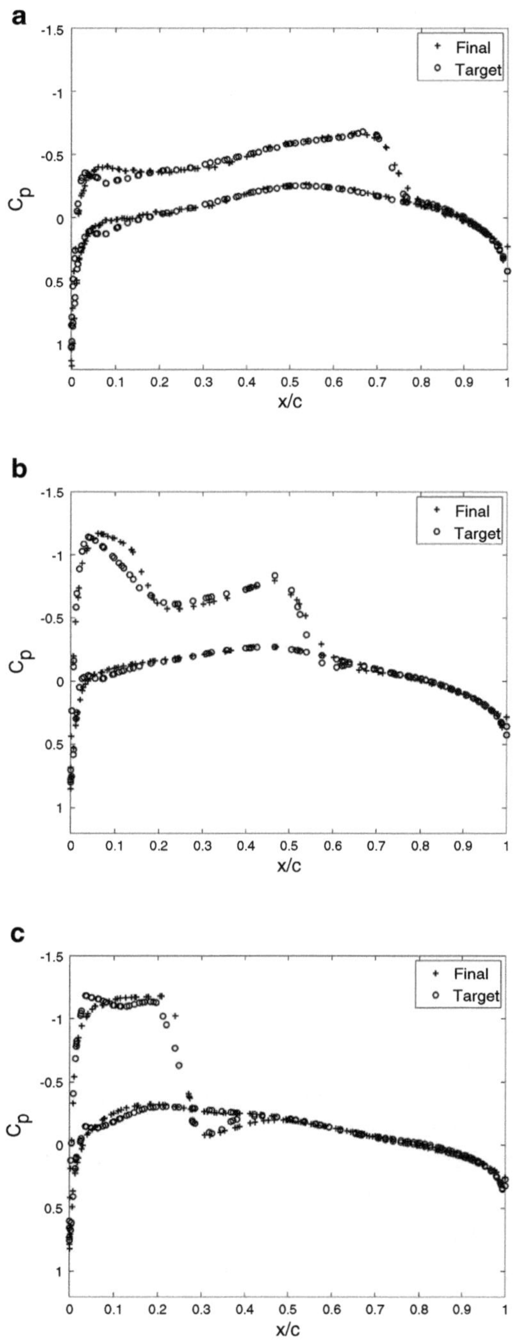

Fig. 21 Final and target pressure distributions along the wing using inverse design. (**a**) Root section, 0 % semi-span (**b**) Mid section: 54.26 % semi-span (**c**) Tip section: 94.32 % semi-span

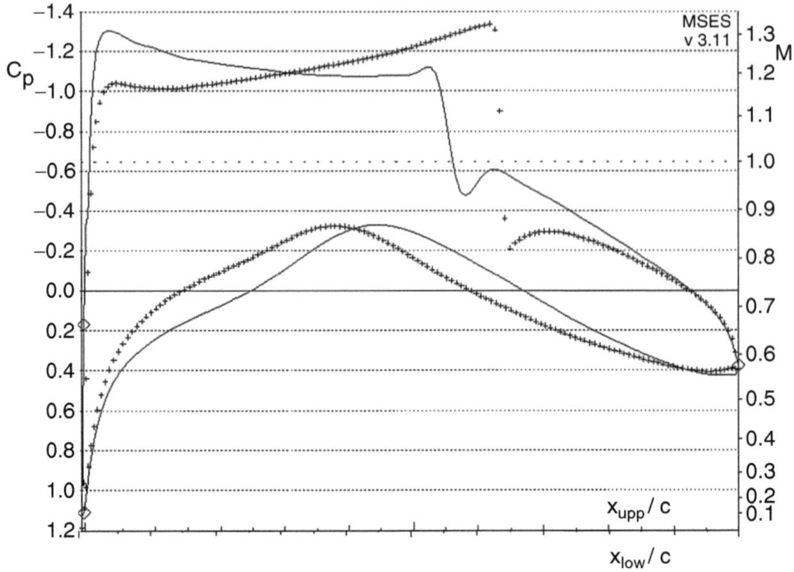

Fig. 22 The pressure for baseline airfoil RAE2822 (*markers*) calculated in MSES inviscid mode and the target pressure distributions (*solid*)

Figure 25 shows the optimized results from RAE 2822 airfoil, inviscid flow solutions. It is found that in inviscid flow, the drag coefficient c_d is reduced by 74 %, the lift coefficient is maintained, and the c_m constraint is held (Table 5).

7.1 Assessment: Hybrid Design

The pros of hybrid design are:

- it can systematically set constraints and cost functions, thus can obtain benefits of both inverse design and direct optimization;
- it offers the engineer know-how advantages (c.f. long list of user options in LINDOP menu);
- iterations with visualized feedback and (thus) a realistic airfoil is obtained at every iteration.

However, there are a number of cons for hybrid design approach. It also has a tough learning curve for *users*, and it is too manual to set/determine too many options, especially for new users. There is no guarantee for convergence, unless the target pressure is a "small" modification of the initial one (e.g., Figures 23, 24). There are some tricks to get convergent solutions. First of all, make small pressure changes (ΔC_p) for each design cycle; second, modify/re-design one surface each

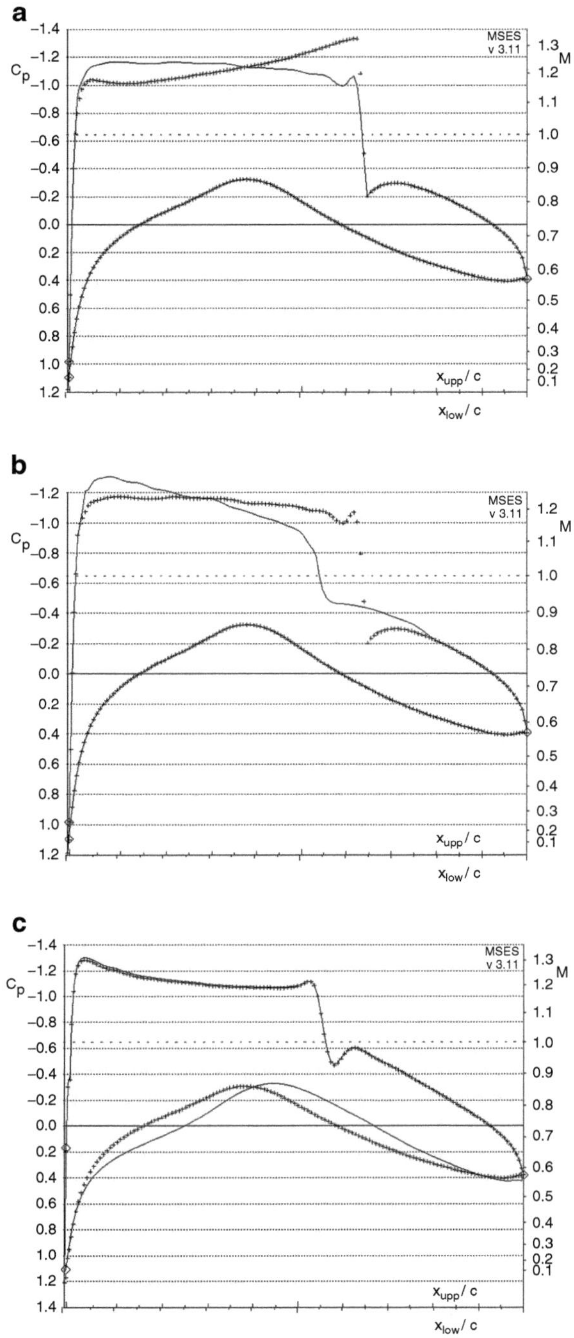

Fig. 23 The design procedures for RAE 2822 in `LINDOP` inviscid mode, modify upper surface *only*, to be continued. (**a**) Fix right endpoint segment (**b**) Fix right endpoint segment (**c**) Fix right endpoint segment

a

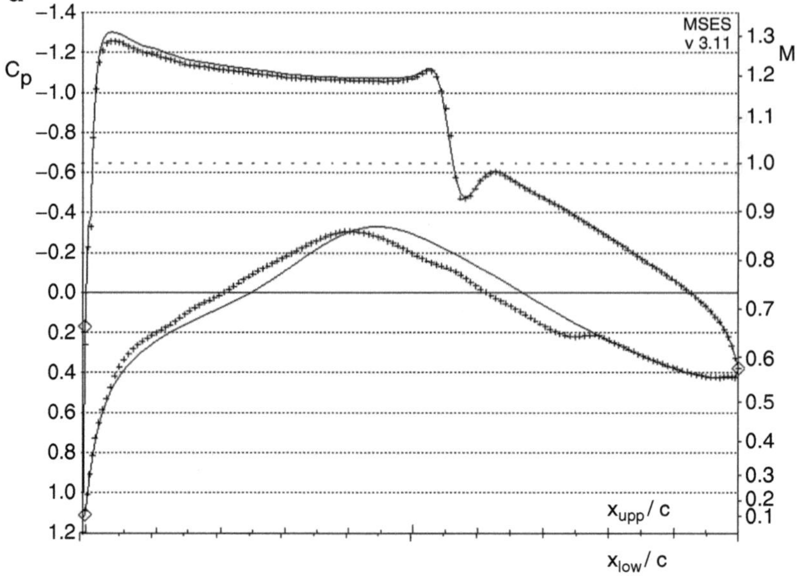

b

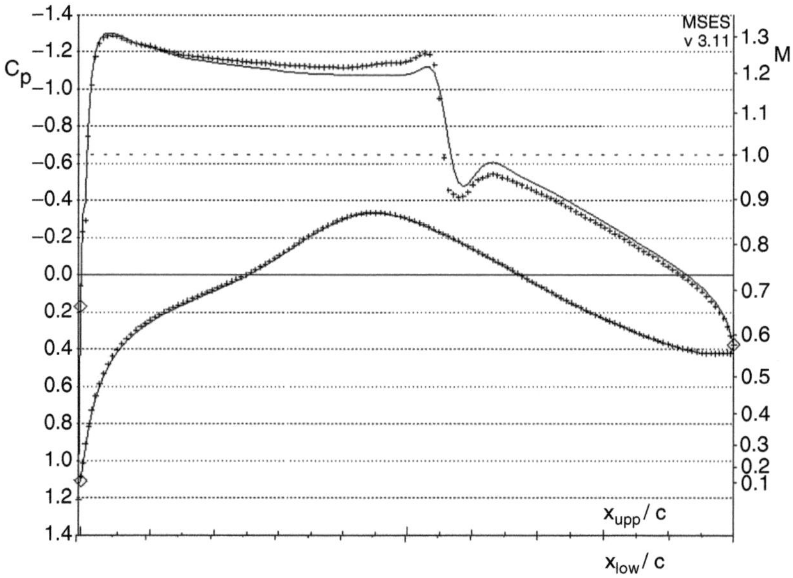

Fig. 24 The design procedures for RAE 2822 `LINDOP` inviscid mode, modify lower surface *only*, completed. (**a**) Fix left endpoint segment (**b**) Fix left endpoint segment

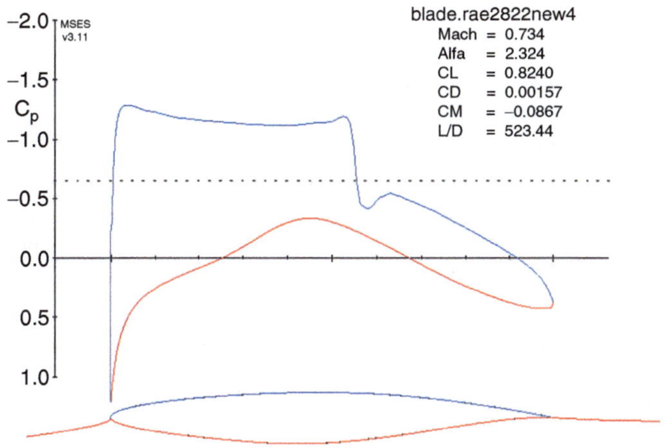

Fig. 25 The optimized design of RAE 2822 in `LINDOP` inviscid mode

Table 5 RAE 2822 optimization results table, computed in `LINDOP` inviscid mode

Airfoil	c_ℓ	c_d	c_m
Baseline RAE2822 - `invis`	0.824	0.00648	−0.1294
Iter-50 -`invis`	0.824	0.00157	−0.0867

time; third, carefully choose order of mixed inverse prescribed shape function and global constraints; fourth, fix the left/right endpoint of freewall segment each time; finally, introduce the target pressure C_p^* from the last cycles when the current C_p is close to C_p^*. All of the tricks and the options which should be determined by users make the hybrid design method tough for new users.

8 Conclusions

This chapter assesses three different design methods for aerodynamic shape design by two test cases. They are not isolated to each other, the "hybrid" design is to combine the first two approaches. It is difficult to say one is superior than the other, each of them has pros and cons. What we can do is to understand the strong and weak points of each method, and use the appropriate and/or combined methods to a specified design problem. Van der Velden called it "cocktails" or combinations of optimizers [45] under the control of the engineer in the loop. This is also stressed in Zhang's PhD thesis [5].

Acknowledgements Part of the computations were performed on resources provided by the Swedish National Infrastructure for Computing (SNIC) at PDC Centre for High Performance Computing (PDC-HPC)

References

1. Sobieszczanski-Sobieski, J.: Multidisciplinary Design Optimization: An Emerging New Engineering Discipline. Kluwer Academic, Boston (1995)
2. Kroo, I.M.: Multidisciplinary Design Optimization: State-of-the-Art, chapter MDO for Large-Scale Design, pp. 22–44. SIAM, Philadelphia (1997)
3. Vassberg, J.C., Jameson, A.: Influence of shape parametrization on aerodynamic shape optimization. Technical report, Brussels, Belgium (2014) Von Karman Institute Lecture-III
4. Castonguay, P., Nadarjah, S.K.: Effect of shape parameterization on aerodynamic shape optimization. In: 45th AIAA Aerospace Sciences Meeting and Exhibit, 2007-59, Reno, Nevada (2007)
5. Zhang, M.: Contributions to Variable Fidelity MDO Framework for Collaborative and Integrated Aircraft Design. Doctoral thesis, Department. Aeronautics, KTH Kungl. Tekniska Högskolan (2015)
6. Griva, I., Nash, S.G., Sofer, A.: Linear and Nonlinear Optimization, 2nd edn. Society for Industrial Applied Mathematics, Philadelphia (2009)
7. Campbell, R.L., Smith, L.A.: A hybrid algorithm for transonic airfoil and wing design. In: AIAA Aerospace Sciences Meeting, 87-2552 (1987)
8. Lamar, J.E.: A vortex-lattice method for the mean camber shapes of trimmed non-coplanar planforms with minimum vortex drag. Technical report, Washington, DC (1976). NASA TN D-8090
9. Barger, R.L., Jr. Brooks, C.W.: A streamline curvature method for design of supercritical and subcritical airfoils. Technical report (1974) NASA TN D-7770
10. Drela, M.: A user's guide to lindop v2.50. Technical report (1996)
11. Drela, M.: Lindop optimization procedures. Technical report, Computational Aerospace Sciences Laboratory, MIT Department of Aeronautics and Astronautics (1993). Technical Report
12. Drela, M.: Newton solution of coupled viscous/inviscid multielement airfoil flows. In: AIAA Aerospace Sciences Meeting, 90-1470 (1990)
13. Jameson, A.: Aerodynamic design via control theory. J. Sci. Comput. **3**, 233–260 (1998)
14. Lions, J.L.: Optimal Control of Systems Governed by Partial Differential Equations. Springer, New York (1971)
15. Palacios, F., Colonno, M.R., Aranake, A.C., Campos, A., Copeland, S.R., Economon, T.D., Lonkar, A.K., Lukaczyk, T.W., Taylor, T.W.R., Alonso, J.: Stanford University Unstructured (SU2): an open-source integrated computational environment for multi-physics simulation and design. In: 51st AIAA Aerospace Sciences Meeting including the New Horizons Forum and Aerospace Exposition, AIAA 2013-0287, Grapevine, 7–10 Jan (2013)
16. Nadarajah, S.K., Jameson, A.: A comparison of the continuous and discrete adjoint approach to automatic aerodynamic optimization. In: 38th Aerospace Sciences Meeting, AIAA 2000-0667 (2000)
17. Palacios, F., Economon, T.D., Wendorff, A.D., Alonso, J.: Large-scale aircraft design using SU2. In: 53st AIAA Aerospace Sciences Meeting, AIAA 2015-1946, Kissimmee, Florida, 5–9 Jan (2015)
18. Zhang, M., Rizzi, A., Nangia, R.: Transonic airfoil and wing design using inverse and direct methods. In: AIAA SciTech, AIAA 2015-1934, Kissimmee, Florida, 5–9 Jan (2015)
19. Dulikravich, G.S., Baker, D.P.: Aerodynamic shape inverse design using a Fourier series method. In: AIAA Aerospace Sciences Meeting, AIAA 90-0185 (1990)
20. Campbell, R.L.: An approach to constrained aerodynamic design with application to airfoils. Technical report
21. Obayashi, S.: Genetic optimization of target pressure distributions for inverse design methods. In: AIAA Aerospace Sciences Meeting, AIAA-95-1649-CP (1995)

22. Streit, T., Wichmann, G., Von Knoblauch Zu Hatzbach, F., Campbell, R.: Implications of conical flow for laminar wing design and analysis. In: 29th AIAA Applied Aerodynamics Conference, AIAA 2011-3808 (2011)
23. Zhang, M., Wang, C., Rizzi, A., Nangia, R.: Hybrid feedback design for subsonic and transonic airfoils and wings. In: AIAA SciTech, AIAA 2014-0414, National Harbor, Maryland, 13–17 Jan (2014)
24. Takanashi, S.: An iterative procedure for three-dimensional transonic wing design by the integral equation method. In: AIAA 2nd Applied Aerodynamic Conference, AIAA-84-2155, Seattle, Washington, 21–23 August 1984
25. Mark Drela, M., Giles, M.B.: Viscous-inviscid analysis of transonic and low Reynolds number airfoils. AIAA J. **25**(10), 1347–1355 (1987)
26. Drela, M.: Design and optimization method for multi-element airfoils. In: AIAA Aerospace Sciences Meeting, 93-0969 (1993)
27. Amoignon, O., Navratil, J., Hradil, J.: Study of parametrization in the project CEDESA. In: AIAA SciTech, 2014-0570, National Harbor, Maryland, (2014)
28. Jakobsson, S., Amoignon, O.: Mesh deformation using radial basis functions for gradient-based aerodynamic shape optimization. Comput. Fluids **36**(6), 1119–1136 (2007)
29. Mohammadi, B., Molho, J.I., Santiago, J.G.: Design of minimal dispersion fluidic channels in a CAD-free framework. In: Center for Turbulence Research, Proceedings of the Summer Program 2000
30. Kenway, G.K.W., Kennedy, G.J., Martins, J.R.R.A.: A CAD-free approach to high-fidelity aerostructural optimization. In: 13th AIAA/ISSMO Multidisciplinary Analysis Optimization Conference, AIAA 2010–9231, Fort Worth, Texas, 13–15 Sept (2010)
31. Tomac, M., Eller, D.: From geometry to CFD grids. an automated approach for conceptual design. Prog. Aerosp. Sci. **47**(11), 589–596 (2011). doi: 10.1016/j.paerosci.2011.08.005
32. Gallier, J.: Curves and surfaces in geometric modeling: Theory and algorithms. Technical report, Department of Computer and Information Science, University of Pennsylvania (2013)
33. Farin, G.E., Hoschek, J., Kim, M.-S.: Handbook of Computer Aided Geometric Design. Elsevier, New York (2002)
34. de Boor, C.: A practical guide to splines. Springer, New York (1978)
35. Melin, T., Amadori, K., Krus, P.: Parametric wing profile description for conceptual design. In: CEAS Meeting, Venice, Italy (2011)
36. Dwight, R.P.: Robust mesh deformation using the linear elasticity equations. In: Proceedings of the Fourth International Conference on Computational Fluid Dynamics, ICCFD, Ghent, 10–14 July 2006
37. Hicks, R., Henne, P.A.: Wing design by numerical optimization. In: AIAA, AIAA 77-1247, Seattle, Washington, 22–24 Aug (1977)
38. Sederberg, T.W., Parry, S.R.: Free-form deformation of solid geometric models. ACM Trans. Math. Softw. **20**(4), 151–160 (1986)
39. Les, P., Wayne, T.: The NURBS Book, 2nd edn. Springer, New York (1997)
40. Schmitt, V., Charpin, F.: Pressure distributions on the ONERA-M6_Wing at transonic Mach numbers, AGARD AR138 (1979)
41. Piquet, J.: Turbulent Flows: Models and Physics. Springer, New York (2001)
42. Lyu, Z., Kenway, G.K.W., Martins, J.R.R.A.: RANS-based aerodynamic shape optimization investigation of the common research model wing. In: 52nd Aerospace Sciences Meeting, AIAA 2014-0567, National Harbor, Maryland, 13–17 Jan 2014. doi: 10.2514/6.2014-0567
43. Jameson, A.: Aerodynamic shape optimization using the adjoint method. Technical report, Brussels, Belgium (2003) Von Karman Institute Lecture-III
44. Anderson, J.D.: Computational Fluid Dynamics: The Basic with Applications, International edition, McGraw-Hill, New York (1995)
45. Van der Velden, A.: The global aircraft shape. In: AGARD-FDP-VKI Special Course at VKI, pp. 9–1–9–11, Rhode-Saint-Genese, April (1994)

Performance Optimization of EBG-Based Common Mode Filters for Signal Integrity Applications

Carlo Olivieri, Francesco de Paulis, Antonio Orlandi, and Slawomir Koziel

Abstract Electromagnetic bandgap structures have been shown to be effective in realizing simple and cheap common mode filters for differential interconnect applications in modern high-speed digital electronics. There are two major advantages offered by this technology. The first is that it relies on the standard planar layout methodology for filter design, applied to either a printed circuit board (PCB) or packaging materials and technology. The second advantage is easy analytical design procedure that requires full wave electromagnetic simulations only at a final stage for the filter geometry refinement to precisely meet given performance specifications. In this chapter, the latter aspect is enhanced by introducing an optimization stage that allows for automated adjustment of geometry parameters of the filter in order to improve its performance in terms of achieving the required central frequency, widening the bandwidth, and increasing the band-notch depth. The optimization approach proposed here combines fast response surface approximation modeling for initial design screening and local derivative-free design improvement using pattern search.

Keywords Optimization • Electromagnetic bandgap structures • Signal integrity • Filter • Common mode

MSC codes: 46N10, 65K10, 78A25 and 74F15

C. Olivieri • F. de Paulis • A. Orlandi (✉)
UAq EMC Laboratory, Department of Industrial and Information Engineering and Economics, University of L'Aquila, L'Aquila, Italy
e-mail: carlo.olivieri@univaq.it; francesco.depaulis@univaq.it; antonio.orlandi@univaq.it

S. Koziel
Engineering Optimization & Modeling Center, School of Science and Engineering, Reykjavik University, Menntavegur 1, 101 Reykjavik, Iceland
e-mail: koziel@ru.is

S. Koziel et al. (eds.), *Simulation-Driven Modeling and Optimization*, Springer Proceedings in Mathematics & Statistics 153,
DOI 10.1007/978-3-319-27517-8_5

1 Introduction

Electromagnetic bandgap (EBG) structures are a sub-class of frequency selective surfaces. Introduced in 1999 [1] for applications in the field of antenna design and for minimization of the coupling between antennas [2–6], their usage has been extended to printed circuit boards (PCBs), mainly in the area of power integrity (PI) [7–16]. High-speed switches in digital systems generate well-known simultaneous switching noise (SSN) that can propagate across the PCB through the cavities made by power planes [17–22]. Moreover, discontinuities along the high-speed interconnects, such as vias and imbalances in differential traces are also a source of noise [23–28].

The EBG structure is a simple and efficient way to minimize this noise. It is implemented in the same technology as used for manufacturing multilayer PCBs, thus without the need of extra components and expensive tools. EBG structures are effective in the GHz range where the lumped capacitors become useless due to their inherited parasitic inductance. Often, mixed-signal systems require isolation of the analog circuitry from the digital section in order to decouple the current return paths. EBGs turn to be attractive noise reduction solutions also in mixed-signal systems [29–31].

The EBG technology attracted attention of many research groups around the world. Its development led to introduction of various types of EBGs, mainly identifiable as the mushroom [13, 15, 32–35] and the planar types [16, 36]. Generally speaking, they consist of specifically designed metal planes with characteristic geometries suitably shaped to form a high impedance surface (HIS) [1]. Widespread utilization as well as optimization of planar EBGs resulted in making this technology flexible and easy to implement. In particular, simple design procedures were developed for effectively sizing the EBG cavity [37, 38], placing it at any level of a multilayer stack-up [39], and minimizing its impact on the IR-Drop of the power distribution network as well as on the signal quality of interconnects referenced to the EBG patterned plane [10, 40–42].

A planar EBG structure, placed within a typical multilayer PCB substrate, affects the propagation of the return current for interconnects being referenced to the patterned plane. Therefore, the impact on the signal integrity (SI) of digital signals has been considered, measured, and predicted [43–46]. Since a close electromagnetic interaction between the signal transmission along this type of interconnects and the resonant behavior of an EBG cavity was found, the coupling mechanisms have been deeply investigated to minimize the unintentional signal degradation. On the other hand, the capability of energy coupling between the signal interconnects and the planar EBGs has been found valuable for realization of efficient common mode filters [27, 47–49].

2 EBGs as Common Mode Filters

In the design of modern link paths for low voltage high-speed differential digital signals, one of the technical challenges is the containment of their common mode (CM) harmonic components [50, 51]. These components have a twofold negative effects: (a) a loss of signal energy due to the differential-to-common mode conversion and, consequently, an implicit attenuation of the intentional differential signal, and (b) EMI radiation when leaving the board assembly through connectors and cables. The origin of these components is always related to certain imbalance (geometrical and/or electrical) of the entire signal path, from the driver to the receiver [27, 52].

In a real-world design, completely removing the asymmetries is impractical or impossible; thus, a suitable solution to reduce the CM harmonics is to filter out the CM portion of the signal. This filtering operation is usually achieved using discrete components, which have some disadvantages: they take up space on the board, generate an additional cost, and are often lead to undesirable attenuation of the intentional differential signal.

An approach similar to the EBG layout technique, based on the periodic interrupted ground plane structure, has been introduced in [53] for the design of a common mode suppression filter. The regular planar EBG has been investigated in [54], where the effects of the patterned plane on both the common mode and the differential mode signal propagation along a differential microstrip line were studied. These principles are applied in [27, 47, 55] where a preliminary filter topology has been developed based on a simple cavity resonator. Later, the proper EBG type of a cavity was introduced for minimizing the layout area required by the filter for a given frequency [56].

A general structure of an EBG CM filter can be explained using the geometry shown in Fig. 1. The relevant geometry parameters are represented by the patch width (a), the gap between the patches (g), and the width of the interconnections between them (w), named "bridges." The filtering effect is achieved due to the resonant behavior of the cavity made by each single EBG cavity and the reference solid plane underneath. In practical applications of a real PCB layout, no vias should be placed within the patch area and no connection should be made between the patches and the reference plane.

The gaps between the patterned plane that are along the signal current return path allow energy coupling between the microstrip/EBG pair and the cavity made of the EBG and the solid plane. The return current goes back to the source flowing mainly under the microstrip. The current, always choosing the path of least impedance, in the case of a discontinuity—such as the gap along its path—follows three different paths. One is through the bridge (at LF due to the highly inductive nature of this part), another is through the patch-to-patch capacitance, and the last one is through the solid plane underneath the EBG patterned plane, as shown in Fig. 2.

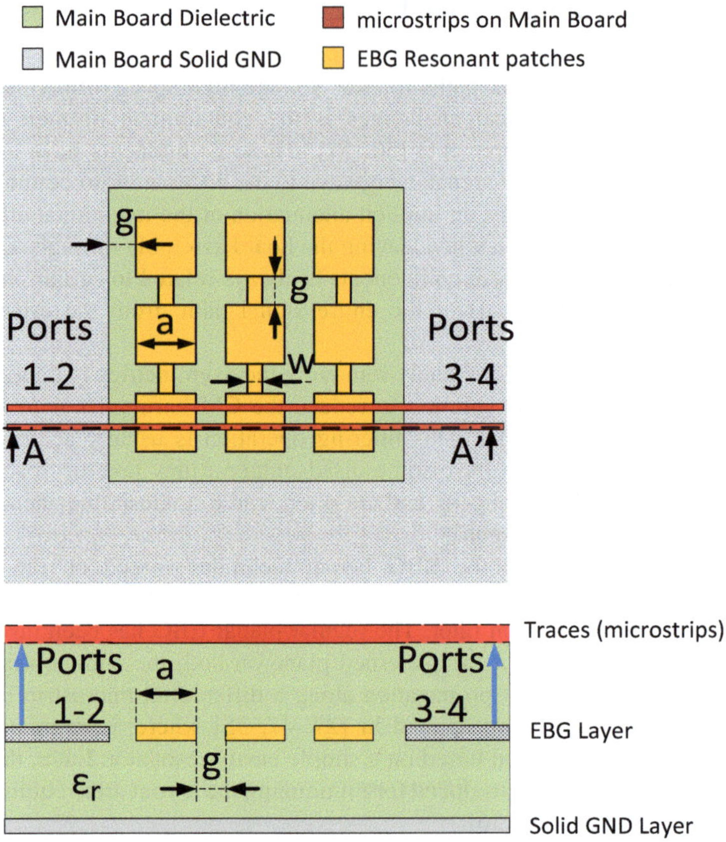

Fig. 1 Basic geometry of an EBG CM filter

The CM filter geometry needs to be appropriately sized to achieve an EBG cavity resonating at the frequency of the harmonic components that should be filtered out. Typically, for EMC purposes, these would be the fundamental and first few harmonics of the intentional differential signal. In [37], the fundamental physics of an EBG cavity is investigated, and a simple design procedure is developed in [38] to design the EBG patches and bridges according to the frequency of the CM strongest harmonic to be filtered out. Moreover, the initial procedure, based on the inductance calculation of bridges and patches as well as evaluation of the EBG equivalent inductance to its solid plane counterpart, is refined as detailed in [57], where the optimal relationships among the design variables are found, depending on the number of patches of the EBG cavity.

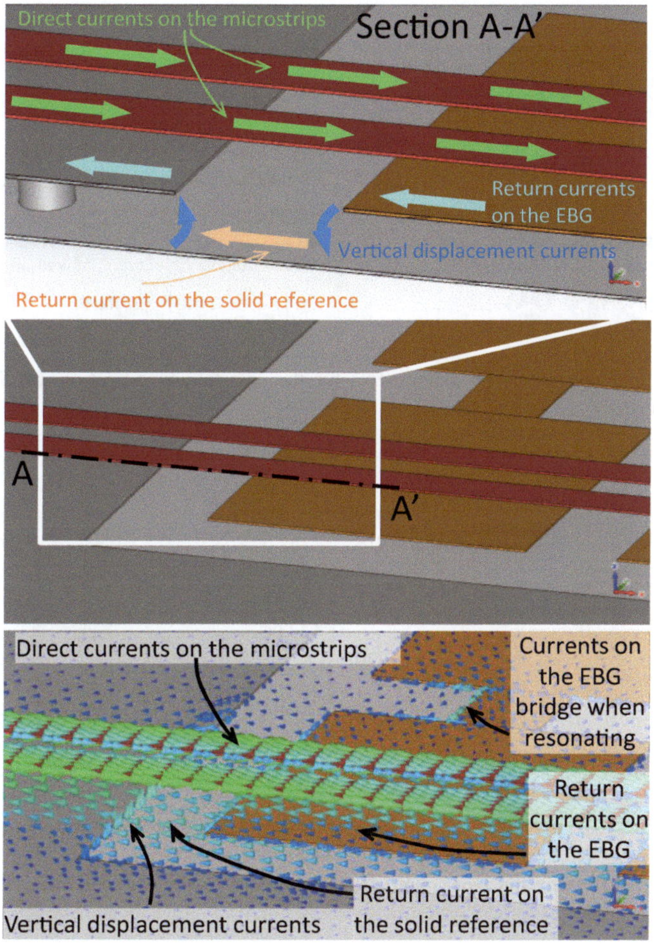

Fig. 2 Identification of the signal and return current path at the gap/bridge location

2.1 On-Board EBG-Based CM Filters

The simplest EBG-based CM filter is laid out on the PCB outermost stack-up layer (the so-called top and bottom layers). It is applied to differential microstrips as shown in Fig. 3. The figure reports the actual layout of a manufactured board which was employed to investigate the crosstalk among the adjacent differential pairs routed on the same EBG filter [58]. However, the embedded CM filters have been shown to be effective also for differential striplines [59], in which case the filter is allocated deep in the stack-up, as shown in Fig. 4. The stripline filter consists of two patterned layers above and below the differential traces, since the return current flows on both the V20 and V22 planes as in Fig. 4b. As an example, the

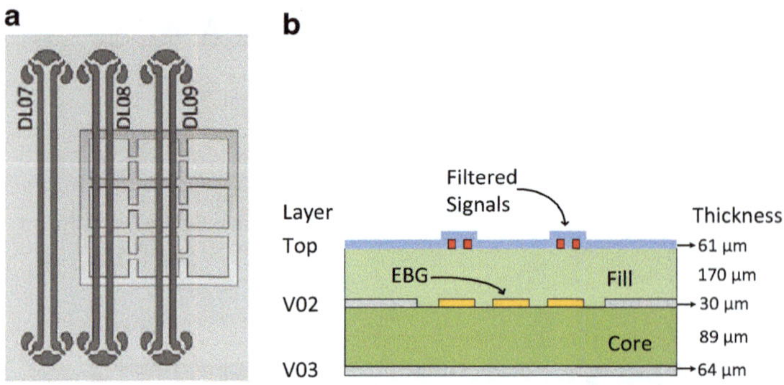

Fig. 3 Design of the microstrip experiments for the crosstalk investigation. (**a**) Top view. (**b**) Cross-section

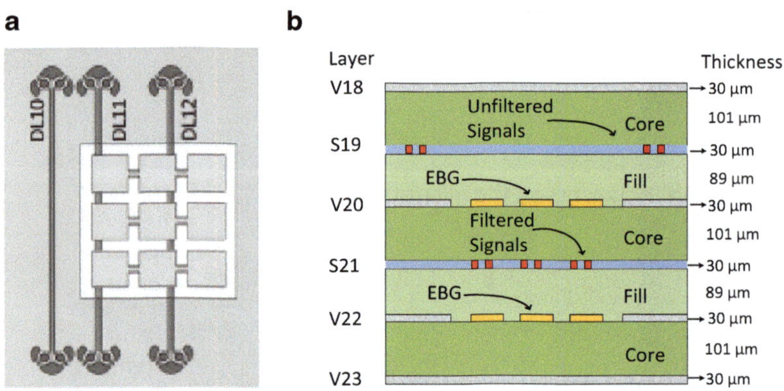

Fig. 4 Design of the stripline experiments for the crosstalk investigation. (**a**) Top view. (**b**) Cross-section

measurement results for the geometries DL08 in Fig. 3 and for the filtered stripline pair DL11 in Fig. 4 are shown in Fig. 5. The notch is clearly seen at the desired frequency of 8 GHz for the specific requirement considered for both the microstrip as well as the stripline EBG filters.

2.2 Removable EBG-Based Common Mode Filters

A different layout strategy was adopted in [60, 61] to provide more flexibility in the filter design. As opposed to the layout described in the previous paragraph, the filter has been removed from the PCB stack-up, and it is modified to be a surface-mount component installed on the top of the PCB. However, the key

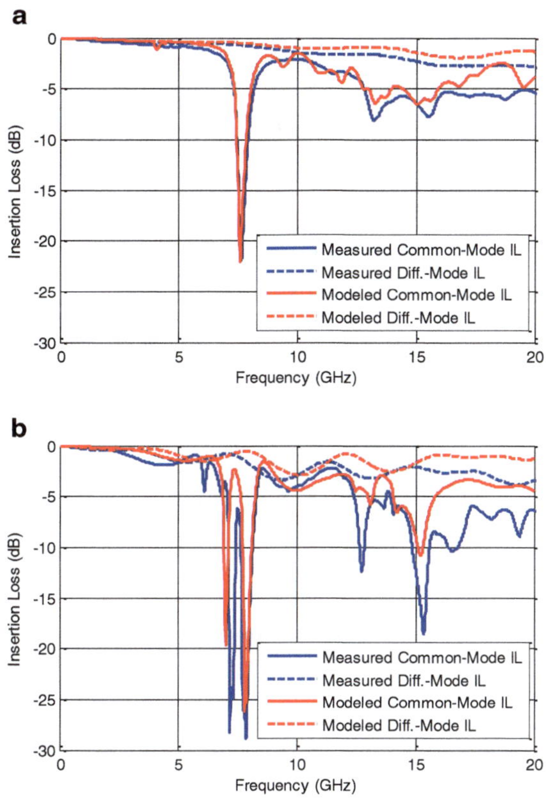

Fig. 5 CM insertion loss for the filtered (**a**) microstrip DL08, and (**b**) striplines DL11

concepts making the EBG filter attractive, such as the use of its standard multilayer laminate technology, easy design procedure, and the reduced costs, are still in place. Moreover, the electromagnetic behavior of the filter remains unchanged with the common mode return currents of the differential pair being responsible for the common mode to EBG cavity mode coupling. The drawback of having a standalone component, as mentioned before, is not a practical issue as long as the filter layout is cheap, it is designed and manufactured similarly to its on-board counterpart, and it ensures no attenuation of the intentional differential signal. The remaining disadvantage of the filter, i.e., utilization of the PCB layout area, can be minimized by employing techniques for its miniaturization; the simplest strategy is to employ a high permittivity material. Again, its larger cost compared to the standard laminates (e.g., FR-4) is not a problem due to the small size of the filter component.

A first version of the removable EBG filter was proposed in [60], where the differential pair runs on the motherboard PCB outer layer. The filter is realized in the auxiliary small PCB using the typical configuration of Fig. 6, with 3 EBG cavities, each made of three patches. The top view of the filter as well as the assembled

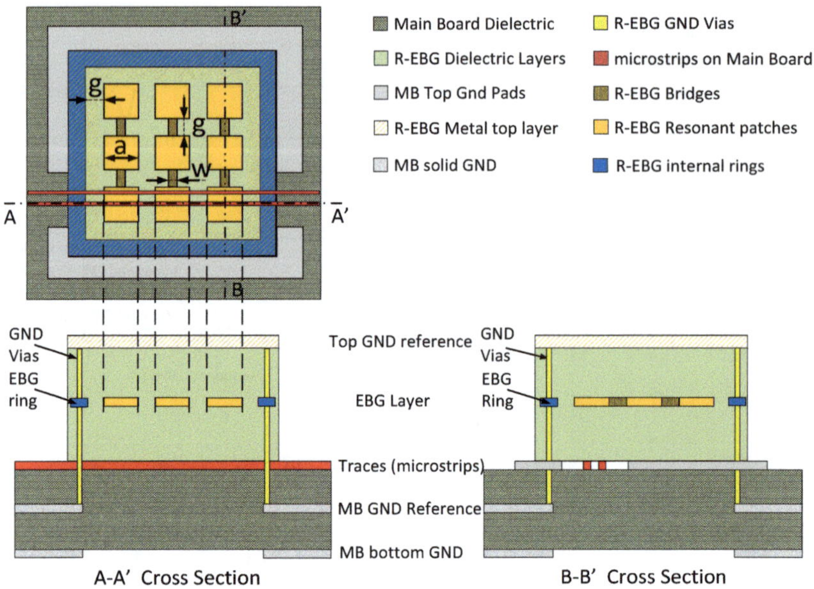

Fig. 6 (**a**) Stack-up obtained from the cut plane along the curve C1. (**b**) Path of the forward and return current at the cut plane along the curve C2. (**c**) Top view of the removable EBG filter

stack-up (the main PCB together with the removable filter) is shown in Fig. 6. The filter is attached to the PCB by means of four corner pads for the current return corresponding to pads on the PCB.

The filter can be realized by a minimum of 3-layer PCB, with the bottom one etched leaving only the four connecting pads. The second layer includes the EBG pattern and the ring; the third is the outermost solid layer for the reference ground that closes the EBG cavity. The PCB area below the EBG (layers L3 and bottom) is voided to allow the return current on L3, once it reaches the EBG area, to flow up toward the EBG through the vias and the ring. Then, the common mode return current flows back and forth between the EBG layer L1 and the top layer, as described in the previous paragraphs.

The S_{cc21} in Fig. 7 shows the predicted filter notch at 8 GHz as well as a lower one at 5.58 GHz. The latter is due to a resonant effect of the ring to ground. Moreover, the figure includes a parametric analysis to investigate the effect of the voided main PCB layers below the PCB area. The four additional models in Fig. 7a are simulated and the corresponding results are shown in Fig. 7b, c. The filter performance degradation is observed when moving the additional EBG reference closer to the filter.

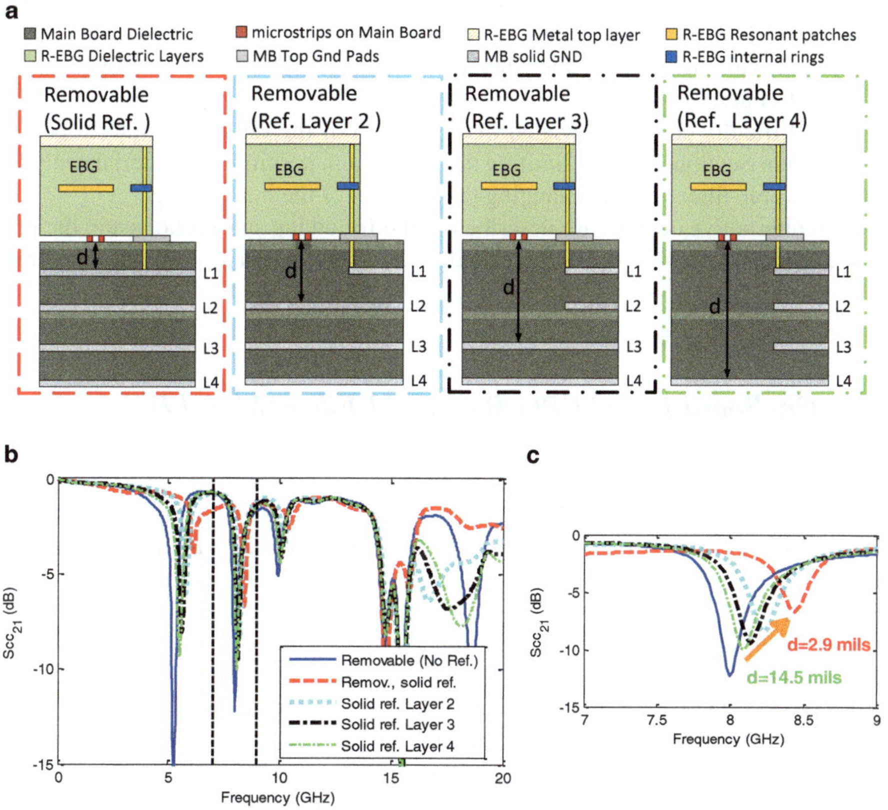

Fig. 7 Variation of the trace-to-solid reference distance. (**a**) Simulation model, (**b**) common mode insertion loss, and (**c**) zoom at the filtering notch between the two *dashed lines* in (**b**) at 7 and 9 GHz with *d* as the trace to reference distance

3 Miniaturized EBG-Based CM Filters

Maintaining small size is an important factor when including an EBG filter in the PCB layout. The present filter topology comes from a miniaturization process described in [56] where the patterned EBG cavity behaves as a resonator having smaller dimension with respect to its solid plane counterpart. Further area reduction can be achieved by manipulating the material properties of the filter. More specifically, increasing dielectric permittivity leads to smaller EBG cavity size at the same required filtering frequency. The main drawback is that materials with higher dielectric permittivity are more expensive, thus the basic design of the on-board EBG filter would require modification of the laminate material for the overall PCB. Although the present PCB technology allows to mix different dielectric layers presented in a single stack-up (i.e., low-loss expensive materials for laying out high frequency RF or high-data-rate-line signals only on a few dielectric layers), the expensive material used for the filter dielectrics should still be employed in the overall PCB area.

This difficulty can be alleviated by utilization of the removable filter topology described in Section 2.2, thus limiting the use of high permittivity materials only for the removable filter without changing the material of the main PCB.

Two alternative designs are described in the following sub-sections and the proposed models are subsequently subjected to the optimization process (cf. Sections 4 and 5).

3.1 On-Board LTCC EBG-Based CM Filter (Model I)

The first design consists of the simplest layout as in Fig. 1 with the use of the high permittivity material on the overall board. Although this model is not quite practical due to relatively large amount of expensive laminate utilized in it, it is considered here as a preliminary illustration example of the optimization procedure.

The material employed is a Low Temperature Co-fired Ceramic (LTCC) typical for aerospace applications. As expected, the stack-up parameters changed with respect to the standard PCB technology employed for the previous filter topologies. In particular, the relative dielectric permittivity of the LTCC is 7.8 with tangent loss of 0.0045. The metal layers are made by a $2.1 \cdot 10^7$ S/m gold allowing 8 μm layer thickness, whereas the employed dielectric has a standard thickness of 137 μm.

The preliminary non-optimized model is shown in Fig. 8 based on the stack-up in Fig. 1. Since the filter is based on an EBG cavity embedded within a larger layout, a key aspect would be to reduce the electromagnetic interference induced across the surrounding signals and vias [62]. Therefore the stitching vias are laid out to minimize the EBG radiation within the same multilayer circuit. The via diameter is set to 130 μm with a distance $s = 400$ μm from the EBG area and 600 μm via-to-via distance.

During the optimization process the variation of the patch and bridge sizes would lead to the overall EBG area resize; the placement of the stitching vias in the simulation model is defined to allow the automatic via number variation according to the overall EBG area.

The EBG parameters involved in the optimization process are the patch width a, the bridge length g, and the bridge width w. These parameters will be varied to achieve the optimization targets as described in the next paragraph.

Fig. 8 (**a**) Layout of the Model I, (**b**) cross-section

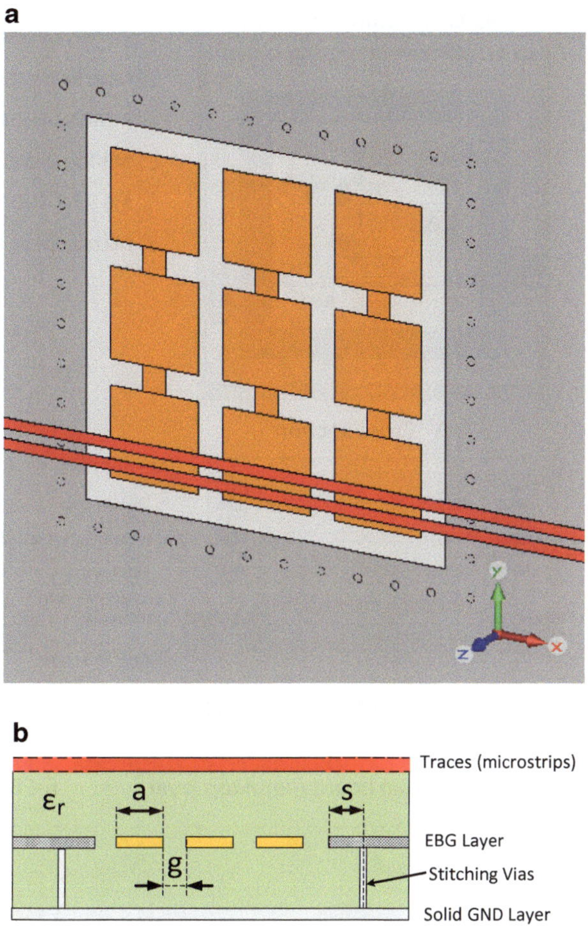

3.2 Removable LTCC EBG-Based CM Filter (Model II)

The second model represents an advancement with respect to the removable EBG filter introduced in Section 2.2. Its architecture is developed to avoid the necessity of using the main board void planes below the filter footprint. To this end, the differential interconnects are moved to the removable component instead of being laid out on the main board. More specifically, the main board microstrips go up inside the removable components through the pads on the main board connected to the pads and vias on the filter substrate, as sketched in Fig. 9. The blue area surrounding the EBG layers acts as the shielding fence to minimize the radiation at the EBG filter resonance; the top solid layer of the filter is designed for the same EMI reduction purpose, thus providing a complete shielding. The finalized layout of the main board external layer as well as of the filter bottom layer is shown in Fig. 10

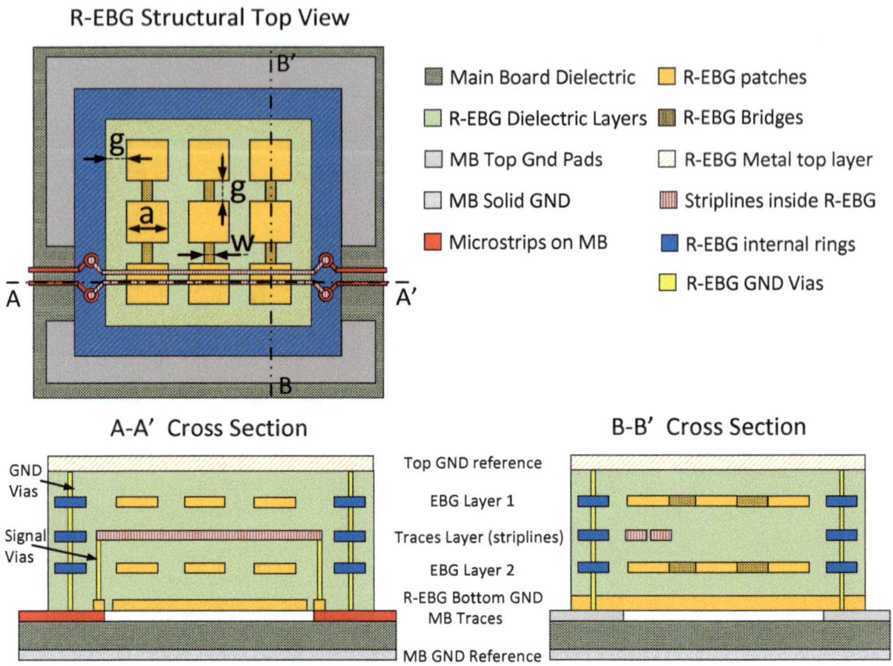

Fig. 9 Details of Model II: top view and cross-sections

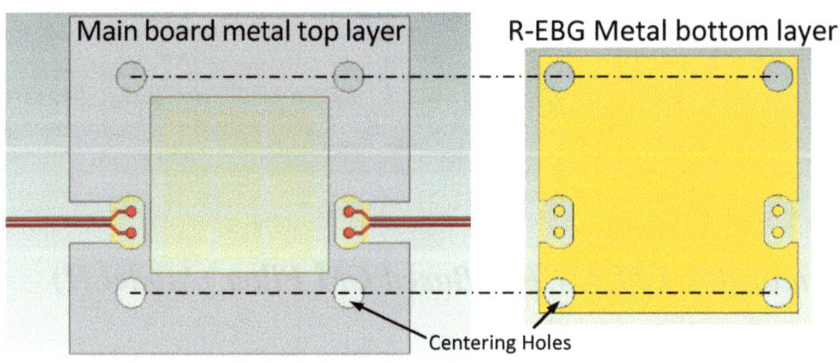

Fig. 10 Details of the external layers to be assembled on the main board and on the removable filter

to highlight the contact points between the main board and the removable filter as well as the four holes necessary for the correct board-to-filter alignment during the assembly process.

Although the layout of this proposed filter topology appears to be much more complex than the filter in Section 3.1, the EBG resonant principles remain the same, thus the variables to be varied to optimize the filter response are still those defined in Section 3.1, the patch width a, the bridge length g, and the bridge width w.

4 Filter Optimization

In this section, we describe the optimization methodology developed and utilized to improve the electrical performance parameters of the EBG filters considered in the previous paragraph. It should be emphasized that the problem at hand is challenging from the numerical point of view because of highly nonlinear responses of the EBG structures that are also very sensitive to geometry parameters. Furthermore, the computational cost of electromagnetic simulations, carried out by using *CST Studio Suite 2015* [63] of the EBG filters, is high so that one of our concerns is to limit the number of simulations as much as possible. On the other hand, the dimensionality of the design space is rather low (typically up to four parameters) which allows us to utilize auxiliary data-driven surrogate models to speed up the design optimization process.

4.1 Problem Formulation

The problem at hand is to adjust the geometry parameters of the EBG filter so that a notch is allocated at a specific design (or center) frequency f_0 (here, 8 GHz) and optimized either to (1) minimize $|S_{21}|$ in a frequency band $f_0 - df \leq f \leq f_0 + df$ (here, $df = 0.1$ GHz), or to (2) increase the bandwidth for which $|S_{21}| \leq -10$ dB.

In more rigorous terms, the problem can be formulated as follows:

$$\boldsymbol{x}^* = \arg\min_{\boldsymbol{x}} U\left(\boldsymbol{R}\left(\boldsymbol{x}\right)\right) \tag{1}$$

where $\boldsymbol{R}$ denotes a response vector of the EM simulation model of the EBG filter (here, S-parameters versus frequency), $\boldsymbol{x}$ is a vector of designable geometry parameters, and U is the objective function. The objective function is defined either as

$$U\left(\boldsymbol{R}\left(\boldsymbol{x}\right)\right) = U\left(\left|S_{21}\left(\boldsymbol{x};f\right)\right|\right) = \max\left\{\left|S_{21}\left(\boldsymbol{x};f\right)\right|_{f_0-df\leq f\leq f_0+df}\right\} \tag{2}$$

for case (1) or

$$\begin{aligned} U\left(\boldsymbol{R}\left(\boldsymbol{x}\right)\right) = U\left(\left|S_{21}\left(\boldsymbol{x};f\right)\right|\right) &= \arg\min_{f}\left\{\left|S_{21}\left(\boldsymbol{x};f\right)\right| \leq -10\ \text{dB}\right\} \\ &- \arg\max_{f}\left\{\ \left|S_{21}\left(\boldsymbol{x};f\right)\right| \leq -10\ \text{dB}\right\} \end{aligned} \tag{3}$$

for case (2).

The geometry parameters are a = patch width, g = bridge length, and w = bridge width forming the geometry parameters vector $\boldsymbol{x} = [a\ g\ w]^T$.

The problem (1) is constrained as follows:

- Lower and upper bounds for geometry parameters $\boldsymbol{l} \leq \boldsymbol{x} \leq \boldsymbol{u}$, and
- Linear inequality constraints $c_k(\boldsymbol{x}) \leq 0$, $k = 1, \ldots, K$.

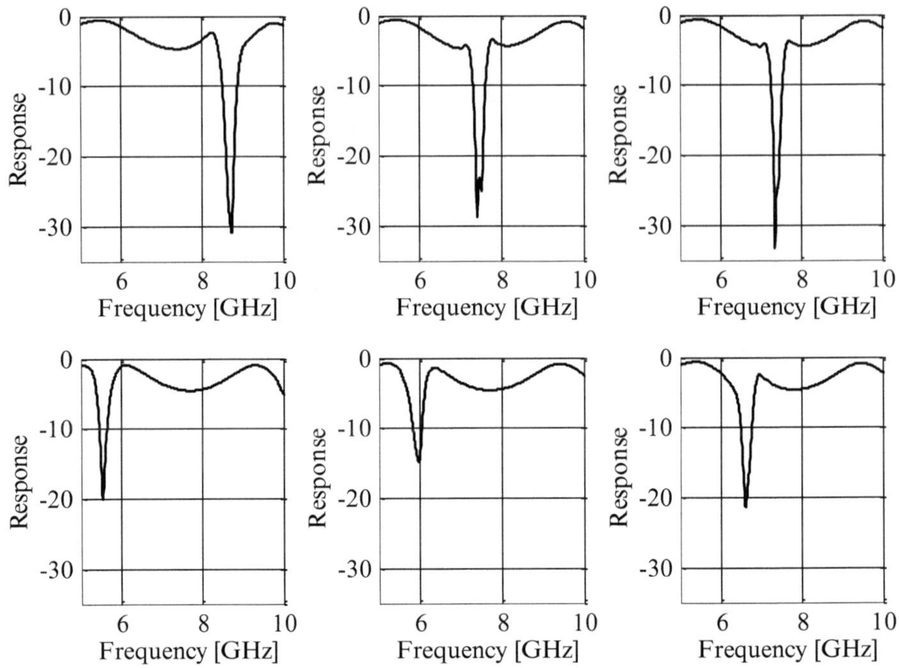

Fig. 11 Responses of the EBG filter at various parameter setups. One can observe that the responses are highly nonlinear (as a function of frequency) and very sensitive to the adjustable parameters of the problem

The constraints are introduced to make the structure physically consistent and able to fulfill the limits of the building technology.

It should be emphasized that the optimization problem is challenging because of sharp narrow responses in the notch region. The typical responses of the filter at various design parameter setup, indicating the difficulty of the problem at hand, are shown in Fig. 11. It can be observed that depending on the initial design, local optimization may fail to find a satisfactory design. Thus, local optimization has to be preceded by a screening stage at which the notch is approximately allocated around f_0 as required.

4.2 Optimization Algorithm

The initial screening mentioned in Section 4.1 cannot be executed through conventional global optimization using, e.g., population-based meta-heuristics due to excessive computational cost associated with such procedures. On the other hand, we are not interested in a precise control of the entire filter response but just in handling its two critical features, i.e., the center frequency and the depth of the notch. It turns out that despite highly nonlinear dependence of the S-parameter responses of the filter on frequency, the aforementioned features of the notch (both its center frequency and depth), change much more linearly with the design variables. Additionally, the number of geometry parameters is small so that is it possible to construct and exploit a data-driven model of the notch features. In this work, Kriging interpolation [64] is utilized for model construction. More specifically, the following procedure is implemented and employed to find a reasonable starting point for further local optimization:

(1) Sample the design space at N locations $\boldsymbol{x}_B{}^{(k)}$, $k = 1, \ldots, N$;
(2) Evaluate the EM model $\boldsymbol{R}$ at all points obtained in Step 1;
(3) Extract center frequencies $f_B{}^{(k)}$ and notch depths $L_B{}^{(k)}$ for all the points;
(4) Construct a Kriging interpolation models $s_f(\boldsymbol{x})$ and $s_L(\boldsymbol{x})$ of the center frequencies and depths as a function of design variables;
(5) Optimize the Kriging models in order to allocate the notch at the required frequency $f_0 = 8$ GHz and increase its depth L.

The objective function used in Step 5 is as follows: $U_L(\boldsymbol{x}) = s_L(\boldsymbol{x}) + \beta \cdot \left[(s_f(\boldsymbol{x}) - f_0)/f_0\right]^2$. Such a formulation allows for increasing the notch depth while centering it at the required operating frequency.

For illustration purposes, Fig. 12 shows the landscapes of the Kriging model (notch depth and its center frequency) for three various values of the patch width a of 1.5, 1.71, and 2.0 mm for the EBG filter of Fig. 8.

Having the notch allocated around the required frequency by optimizing the Kriging interpolation models, local optimization is executed. Here, a pattern search algorithm [65] is utilized because of low-dimensionality of the search space. In case of a larger number of parameters more efficient methods would have to be used.

5 Numerical Results

In this section, we provide optimization results of the two EBG filters considered in this chapter. Optimization was executed using the methodology described in Section 4.

5.1 *On-Board LTCC EBG-Based CM Filter (Model I)*

In case of Model I, only the lower $\boldsymbol{l}$ and upper $\boldsymbol{u}$ bounds for design variables were set as follows: $\boldsymbol{l} = [0.836\ 0.15\ 0.15]^T$ mm and $\boldsymbol{u} = [2.0\ 1.0\ 1.0]^T$ mm. There was no need to execute the initial screening because the notch was allocated sufficiently close to the center frequency of 8 GHz at the initial design $\boldsymbol{x}^{init} = [1.4000\ 0.4036\ 0.3750\ 0.4030]^T$. The local search was only run for case (2) of Section 4.1 (bandwidth maximization) and resulted in the final design $\boldsymbol{x}^* = [1.3444\ 0.0.3969\ 0.02789]^T$.

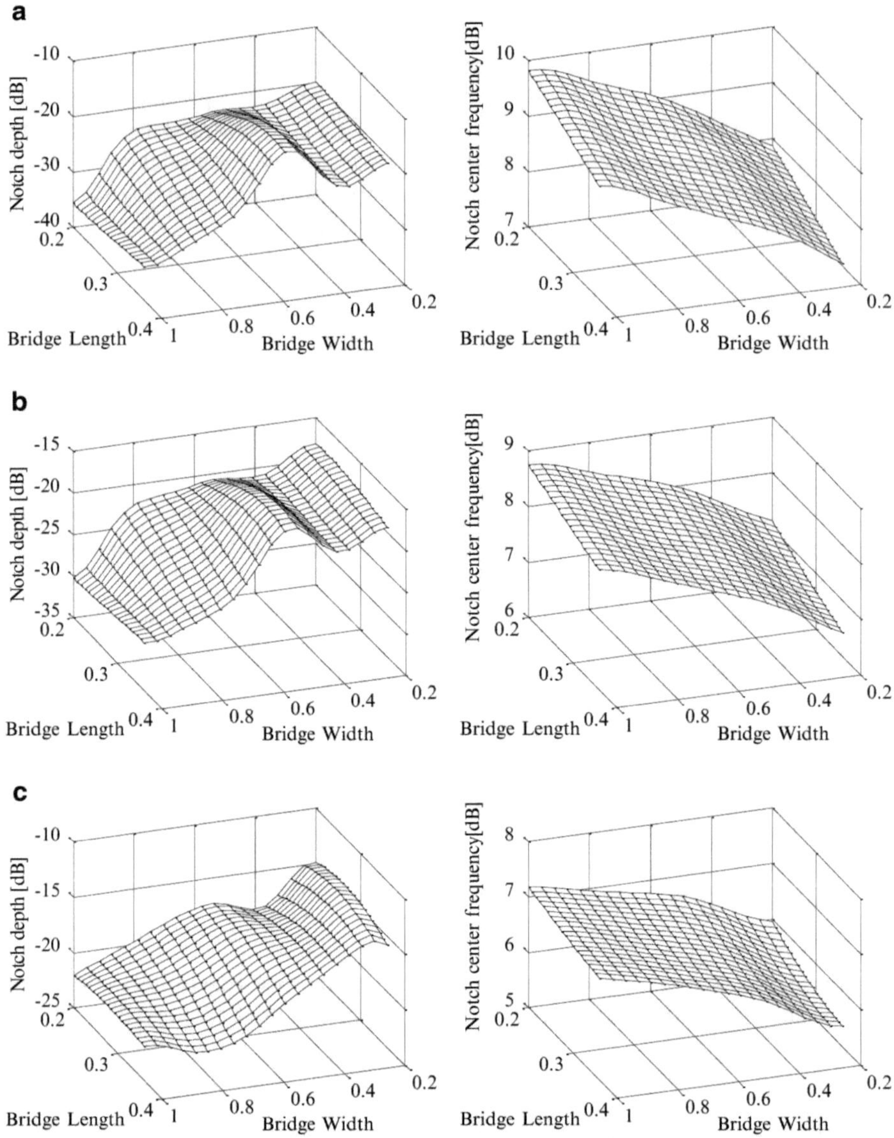

Fig. 12 Two-dimensional cuts of the Kriging interpolation models $s_f(\mathrm{x})$ and $s_L(\mathrm{x})$ for the EBG filter of Fig. 8, corresponding to patch width $a = 1.5$ (**a**), 1.71 (**b**), and 2.0 (**c**)

Fig. 13 shows the responses of the filter at the initial and at the final design. The design cost is 45 evaluations of the EM filter model. It can be observed that both the notch depth and the bandwidth were greatly improved compared to the initial design. Also, the bandwidth at the final design is centered at the frequency $f_0 = 8$ GHz as required.

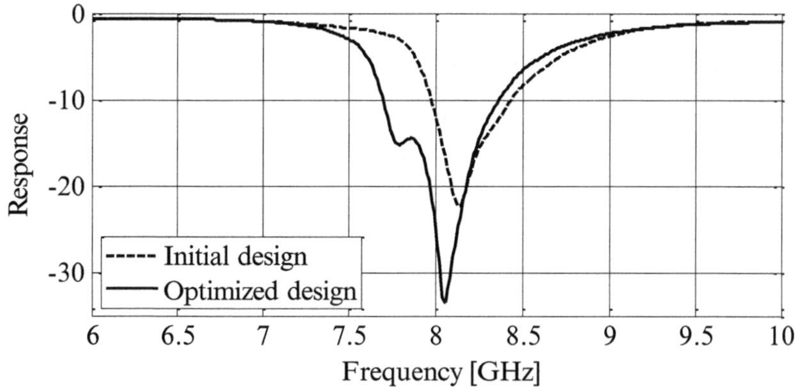

Fig. 13 EBG filter (Model I) responses at the initial and the optimized designs

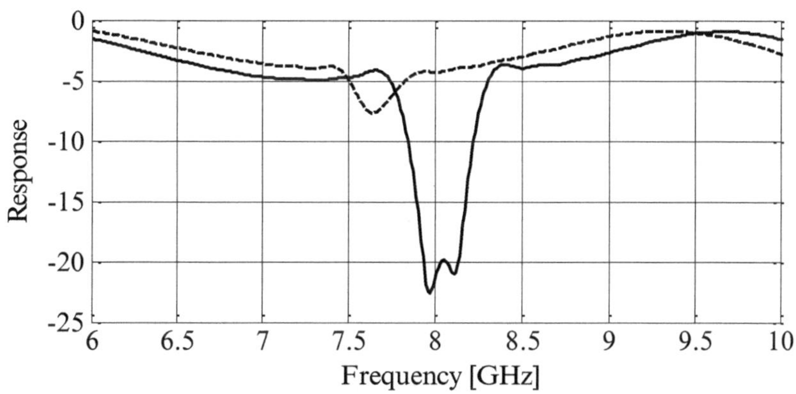

Fig. 14 Responses of the EBG filter (Model II) at the initial design *dashed line* and at the design obtained at the first stage of the optimization process (screening) *solid line*

5.2 Removable LTCC EBG-Based CM Filter (Model II)

In this case of Model II, the following lower and upper bounds for design variables were set: $\boldsymbol{l} = [0.836\ 0.15\ 0.15]^T$ mm and $\boldsymbol{u} = [2.0\ 1.0\ 1.0]^T$ mm. Additionally, the following two inequality constraints were defined: $2{*}w + 2{*}g \le 6.385$ and $w \le a$. The result obtained in the screening stage is $\boldsymbol{x}^{init} = [1.5000\ 0.2733\ 0.3489]^T$. The EM-simulated response is shown in Fig. 14. The design is already very good both in terms of the notch frequency and depth. Dashed line shows the initial design $[2.500\ 0.400\ 0.175]^T$, which is very poor, especially in terms of the notch depth. The cost of the screening stage was 48 evaluations of the EM filter model required to set up the Kriging interpolation model.

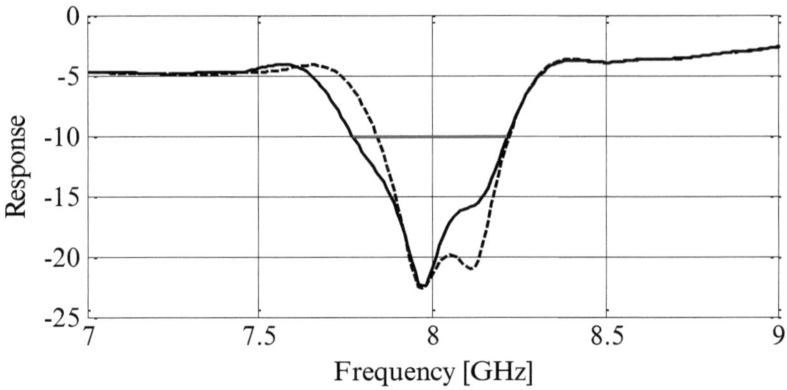

Fig. 15 Local optimization of the EBG filter (Model II) for case (1) (bandwidth enhancement): *dashed line* initial design from screening, *solid line* final design

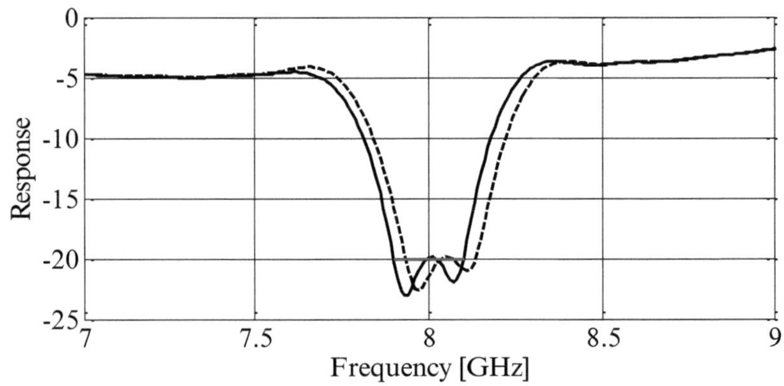

Fig. 16 Local optimization of the EBG filter (Model II) for case (2) (notch depth): *dashed line* initial design from screening, *solid line* final design

Figures 15 and 16 show the filter responses at the two designs obtained by local optimization starting from $\boldsymbol{x}^{init}$, the first one for case (1) (bandwidth enhancement) and case (2) (notch depth enhancement): $\boldsymbol{x}^{*1} = [1.5000\ 0.2733\ 03558]^T$ and $\boldsymbol{x}^{*2} = [1.5000\ 0.2733\ 03451]^T$. The -10 dB bandwidth obtained in the first case is 442 MHz and it is wider than for the second design (365 MHz). On the other hand, the second design exhibits notch depth of -20 dB and better for the frequency range of 7.9–8.1 GHz, which is not the case for the first design. The cost of the local optimization was only 15 and 19 EM model evaluations for case (1) and case (2), respectively, which is because the final designs were close to the one obtained at the screening stage.

6 Conclusions

In the chapter, an efficient numerical procedure for EM-simulation-driven design optimization of EBG-based filters has been presented. The design objectives are to obtain the band-notch at a specified center frequency and increase the notch bandwidth and depth, and, consequently, the bandwidth of the common mode insertion loss. The optimization algorithm is applied to two different topologies, initially designed employing a previously published analytical procedure. The first case is based on a typical PCB configuration where the filter is laid out within the PCB stack-up. The second filter structure consists of a more complex topology with the filter modified with respect to the typical on-board case. At the conceptual level, the goal is to develop a miniaturized standalone component (still designed with the typical PCB/package planar layout) to be removed and substituted if necessary once the electronics of the system changes, i.e. when the data rate and thus the common mode harmonic components or simply the design requirements are modified.

The proposed optimization algorithm can be applied at the design closure stage to fine-tune the geometry parameters of the EBG filters. It allows for automation and reduction of the computational cost of the simulation-driven design, previously realized using inefficient and error-prone hands-on procedures involving parameter sweeps. The future work will include further developments of the algorithmic frameworks for EBG filter optimization as well as its application to various practical design cases.

References

1. Sievenpiper, D., Zhang, L., Broas, R.F.J., Alexopulos, N.G., Yablonovitch, E.: High-impedance electromagnetic surfaces with a forbidden frequency band. IEEE Trans. Microw. Theory Tech. **47**(11), 2059–2073 (1999)
2. Tan, M.N.M., Ali, M.T., Subahir, S., Rahman, T.A., Rahim, S.K.A.: Backlobe reduction using mushroom-like EBG structure. In: Proceedings of IEEE Symposium on Wireless Technology and Applications (ISWTA), pp. 206–209, 23–26 September 2012
3. Chen, X., Su, Z.J., Li, L., Liang, C.H.: Radiation pattern improvement in closely-packed array antenna by using mushroom-like EBG structure. In: Radar Conference 2013, IET International, pp. 1–3, 14–16 April 2013
4. Neo, C., Lee, Y.H.: Patch antenna enhancement using a mushroom-like EBG structures. In: Antennas and Propagation Society International Symposium (APSURSI), 2013 IEEE, pp. 614–615, 7–13 July 2013
5. Coulombe, M., Koodiani, S.F., Caloz, C.: Compact elongated mushroom (EM)-EBG structure for enhancement of patch antenna array performances. IEEE Trans. Antennas Propag. **58**(4), 1076–1086 (2010)
6. Azad, M.Z., Ali, M.: Novel wideband directional dipole antenna on a mushroom like EBG structure. IEEE Trans. Antennas Propag. **56**(5), 1242–1250 (2008)
7. Qin, J., Ramahi, O.M., Granatstein, V.: Novel planar electromagnetic bandgap structures for wideband noise suppression and EMI reduction in high speed circuits. IEEE Trans. Electromagn. Compat. **49**(3), 661–669 (2007)

8. Wu, T.L., Wang, C.C., Lin, Y.H., Wang, T.K., Chang, G.: A novel power plane with super-wideband elimination of ground bounce noise on high speed circuits. IEEE Microw. Wireless Compon. Lett. **15**(3), 174–176 (2005)
9. Shapharnia, S., Ramahi, O.M.: Electromagnetic interference (EMI) reduction from printed circuit boards (PCB) using electromagnetic bandgap structures. IEEE Trans. Electromagn. Compat. **46**(4), 580–587 (2004)
10. de Paulis, F., Nisanci, M.N., Orlandi, A.: Practical EBG application to multilayer PCB: Impact on power integrity. IEEE Electromagn. Compat. Mag. **1**(3), 60–65 (2012)
11. Wu, T.L., Lin, Y.H., Wang, T.K., Wang, C.C., Chen, S.T.: Electromagnetic bandgap power/ground planes for wideband suppression of ground bounce noise and radiated emission in highspeed circuits. IEEE Trans. Microw. Theory Tech. **53**(9), 2935–2942 (2005)
12. Kamgaing, T., Ramahi, O.M.: A novel power plane with integrated simultaneous switching noise mitigation capability using high impedance surface. IEEE Microw. Wireless Compon. Lett. **13**(1), 21–23 (2003)
13. Abhari, R., Eleftheriades, G.V.: Metallo-dielectric electromagnetic bandgap structures for suppression and isolation of the parallel-plate noise in high-speed circuits. IEEE Trans. Microw. Theory Tech. **51**(6), 1629–1639 (2003)
14. Tavallaee, M. Iacobacci, Abhari, R.: A new approach to the design of power distribution networks containing electromagnetic bandgap structures. Electr. Perform. Elect. Packag. (EPEP) conference, 43–46 (2006)
15. Kamgaing, T., Ramahi, O.M.: Design and modeling of high impedance electromagnetic surfaces for switching noise suppression in power planes. IEEE Trans. Electromagn. Compat. **47**(3), 479–489 (2005)
16. Kim, K.H., Shutt-Ainé, J.E.: Analysis and modeling of hybrid planar-type electromagnetic-bandgap structures and feasibility study on power distribution network applications. IEEE Trans. Microw. Theory Tech. **56**(1), 178–186 (2008)
17. Swaminathan, M., Engin, A.E.: Power Integrity Modeling and Design for Semiconductors and Systems. Prentice Hall, Boston, USA (2008)
18. Lei, G.-T., Techentin, R.W., Gilbert, B.K.: High frequency characterization of power/ground-plane structures. IEEE Trans. Microw. Theory Tech. **47**, 562–569 (1999)
19. Berghe, S.V., Olyslager, F., de Zutter, D., Moerloose, J.D., Temmerman, W.: Study of the ground bounce caused by power plane resonances. IEEE Trans. Electromagn. Compat. **40**(2), 111–119 (1998)
20. Cui, W., Fan, J., Ren, Y., Shi, H., Drewniak, J.L., DuBroff, R.E.: DC power-bus noise isolation with power-plane segmentation. IEEE Trans. Electromagn. Compat. **45**(2), 436–443 (2003)
21. Na, N., Jinseong, J., Chun, S., Swaminathan, M., Srinivasan, J.: Modeling and transient simulation of planes in electronic packages. IEEE Trans. Adv. Packag. **23**(3), 340–352 (2000)
22. Xu, M., Hubing, T.H., Chen, J., Van Doren, T.P., Drewniak, J.L., DuBroff, R.E.: Power-bus decoupling with embedded capacitance in printed circuit board design. IEEE Trans. Electromagn. Compat. **45**(1), 22–30 (2003)
23. Huang, W.-T., Lu, C.-H., Lin, D.-B.: The optimal number and location of grounded vias to reduce crosstalk. Prog. Electromagn. Res. **95**, 241–266 (2009)
24. Wu, B., Tsang, L.: Full-wave modeling of multiple vias using differential signaling and shared antipad in multilayered high speed vertical interconnects. Prog. Electromagn. Res. **97**, 129–139 (2009)
25. de Paulis, F., Zhang, Y.-J., Fan, J.: Signal/power integrity analysis for multilayer printed circuit boards using cascaded S-parameters. IEEE Trans. Electromagn. Compat. **52**(4), 1008–1018 (2010)
26. Wu, B., Tsang, L.: Full-wave modeling of multiple vias using differential signaling and shared antipad in multilayered high speed vertical interconnects. Prog. Electromagn. Res. **97**, 129–139 (2009)
27. de Paulis, F., Archambeault, B., Connor, S., Orlandi, A.: Electromagnetic band gap structure for common mode filtering of high speed differential signals. In: Proceedings of IEC DesignCon 2011, Santa Clara, USA, 31 January–3 February 2011

28. Ricchiuti, V., de Paulis, F., Orlandi, A.: An equivalent circuit model for the identification of the stub resonance due to differential vias on PCB. In: Proceedings of IEEE Workshop on Signal Propagation on Interconnects 2009, SPI '09, Strasbourg, France, 12–15 May 2009
29. Choi, J., Govind, V., Mandrekar, R., Janagama, S., Swaminathan, M.: Noise reduction and design methodology for the mixed-signal systems with alternating impedance electromagnetic bandgap (AI-EBG) structure. In: International Microwave Symposium Digest, Long Beach, CA, pp. 645–651, June 2005
30. Kim, T.H., Chung, D., Engin, E., Yun, W., Toyota, Y., Swaminathan, M.: A novel synthesis method for designing electromagnetic bandgap (EBG) structures in packaged mixed signal systems. In: Proceedings of 56th Electronic Components and Technology Conference, pp. 1645–1651, 2006
31. Rajo-Iglesias, E., Caiazzo, M., Inclán-Sánchez, L., Kildal, P-.S.: Comparison of bandgaps of mushroom-type EBG surface and corrugated and strip-type soft surfaces. IET Microw. Antennas Propag. **1**(1), 184–189 (2007)
32. Liang, L., Liang, C.H., Chen, L., Chen, X.: A novel broadband EBG using cascaded mushroom-like structure. Microw. Opt. Technol. Lett. **50**, 2167–2170 (2008)
33. Kamgaing, T., Ramahi, O.M.: Multiband electromagnetic-bandgap structures for applications in small form-factor multichip module packages. IEEE Trans. Microw. Theory Tech. **56**(10), 2293–2300 (2008)
34. Wu, T.L., Fan, J., de Paulis, F., Wang, C.D., Ciccomancini, A., Orlandi, A.: Mitigation of noise coupling in multilayer high-speed PCB: State of the art modeling methodology and EBG technology. IEICE Trans. Commun. **E93-B**(7), 1678–1689 (2010)
35. Wang, C.-D., Yu, Y.-M., de Paulis, F., Scogna, A.C., Orlandi, A., Chiou, Y.-P., Wu, T.-L.: Bandwidth enhancement based on optimized via location for multiple vias EBG power/ground planes. IEEE Trans. Compon. Packag. Manuf. Technol. **2**(2), 332–341 (2012)
36. Oh, S.S., Kim, J.M., Kwon, J.H., Yook, J.G.: Enhanced power plane with photonic bandgap structures for wide band suppression of parallel plate resonances. In: IEEE International Symposium on Antennas and Propagation, vol. 2B, pp. 655–658, July 2005
37. de Paulis, F., Orlandi, A.: Accurate and efficient analysis of planar electromagnetic band-gap structures for power bus noise mitigation in the GHz band. Prog. Electromagn. Res. B **37**, 59–80 (2012)
38. Raimondo, L., de Paulis, F., Orlandi, A.: A simple and efficient design procedure for planar electromagnetic bandgap structures on printed circuit boards. IEEE Trans. Electromagn. Compat. **53**(2), 482–490 (2011)
39. de Paulis, F., Raimondo, L., Orlandi, A.: Impact of shorting vias placement on embedded planar electromagnetic bandgap structures within multilayer printed circuit boards. IEEE Trans. Microw. Theory Tech. **58**(7), 1867–1876 (2010)
40. de Paulis, F., Raimondo, L., Orlandi, A.: IR-Drop analysis and thermal assessment of planar electromagnetic band-gap structures for power integrity applications. IEEE Trans. Adv. Packag. **33**(3), 617–622 (2010)
41. Di Febo, D., Nisanci, M.H., de Paulis, F., Orlandi, A.: Impact of planar electromagnetic band-gap structures on IR-DROP and signal integrity in high speed printed circuit boards. In: Proceedings at IEEE International Symposium on EMC – EMC Europe 2012, Rome, Italy, 17–21 September 2011
42. Nisanci, M.N., de Paulis, F., Di Febo, D., Orlandi, A.: Practical EBG application to multilayer PCB: Impact on signal integrity. IEEE Electromagn. Compat. Mag. **2**(2), 82–87 (2013)
43. Scogna, A.C., Orlandi, A., Ricchiuti, V.: Signal and power integrity performances of striplines in presence of 2D EBG planes. In: Proceedings of IEEE Workshop on Signal Propagation and Interconnects, Avignon, France, May 2008
44. de Paulis, F., Orlandi, A.: Signal integrity analysis of single-ended and differential striplines in presence of EBG planar structures. IEEE Microw. Wireless Compon. Lett. **19**(9), 554–556 (2009)

45. de Paulis, F., Orlandi, A., Raimondo, L., Antonini, G.: Fundamental mechanisms of coupling between planar electromagnetic bandgap structures and interconnects in high-speed digital circuits—Part I: Microstrip lines. Presented at the Electromagnetic Compatibility Europe Workshop, Athens, Greece, 11–12 June 2009
46. de Paulis, F., Raimondo, L., Orlandi, A.: Signal integrity analysis of embedded planar EBG structures. In: Proceedings of Asia-Pacific EMC 2010, Beijing, China, 12–16 April 2010
47. de Paulis, F., Raimondo, L., Connor, S., Archambeault, B., Orlandi, A.: Design of a common mode filter by using planar electromagnetic bandgap structures. IEEE Trans. Adv. Packag. **33**(4), 994–1002, 2010
48. de Paulis, F., Raimondo, L., Connor, S., Archambeault, B., Orlandi, A.: Compact configuration for common mode filter design based on electromagnetic band-gap structures. IEEE Trans. Electromagn. Compat. **54**(3), 646–654 (2012)
49. de Paulis, F., Raimondo, L., Di Febo, D., Archambeault, B., Connor, S., Orlandi, A.: Experimental validation of common-mode filtering performances of planar electromagnetic band-gap structures. In: Proceedings of IEEE International Symposium on Electromagnetic Compatibility, Ft. Lauderdale, USA, 25–30 July 2010
50. Archambeault, B.: PCB Design for Real-World EMI Control. Kluwer Academic Publisher, Norwell, MA (2002)
51. Connor, S., Archambeault, B., Mondal, M.: The impact of common mode currents on signal integrity and EMI in high-speed differential data links. In: Proceedings of IEEE International Symposium on Electromagnetic Compatibility, pp. 1–5, 18–22 August 2008
52. Jaze, A., Archambeault, B., Connor, S.: Differential mode to common mode conversion on differential signal vias due to asymmetric GND via configurations. In: Proceedings of IEEE International Symposium on Electromagnetic Compatibility, pp. 735–740, 5–9 August 2013
53. Liu, W.T., Tsai, C.H., Han, T.W., Wu, T.L.: An embedded common-mode suppression filter for GHz differential signals using periodic defect ground plane. IEEE Microw. Wireless Compon. Lett. **18**(4), 248–250 (2008)
54. de Paulis, F., Orlandi, A., Raimondo, L., Archambeault, B., Connor, S.: Common mode filtering performances of planar EBG structures. In: Proceedings of IEEE International Symposium on Electromagnetic Compatibility, pp. 86–90, 17–21 August 2009
55. de Paulis, F., Raimondo, L., Di Febo, D., Orlandi, A.: Routing strategies for improving common mode filter performances in high speed digital differential interconnects. In: Proceedings of IEEE Workshop on Signal Propagation on Interconnects 2011, SPI '11, Naples, Italy, 8–11 May 2011
56. de Paulis, F., Archambeault, B., Nisanci, M.H., Connor, S., Orlandi, A.: Miniaturization of common mode filter based on EBG patch resonance. In: Proceedings of IEC DesignCon 2012, Santa Clara, USA, 30 January–2 February 2012
57. Nisanci, M.H., de Paulis, F., Orlandi, A., Archambeault, B., Connor, S.: Optimum geometrical parameters for the EBG-based common mode filter design. In: Proceedings at 2012 IEEE Symposium on Electromagnetic Compatibility, Pittsburgh, PA, USA, 5–10 August 2012
58. de Paulis,F., Cracraft, M., Di Febo, D., Nisanci, M.H., Connor, S., Archambeault, B., Orlandi, A.: EBG-based common-mode microstrip and stripline filters: Experimental investigation of performances and crosstalk. IEEE Trans. Electromagn. Compat. **57**(5), 996–1004 (2015)
59. de Paulis, F., Cracraft, M., Olivieri, C., Connor, S., Orlandi, A., Archambeault, B.: EBG-based common-mode stripline filters: Experimental investigation on interlayer crosstalk. IEEE Trans. Electromagn. Compat. **57**(6), 1416–1424 (2015)
60. de Paulis, F., Nisanci, M.H., Di Febo, D., Orlandi, A., Connor, S., Cracraft, M., Archambeault, B.: Standalone removable EBG-based common mode filter for high speed differential signaling. In: Proceedings of IEEE International Symposium on Electromagnetic Compatibility, Raleigh NC (USA), pp. 244–249, 3–8 August 2014

61. Varner, M.A., de Paulis, F., Orlandi, A., Connor, S., Cracraft, M., Archambeault, B., Nisanci, M.H., Di Febo, D.: Removable EBG-based common-mode filter for high-speed signaling: Experimental validation of prototype design. IEEE Trans. Electromagn. Compat. **57**(4), 672–679 (2015)
62. Kodama, C., O'Daniel, C., Cook, J., de Paulis, F., Cracraft, M., Connor, S., Orlandi, A.,. Wheeler, E: Mitigating the threat of crosstalk and unwanted radiation when using electromagnetic bandgap structures to suppress common mode signal propagation in PCB differential interconnects. In: Proceedings of IEEE International Symposium on Electromagnetic Compatibility, Dresden, pp. 622–627, 16–22 August 2015
63. Computer Simulation Technology, *CST Studio Suite 2015*, available at www.cst.com
64. Queipo, N.V., Haftka, R.T., Shyy, W., Goel, T., Vaidynathan, R., Tucker, P.K.: Surrogate-based analysis and optimization. Prog. Aerosp. Sci. **41**(1), 1–28 (2005)
65. Koziel, S.: Computationally efficient multi-fidelity multi-grid design optimization of microwave structures. Appl. Comput. Electromagn. Soc. J. **25**(7), 578–586 (2010)

Unattended Design of Wideband Planar Filters Using a Two-Step Aggressive Space Mapping (ASM) Optimization Algorithm

Marc Sans, Jordi Selga, Ana Rodríguez, Paris Vélez, Vicente E. Boria, Jordi Bonache, and Ferran Martín

Abstract This chapter deals with the automated and unattended design of planar wideband bandpass filters by means of aggressive space mapping (ASM) optimization. The approach can be applied to bandpass filters based on semi-lumped element resonators (e.g., stepped impedance resonators, ring resonators, etc.) coupled through admittance inverters (implemented with quarter-wavelength transmission lines). It will be explained how the filter layout is automatically generated from filter specifications, i.e., central frequency, fractional bandwidth, in-band ripple, and order, without the need of any external aid to the design process. For this purpose, a novel optimization algorithm based on two independent ASM processes will be fully described. The proposed automatic design procedure will be detailed and validated through its application to generate several filter layouts starting from different sets of practical specifications.

Keywords Wideband bandpass filters • Balanced filters • Optimization • Unattended design • Circuit synthesis • Space mapping (SM) • Step impedance resonators (SIRs)

MSC codes: 94-01, 94-02 y 78-01

M. Sans • J. Selga • P. Vélez • J. Bonache • F. Martín
Departament d'Enginyeria Electrònica, GEMMA/CIMITEC, Universitat Autònoma de Barcelona, 08193 Bellaterra, Barcelona, Spain

A. Rodríguez • V.E. Boria (✉)
Departamento de Comunicaciones, iTEAM, Universitat Politècnica de València, 46022 Valencia, Spain
e-mail: vboria@dcom.upv.es

S. Koziel et al. (eds.), *Simulation-Driven Modeling and Optimization*, Springer Proceedings in Mathematics & Statistics 153,
DOI 10.1007/978-3-319-27517-8_6

1 Introduction

The synthesis of planar microwave circuits able to satisfy a set of given specifications is a subject of interest for microwave engineers. Despite the fact that most commercially available computer aided design (CAD) tools and electromagnetic solvers include optimizers, obtaining the circuit topologies that satisfy the design requirements is not always straightforward. This design difficulty increases with circuit complexity, and convergence to the optimum solution is not always guaranteed (for instance, due to limitations related to local minima), unless the seeding layout is already very close to the one providing the target response.

This chapter is focused on a specific type of planar circuits, of interest in many different microwave applications, whose design requires significant computational effort: high-order (and hence highly selective) wideband bandpass filters. There are many approaches for the design of wideband planar microwave filters [1, 2]. The interest in this chapter is the design of wideband bandpass filters based on semi-lumped (i.e., electrically small and planar) resonant elements coupled through admittance inverters [3] (see the generalized bandpass filter network in Fig. 1). In principle, these filters can be designed by forcing the planar resonant elements to exhibit the fundamental resonance at the filter central frequency, f_0, and the impedance slope at the value of the corresponding LC resonant tank of the generalized bandpass filter network. With such network, standard filter responses (e.g., Butterworth, Chebyshev) can ideally be achieved.[1] However, deviations from the ideal responses are caused by the limited functionality of the inverters (implemented by means of quarter-wavelength transmission lines at f_0) and by the distributed effects of the planar resonators at sufficiently high frequencies. These deviations are more pronounced if the filter is broadband, mainly because the required phase shift (90°) of the inverters is not preserved over the whole filter pass band. Thus, broadband planar filters designed by implementing the inverters with quarter-wavelength transmission line sections typically exhibit a narrower bandwidth than the required one (nominal bandwidth), unless such bandwidth degradation is compensated at the design stage. One possibility to compensate for this narrowing effect is to over-dimension the filter bandwidth [4]. However, this tends to affect

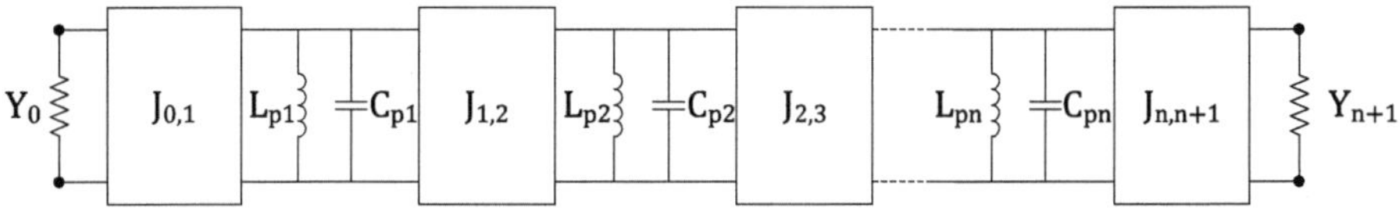

Fig. 1 Generalized bandpass filter network based on shunt LC resonators coupled through admittance inverters. From [5]; reprinted with permission

[1]There are available expressions that provide the element values of the resonators from the filter order, central frequency, bandwidth, and response type (see [3]).

the reflection coefficient (ripple level in Chebyshev filters), and hence this is not an optimum solution. In the present chapter, a systematic design procedure, able to provide the filter layout satisfying the required specifications, and first reported in [5], is studied in detail. Moreover, it will be shown that, once specifications and order are introduced, the final layout of the filter is obtained without any action by the user, namely, filter layout is generated following a completely unattended scheme. Filter design is a two-step process. In the first step, the filter schematic satisfying the specifications (optimum schematic) is obtained. This equivalent circuit consists in the generalized network of Fig. 1 with the admittance inverters replaced with quarter-wavelength transmission lines (not necessarily at f_0) and the lumped LC parallel tanks substituted by a reactive element network describing the semi-lumped resonators. Once the optimum filter schematic is inferred, the second step consists in determining the filter layout described by the optimum schematic.

Both design steps (optimum schematic and layout generation) are based on the aggressive space mapping (ASM) methodology [6], a technique that uses quasi-Newton type iteration to obtain the optimum solution. The two-step design process can in principle be applied to the automated synthesis of any type of wideband bandpass filter implemented by means of semi-lumped resonators coupled through admittance inverters. Planar resonant elements such as split rings, stepped impedance resonators (SIRs), and combinations of inductive/capacitive stubs and capacitive patches, among others, can be also considered as semi-lumped resonators for filter design based on the approach reported in this chapter.

Chapter organization is as follows. In Section 2, the general formulation of ASM is presented. Section 3 is focused on the first design step, namely on the determination of the optimum filter schematic. Thus, the first ASM iterative algorithm will be presented through a guide example. The second design step (and hence the second ASM), providing the filter layout, will be presented in Section 4. In Section 5, further examples will be reported. Finally, in Section 6, the application of the two-step ASM algorithm will be applied to the design of wideband balanced filters.

2 General Formulation of ASM

Among the considered techniques for microwave circuit synthesis and optimization, space mapping (SM), first proposed by Bandler et al. in 1994 [7], has revealed to be a powerful and efficient approach. Since this seminal work, several variants of SM have been proposed and applied to the synthesis and optimization of many different microwave components, including not only planar circuits [6, 8, 9], but also waveguide-based components [10–12]. The interest in this chapter is on the so-called ASM [6], an approach that uses quasi-Newton type iteration to find the optimum solution of the considered problem, as mentioned before. ASM uses two simulation spaces [6, 7, 13]: (1) the optimization space, $\mathbf{X_c}$, where the variables are linked to a coarse model, which is simple and computationally efficient, although not accurate, and (2) the validation space, $\mathbf{X_f}$, where the variables are linked to a fine model, typically more complex and CPU intensive, but significantly more precise. In each space, a vector containing the different model parameters can be defined. Let us call such vectors $\mathbf{x_f}$ and $\mathbf{x_c}$ for the fine and coarse model spaces, respectively, and let us designate by $\mathbf{R_f}(\mathbf{x_f})$ and $\mathbf{R_c}(\mathbf{x_c})$ their corresponding responses. The goal in ASM is to minimize the following error function:

$$\mathbf{f}(\mathbf{x_f}) = \mathbf{P}(\mathbf{x_f}) - \mathbf{x_c^*} \tag{1}$$

where $\mathbf{x_c}^*$ is a vector containing the target parameters in the coarse model (so that $\mathbf{R_c}(\mathbf{x_c}^*)$ is the target response), and $\mathbf{P}(\mathbf{x_f})$ is a mapping function that gives (via a parameter extraction procedure) the corresponding coarse model parameters that provide the same response of the fine model parameters.

Let us assume that $\mathbf{x_f}^{(j)}$ is the *j*-th approximation to the solution in the validation space, and $\mathbf{f}^{(j)}$ is the corresponding error function to this solution. The next vector of the iterative process $\mathbf{x_f}^{(j+1)}$ is obtained by a quasi-Newton iteration according to

$$\mathbf{x_f}^{(j+1)} = \mathbf{x_f}^{(j)} + \mathbf{h}^{(j)} \tag{2}$$

where $\mathbf{h}^{(j)}$ is given by:

$$\mathbf{h}^{(j)} = -\left(\mathbf{B}^{(j)}\right)^{-1}\mathbf{f}^{(j)} \tag{3}$$

and $\mathbf{B}^{(j)}$ is an approach to the Jacobian matrix, which is updated according to the Broyden formula [6]:

$$\mathbf{B}^{(j+1)} = \mathbf{B}^{(j)} + \frac{\mathbf{f}^{(j+1)}\mathbf{h}^{(j)T}}{\mathbf{h}^{(j)T}\mathbf{h}^{(j)}} \tag{4}$$

In (4), $\mathbf{f}^{(j+1)}$ is obtained by evaluating (1), using a certain parameter extraction method providing the coarse model parameters from the fine model parameters, and the super-index T stands for transpose.

A typical scenario in ASM optimization is the determination of the layout of a certain microwave circuit (e.g., a filter) described by a lumped element equivalent circuit model. In this case, the coarse model parameters are constituted by the set of lumped elements describing the equivalent circuit, and the response (optimization space) is, for instance, the set of S-parameters, that can be inferred from the electrical analysis of the circuit model. The fine model parameters are a set of geometrical values defining the layout geometry, and the response in the validation space is also given by the S-parameters, typically inferred from electromagnetic simulation by means of commercial solvers. Note that the substrate parameters, necessary for the electromagnetic simulations, i.e., thickness, dielectric constant, and loss tangent, are not considered as optimization variables. The second step in the ASM-based optimization method studied in this chapter is very similar to the example of the present paragraph (despite the fact that the considered filters are described by a combination of lumped and distributed components). Namely, the filter layout is synthesized from the circuit schematic in the second ASM algorithm. However, the main relevant aspect of the two-step ASM iterative process concerns the first ASM stage, where the schematic satisfying the filter requirements is determined from the nominal specifications [5]. In the next two sections, these two ASM algorithms are carefully and independently analyzed, using for that purpose a guide example for better understanding.

3 First ASM Algorithm: Determination of the Optimum Filter Schematic

The objective of the first ASM algorithm is the determination of the filter schematic able to satisfy the filter requirements (specifications). It can be applied to any type of bandpass filter described by a set of semi-lumped resonators coupled through admittance inverters, where the replacement of the inverters with quarter-wavelength transmission lines in the theoretical model, which results from direct application of the design formulas [3], degrades the bandwidth. Thus, the objective is to automatically find the filter schematic providing the target response. Such theoretical model will be designated as optimum filter schematic from now on. It is important to note that each specific filter requires a particular ASM algorithm, and for that reason, a representative case example is considered throughout this section and the next one.[2] Let us thus first present the considered wideband bandpass filters, including their equivalent circuit and topology, and then we will describe the first ASM algorithm for the determination of the optimum filter schematic.

The topology and schematic of the case example filters are depicted in Fig. 2 [14]. The shunt resonators are implemented through a combination of SIRs and grounded stubs. The SIRs provide transmission zeros (at frequencies designated as f_z) above the central filter frequency, f_0, which are useful for spurious suppression and for achieving a pronounced fall-off above the upper band edge. Moreover, with the parallel combination of SIRs and inductive stubs, the susceptance slopes at the filter central frequency can be made small, resulting in broad fractional bandwidths [14]. In the filter schematic, the resonators L_{ri}-C_{ri} describe the SIRs, the inductances L_{pi} account for the grounded stubs (i denotes the filter stage), and the quarter-wavelength transmission line sections correspond to the admittance inverters in the canonical prototype network (Fig. 1).

As mentioned before, at the schematic level, deviations from the target response (given by the network of Fig. 1 that results from specifications and transformation from the low-pass filter prototype) are due to the limited functionality of the

[2] Nevertheless, the reported ASM algorithm can be easily adapted to different type of filters (i.e., considering different semi-lumped resonators).

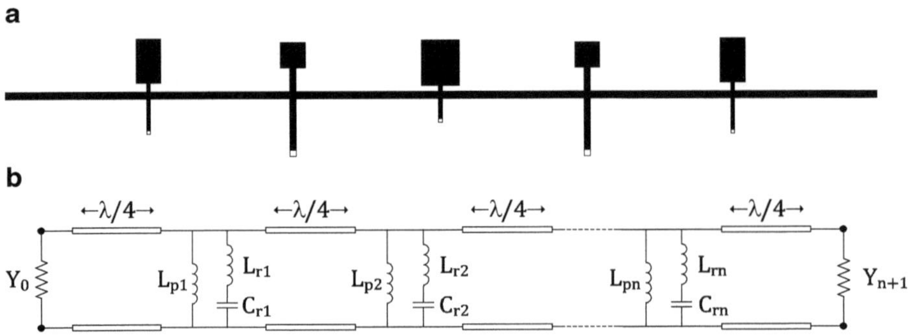

Fig. 2 Typical topology (**a**) and circuit schematic (**b**) of the wideband bandpass filters considered as case example. The topology corresponds to an order-5 filter. From [5]; reprinted with permission

inverters and to the fact that the considered resonators (Fig. 2b) are not purely parallel resonant LC tanks, but inductors connected to series resonators in shunt configuration. However it does not mean that the intended filter response (or at least a very good approximation to it in the region of interest) cannot be achieved with a certain schematic (described by the circuit of Fig. 2b). The hypothesis is that there exists a set of specifications, different than the target, that provide a filter schematic satisfying the target specifications [5]. This schematic is obtained by replacing the ideal admittance inverters of the canonical network with quarter-wavelength (not necessarily at f_0) transmission line sections, and the LC parallel resonators with the resonators L_{ri}-C_{ri} connected in parallel to the inductances L_{pi}, with the reactance values necessary to obtain the reactance slope and central frequency of the given specifications (different from the target).[3] This means that it is necessary to tailor the parameters of the circuit schematic of Fig. 2b, that is, the reactive parameters (L_{ri}, C_{ri}, and L_{pi}) and the electrical lengths (at f_0) of the transmission line sections. Let us now present an ASM-based algorithm that automatically re-calculates these parameters in order to satisfy the filter specifications, thus providing the optimum filter schematic.

Let us consider that the filter order, n, is known. The order is determined by the required selectivity. However, it is important to bear in mind that the responses obtained by the proposed filters are more selective than the Chebyshev responses at the upper transition band (due to the transmission zeros), but somehow less selective at the lower transition band (for the same specifications, i.e., central frequency, fractional bandwidth, and ripple). Therefore, although the order can in principle be estimated by considering the standard Chebyshev response, it might be necessary to increase it if the resulting response does not satisfy the selectivity requirements below the pass band. Nevertheless, the ASM algorithm to determine the optimum

[3]The additional condition to univocally determine the three element values of the resonators is the transmission zero frequency, set to a fixed value.

filter schematic is very fast, and hence a new optimum filter schematic with a higher order (and hence higher selectivity) can be easily inferred.

By considering a quasi-Chebyshev response,[4] the filter specifications are the central frequency, f_0, the fractional bandwidth, *FBW*, and the in-band ripple level L_{Ar} (or minimum return loss level). On the other hand, as many transmission zeros as SIRs (and hence order) can be forced. However, it is convenient to set all the transmission zeros to $f_z = 2f_0$, since this is an efficient strategy to achieve spurious suppression, and to improve filter selectivity above the upper band edge [4, 14].

From the well-known impedance and frequency transformations from the low-pass filter prototype [3], and assuming a Chebyshev response, the reactive elements of the shunt resonators of the network of Fig. 2 (L_{ri}, C_{ri}, and L_{pi}) can be easily inferred. The three conditions to unequivocally determine L_r, C_r, and L_p are:

(1) the filter central frequency, given by

$$f_0 = \frac{1}{2\pi\sqrt{(L_{ri} + L_{pi})\, C_{ri}}} \tag{5}$$

(2) the transmission zero frequency

$$f_z = \frac{1}{2\pi\sqrt{L_{ri} C_{ri}}} \tag{6}$$

(3) and the susceptance slope at f_0 (dependent on the filter stage):

$$b_i = 2\pi f_0 \frac{C_{ri}(L_{ri} + L_{pi})^2}{L_{pi}^2} \tag{7}$$

In (5) and (7), the left-hand side terms are the resonance frequency and susceptance slope, respectively, of the LC resonant tanks in the circuit of Fig. 1 corresponding to the required Chebyshev response. Without loss of generality, the admittance of the inverters is set to $J = 0.02$ S.[5]

Let us consider for the case example of an order-5 ($n = 5$) Chebyshev response with $f_0 = 2.4$ GHz, $FBW = 40$ % (corresponding to a 43.96 % −3-dB fractional

[4] As mentioned, the filter responses are similar, but not identical, to the standard Chebyshev responses.

[5] For a given filter response, there is not a unique solution for the network of Fig. 1. However, if the admittance of the inverters is set to a certain value (typically $J = 0.02$ S, as considered in the guide example), then the element values of the resonators are univocally determined. This is a usual procedure, although sometimes the resonator elements are all fixed to the same value, and the resulting admittance of the inverters is univocally determined by the design equations.

Table 1 Element values of the shunt resonators [5]

Stage	L_p (nH)	L_r (nH)	C_r (pF)
1,5	1.3202	0.4401	2.4983
2,4	1.3226	0.4409	2.4937
3	0.8164	0.2721	4.0400

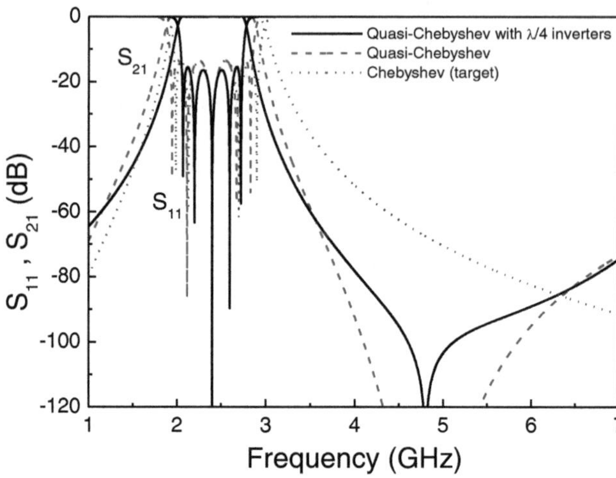

Fig. 3 Quasi-Chebyshev response of the filter that results by using the element values of Table 1 and ideal admittance inverters (*dashed line*), compared to the filter response that results by replacing the ideal inverters with quarter-wavelength transmission lines (*black solid line*), and with the ideal Chebyshev (target) response. From [5]; reprinted with permission

bandwidth[6]), and $L_{Ar} = 0.2$ dB [5]. From (5–7), the element values of the shunt resonators are found to be those indicated in Table 1.

The quasi-Chebyshev filter response (i.e., the one inferred from the schematic of Fig. 2b, but with ideal admittance inverters), depicted in Fig. 3, is similar to the ideal Chebyshev response in the pass band region, and it progressively deviates from it as frequency approaches f_z, as expected. The discrepancies are due to the fact that the shunt resonator is actually a combination of a grounded series resonator (providing the transmission zero) and a grounded inductor. The quasi-Chebyshev response satisfies the specifications to a rough approximation. Hence the target is considered to be the ideal Chebyshev response, except for the transmission zero frequency.

[6]Note that for Chebyshev bandpass filters the fractional bandwidth is given by the ripple level and is hence smaller than the once given by −3-dB level. However, in this chapter, the −3-dB fractional bandwidth is considered, since the ripple level is not constant in the optimization process (to be described). Thus, from now on, this −3-dB fractional bandwidth is designated as *FBW*, rather than FBW_{-3dB} (as usual), for simplicity, and to avoid an excess of subscripts in the formulation.

Let us now replace the ideal admittance inverters with quarter-wavelength transmission lines. The resulting response is further modified, as revealed by the significant bandwidth reduction (see Fig. 3). These results indicate that the three-element resonators and the limited functionality of the quarter-wavelength transmission lines (acting as admittance inverters) degrade the filter bandwidth, as anticipated before, and point out the need to recalculate the element values and electrical lengths of the line sections of the filter schematic of Fig. 2b, in order to satisfy the specifications. To this end, a new ASM concept that carries out the optimization at the schematic level has been proposed [5], and it is detailed in the following paragraphs.

The main hypothesis in the development of the iterative ASM algorithm able to provide the optimum filter schematic is to assume that there is a set of filter specifications, different than the target, that leads to a filter schematic (inferred by substituting the ideal admittance inverters with quarter-wavelength transmission lines), whose response satisfies the target specifications. Let us now try to define the optimization (coarse model) space and the validation (fine model) space in the proposed ASM iterative scheme. The first one is constituted by the set of specifications, f_0, FBW, L_{Ar}, being its response the same from the ideal Chebyshev prototype—target response—depicted in Fig. 3. The validation space is constituted by the same variables, but their response is inferred from the schematic of Fig. 2b, with element values calculated as specified above, and quarter-wavelength transmission lines at f_0, where f_0 is the considered value of this element in the validation space (not necessarily the target filter central frequency). The variables of each space are differentiated by a subscript. Thus, the corresponding vectors in the coarse and fine models are written as $\mathbf{x_c} = [f_{0c}, FBW_c, L_{Arc}]$ and $\mathbf{x_f} = [f_{0f}, FBW_f, L_{Arf}]$, respectively. The optimum coarse model solution (target specifications) is expressed as $\mathbf{x_c}^* = [f_{0c}^*, FBW_c^*, L_{Arc}^*]$. Notice that the transmission zero frequency, necessary to unequivocally determine the element values of the shunt resonators, is set to $f_z = 2f_0$, as indicated before. Hence f_z is not a variable in the optimization process.

Following the standard procedure in ASM, the first step before starting the iterative process is to make an estimation of the initial vector in the validation space, $\mathbf{x_f}^{(1)}$. Since the variables in both spaces are the same ones, the most canonical (and simplest) procedure is to consider $\mathbf{x_f}^{(1)} = \mathbf{x_c}^*$. From $\mathbf{x_f}^{(1)}$, the response of the fine model space is obtained (using the schematic with quarter-wavelength transmission lines), and from it, the parameters of the coarse model can directly be extracted by inspection of that response, i.e., $\mathbf{x_c}^{(1)} = \mathbf{P}(\mathbf{x_f}^{(1)})$. Applying (1), the first error function can be obtained. To iterate the process (obtaining $\mathbf{x_f}^{(2)}$ from (2), using (3)), the Jacobian matrix must be initiated. To this end, the parameters of the fine model, f_{0f}, FBW_f, L_{Arf}, are slightly perturbed, and the effects of such perturbations on the coarse model parameters, f_{0c}, FBW_c, L_{Arc}, are inferred. Thus, the first Jacobian matrix is given by:

$$\mathbf{B} = \begin{pmatrix} \frac{\delta f_{0c}}{\delta f_{0f}} & \frac{\delta f_{0c}}{\delta FBW_f} & \frac{\delta f_{0c}}{\delta L_{Arf}} \\ \frac{\delta FBW_c}{\delta f_{0f}} & \frac{\delta FBW_c}{\delta FBW_f} & \frac{\delta FBW_c}{\delta L_{Arf}} \\ \frac{\delta L_{Arc}}{\delta f_{0f}} & \frac{\delta L_{Arc}}{\delta FBW_f} & \frac{\delta L_{Arc}}{\delta L_{Arf}} \end{pmatrix} \tag{8}$$

Once the first Jacobian matrix is obtained, the process can be iterated until convergence is obtained. At each iteration, the elements of the coarse space vector, $\mathbf{x_c}^{(j)}$, are compared to the target (filter specifications), $\mathbf{x_c}^*$, and the error function is calculated according to:

$$\|f_{norm}\| = \sqrt{\left(1 - \frac{f_{0c}}{f_{0c}^*}\right)^2 + \left(1 - \frac{FBW_c}{FBW_c^*}\right)^2 + \left(1 - \frac{L_{Arc}}{L_{Arc}^*}\right)^2} \tag{9}$$

The scheme of the proposed ASM algorithm is depicted in Fig. 4.

Applying the first ASM algorithm to the considered example ($\mathbf{x_c}^* = [f_{0c}^*, FBW_c^*, L_{Arc}^*] = [2.4\ \text{GHz}, 43.96\ \%, 0.2\ \text{dB}]$), the error function rapidly decreases, so that the error is smaller than 1.2 % after $N = 13$ iterations. The evolution of the error function is depicted in Fig. 5. The fine model parameters for $N = 13$ are $\mathbf{x_f}^{(13)} = [f_{0f}^{(13)}, FBW_f^{(13)}, L_{Arf}^{(13)}] = [2.4690\ \text{GHz}, 65.53\ \%, 0.4413\ \text{dB}]$, and the coarse model parameters are $\mathbf{x_c}^{(13)} = [f_{0c}^{(13)}, FBW_c^{(13)}, L_{Arc}^{(13)}] = [2.3999\ \text{GHz}, 43.75\ \%, 0.1978\ \text{dB}]$. Note that $\mathbf{x_f}^{(13)}$ is appreciably different than $\mathbf{x_c}^*$. The optimum filter schematic is the one given by the last fine model response (which provides

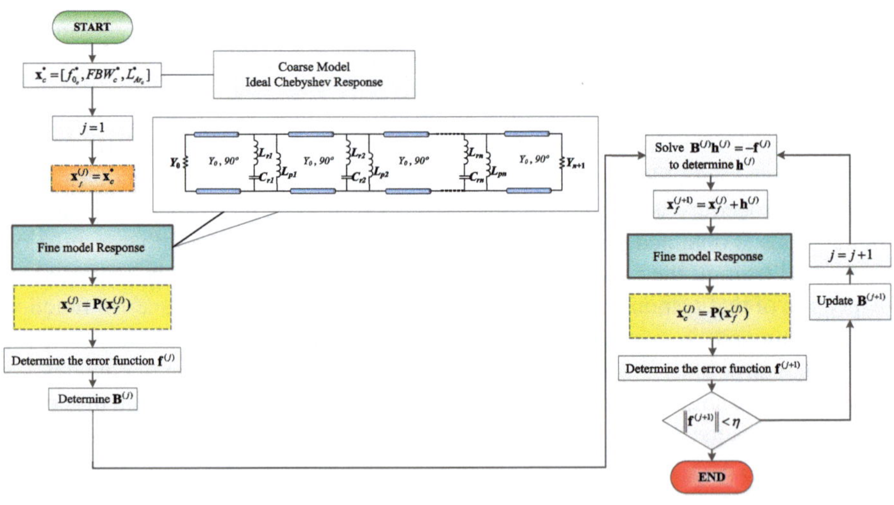

Fig. 4 Flow diagram of the first ASM algorithm. From [5]; reprinted with permission

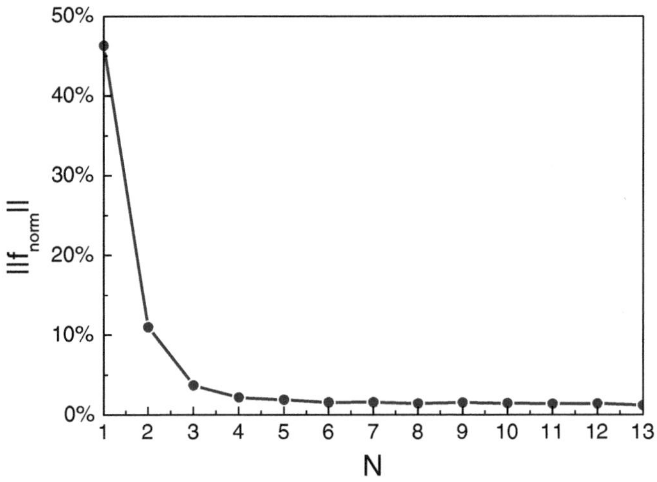

Fig. 5 Evolution of the error function of the first ASM algorithm for the considered example. From [5]; reprinted with permission

Table 2 Element values of the shunt resonators for the optimum filter schematic [5]

Stage	L_p (nH)	L_r (nH)	C_r (pF)
1,5	1.6090	0.5363	1.9368
2,4	2.1196	0.7065	1.4703
3	1.0685	0.3562	2.9168

an error below a predefined value). The elements of the shunt resonators for this optimum filter schematic are indicated in Table 2, whereas the 50-Ω line sections are quarter-wavelength transmission lines at $f_{0f}^{(13)} \neq f_0^* = 2.4$ GHz.

The response of the optimum schematic is compared to the target response in Fig. 6. The agreement in terms of central frequency, bandwidth, and in-band ripple is very good as indicates the small error function that results after 13 iterations. Nonetheless, the position of the reflection zero frequencies is different in both responses. The reason is that these frequency positions are not goals in the optimization process. Unavoidably, it is not possible to perfectly match the Chebyshev (target) response by replacing the ideal admittance inverters with transmission line sections, and the LC shunt resonators of Fig. 1 with the resonators of the schematic of Fig. 2b. Nevertheless the synthesized circuit fulfills the target specifications, and hence it is the optimum filter schematic. This schematic is used as the starting point in the ASM algorithm developed to obtain the filter layout, to be described in the next section.

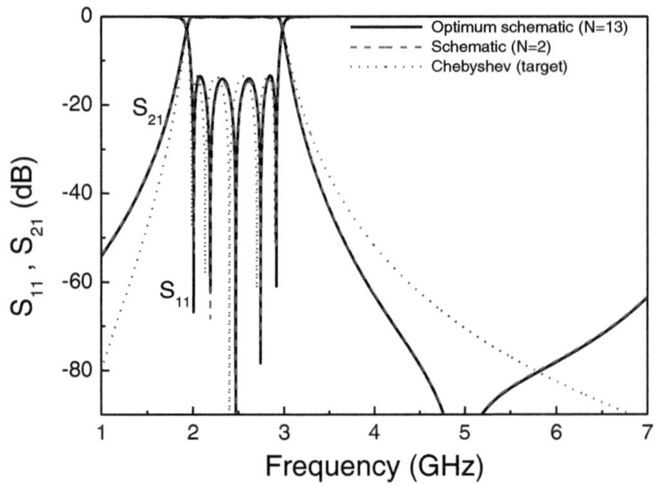

Fig. 6 Response of the optimum filter schematic, derived from the ASM algorithm, compared with the Chebyshev target response. The response at the second iteration ($N = 2$), also included, is very close to the final solution ($N = 13$). From [5]; reprinted with permission

4 Second ASM Algorithm: Determination of the Filter Layout

To determine the final layout from the optimum filter schematic, a second ASM algorithm is considered. Each unit cell is synthesized from the element values of the shunt resonator and the characteristic impedance of the cascaded $\lambda/4$ (at $f_{0f}^{(13)}$) transmission lines independently. This second ASM process involves three stages: (1) determination of the resonator layout, (2) determination of the line width, and (3) optimization of the line length. Let us now discuss in detail the implementation of these three specific ASM algorithms.

4.1 Synthesis of the Resonators

In order to obtain the layout of the filter resonators, composed by the SIRs connected in parallel to the grounded stubs, a specific ASM iterative process is applied. The variables in the optimization space are the resonator elements, i.e., $\mathbf{x}_c = [L_p, L_r, C_r]$, and the coarse model response is obtained through circuit simulation. The validation space is constituted by a set of variables defining the resonator layout. In order to deal with the same number of variables in both spaces, the widths of the grounded stubs, W_{Lp}, as well as the widths of the low- and high-impedance transmission line sections of the SIRs, W_{Cr} and W_{Lr}, respectively, are set to fixed values. Specifically, the values of W_{Cr} and

Table 3 Geometry parameters of the synthesized order-5 filter [5]

Stage	l_{Lr} (mm)	l_{Cr} (mm)	l_{Lp} (mm)	l_{cell} (mm)	W_{Cr} (mm)	W_{Lp} (mm)
1,5	0.7062	3.6710	2.5955	11.4003	2	0.3
2,4	1.9438	2.1037	4.2334	11.2583	2	0.5
3	0.4725	3.7461	1.6629	11.4786	3	0.3

$W_{Lr} = W_{Lp}$ indicated in Tables 3 and 5 have been considered. There is some flexibility to choose these widths, but it is convenient to consider wide capacitive and narrow inductive sections in order to reduce the length of the SIRs and inductive stubs. Nevertheless, extreme widths of the capacitive sections are not recommended, since they can introduce transversal resonances in the frequency region of interest (i.e., up to frequencies above f_z) [15]. The widths of the inductive sections must be set to values above the tolerance limits (representing a good balance between SIR/stub dimensions and fabrication accuracy). Thus, the variables in the validation space are the remaining dimensions of the shunt resonators, that is, the length of the grounded stub, l_{Lp}, and the lengths l_{Lr} and l_{Cr} of the high- and low-impedance transmission line sections, respectively, of the SIR (i.e., $\mathbf{x_f} = [l_{Lp}, l_{Lr}, l_{Cr}]$). The fine model response is obtained through electromagnetic simulation of the layout, inferred from the fine model variables plus the fixed dimensions (specified above) and substrate parameters. Let us consider in the present guide example the substrate parameters of the *Rogers RO3010* with thickness $h = 635$ μm and dielectric constant $\varepsilon_r = 10.2$.

Following the general scheme of the ASM optimization described in Section 2, to initiate the algorithm it is necessary to obtain an initial layout for the SIR and shunt stub. This is obtained from the well-known (and simple) approximate formulas providing the inductance and capacitance of a narrow and wide, respectively, electrically small transmission line section [16]. Isolating the lengths, the following results are obtained:

$$l_{Lr} = \frac{L_r v_{ph}}{Z_h} \tag{10a}$$

$$l_{Cr} = C_r v_{pl} Z_l \tag{10b}$$

$$l_{Lp} = \frac{L_p v_{ph}}{Z_h} \tag{10c}$$

where v_{ph} and v_{pl} are the phase velocities of the high- and low-impedance transmission lines sections, respectively, and Z_h and Z_l are the corresponding characteristic impedances. Once the initial layout (i.e., $\mathbf{x_f}^{(1)}$) is determined, the circuit elements can be extracted from the electromagnetic response using (5–7). This provides $\mathbf{x_c}^{(1)} = \mathrm{P}(\mathbf{x_f}^{(1)})$, and using (1), the first error function can be inferred. To iterate the process using (2), with $\mathbf{h}^{(1)}$ derived from (3), a first approximation of the Jacobian matrix is needed. Following a similar approach to the one explained in Section 3, the lengths l_{Lp}, l_{Lr}, l_{Cr} are slightly perturbed, and the values of L_p, L_r, and C_r resulting after each perturbation are obtained from parameter extraction. The first Jacobian matrix can thus be expressed as:

$$\mathbf{B} = \begin{pmatrix} \frac{\delta L_r}{\delta l_{Lr}} & \frac{\delta L_r}{\delta l_{Cr}} & \frac{\delta L_r}{\delta l_{Lp}} \\ \frac{\delta C_r}{\delta l_{Lr}} & \frac{\delta C_r}{\delta l_{Cr}} & \frac{\delta C_r}{\delta l_{Lp}} \\ \frac{\delta L_p}{\delta l_{Lr}} & \frac{\delta L_p}{\delta l_{Cr}} & \frac{\delta L_p}{\delta l_{Lp}} \end{pmatrix} \tag{11}$$

By applying this procedure to the resonator elements of each filter stage, the corresponding layouts of the SIRs and grounded stubs are determined.

4.2 Determination of the Line Width

The initial line width is estimated from the formulas provided in several microwave textbooks (for instance, [17]). Once the initial width is estimated, the specific ASM algorithm developed to determine the line width is applied. In such one-variable ASM scheme, the initial Jacobian matrix (actually just composed of one element) is inferred by perturbing the line width and obtaining the characteristic impedance through electromagnetic simulation (i.e., the fine model variable is the line width, W, whereas the variable of the coarse model is the characteristic impedance).

4.3 Optimization of the Line Length (Filter Cell Synthesis)

As previously mentioned, the length of the lines cascaded to the resonant elements is optimized by considering the whole filter cell. Let us define l_{cell} as the length of the cell excluding the width of the grounded stubs, W_{Lp}, (roughly corresponding to $\lambda/4$ at $f_{0f}^{(13)}$). To determine l_{cell}, a single parameter ASM optimization is applied to the filter cell (filter cell synthesis), where the initial value of l_{cell} is inferred from the well-known formula [15] that gives the line length as a function of the required phase (90°) and frequency ($f_{0f}^{(13)}$). At this stage, the ASM optimization consists of varying the length of the lines cascaded to the resonator until the required phase at $f_{0f}^{(13)}$ (i.e., 90°) per filter cell is achieved (the other geometrical parameters of the cell are kept unaltered). The phase is directly inferred from the frequency response of the cell obtained from electromagnetic simulation at each iteration step.

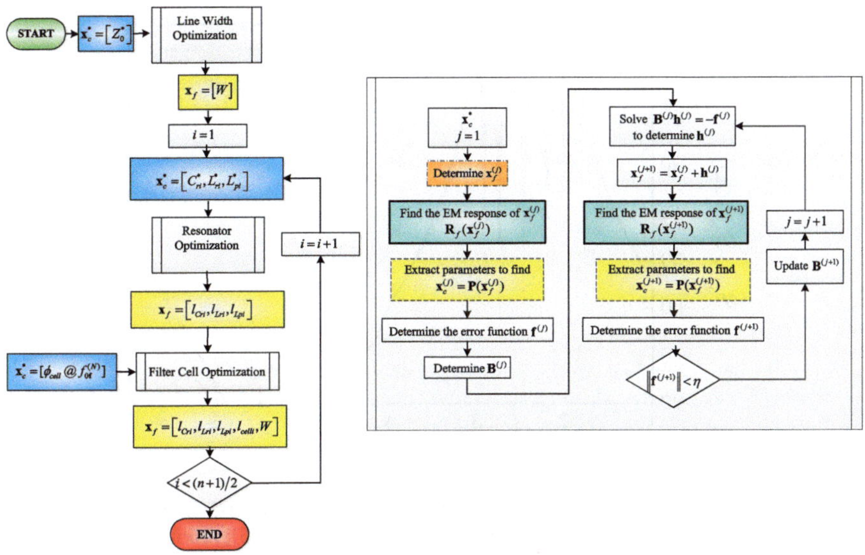

Fig. 7 Flow diagram of the second ASM algorithm. The sub-process depicted at the right of the figure represents a typical ASM algorithm used in each optimization process (particularly the indicated one is for the resonator optimization). Notice that the loop must be executed $(n+1)/2$ times, where n is the filter order. The order is assumed to be odd, since for odd Chebyshev response the cells i and $n+1-i$ are identical. However, this does not affect line width optimization since line width is identical for all filter stages. From [5]; reprinted with permission

Once each filter cell has been synthesized, the cells are simply cascaded to generate the final filter layout (coupling between adjacent resonators is not taken into account since the results reveal that this effect is not significant). The scheme of the complete ASM process able to automatically provide the layout from the optimum filter schematic, and consisting of three independent quasi-Newton iterative algorithms, is depicted in Fig. 7. Using the element values of Table 2, corresponding to the optimum filter schematic of the example reported in Section 3, where the lines present between adjacent resonators exhibit a characteristic impedance of $Z_o = 50\ \Omega$ and an electrical length of 90° at $f_{0f}^{(13)} = 2.4690$ GHz, the second ASM algorithm was applied to automatically generate the filter layout (which is actually the one depicted in Fig. 2). The dimensions are summarized in Table 3, except the line width, which does not depend on the filter stage, i.e., $W = 0.6055$ mm. Notice that the cell length slightly varies from cell to cell. This variation is due to the phase effects produced by the different resonators, and justifies the need to optimize the length of the lines by considering the complete filter cell (as described in the preceding paragraph).

The electromagnetic simulation (excluding losses) of the synthesized filter is compared to the response of the optimum filter schematic and to the target (ideal Chebyshev) response in Fig. 8. The agreement between the lossless electromagnetic simulation and the response of the optimum filter schematic (where losses are

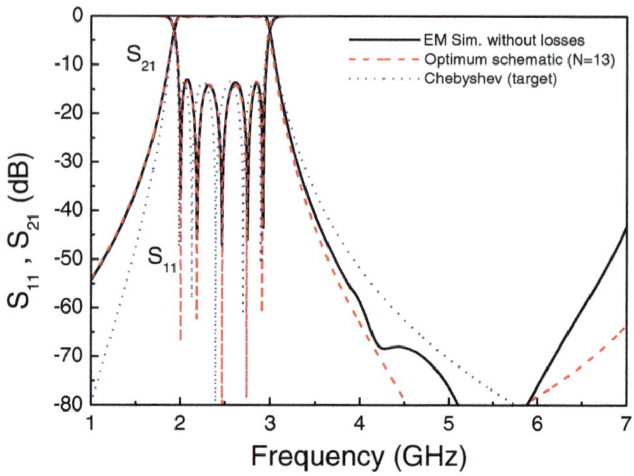

Fig. 8 Lossless electromagnetic simulation of the synthesized order-5 filter, compared to the response of the optimum filter schematic and target response. From [5]; reprinted with permission

excluded) is very good, pointing out the validity of the second ASM synthesis method. The fabricated filter and the measured frequency response are depicted in Fig. 9. The measured response is in reasonable agreement with the lossy electromagnetic simulation, and reveals that filter specifications are satisfied to a good approximation. Slight discrepancies between the measured response and the lossy electromagnetic simulation can be mainly attributed to inaccuracies in the dielectric constant provided by the substrate supplier, although they can be also influenced by fabrication related tolerances, substrate anisotropy, and foil roughness. Nevertheless, these effects are not considered in the optimization process, because the aim is to automatically obtain the filter layout providing a lossless electromagnetic simulation able to satisfy the specifications.

It is worth highlighting that layout generation with the reported two-step ASM procedure (synthesis of the filter schematic and synthesis of the filter layout) is obtained following a completely unattended scheme. External action is only required in the first step, to provide the filter specifications, which are the input variables of the proposed two-step ASM algorithm.

5 Further Examples of Filter Synthesis

In order to demonstrate the potential of the two-step ASM algorithm for the synthesis of filters based on SIRs and shunt inductive stubs, let us now apply the developed tool to the synthesis of a higher order filter with the following specifications: order $n = 7$, central frequency $f_0 = 3.0$ GHz, fractional bandwidth $FBW = 37.2$ % (−3 dB fractional bandwidth), and ripple level $L_{Ar} = 0.12$ dB.

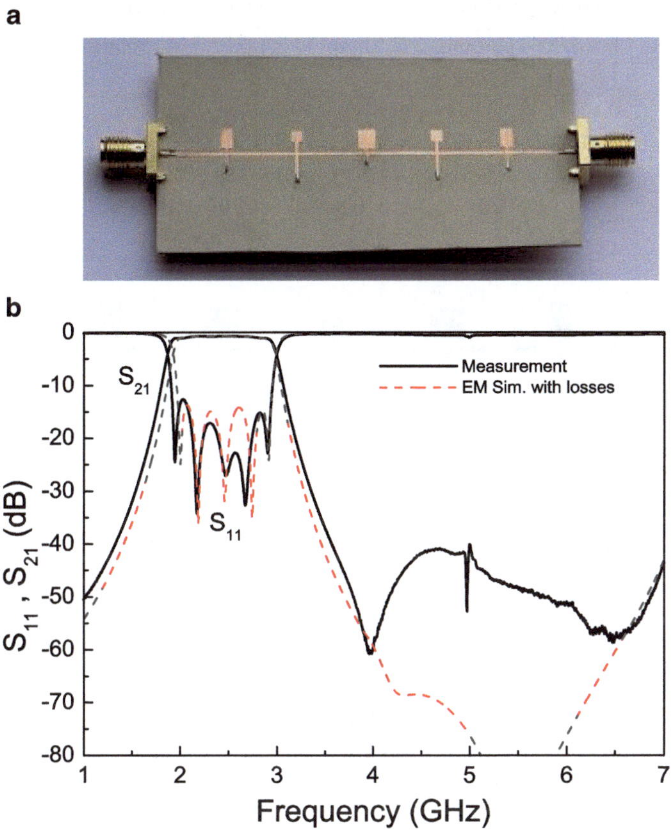

Fig. 9 Photograph of the fabricated order-5 filter (**a**) and measured response compared to the lossy electromagnetic simulation (**b**). The layout of the fabricated filter is the one depicted in Fig. 2. From [5]; reprinted with permission

Table 4 Element values of the shunt resonators for the optimum filter schematic of the 7th order filter

Stage	L_p (nH)	L_r (nH)	C_r (pF)
1,7	1.6280	0.5427	1.2431
2,6	1.1377	0.3792	1.7788
3,5	0.8442	0.2814	2.3974
4	1.0073	0.3358	2.0092

Application of the first ASM algorithm gives the schematic with element values indicated in Table 4 and 90° (at $f_{0f}^{(8)} = 3.0638$ GHz) transmission line sections. Convergence has been achieved after $N = 8$ iterations, when the error function is as small as 0.1 %.

Application of the second ASM algorithm, considering the substrate used for the filter of the previous section (*Rogers RO3010* with thickness $h = 635$ μm and

Table 5 Geometry parameters of the synthesized order-7 filter

Stage	l_{Lr} (mm)	l_{Cr} (mm)	l_{Lp} (mm)	l_{cell} (mm)	W_{Cr} (mm)	W_{Lp}(mm)
1,7	1.0493	1.9982	2.6430	9.0701	2.0	0.3
2,6	0.4612	3.2080	1.7949	9.1020	2.0	0.3
3,5	0.4407	2.7972	1.2780	9.2243	3.0	0.3
4	0.5882	2.3437	1.5738	9.1482	3.0	0.3

a

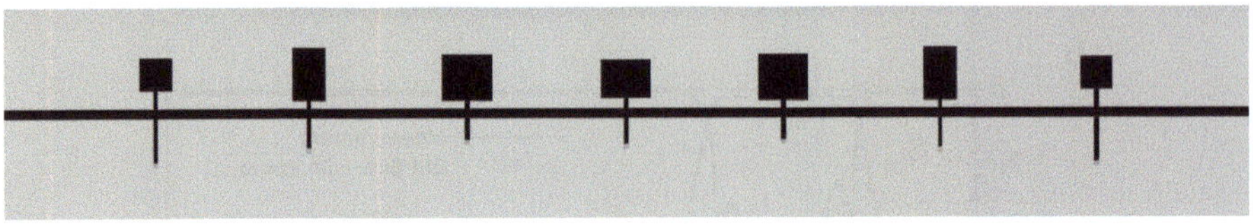

b

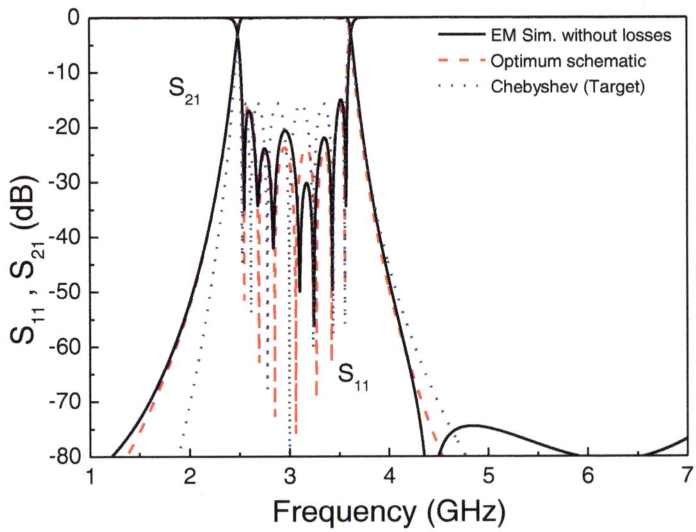

Fig. 10 Layout of the synthesized order-7 filter (**a**), and lossless electromagnetic simulation compared to the response of the optimum filter schematic and target response (**b**)

dielectric constant $\varepsilon_r = 10.2$), provides the filter geometry indicated in Table 5. Figure 10 shows the layout of the filter and the lossless electromagnetic simulation, compared to the optimum filter schematic and target responses. The fabricated filter is depicted in Fig. 11, together with the measured response and the lossy electromagnetic simulation. Again, very good agreement between the different responses can be appreciated, and the filter response satisfies the considered specifications.

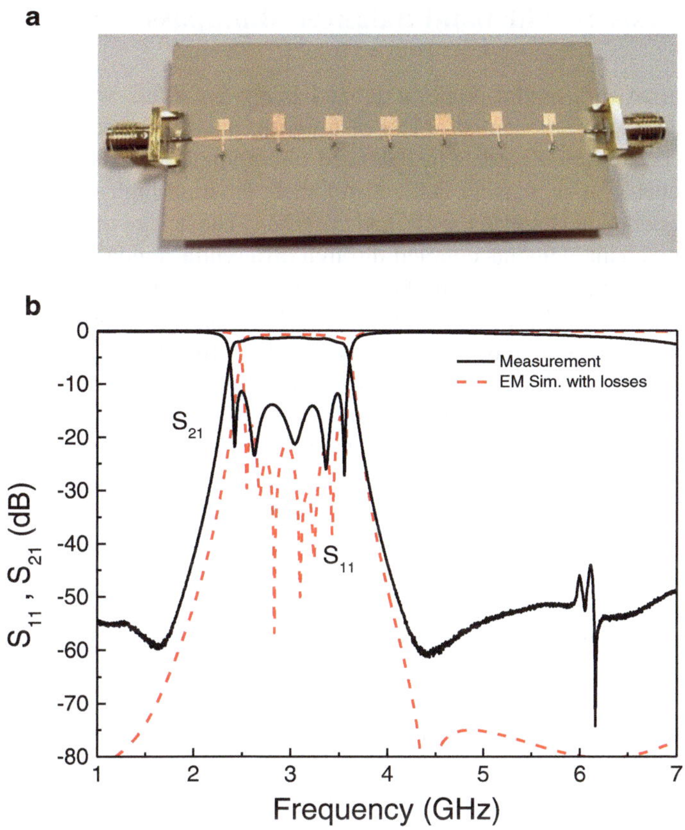

Fig. 11 Photograph of the fabricated order-7 filter (**a**) and measured response compared to the lossy electromagnetic simulation (**b**)

This additional example of an order-7 filter with different specifications (as compared to the 5th order filter reported before) and other synthesized higher order filters reported in [5] demonstrate that the two-step ASM algorithm discussed in this chapter is a powerful tool to automatically provide the layout of the considered type of filters. In principle, the reported two-step ASM tool can be applied to any type of filter consisting on semi-lumped resonators coupled through admittance inverters. For instance, application to wideband bandpass filters based on open complementary split ring resonators (OCSSRs) coupled through admittance inverters, first reported in [18], can be envisaged. It is also possible to apply the reported two-step ASM technique to the synthesis of wideband balanced bandpass filters. Indeed, this is the subject of the next section.

6 Synthesis of Wideband Balanced Bandpass Filters

The design of differential-mode wideband bandpass filters with common-mode noise suppression has been an object of growing interest in recent years [19–27]. These filters are key elements in balanced circuits and systems (e.g., high-speed digital circuits), of increasing demand for their inherent high immunity to noise, electromagnetic interference (EMI), and crosstalk. The design of balanced filters by means of resonant elements coupled through differential-mode quarter-wavelength admittance inverters suffers from the same limitations than the single-ended counterparts, i.e., bandwidth degradation [24]. Hence the application of the two-step ASM scheme presented before is fully justified for the automated design of these differential-mode bandpass filters. As an illustrative example, the filter type considered in this section is based on mirrored SIRs coupled through admittance inverters [24]. The typical layout is depicted in Fig. 12a. The central metallic patches of the mirrored SIRs introduce common-mode transmission zeros, useful for the suppression of that mode in the differential filter pass band.

The mirrored SIRs are described by means of a combination of capacitances and inductances, as indicated in the schematic of Fig. 12b. Note that the symmetry plane is an electric wall for the differential mode, and hence the capacitances C_{zi} do not play an active role for that mode (they are grounded). Thus, the equivalent circuit schematic for the differential mode is the canonical circuit of a bandpass filter (Fig. 12c) [5]. Conversely, the symmetry plane for the common-mode is a magnetic wall (open circuit) and the equivalent circuit schematic is the one depicted in Fig. 12d. The resonators L_{pi}-C_{zi} provide transmission zeros that are useful for the suppression of the common-mode in the region of interest (differential filter pass band). According to the schematics of Fig. 12c–d, the position of the transmission zeros does not affect the differential-mode response.

For the synthesis of the filter, the two-step process described above can be applied after suitable modifications. Notice that for the determination of the circuit schematic for the differential mode, the capacitances C_{zi} do not play a role. Moreover, for the differential mode, the shunt resonators only depend on two elements, and hence they are determined from the filter central frequency and reactance slope. An important difference, as compared to the first ASM algorithm reported in Section 3, is that in this case the resonators are considered to be identical, (and thus $L_{pi} = L_p$ and $C_{pi} = C_p$ for $i = 1,2,\ldots,n$), whereas the admittance of the inverters depends on the device stage. The reason is that, from a topological point of view, it is convenient to implement the filter layout with identical mirrored SIRs, since the admittance inverters can thus be implemented by transmission line sections parallel to the line axis.

Let us consider the following differential filter specifications: $n = 3$, $f_0 = 2.4$ GHz, $FBW = 40$ % (corresponding to a 52.98 % −3 dB fractional bandwidth, considered in the optimization), and $L_{Ar} = 0.15$ dB. The ideal Chebyshev response is depicted in Fig. 13. Such response is achieved by considering $L_p = 1.1637$ nH, $C_p = 3.7779$ pF, and ideal admittance inverters with

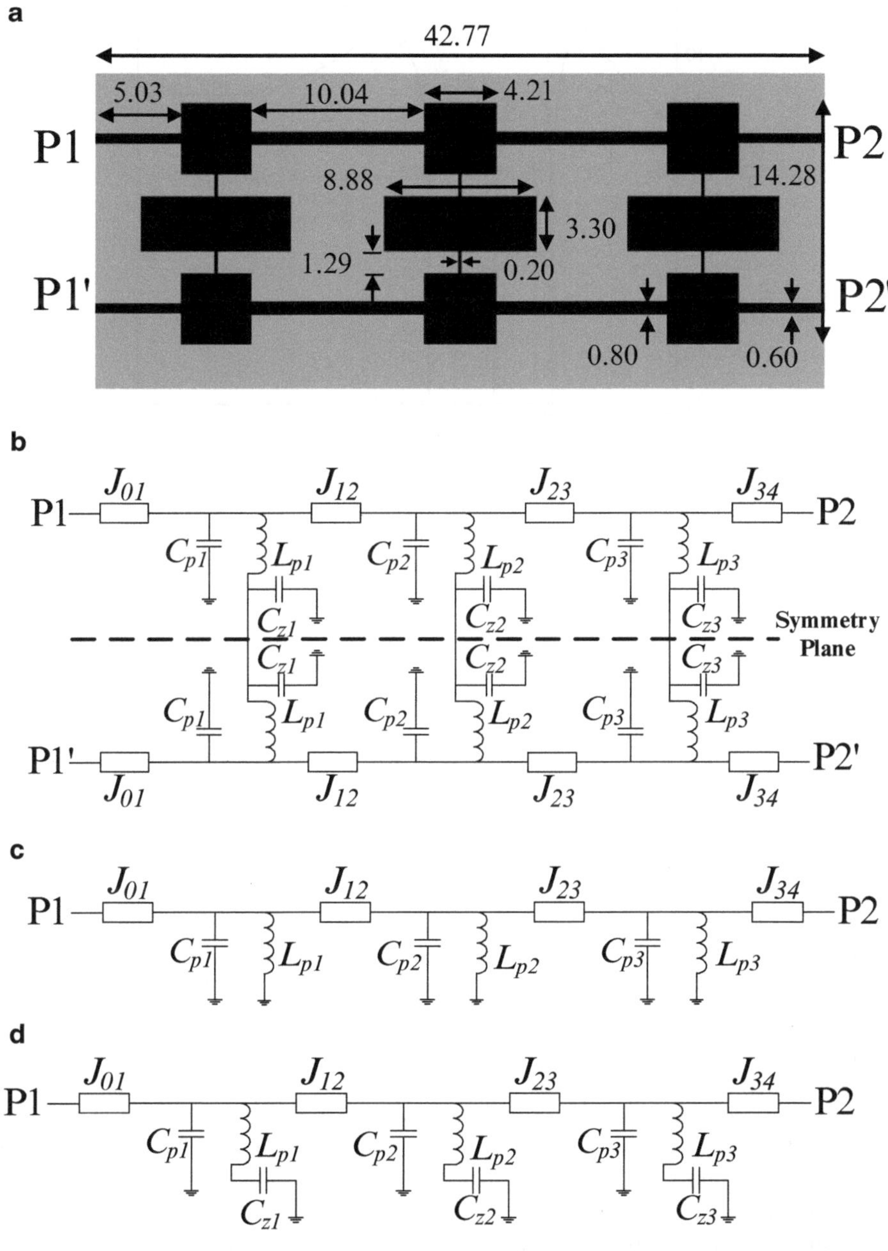

Fig. 12 Typical topology (order-3) of the considered balanced wideband bandpass filters (**a**), circuit schematic (**b**), and circuit schematic for the differential (**c**) and common (**d**) modes. Dimensions, in millimeters, correspond to the designed prototype

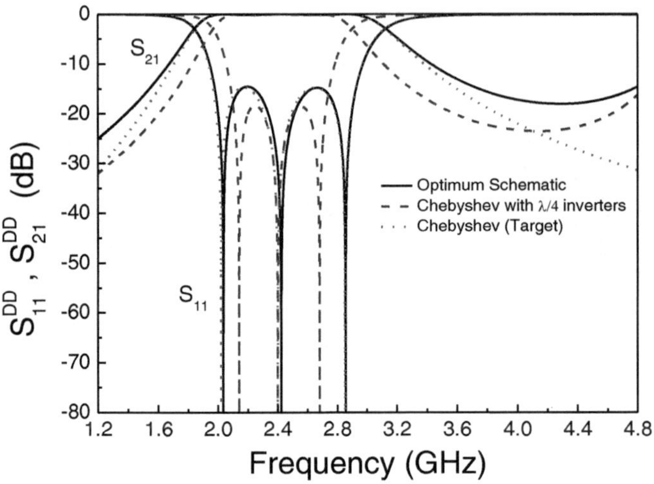

Fig. 13 Ideal Chebyshev response of the differential bandpass filter, response that results by replacing the ideal inverters with quarter-wavelength transmission lines, and response of the optimum schematic that results after applying the first ASM algorithm

$J_{0,1} = J_{3,4} = 0.02$ S, $J_{1,2} = J_{2,3} = 0.0199$ S. The application of the first ASM algorithm to the considered example provides an error function smaller than 3.2 % after $N = 4$ iterations. The resulting fine and coarse model parameters are $\mathbf{x_f}^{(4)} = [f_{0f}{}^{(4)}, FBW_f{}^{(4)}, L_{Arf}{}^{(4)}] = [2.423\text{ GHz}, 65.78\ \%, 0.332\text{ dB}]$ and $\mathbf{x_c}^{(4)} = [f_{0c}{}^{(4)}, FBW_c{}^{(4)}, L_{Arc}{}^{(4)}] = [2.3999\text{ GHz}, 0.5344\ \%, 0.1546\text{ dB}]$, and the resulting response is depicted in Fig. 13. The element values of the resonators are $L_p = 2.8195$ nH, $C_p = 1.5302$ pF, and the admittance of the quarter-wavelength (at $f_{0f}{}^{(4)}$) transmission line sections are $J_{0,1} = J_{3,4} = 0.02$ S, $J_{1,2} = J_{2,3} = 0.0223$ S. To complete the schematic of Fig. 12b, the capacitances C_{zi} must be set to a certain value. In this example, all the transmission zero frequencies have been set to the same value, i.e., $f_z = 1.1f_0$. This gives a good common-mode rejection ratio (CMRR) in the whole differential filter pass band. With this value of f_z, the central patch capacitances are found to be $C_z = 2.3302$ pF.

The second ASM algorithm, necessary to determine the filter layout, is similar to the one reported in Section 4. The resulting layout is the one depicted in Fig. 12. The lossless electromagnetic response is very close to the response of the schematic and hence to the target response (see Fig. 14). The fabricated device (depicted in Fig. 15) exhibits a frequency response in very good agreement to the lossy electromagnetic simulation (see Fig. 15b). As for the single-ended filter synthesis reported before, the layout of the balanced filter has been determined from the specifications without the need of any further action during the optimization process. The results of this section demonstrate that the two-step ASM scheme analyzed in this chapter is also useful for the synthesis of common-mode suppressed balanced filters based on resonators coupled through admittance inverters.

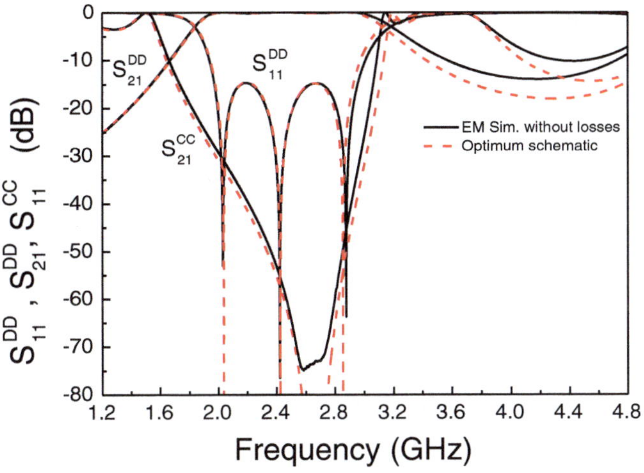

Fig. 14 Lossless electromagnetic simulation of the synthesized order-3 balanced filter, compared to the response of the optimum filter schematic

7 Conclusions

In summary, a method for the automated and unattended design of single-ended and balanced wideband bandpass filters based on shunt resonators coupled through admittance inverters has been reviewed. It is based on ASM optimization and is divided in two steps: a first ASM algorithm, devoted to determine the filter schematic able to satisfy the specifications, and a second ASM algorithm, that automatically determines the filter layout. With this approach, the bandwidth degradation typical of wideband filters based on resonators coupled through admittance inverters is solved, and the filter design does not require any external action in the whole process. Several examples of filter design have been reported to demonstrate the viability and flexibility of the method. Moreover, the applicability of this method for efficient solving of numerically expensive problems has been revealed.

Acknowledgements The results presented in this chapter have been generated thanks to the support of several projects by MINECO-Spain (projects TEC2013-47037-C5-1-R, TEC2013-40600-R, TEC2013-49221-EXP) and *Generalitat de Catalunya* (project 2014SGR-157). This work has also been supported by FEDER funds. Ferran Martín is in debt to *Institució Catalana de Recerca i Estudis Avançats*, who awarded him with an ICREA Academia Prize.

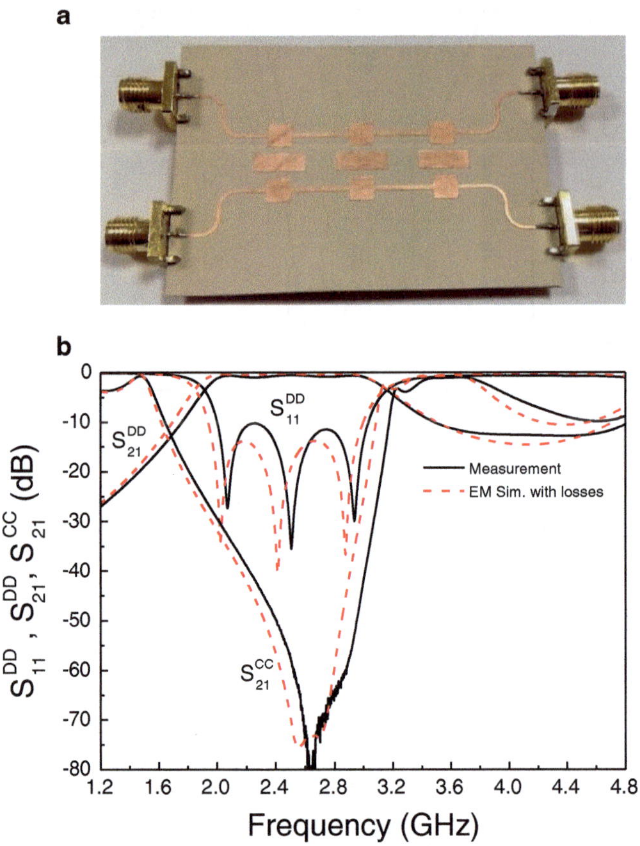

Fig. 15 Photograph of the fabricated order-3 balanced filter (**a**) and measured response compared to the lossy electromagnetic simulation (**b**). The layout of the fabricated filter is the one depicted in Fig. 12

References

1. Sun, S., Zhu, L.: Multimode resonator-based bandpass filters. Microw. Mag. **10**(2), 88–98 (2009)
2. Hao, Z.-C., Hong, J.-S.: Ultrawideband filter technologies. Microw. Mag. **11**(4), 56–68 (2010)
3. Hong, J.S., Lancaster, M.J.: Microstrip Filters for RF/Microwave Applications. Wiley, New York, NY, USA (2001)
4. Selga, J., Sans, M., Rodríguez, A., Bonache, J., Boria, V., Martín, F.: Automated synthesis of planar wideband bandpass filters based on stepped impedance resonators (SIRs) and shunt stubs through aggressive space mapping (ASM). In: IEEE MTT-S International Microwave Symposium, Tampa, FL (USA), June 2014
5. Sans, M., Selga, J., Rodríguez, A., Bonache, J., Boria, V.E., Martín, F.: Design of planar wideband bandpass filters from specifications using a two-step aggressive space mapping (ASM) optimization algorithm. IEEE Trans. Microw. Theory Techn. **62**, 3341–3350 (2014)

6. Bandler, J.W., Biernacki, R.M., Chen, S.H., Hemmers, R.H., Madsen, K.: Electromagnetic optimization exploiting aggressive space mapping. IEEE Trans. Microw. Theory Techn. **43**, 2874–2882 (1995)
7. Bandler, J.W., Biernacki, R.M., Chen, S.H., Grobelny, P.A., Hemmers, R.H.: Space mapping technique for electromagnetic optimization. IEEE Trans. Microw. Theory Techn. **42**, 2536–2544 (1994)
8. Bakr, M.H., Bandler, J.W., Madsen, K., Rayas-Sánche, J.E., Søndergaard, J.: Space-mapping optimization of microwave circuits exploiting surrogate models". IEEE Trans. Microw. Theory Tech. **43**, 2297–2306 (2000)
9. Koziel, S., Cheng, Q.S., Bandler, J.W.: Space mapping. IEEE Microw. Mag. **9**, 105–122 (2008)
10. Bandler, J.W., Biernacki, R.M., Chen, S.H., Omeragic, D.: Space mapping optimization of waveguide filters using finite element and mode-matching electromagnetic simulators. In: IEEE MTT-S International Microwave Symposium, Denver, CO (USA), June 1997
11. Bandler, J.W., Cheng, Q.S., Hailu, D.M., Nikolova, N.K.: A space-mapping design framework. IEEE Trans. Microw. Theory Techn. **53**, 2601–2610 (2004)
12. Morro, J.V., Soto, P., Esteban, H., Boria, V.E., Bachiller, C., Taroncher, M., Cogollos, S., Gimeno, B.: Electromagnetic optimization exploiting aggressive space mapping. IEEE Trans. Microw. Theory Techn. **53**, 1130–1142 (2005)
13. Bandler, J.W., Cheng, Q.S., Dakroury, S.A., Mohamed, A.S., Bakr, M.H., Madsen, K., Søndergaard, J.: Space mapping: the state of the art. IEEE Trans. Microw. Theory Techn. **52**, 337–361 (2004)
14. Bonache, J., Gil, I., García-García, J., Martín, F.: Compact microstrip band-pass filters based on semi-lumped resonators. IET Microw. Antennas Propag. **1**, 932–936 (2007)
15. Naqui, J., Durán-Sindreu, M., Bonache, J., Martín, F.: Implementation of shunt connected series resonators through stepped-impedance shunt stubs: analysis and limitations. IET Microw. Antennas Propag. **5**, 1336–1342 (2011)
16. Pozar, D.M.: Microwave Engineering. Addison Wesley, New York, USA (1990)
17. Bahl, I., Barthia, P.: Microwave Solid State Circuit Design. John Wiley, New York (1998)
18. Vélez, P., Naqui, J., Durán-Sindreu, M., Bonache, J., Martín, F.: Broadband microstrip bandpass filter based on open complementary split ring resonators. Int. J. Antennas Propag. **2012**, 6 pp. (2012). doi:10.1155/2012/174023. Article ID 174023
19. Lim, T.B., Zhu, L.: A differential-mode wideband bandpass filter on microstrip line for UWB applications. IEEE Microw. Wireless Compon. Lett. **19**, 632–634 (2009)
20. Abbosh, A.M.: Ultrawideband balanced bandpass filter. IEEE Microw. Wireless Compon. Lett. **21**, 480–482 (2011)
21. Zhu, H.T., Feng, W.J., Che, W.Q., Xue, Q.: Ultra-wideband differential bandpass filter based on transversal signal-interference concept. Electron. Lett. **47**, 1033–1035 (2011)
22. Wu, X.-H., Chu, Q.-X.: Compact differential ultra-wideband bandpass filter with common-mode suppression. IEEE Microw. Wireless Compon. Lett. **22**, 456–458 (2012)
23. Vélez, P., Naqui, J., Fernández-Prieto, A., Durán-Sindreu, M., Bonache, J., Martel, J., Medina, F., Martín, F.: Differential bandpass filter with common mode suppression based on open split ring resonators and open complementary split ring resonators. IEEE Microw. Wireless Compon. Lett. **23**, 22–24 (2013)
24. Vélez, P., Durán-Sindreu, M., Bonache, J., Fernández-Prieto, A., Martel, J., Medina, F., Martín, F.: Differential bandpass filters with common-mode suppression based on stepped impedance resonators (SIRs). In: IEEE MTT-S International Microwave Symposium, Seattle (USA), June 2013
25. Wang, X.-H., Zhang, H., Wang, B.-Z.: A novel ultra-wideband differential filter based on microstrip line structures. IEEE Microw. Wireless Compon. Lett. **23**, 128–130 (2013)
26. Shi, J., Shao, C., Chen, J.-X., Lu, Q.-Y., Peng, Y., Bao, Z.-H.: Compact low-loss wideband differential bandpass filter with high common-mode suppression. IEEE Microw. Wireless Compon. Lett. **23**(9), 480–482 (2013)
27. Vélez, P., Naqui, J., Fernández-Prieto, A., Bonache, J., Mata-Contreras, J., Martel, J., Medina, F., Martín, F.: Ultra-compact (80 mm^2) differential-mode ultra-wideband (UWB) bandpass filters with common-mode noise suppression. IEEE Trans. Microw. Theory Tech. **63**(4), 1272–1280 (2015)

Two-Stage Gaussian Process Modeling of Microwave Structures for Design Optimization

J. Pieter Jacobs and Slawomir Koziel

Abstract Accurate models that can be rapidly evaluated are indispensable in microwave engineering. Kernel-based machine learning methods applied to the modeling of microwave structures have recently attracted attention; these include support vector regression, Bayesian support vector regression, and Gaussian process regression (GPR). In this chapter, we apply an extended methodology based on GPR, namely two-stage GPR, to the modeling of microwave antennas and filters. At the core of the method lies variable-fidelity electromagnetic simulations. In the first stage, a mapping between electromagnetic models (simulations) of low and high fidelity is learned, which allows for significantly reducing the computational effort necessary to set up the high-fidelity training data sets for the actual surrogate models (second stage), with negligible loss in predictive power. We apply the two-stage models to design optimization involving several examples of antennas and microstrip filters.

Keywords Gaussian process regression • Computer-aided design (CAD) • Electromagnetic (EM) simulation • Microwave engineering • Space mapping • Surrogate-based optimization • Surrogate modeling

MSC codes: 65D17, 93A30, 74P99, 78M50

J.P. Jacobs (✉)
Department of Electrical, Electronic and Computer Engineering, Centre for Electromagnetism, University of Pretoria, Pretoria 0002, South Africa
e-mail: jpjacobs@up.ac.za

S. Koziel
Engineering Optimization & Modeling Center, School of Science and Engineering, Reykjavik University, Menntavegur 1, 101 Reykjavik, Iceland
e-mail: koziel@ru.is

S. Koziel et al. (eds.), *Simulation-Driven Modeling and Optimization*, Springer Proceedings in Mathematics & Statistics 153,
DOI 10.1007/978-3-319-27517-8_7

1 Introduction

Microwave engineering depends extensively on full-wave electromagnetic simulations as they permit highly accurate evaluation of microwave structures such as antennas, filters, and circuit components. However, high-fidelity simulations may pose significant computational demands. Hence the use of accurate electromagnetic simulations to solve especially tasks involving multiple analyses, such as statistical analysis, yield-driven design, or parametric design optimization, might become infeasible under certain conditions. Consider, for instance, global optimization using metaheuristics such as genetic algorithms [1, 2] that might require thousands of full-wave analyses of possible geometries of the structure to be optimized. In such cases, the use of fast and yet accurate models of the microwave structures being analyzed (so-called surrogates) becomes indispensable. Identified on a training set consisting of a limited number of input–output pairs, a model of this sort can generalize over the input space and therefore quickly obtain the desired performance characteristics for inputs not previously presented to it.

A highly effective approach for constructing surrogate models of antenna structures is Gaussian process regression (GPR) [3]. In particular, GPR has been shown to be a very successful tool for modeling antenna performance characteristics such as input reflection coefficient against frequency [4, 5]. Other kernel-based machine learning methods that have been used for antenna modeling include standard support vector regression, e.g., [6], and the more expressive, GPR-based Bayesian support vector regression [7].

A Gaussian process (GP) is a stochastic process that can be viewed as the generalization of the Gaussian probability distribution to functions. The Gaussian nature of the distribution leads to tractable calculations when learning and inference need to be performed. Gaussian processes are generally easier to implement and interpret than neural networks—a reason is that training of far fewer parameters (in the order of the dimension of the input vectors) is required compared to the number of weights in, for example, a multi-layer perceptron neural network.

An important limitation of approximation-based modeling methods such as GPR is the high cost of gathering the high-fidelity data necessary to train the model for sufficient predictive accuracy. Here we address this problem by using variable-fidelity electromagnetic simulations within a two-stage modeling scheme [8, 9]: in the first stage, we use full-wave simulations to generate by a low-fidelity (coarse) training data set of n points, and $n_{\text{aux}} < n$ points of the corresponding (computationally expensive) high-fidelity (fine) training set. We then train a model that maps low-fidelity training targets ($\text{Re}\{S_{11}\}$ or $\text{Im}\{S_{11}\}$) to the high-fidelity ones, and use it to predict the remaining $n - n_{\text{aux}}$ high-fidelity targets that were not simulated. The n_{aux} simulated high-fidelity targets and the n-n_{aux} predicted ones—together with the input vectors—then yield the n-point "approximate" high-fidelity training set. The second stage entails the construction of a final GPR model using the latter training set. Exploiting the knowledge embedded in the low-fidelity simulations in this way enables significant reductions in model setup cost without compromising accuracy.

Data selection for microwave modeling problems in the past has been accomplished through adaptive sampling techniques embedded within optimization contexts that aim to reduce the number of samples necessary to achieve the desired modeling accuracy by iterative identification of the model and adding new training samples based on the actual model error at selected locations [10] or expected error values (statistical infill criteria, e.g., [11]). We note that our focus—in contrast to [10, 11] that are local models—is on global or library type models that give reliable predictions over the entire input space, and that can be used for a variety of applications (e.g., optimization, statistical analysis). An alternative approach to reducing surrogate model setup cost was demonstrated in [12], where only the support vectors of an initial (global) BSVR model trained on low-fidelity data were selected for high-fidelity simulation. A 31-to-48 % reduction in computational expense could be achieved without compromising predictive ability.

The novelty of the methodology described in this chapter lies in that it maps the correlations between physically related coarse and fine simulation models of the same antenna via an auxiliary model, blending full-wave simulation data at two fidelity levels into one final surrogate model by means of training on the above "approximate" high-fidelity data set. This stands in contrast to [4, 5], where conventional GPR models were trained on data sets obtained in full from expensive high-fidelity data; there was no attempt to reduce the costs associated with acquiring this data, even though this contributed by far the bulk of the model setup costs.

The two-stage approach is demonstrated using both microstrip antenna and filter examples (cf. Sections 4 and 5, respectively). We also evaluate the accuracy of our surrogates by using them within a space-mapping (SM) optimization framework. These sections are preceded by an overview in Section 2 of standard GPR, and a discussion of two-stage GPR in Section 3. The chapter is concluded by some summary remarks (Section 6).

2 Fundamentals of Standard Gaussian Process Regression

This section summarizes the basic tenets of GPR along the lines of [3], and explains how these equations map to practical modeling (using microwave filter modeling as example).

The multivariate Gaussian probability distribution is fundamental to GPR. Consider n continuous random variables $f_1, \ldots, f_n$ with joint probability $p(f_1, \ldots, f_n)$, or equivalently $p(\mathbf{f})$, where $\mathbf{f} = [f_1 \ \ldots \ f_n]$. Assume that variables $\mathbf{f}$ are distributed according to the multivariate Gaussian distribution [3]:

$$p\left(\mathbf{f}\middle|\mathbf{m}, A\right) = (2\pi)^{-n/2}|A|^{-1/2}\exp\left(-\frac{1}{2}(\mathbf{f}-\mathbf{m})^T A^{-1}(\mathbf{f}-\mathbf{m})\right) = N(\mathbf{m}, A) \quad (1)$$

with $\mathbf{f}$ a multi-dimensional "point" under the distribution; $\mathbf{m}$ the mean vector of length n; and A the covariance matrix of size $n \times n$ determining the shape of the distribution.

Consider now standard GPR. Of interest is learning a mapping between filter geometry dimensions and frequency, and $|S_{21}|$ (or Re/Im$\{S_{21}\}$; for conciseness we will refer to $|S_{21}|$ throughout). The first step is to collect a training data set of n input–output pairs, $\{(\mathbf{x}_i, y_i) \mid i = 1, \ldots, n\}$. Input vectors $\mathbf{x}_i$ are of dimension P, while the target responses y_i are scalars. Specifically, each input vector $\mathbf{x}_i$ corresponds to a set of adjustable filter geometry parameters and a frequency value within the band of interest, while each output scalar y_i is the corresponding $|S_{21}|$ value. Also selected is a test data set of n^* input–output pairs $\{(\mathbf{x}_i^*, y_i^*) \mid i = 1, \ldots, n^*\}$: input vectors $\mathbf{x}_i^*$ consisting of previously unseen geometry-values-and-frequency for which $|S_{21}|$ needs to be predicted, and output scalars y_i^* that are known associated values of $|S_{21}|$, computed for evaluating the model's predictions.

Under GPR, the n training output scalars (associated with the n input vectors $\mathbf{x}_i$) are modeled by random variables $[f_1 \ \ldots \ f_n]^T = [f(\mathbf{x}_1) \ \ldots \ f(\mathbf{x}_n)]^T$, and the n^* test output scalars by random variables $[f_1^* \ \ldots \ f_n^*]^T = [f(\mathbf{x}_1^*) \ \ldots \ f(\mathbf{x}_n^*)]^T$, where $f(\mathbf{x})$ is a Gaussian process (GP). A GP is a stochastic process that is the result of generalization of the Gaussian probability distribution (1) to functions. The latter implies the mean vector $\mathbf{m}$ becoming infinitely long, resulting in a mean *function* $m(\mathbf{x})$; and the two-dimensional covariance matrix A becoming infinitely large, with entries given by a covariance function $k(\mathbf{x}, \mathbf{x}')$. $f(\mathbf{x})$, which corresponds to an infinitely long vector, can be seen as a "point" under this distribution. The mean function is defined in the standard manner as

$$m(\mathbf{x}) = E[f(\mathbf{x})] \tag{2}$$

while the covariance function, which gives the covariance between outputs $f(\mathbf{x})$ and $f(\mathbf{x}')$ in terms of the associated inputs $\mathbf{x}$ and $\mathbf{x}'$, is defined as [3]

$$k\left(f(\mathbf{x}), f(\mathbf{x}')\right) = k\left(\mathbf{x}, \mathbf{x}'\right) = E\left[\left(f(\mathbf{x}) - m(\mathbf{x})\right)\left(f(\mathbf{x}') - m(\mathbf{x}')\right)\right] \tag{3}$$

where $E(X)$ is the expected value of the random variable X. It should be noted that actual computation of covariance functions takes place through (7) and (8), as will be explained below. Hence the GP $f(\mathbf{x})$ is a set consisting of an infinite number of random variables, of which any finite subset, for example, the training outputs $\mathbf{f} = [f_1 \ \ldots \ f_n]$, has a jointly Gaussian distribution by virtue of the general properties of the multivariate Gaussian distribution [3].

Predictions in GPR are carried out using standard probability rules applied to Gaussian multivariate distributions. A jointly Gaussian distribution (1) with zero mean is assumed over the n training outputs and the n^* test outputs ($n + n^*$ random variables in total). This is referred to as the *prior* distribution, and can be written as

$$\begin{bmatrix} \mathbf{f} \\ \mathbf{f}^* \end{bmatrix} \sim N\left(\mathbf{0}, \begin{bmatrix} K(X,X) & K(X,X^*) \\ K(X^*,X) & K(X^*,X^*) \end{bmatrix}\right) \tag{4}$$

(4) indicates that the random variables contained in the vector $[\mathbf{f} \ \mathbf{f}^*]^T$ have a multivariate jointly Gaussian distribution with zero mean and covariance matrix [•]. In (4), matrices X and X* contain the training and test input vectors, respectively;

and $K(X, X^*)$ is an $n \times n^*$ sub-matrix of covariances evaluated between all possible pairs of n training and n^* test outputs—for example, $K_{12} = k(f(\mathbf{x}_1), f(\mathbf{x}_2^*)) = k(\mathbf{x}_1, \mathbf{x}_2^*)$ (other sub-matrices in (4) are set up in a similar manner).

Since the training outputs $\mathbf{y}$ are known, the distribution of the test outputs conditioned on the training outputs $\mathbf{y}$ can be computed, yielding the *posterior* distribution, a multivariate Gaussian with mean vector $\mathbf{p}$ and covariance matrix Σ [3]

$$\mathbf{p} = K\left(X^*, X\right) K(X, X)^{-1} \mathbf{y} \tag{5}$$

$$\Sigma = K\left(X^*, X^*\right) - K\left(X^*, X\right) K(X, X)^{-1} K\left(X, X^*\right) \tag{6}$$

The mean vector $\mathbf{p}$ contains the $|S_{21}|$ predictions, i.e., the most likely values of the test outputs associated with the test input vectors in X^*. In other words, $\mathbf{p} = [|S_{21}|_1 \; |S_{21}|_2 \; \ldots \; |S_{21}|_{n^*}]$ where $|S_{21}|_1$ is the prediction for test input vector $\mathbf{x}_1^*$, $|S_{21}|_2$ is the prediction for $\mathbf{x}_2^*$, and so forth. The diagonal of Σ determines the corresponding predictive variances, which can be viewed as the confidence of the model in its predictions.

The covariance functions determine the covariance matrices in the prior and posterior probability distributions, and hence are critical in determining the shapes of these distributions and the GPs that will be favored by them. In what follows we consider two well-known covariance functions for calculating the covariance between outputs $f(\mathbf{x}_i)$ and $f(\mathbf{x}_j)$. The first is the squared-exponential covariance function with automatic relevance determination (ARD) [3],

$$k_{SE}\left(\mathbf{x}_i, \mathbf{x}_j\right) = \sigma_f^2 \exp\left(-\frac{1}{2}\sum_{k=1}^{P} \frac{\left(x_{i.k} - x_{j.k}\right)^2}{\tau_k^2}\right) \tag{7}$$

where $x_{i,k}$ and $x_{j,k}$ are the kth components of input vectors $\mathbf{x}_i$ and $\mathbf{x}_j$, respectively ($k = 1, \ldots, P$); $\tau_k > 0$ is the length-scale parameter that corresponds to component k of the two input vectors; and σ_f^2 is the signal variance. The second covariance function is the rational quadratic function with ARD. This covariance function can be viewed as a scaled mixture of squared-exponential functions with different length scales [3]:

$$k_{RQ}\left(\mathbf{x}_i, \mathbf{x}_j\right) = \sigma_f^2\left(1 + \frac{1}{2^\alpha}\sum_{k=1}^{P} \frac{\left(x_{i.k} - x_{j.k}\right)^2}{\tau_k^2}\right)^{-\alpha} \tag{8}$$

In (8), $\alpha > 0$ is the shape parameter, with the remaining symbols defined as for (1). Together, ${\sigma_f}^2$ and τ_k in (1) and (2)—as well as α in the case of (8)—constitute the hyperparameters of the covariance function.

Training in GPR entails finding the set of hyperparameters that minimizes the negative log marginal likelihood; this is usually accomplished by means of gradient-based search. The log marginal likelihood can be expressed as [3]

$$\log p\left(\mathbf{y}\middle|X\right) = -\frac{1}{2}\mathbf{y}^T K^{-1}\mathbf{y} - \frac{1}{2}\log|K| - \frac{n}{2}\log 2\pi \tag{9}$$

In the above, K is shorthand for the $n \times n$ matrix $K(X, X)$, and other symbols are as defined above. Once optimized, the magnitude of hyperparameter τ_k will reflect the relative importance of the kth input vector component, as large values of τ_k will ensure an insignificant contribution of that component to the covariance. This property is the above-mentioned ARD.

The computational cost of GPR is $O(n^3)$ due to the inversion of $K(X, X)$ which is of dimension $n \times n$.

3 Two-Stage Gaussian Process Regression

In this section we describe our two-stage GPR approach using filter modeling as practical example. Our objective is to construct highly accurate GPR surrogate models $\boldsymbol{R}_s$ that map geometry (design) variables and frequency to either Re$\{S_{21}\}$, Im$\{S_{21}\}$, or $|S_{21}|$ (in some cases, better results might be obtained if the third quantity is constructed from separate models of the first two). For the sake of conciseness, we will only refer to $|S_{21}|$ in the description below.

Assume that in order to ensure sufficient model accuracy, an n-element set of high-fidelity training data (simulated using a finely discretized mesh) is required:

$$D_{\text{fine}} = \left\{ (\mathbf{x}_i, y_{\text{fine},i}) \;\middle|\; i = 1, \ldots, n \right\} \tag{10}$$

with P-dimensional input vectors

$$\mathbf{x}_i = \left[\mathbf{u}_i^T \, f_{oi} \right]^T = \left[u_{1i} \, u_{2i} \ldots u_{Mi} \, f_{oi} \right]^T \tag{11}$$

and scalar targets $y_{\text{fine},i} = |S_{21}|_{\text{fine},i}$. The design vector $\mathbf{u}_i = [u_{1i} \, u_{2i} \; \ldots \; u_{Mi}]^T$ consists of M geometry variables of the filter and f_{oi} is the ith frequency sample in the frequency band of interest; hence $P = M + 1$.

Generating D_{fine} however may be prohibitively expensive. This can be addressed by adopting a two-stage modeling approach. It aims to construct a final GPR model that is based on a fraction of the high-fidelity simulations required to set up D_{fine} but is almost as accurate as a GPR model trained on the actual D_{fine}. The details of the two modeling stages are described below.

3.1 Two-Stage GPR: First Stage

In this stage, a separate auxiliary GPR model $\boldsymbol{R}_{aux}$ is used to "approximate" the expensive fine training data set D_{fine} by a relatively inexpensive data set $D_{\text{fine,approx}}$ of the same size.

Initially, we inexpensively simulate the n input vectors of D_{fine} using a coarse discretization, yielding the data set

$$D_{\text{coarse}} = \left\{ (\mathbf{x}_i, y_{\text{coarse},i}) \;\middle|\; i = 1, \ldots, n \right\}, \tag{12}$$

with $\mathbf{x}_i$ as before and $y_{\text{coarse},i} = |S_{21}|_{\text{coarse},i}$. We also simulate at high fidelity a (small) randomly selected subset of D_{fine} consisting of $n_{aux} < n$ points. Using this subset of D_{fine}, we construct a training set D_{aux} for $\boldsymbol{R}_{aux}$ as follows:

$$D_{\text{aux}} = \left\{ (\mathbf{x}_{\text{aux},k}, y_{\text{fine},k}) \;\middle|\; k = 1, \ldots, n_{\text{aux}} \right\} \tag{13}$$

where the training input vector (of dimension $\mathrm{M} + 2$)

$$\mathbf{x}_{\text{aux},k} = \left[u_{1k}\; u_{2k} \ldots u_{Mk}\; f_{ok}\; |S_{21}|_{\text{coarse},k} \right]^T \tag{14}$$

is of the form of (7) augmented by the associated coarse $|S_{21}|$ target value from D_{coarse}, and the target $y_{\text{fine},k} = |S_{21}|_{\text{fine},k}$ is the corresponding $|S_{21}|$ value from the above subset of D_{fine} (it may be noted that D_{coarse} and D_{fine} have the same set of input vectors; the only difference lies in the meshing density with which the targets have been obtained). A mapping is thus learned between coarse and fine $|S_{21}|$ simulations using training data that correspond to n_{aux} specific instances of sets of design variables and frequency; the first $M + 1$ elements of the input vector $\boldsymbol{u}_{\text{aux},k}$ can be viewed as unique identifiers of the $|S_{21}|$ values. The mapping represents the correlations between the coarse and fine model responses. Due to the fact that these models are physically related by virtue of being evaluated using the same EM solver, the mapping learned for a limited number of fine training points is likely to be preserved across the full design space.

Following training, $\boldsymbol{R}_{aux}$ is used to predict the $n - n_{aux}$ fine $|S_{21}|$ values that were not simulated from their coarsely simulated counterparts; we refer to these predicted targets as $y_{\text{pred},k} = |S_{21}|_{\text{pred},k}$, $k = (n_{\text{aux}} + 1), \ldots, n$. The n_{aux} full-wave simulated fine $|S_{21}|$ target values in conjunction with the $n - n_{aux}$ predicted ones yield—along with input vectors consisting of geometry parameters and frequency of the form (7)—an n-point "approximate" fine training data set for $\boldsymbol{R}_s$,

$$D_{\text{fine,approx}} = \left\{ \begin{array}{c} (\mathbf{x}_k, y_{\text{fine},k}) \;\middle|\; k = 1, \ldots, n_{\text{aux}} \\ (\mathbf{x}_k, y_{\text{pred},k}) \;\middle|\; k = (n_{\text{aux}} + 1), \ldots, n \end{array} \right\} \tag{15}$$

Acquiring the targets $y_{\text{pred},k}$ via model predictions as opposed to direct full-wave simulations can result in significant savings in computational costs, as described below.

3.2 Two-Stage GPR: Second Stage

Here we use $D_{\text{fine,approx}}$ instead of the (unavailable) full D_{fine} as training set for $\boldsymbol{R}_s$, the desired final surrogate that maps design variables and frequency to $|S_{21}|$ by means of "standard" GPR (cf. Section 2). In Sections 4 and 5 we show that these surrogates are sufficiently accurate to be used effectively for antenna and filter optimization using space mapping.

It may be emphasized that reducing the number of simulated high-fidelity training points without compromising predictive accuracy is possible because we exploit the knowledge embedded in the low-fidelity model, in particular via the mapping learned in the first stage that identifies correlations between the low- and high-fidelity simulation data.

4 Modeling and Optimization of Antennas Using Two-Stage GPR

In this section, we present examples illustrating how two-stage GPR models for the reflection coefficients of planar slot antennas can be set up based on substantially reduced finely discretized full-wave simulations. We then use these models for design optimization to illustrate their robustness. We consider three examples of antennas with highly non-linear $|S_{11}|$ responses as a function of geometry parameters and frequency: a narrowband coplanar waveguide (CPW)-fed slot dipole antenna, an ultra-wideband (UWB) CPW-fed T-shaped slot antenna, and a dielectric resonator antenna.

4.1 Slot Dipole Antenna (Antenna 1)

Figure 1 shows the geometry of a CPW-fed slot dipole antenna on a dielectric substrate. The design vector was $\mathbf{u} = [W\ L]^T$, with the design variable space being defined by the center and size vectors $\mathbf{u}^0 = [7.5\ 39]^T$ mm and $\boldsymbol{\delta} = [2.5\ 11]^T$ mm

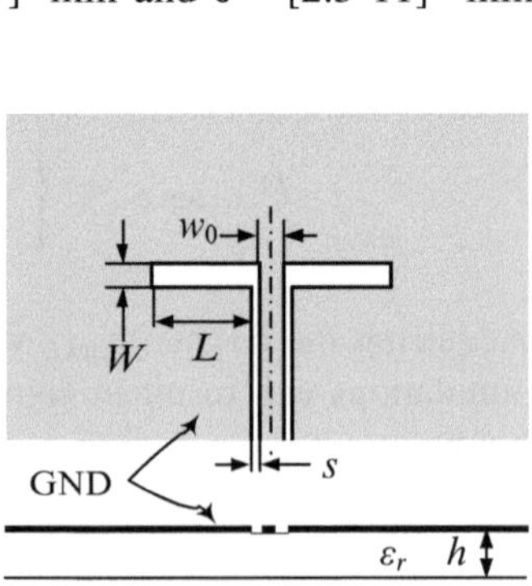

Fig. 1 Geometry of CPW-fed slot dipole antenna (Antenna 1). The ground plane (GND) is of infinite lateral extent

such that the variable ranges were $\mathbf{u}^0 \pm \boldsymbol{\delta}$ mm. Other dimensions and parameters were $w_0 = 4.0$ mm, $s = 0.5$ mm, $h = 1.6$ mm, and $\varepsilon_r = 4.4$. Of interest was S_{11} over the frequency band 2.0–2.7 GHz.

Training data input vectors for $\boldsymbol{R}_s$ were defined by randomly selecting 91 geometries from the input space using Latin hypercube sampling (LHS), with three frequencies per geometry uniformly randomly sampled from the above frequency band; in general each geometry had a different set of frequencies. The total number of training input vectors was $n = 91 \times 3 = 273$; they had the form $\{\mathbf{x}_i = [\mathbf{u}_i^T\ f_{oi}]^T \mid i = 1, \ldots, n\}$, with f_{oi} a frequency value within the range of interest. Test input vectors were obtained from 100 new LHS geometries, with 71 equally spaced frequencies per geometry yielding $n_* = 7100$.

The above training input vectors were simulated using CST Microwave Studio [13] on a dual-core 2.33 GHz Intel CPU with 2 GB RAM at a fine mesh density (~130,000 mesh cells, simulation time 12 min) resulting in the full high-fidelity training data set D_{fine}, and at a coarse mesh density (~5000 mesh cells, simulation time 30 s) to give D_{coarse}. (We refer to the CST simulations at the fine mesh density as the high-fidelity model $\boldsymbol{R}_f$, and the simulations at the coarse density as the low-fidelity model $\boldsymbol{R}_c$). The test inputs were only simulated at the fine mesh density, yielding the test data set D_{test} used to evaluate the predictions of $\boldsymbol{R}_s$.

For the first stage of our method we constructed training sets D_{aux} by randomly selecting n_{aux} data points from D_{fine}, and then trained a model $\boldsymbol{R}_{\text{aux}}$ as described in Section 3.1 that was used to estimate the rest of the high-fidelity target values in D_{fine}. This was repeated for $n_{\text{aux}}/n \times 100\ \% \in \{70\ \%, 60\ \%, 50\ \%, 40\ \%, 30\ \%, 20\ \%, 10\ \%\}$. Table 1 gives the predictive errors of $\boldsymbol{R}_{\text{aux}}$ on the remaining n - n_{aux} training points in D_{fine}. The results indicate that the remaining training targets could be predicted with reasonable accuracy by $\boldsymbol{R}_{\text{aux}}$, likely due to the fact that values of Re/Im$\{S_{11}\}_{\text{coarse},k}$ in the training input vectors (13) were well correlated with the targets Re/Im$\{S_{11}\}_{\text{fine},k}$. Fig. 2a gives, for a sample geometry, fine and coarse responses of Re$\{S_{11}\}$ and Im$\{S_{11}\}$ against frequency that are indicative of typical discrepancies for Antenna 1. It may be noted that the overall "shapes" of the

Table 1 Predictive errors[a] of auxiliary antenna models $\boldsymbol{R}_{\text{aux}}$ on remaining n - n_{aux} fine training data points

	RMSE [%]					
	Antenna 1 ($n = 273$)		Antenna 2 ($n = 3348$)		Antenna 3 ($n = 1600$)	
$n_{\text{aux}}/n \times 100$ %	Re$\{S_{11}\}$	Im$\{S_{11}\}$	Re$\{S_{11}\}$	Im$\{S_{11}\}$	Re$\{S_{11}\}$	Im$\{S_{11}\}$
70	0.416	0.356	1.35	1.16	0.631	0.634
60	0.401	0.916	1.34	1.22	0.672	0.633
50	0.304	0.604	1.51	1.26	0.754	0.842
40	0.235	0.561	1.27	1.24	0.863	0.805
30	0.219	0.612	1.54	1.51	1.07	0.945
20	0.444	0.648	1.63	1.65	1.48	1.261
10	0.947	1.834	2.09	2.11	3.04	1.948

[a]Root mean square error normalized to the target range, expressed as percentage

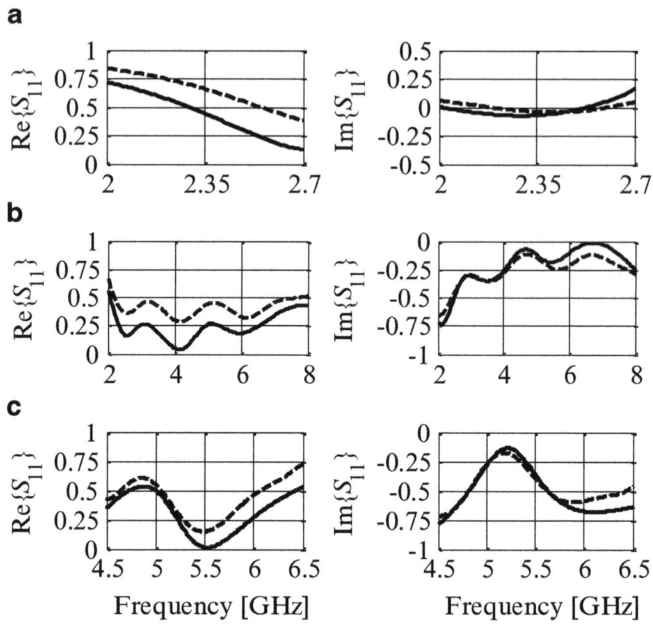

Fig. 2 Typical fine (*solid line*) and coarse (*dashed line*) responses for Re$\{S_{11}\}$ and Im$\{S_{11}\}$ against frequency for (**a**) Antenna 1: $\mathbf{x} = [7.9107\ 35.4954]^T$ mm, (**b**) Antenna 2: $\mathbf{x} = [37.5046\ 25.5961\ 4.0234\ 14.8285]^T$ mm, and (**c**) Antenna 3: $\mathbf{x} = [7.7187\ 14.8472\ 8.8390\ 0.0149\ 1.9376\ 8.1535\ 8.9098]^T$ mm

Table 2 Predictive errors of surrogate antenna models $\boldsymbol{R}_s$ on fine test data

	RMSE [%]					
	Antenna 1 ($n = 273$)		Antenna 2 ($n = 3348$)		Antenna 3 ($n = 1600$)	
$n_{aux}/n \times 100$ %	Re$\{S_{11}\}$	Im$\{S_{11}\}$	Re$\{S_{11}\}$	Im$\{S_{11}\}$	Re$\{S_{11}\}$	Im$\{S_{11}\}$
100 ($\boldsymbol{R}_{s,\text{full}}$)	1.39	1.29	1.95	1.78	0.854	0.753
30	1.27	1.32	2.36	2.28	1.29	1.08
20	1.34	1.31	2.56	2.46	1.86	1.45
10	1.55	1.38	2.81	2.65	3.85	2.35

coarse and fine model responses as functions of frequency are similar—the major misalignment relates to the level of the responses. This indicates relatively good correlation between both models, giving support for the notion of exploiting this correlation for coarse model enhancement even if a limited number of fine model training data points are used.

Next we constructed "approximate" fine training sets $D_{\text{fine,approx}}$ (cf. Section 3.1) for cases where the savings in finely discretised training points were highly significant, i.e., $n_{aux}/n \times 100\ \% \in \{30\ \%, 20\ \%, 10\ \%\}$, and trained GPR models $\boldsymbol{R}_s$ in each case. The predictive errors of these models on the test data set D_{test} are listed in Table 2. For comparison, the predictive error for the case where the full

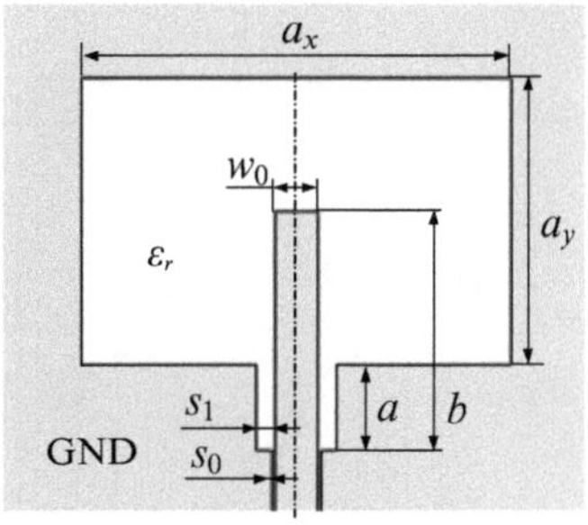

Fig. 3 Geometry of UWB CPW-fed T-shaped slot antenna (Antenna 2; top view). The ground plane (GND) is of infinite lateral extent

D_{fine} was used as training data ($n_{\text{aux}}/n \times 100\ \% = 100\ \%$) is also given—we refer to this model as $\boldsymbol{R}_{s,\text{full}}$. Predictive accuracies were good, especially given the relatively small proportions of high-fidelity data present in the "approximate" fine training data sets.

4.2 UWB T-Shaped Slot Antenna (Antenna 2)

Fig. 3 shows the layout of a CPW-fed antenna with T-shaped slot [14]. The design vector was $\mathbf{u} = [a_x\ a_y\ a\ b]^T$, and the design space was delimited by center vector $\mathbf{u}^0 = [40\ 27.5\ 7\ 20]^T$ mm and size vector $\boldsymbol{\delta} = [5\ 7.5\ 5\ 10]^T$ mm (other dimensions were $w_0 = 4.0$ mm, $s_0 = 0.3$ mm, and $s_1 = 1.7$ mm; the dielectric substrate had height $h = 0.813$ mm and dielectric constant $\varepsilon_r = 3.38$). Of interest was the frequency band 2–8 GHz.

The training data represented 270 geometries obtained by LHS, with 12 randomly selected frequencies per geometry ($n = 3348$). Test data were made up of 50 new LHS geometries, with 121 frequencies that were equally spaced per geometry.

To obtain D_{fine} we simulated [13] the training input vectors at a fine mesh density (~2,962,000 mesh cells, simulation time 21 min); coarse mesh density simulations (~44,500 mesh cells, simulation time 20 s) yielded D_{coarse}.

Models $\boldsymbol{R}_{\text{aux}}$ and $\boldsymbol{R}_s$ were set up in a manner similar to that described in Section 4.1, and Tables 1 and 2 give the relevant predictive errors, again showing that both types of models had good predictive capabilities. In particular, predictive accuracies for the $\boldsymbol{R}_s$ models appeared to be good given the relatively small fractions of high-fidelity data present in the "approximate" fine training data sets. Fig. 2b gives fine and coarse responses of Re$\{S_{11}\}$ and Im$\{S_{11}\}$ against frequency for a sample geometry that are representative of the fine/coarse discrepancies observed for this antenna.

4.3 Dielectric Resonator Antenna (Antenna 3)

Fig. 4 shows the antenna geometry [15]. The design vector was $\mathbf{u} = [a_x\ a_y\ a_z\ a_c\ u_s\ w_s\ y_s]^T$, where a_x, a_y, and a_z are dimensions of the dielectric resonator (DR) brick,

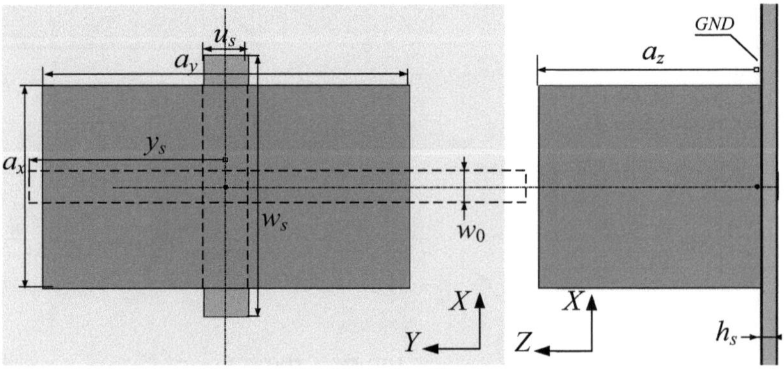

Fig. 4 Geometry of dielectric resonator antenna (Antenna 3): (**a**) top and (**b**) side views

a_c stands for the shift of the DR center in the Y-direction relative to the slot center, u_s is the slot width, w_s is the slot length, and y_s is the length of the microstrip stub. The relative dielectric constant and loss tangent of the dielectric resonator were 10 and 0.0001. The substrate consisted of 0.5 mm thick RO4003C laminate [15], and the metallization of the trace and ground was 50 μm copper. The design variable space was described by the center vector $\mathbf{u}^0 = [8\ 14\ 8\ 1\ 2\ 9\ 8]^T$ mm and size vector $\boldsymbol{\delta} = [1\ \ 1\ \ 1\ \ 1\ \ 1\ \ 1]^T$ mm; other dimensions were $w_0 = 1.15$ mm and $h_s = 0.5$ mm. Simulations included the frequency band 4.5–6.5 GHz.

Training data were obtained from 400 LHS geometries, and there were four randomly selected frequencies per geometry ($n = 1600$). Test data were constituted by 50 new LHS geometries with 121 equally spaced frequencies per geometry.

We simulated the training input vectors in CST Microwave Studio at a fine mesh density (~500,000 mesh cells, simulation time 12.5 min) to obtain D_{fine}, and at a coarse mesh density (~15,000 mesh cells, simulation time 30 s) to obtain D_{coarse}.

$\boldsymbol{R}_{\text{aux}}$ and $\boldsymbol{R}_s$ were set up in a manner similar to those for Antennas 1 and 2. Tables 1 and 2 list the relevant predictive errors. On the whole these were good given that this antenna had seven design variables. Fig. 2c gives fine and coarse responses of Re$\{S_{11}\}$ and Im$\{S_{11}\}$ against frequency for a sample geometry that are representative of the fine/coarse disagreement for this antenna.

4.4 Application Examples: Antenna Optimization

We apply $\boldsymbol{R}_s$ within a space-mapping algorithm aimed at optimizing the input characteristics of the antenna structures considered in Sections 4.1–4.3. This is intended as a way to validate the proposed modeling methodology where our GPR surrogates $\boldsymbol{R}_s$ are trained using the "approximate" high-fidelity training set $D_{\text{fine,approx}}$ rather than the "original" training set D_{fine} obtained in full through direct simulations. Optimization results are compared to results obtained by using $\boldsymbol{R}_{s\text{,full}}$ (i.e., $\boldsymbol{R}_s$ trained on D_{fine}).

We note that the GPR surrogates considered in this chapter are intended to be multiple-purpose library models. Antenna optimization with respect to various sets of design specifications is one example of a typical application task. Another could be yield-driven optimization or statistical analysis.

Here, we consider antenna optimization where the initial design is the center of the region of interest $\mathbf{x}^{(0)}$. The design process starts from directly optimizing the GPR model. Because of modeling error that is nonzero, an iterative design refinement procedure is used based on space-mapping technology [16, 24, 25]

$$\mathbf{x}^{(i+1)} = \arg\min_{\mathbf{x}} U\left(\mathbf{R}_{su}^{(i)}(\mathbf{x})\right) \tag{16}$$

where $\boldsymbol{R}_{su}{}^{(i)}$ is a surrogate model obtained by output space mapping [16]. The surrogate model setup is carried out using an evaluation of $\boldsymbol{R}_f$ at $\mathbf{x}^{(i)}$. U implements design specifications. For conciseness considerations, we simply use the symbol $\boldsymbol{R}_{co}$ below to denote either of $\boldsymbol{R}_{s.full}$ or $\boldsymbol{R}_s$, which can be considered the "coarse" models in the space-mapping context. The surrogate model is then defined as

$$\mathbf{R}_{su}^{(i)}(\mathbf{x}) = \mathbf{R}_{co}(\mathbf{x}) + \mathbf{d}^{(i)} \tag{17}$$

with

$$\mathbf{d}^{(i)} = \mathbf{R}_f\left(\mathbf{x}^{(i)}\right) - \mathbf{R}_{co}\left(\mathbf{x}^{(i)}\right) \tag{18}$$

The additive correction term $\boldsymbol{d}^{(i)}$ is computed so that zero-order consistency (i.e., $\boldsymbol{R}_{su}{}^{(i)}(\mathbf{x}^{(i)}) = \boldsymbol{R}_f(\mathbf{x}^{(i)})$) between the surrogate and the high-fidelity model $\boldsymbol{R}_f$ [17] is ensured at the current design $\mathrm{x}^{(i)}$. In practice, because of the good initial accuracy of the GPR surrogates, one or two iterations of the algorithm (16) usually suffices with respect to yield an optimized design. It should be noted that the cost of each iteration (16) in effect corresponds to a single evaluation of the high-fidelity model (the expense of optimizing the surrogate itself can be neglected as compared to the evaluation of the high-fidelity model).

Fig. 5 shows, for all three antennas, the responses of models $\boldsymbol{R}_{s,\mathrm{full}}$ and $\boldsymbol{R}_f$ (the latter being direct high-fidelity CST simulations) at the initial designs, as well as the response of $\boldsymbol{R}_f$ at the final designs. Fig. 6 likewise gives the responses of the GPR models $\boldsymbol{R}_s$ trained on the "approximate" high-fidelity data set $D_{\mathrm{fine,approx}}$ (here with $n_{\mathrm{aux}}/n \times 100\ \% = 20\ \%$) and the $\boldsymbol{R}_f$ model responses at $\boldsymbol{x}^{(0)}$; and the response of $\boldsymbol{R}_f$ at the final designs. Table 3 provides a summary of the corresponding numerical results. These indicate that the design quality and cost expressed in terms of number of $\boldsymbol{R}_f$ evaluations are very similar for the GPR models obtained using the original and approximate high-fidelity training data sets. In the case of Antennas 1 and 2, the optimization cost corresponds to three $\boldsymbol{R}_f$ evaluations. For Antenna 3, $\boldsymbol{R}_{s,\mathrm{full}}$ exhibits better performance with only one refinement iteration necessary to yield an optimized design (compared to three iterations for $\boldsymbol{R}_s$). Table 3 also shows the optimization results with the $\boldsymbol{R}_s$ models trained on $D_{\mathrm{fine,approx}}$ where only 10 % of the

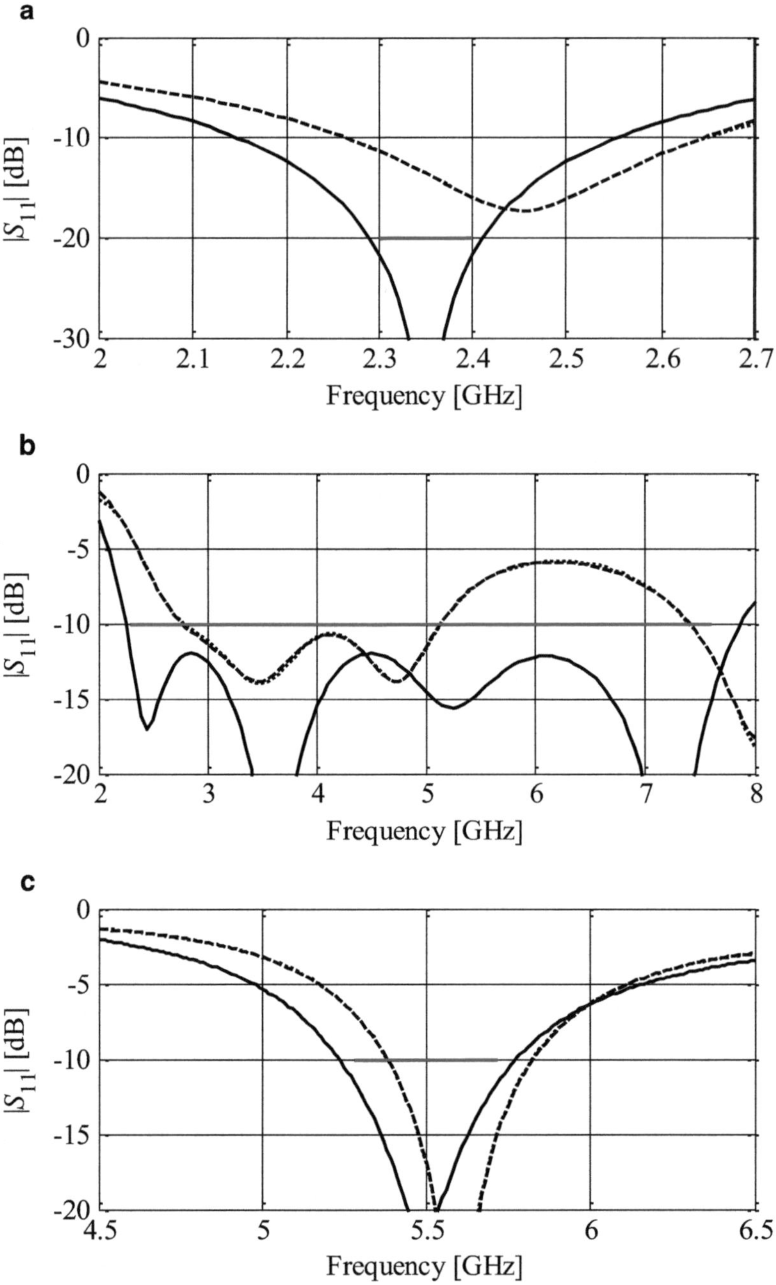

Fig. 5 Optimization results: responses of $\boldsymbol{R}_s$, full (*dotted line*) and $\boldsymbol{R}_f$ (*dashed line*) at the initial design, and $\boldsymbol{R}_f$ at the optimized design (*solid line*) for (**a**) Antenna 1, (**b**) Antenna 2, and (**c**) Antenna 3. Design specifications marked with *horizontal solid line*. GPR model responses (computed from separate models for Re$\{S_{11}\}$ and Im$\{S_{11}\}$) are hardly distinguishable from the corresponding high-fidelity simulation ($\boldsymbol{R}_f$) response

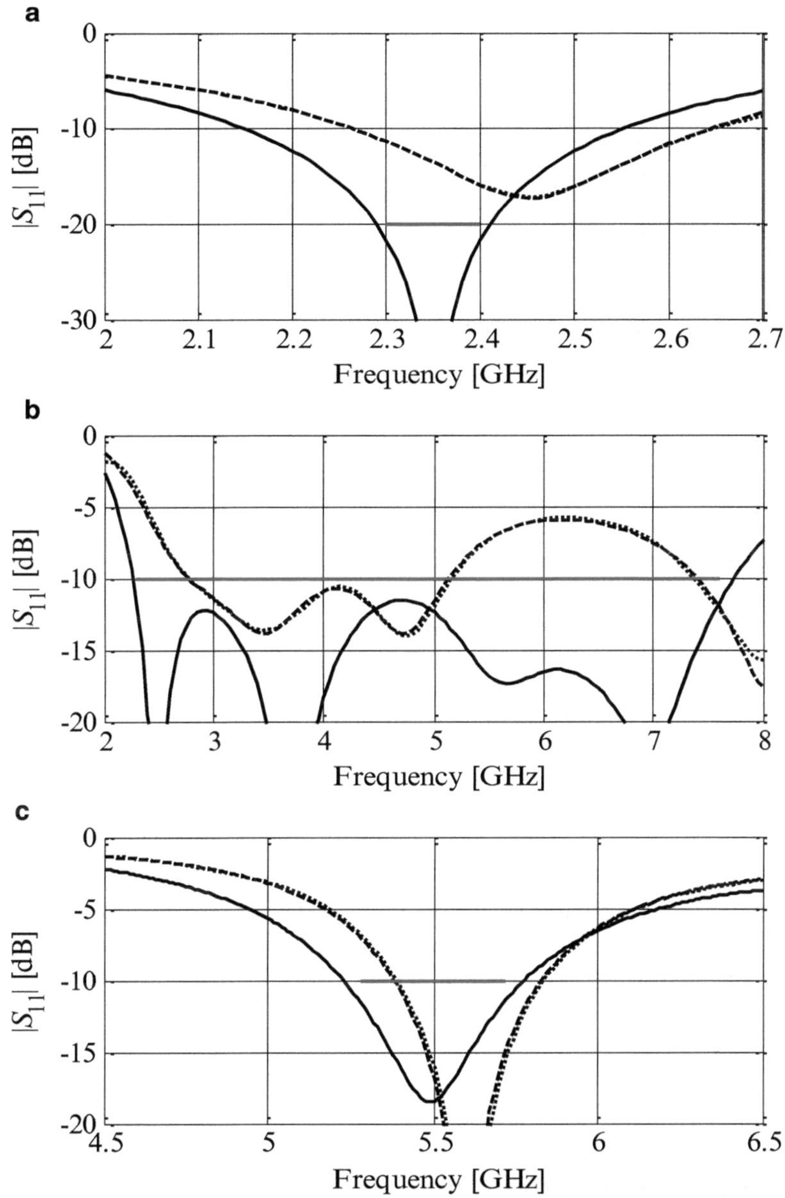

Fig. 6 Optimization results: responses of $\boldsymbol{R}_s$ (*dotted line*) and $\boldsymbol{R}_f$ (*dashed line*) at the initial design, and $\boldsymbol{R}_f$ at the optimized design (*solid line*) for (**a**) Antenna 1, (**b**) Antenna 2, and (**c**) Antenna 3. Design specifications marked with *horizontal solid line*. Note that the GPR model responses are very close to the high-fidelity model ($\boldsymbol{R}_f$) response. The $\boldsymbol{R}_s$ models represented here were trained on $D_{\text{fine,approx}}$ utilizing 20 % data points simulated at high fidelity

Table 3 Antenna optimization results

Antenna	Model	max $\|S_{11}\|$ at final design (dB)[a]	Optimization cost[b]
1	$\boldsymbol{R}_{s,full}$	−21.7	3
	$\boldsymbol{R}_s$ (20 %)[c]	−21.6	3
	$\boldsymbol{R}_s$ (10 %)[c]	−21.5	3
2	$\boldsymbol{R}_{s,full}$	−12.0	3
	$\boldsymbol{R}_s$ (20 %)[c]	−11.5	3
	$\boldsymbol{R}_s$ (10 %)[c]	−11.4	3
3	$\boldsymbol{R}_{s,full}$	−11.5	2
	$\boldsymbol{R}_s$ (20 %)[c]	−11.3	4
	$\boldsymbol{R}_s$ (10 %)[c]	−11.2	4

[a]max $|S_{11}|$ at the frequency band of interest: 2.3–2.4 GHz (Antenna 1), 2.3–7.6 GHz (Antenna 2), and 5.28–5.72 GHz (Antenna 3)
[b]Number of $\boldsymbol{R}_f$ evaluations including evaluation at the initial design
[c]20 % refers to the model trained on $D_{\text{fine,approx}}$ utilizing 20 % data points that are actually simulated at high fidelity (10 % accordingly)

data were high-fidelity-simulated points. Regardless of the fact that these models are generally less accurate than the 20 % versions (cf. Table 2), they apparently are still reliable enough—in combination with the particular surrogate-based optimization technique (16–18)—to optimize our antenna structures. The quality of the final designs as well as the corresponding design costs therefore is essentially the same for the 10- and 20 % high-fidelity-simulated GPR models.

For the sake of comparison, we also optimized the three antennas using a conventional (i.e., not surrogate-based) method, namely a state-of-the-art pattern-search algorithm [18, 19] that directly relied on high-fidelity full-wave simulations for its objective function evaluations. Maximum $|S_{11}|$ values at the final designs obtained for Antennas 1, 2, and 3 (−21.6, −11.6, and −10.4 dB, respectively) were similar to those obtained using our GPR models and the above space-mapping procedure—however, the computational expense for the conventional optimization was at least an order of magnitude larger (i.e., 40, 148, and 117 $\boldsymbol{R}_f$ evaluations for Antennas 1, 2, and 3, respectively). This confirms that fast and accurate surrogates are indispensable in the antenna design process, particularly if they can be set up at relatively low computational cost.

5 Modeling and Optimization of Filters Using Two-Stage GPR

In this section, numerical verification of the two-stage GPR modeling technique using two examples of microstrip filters is presented. As before, the GPR surrogates are also applied for design optimization purposes.

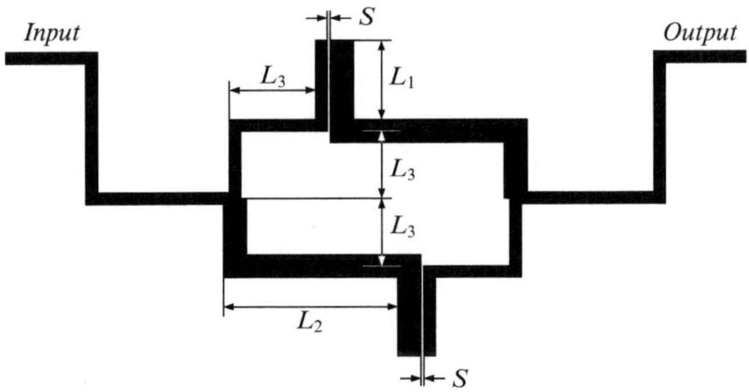

Fig. 7 CCDBR filter: geometry [20]

5.1 Capacitively Coupled Dual-Behavior Resonator (CCDBR) Microstrip Bandpass Filter (Filter 1)

The first example is the second-order capacitively coupled dual-behavior resonator (CCDBR) microstrip filter [20] shown in Fig. 7. The filter structure is described by three design variables as $\mathbf{u} = [L_1\, L_2\, L_3]^T$. The microstrip line widths are 0.25 and 0.5 mm, whereas the line spacing $S = 0.05$ mm. The substrate parameters are $h = 0.254$ mm and $\varepsilon_r = 9.9$. The GPR surrogate model is set up in the interval $[\mathbf{u}^0 - \boldsymbol{\delta},\ \mathbf{u}^0 + \boldsymbol{\delta}]$ with $\mathbf{u}^0 = [3\ 5\ 1.5]^T$ mm and $\boldsymbol{\delta} = [1\ 1\ 0.5]^T$ mm. The objective is to model the transmission coefficient $|S_{21}|$ for the frequency range of 2–6 GHz. The rational quadratic covariance function (8) was used during both stages of our two-stage modeling approach.

The training data input vectors for creating the two-stage GPR surrogate model $\boldsymbol{R}_s$ were allocated using LHS [21]. Twelve random frequencies were associated with each training geometry within the frequency range of interest. Consequently, a different set of frequencies was effectively assigned for each geometry. Furthermore, the total number of training vectors was $n = 600 \times 12 = 7200$. The training input vectors had the form $\{\mathbf{x}_i = [\mathbf{u}_i^T\, f_{oi}]^T \mid i = 1, \ldots, n\}$ with f_{oi} being the frequency values. The surrogate models were tested using a split-sample method with 50 separate testing geometries, also obtained with LHS, however, using 81 frequencies per geometry, distributed uniformly on the frequency scale ($n^* = 4050$).

The training data set D_{fine} was acquired by means of FEKO [22] simulations of the high-fidelity model $\boldsymbol{R}_f$. It consists of n input–output pairs, $\{(\mathbf{x}_i, y_i) \mid i = 1, \ldots, n\}$, with $\mathbf{x}_i = [\mathbf{u}_i^T\, f_{oi}]^T = [L_1\, L_2\, L_3\, f_{oi}]^T$, and $y_i = \text{Re}\{S_{21}\}_{\text{fine},i}$ or $\text{Im}\{S_{21}\}_{\text{fine},i}$. Similarly, the training set D_{coarse} was acquired through coarse-discretization FEKO simulations of the low-fidelity model $\boldsymbol{R}_c$. Total mesh numbers for $\boldsymbol{R}_f$ and $\boldsymbol{R}_c$ were 614 (evaluation time 6 seconds per frequency) and 130 (evaluation time 0.3 seconds per frequency), respectively. The testing data set D_{test} was obtained—for the sake of evaluating the predictive power of the surrogate—from fine-discretization EM simulations.

Table 4 Predictive errors[a] of auxiliary filter models $\boldsymbol{R}_{\text{aux}}$ on remaining n - n_{aux} fine training data points

	RMSE [%]				
	Filter 1		Filter 2		Filter 3
$n_{\text{aux}}/n \times 100$ %	Re$\{S_{21}\}$	Im$\{S_{21}\}$	Re$\{S_{21}\}$	Im$\{S_{21}\}$	$\lvert S_{21}\rvert$
40	1.45	1.47	1.79	2.03	3.17
30	1.77	1.63	1.73	1.98	3.43
20	1.87	1.97	2.74	2.23	3.75

[a]Normalized root mean square error (RMSE), expressed as a percentage of the target value range

The two-stage GPR surrogate model is constructed separately for Re$\{S_{21}\}$ and Im$\{S_{21}\}$, however, the description below is—for the sake of brevity—only provided for Re$\{S_{21}\}$. The first stage of the process involves construction of the training set D_{aux}, which is realized by randomly selecting n_{aux} data points from the original set D_{fine}. The auxiliary surrogate $\boldsymbol{R}_{\text{aux}}$ is then set up as described in Section 3.1. More specifically, the training set D_{aux} consists of n_{aux} input–output pairs $\{(\mathbf{x}_{aux,k}, y_{\text{fine},k}) \mid k = 1, \ldots, n_{aux}\}$, with $\mathbf{x}_{aux,k} = [\mathbf{u}_k{}^T\ f_{ok}\ \text{Re}\{S_{21}\}_{\text{coarse},k}]^T = [L_1\, L_2\, L_3\ f_{ok}$ $\text{Re}\{S_{21}\}_{\text{coarse},k}]^T$ and $y_{\text{fine},k} = \text{Re}\{S_{21}\}_{\text{fine},k}$. Upon accomplishing the model training, it is used to estimate the rest of the high-fidelity target values in D_{fine} by finding the mean of the posterior distribution (cf. Eq. (15))—i.e., yielding $n - n_{aux}$ $\text{Re}\{S_{21}\}_{\text{pred}}$ values. These steps are repeated for $n_{\text{aux}}/n \times 100\ \% \in \{40\ \%, 30\ \%, 20\ \%\}$, and the predictive errors of $\boldsymbol{R}_{\text{aux}}$ (i.e., the root mean square values of the n - n_{aux} residuals of $\text{Re}\{S_{21}\}_{\text{pred}}$ and $\text{Re}\{S_{21}\}_{\text{fine}}$) are listed in Table 4 for each case.

In the second stage, the "approximate" fine training data sets $D_{\text{fine,approx}}$ (as described by Eq. (15), Section 3.1) corresponding to $n_{\text{aux}}/n \times 100\ \% \in \{40\ \%, 30\ \%, 20\ \%\}$ have been constructed, and the GPR models $\boldsymbol{R}_s$ have been trained on each set. The n-point "approximate" fine training set corresponding to a specific n_{aux} value was

$$D_{\text{fine,approx}} = \left\{ \begin{array}{c|c} \left([L_{1k}\ L_{2k}\ L_{3k}\ f_{ok}]\ ,\ Re\{S_{21}\}_{\text{fine},k}\right) & k = 1, \ldots, n_{\text{aux}} \\ \left([L_{1k}\ L_{2k}\ L_{3k}\ f_{ok}]\ ,\ Re\{S_{21}\}_{\text{pred},k}\right) & k = (n_{\text{aux}} + 1), \ldots, n \end{array} \right\}$$

where $n = 7200$ is the total number of the training points.

The predictive errors for the surrogates constructed for various n_{aux}/n are listed in Table 5. The test set, which was independent of the training set, was given by

$$D_{\text{test}} = \left\{ \left[L_{1j}\ L_{2j}\ L_{3j}\ f_{oj}\right]^T,\ Re\{S_{21}\}_{\text{fine},j} \ \middle|\ j = 1, \ldots, n^* \right\}$$

with $n^* = 4050$.

Apart from the fractional data sets ($n_{\text{aux}}/n \times 100\ \% < 100\ \%$), the predictive error for the case corresponding to the full D_{fine} utilized as the training data

Table 5 Predictive errors of surrogate filter models $\boldsymbol{R}_s$ on fine test data

	RMSE [%]				
	Filter 1		Filter 2		Filter 3
$n_{aux}/n \times 100$ %	Re{S_{21}}	Im{S_{21}}	Re{S_{21}}	Im{S_{21}}	$\|S_{21}\|$
100 ($\boldsymbol{R}_{s.\text{full}}$)[a]	3.11	3.22	2.56	2.93	3.92
40	3.14	3.31	3.12	3.37	4.12
30	3.18	3.38	2.99	3.33	4.22
20	3.24	3.36	3.54	3.53	4.39

[a] $\boldsymbol{R}_{s.\text{full}}$ is the benchmark that we compare to, i.e., standard GPR using D_{fine} as training data (the fine training data obtained in full via EM simulations)

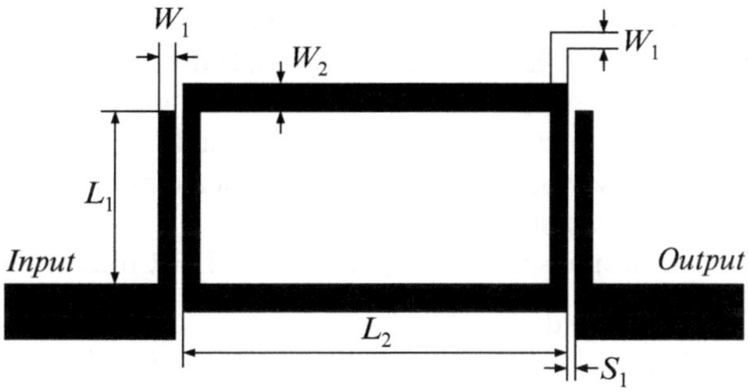

Fig. 8 Second-order ring resonator bandpass filter: geometry [23]

($n_{aux}/n \times 100\ \% = 100\ \%$) is also indicated. We refer to this model as $\boldsymbol{R}_{s,\text{full}}$. It can be observed that the predictive accuracies are good given the relatively small proportions of high-fidelity data present in the "approximate" fine training data sets.

5.2 *Open-Loop Ring Resonator (OLRR) Bandpass Filter (Filter 2)*

The second example is the second-order ring resonator bandpass filter [23] shown in Fig. 8. The structure is described by five design parameters $\mathbf{u} = [L_1\ L_2\ S_1\ W_1\ W_2]^T$. The substrate parameters are $h = 1.52$ mm and $\varepsilon_r = 4.32$. The region of interest for the surrogate model construction is $[\mathbf{u}^0 - \boldsymbol{\delta}, \mathbf{u}^0 + \boldsymbol{\delta}]$ with $\mathbf{u}^0 = [20\ 22\ 0.2\ 0.8\ 1.7]^T$ mm and $\boldsymbol{\delta} = [2\ 2\ 0.1\ 0.1\ 0.1]^T$ mm. The objective is to model the transmission coefficient $|S_{21}|$ for the frequency range of 1–3 GHz.

The overall setup for the models $\boldsymbol{R}_{\text{aux}}$ and $\boldsymbol{R}_s$ was similar to that used for Filter 1. However, the squared-exponential covariance function (7) was used in this case. Tables 4 and 5 show the predictive errors for the models. It can be concluded from these results that both types of models had good predictive capabilities. Similarly as for Filter 1, predictive accuracies of the final surrogate $\boldsymbol{R}_s$ are good, particularly given greatly reduced number of the high-fidelity training points compared to the "full" set.

The training data set selected for this example consists of 400 geometries allocated using LHS with six randomly selected frequencies per geometry. Thus, the total number of training points is $n = 2400$. A separate set of 50 random geometries (with 81 uniformly distributed frequencies per geometry) is utilized in the testing stage. The filter models are evaluated in FEKO [22] using the following setup: 828 mesh cells for $\boldsymbol{R}_f$ (simulation time 8 seconds per frequency), and 64 mesh cells for $\boldsymbol{R}_c$ (simulation time about 0.1 seconds per frequency).

5.3 Application Examples: Filter Optimization

The two filter structures considered in Section 5.2 were optimized—for the sake of additional verification—using the final GPR surrogate models $\boldsymbol{R}_s$. Note that the GPR surrogates considered in this chapter are intended to be multiple-purpose library models. In particular, such models could be utilized for filter optimization with respect to various sets of design specifications or robust (yield-driven) optimization as well as statistical analysis.

The design specifications for the CCDBR filter (Filter 1) are the following:

- $|S_{21}| \geq -3$ dB for 3.8–4.2 GHz
- $|S_{21}| \leq -20$ dB for 2–3.2 GHz and for 4.8–6 GHz

The initial design is $\mathbf{u}_{init} = [3.5\ 4.5\ 1.5]^T$ mm.

The design specifications for the second-order ring resonator filter (Filter 2) are:

- $|S_{21}| \geq -1$ dB for 1.8–2.2 GHz
- $|S_{21}| \leq -20$ dB for 1–1.55 GHz and for 2.45–3 GHz

The initial design is $\mathbf{u}_{init} = [18.0\ 22.0\ 0.2\ 0.8\ 1.7]^T$ mm.

The first stage of the process is to optimize the surrogate model $\boldsymbol{R}_s$. The second stage is an iterative design refinement procedure, necessary due to a nonzero error of $\boldsymbol{R}_s$. This stage is executed using the space-mapping algorithm described in Section 4.4 (Eqs. (16–18)). Due to a good initial accuracy of the GPR surrogates, one or two iterations of the algorithm (16) are usually sufficient to yield an optimized design. The optimization results are presented in Table 6, as well as in Figs. 9 and 10.

Figure 9 shows, for Filters 1 and 2, the responses of models $\boldsymbol{R}_{s,\text{full}}$ and $\boldsymbol{R}_f$ (i.e., direct high-fidelity FEKO simulations) at the initial designs, and the high-fidelity model response at the final designs. Similarly, Fig. 10 shows the GPR surrogate model responses trained on the "approximate" high-fidelity data set $D_{\text{fine,approx}}$ (here with $n_{\text{aux}}/n \times 100\ \% = 20\ \%$) and the $\boldsymbol{R}_f$ model responses at $\boldsymbol{x}^{(0)}$, and the response of $\boldsymbol{R}_f$ at the final designs.

Table 6 gathers the numerical results. It should be emphasized that the differences between the design quality and cost (the latter expressed in terms of number of evaluations of the high-fidelity model $\boldsymbol{R}_f$) are very small for the GPR models obtained using the original and approximate high-fidelity training data sets. Furthermore, the average design cost is about three evaluations of the high-fidelity model and it is similar in all cases, regardless of the n_{aux}/n ratio.

Table 6 Filter optimization results

Filter	Model	Minimax specification error at final design (dB)[a]	Optimization cost[b]
1	$\boldsymbol{R}_{s,full}$	−2.4	3
	$\boldsymbol{R}_s$ (40 %)[c]	−2.3	4
	$\boldsymbol{R}_s$ (30 %)[c]	−2.2	3
	$\boldsymbol{R}_s$ (20 %)[c]	−2.4	3
2	$\boldsymbol{R}_{s,full}$	−0.0	4
	$\boldsymbol{R}_s$ (40 %)[c]	−0.1	4
	$\boldsymbol{R}_s$ (30 %)[c]	−0.0	4
	$\boldsymbol{R}_s$ (20 %)[c]	−0.1	4

[a]Maximum violation of $|S_{21}|$ specifications at the frequency bands of interest
[b]Number of evaluations of the high-fidelity model $\boldsymbol{R}_f$, including evaluation at the initial design
[c]40 % refers to the model trained on $D_{\text{fine,approx}}$ utilizing 40 % data points that are actually simulated at high fidelity (30 and 20 % accordingly)

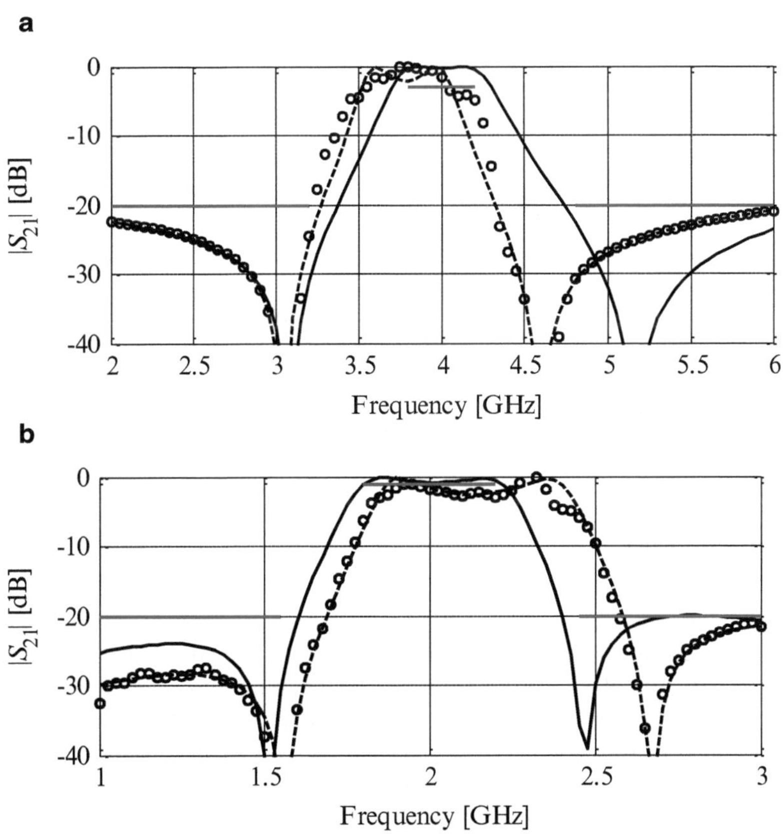

Fig. 9 Optimization results for the three filter structures: responses of $\boldsymbol{R}_{s,\text{full}}$ (*open circle*) and $\boldsymbol{R}_f$ (*dashed line*) at the initial design, and $\boldsymbol{R}_f$ at the optimized design (*solid line*) for (**a**) Filter 1, (**b**) Filter 2. *Horizontal solid lines* denote design specifications

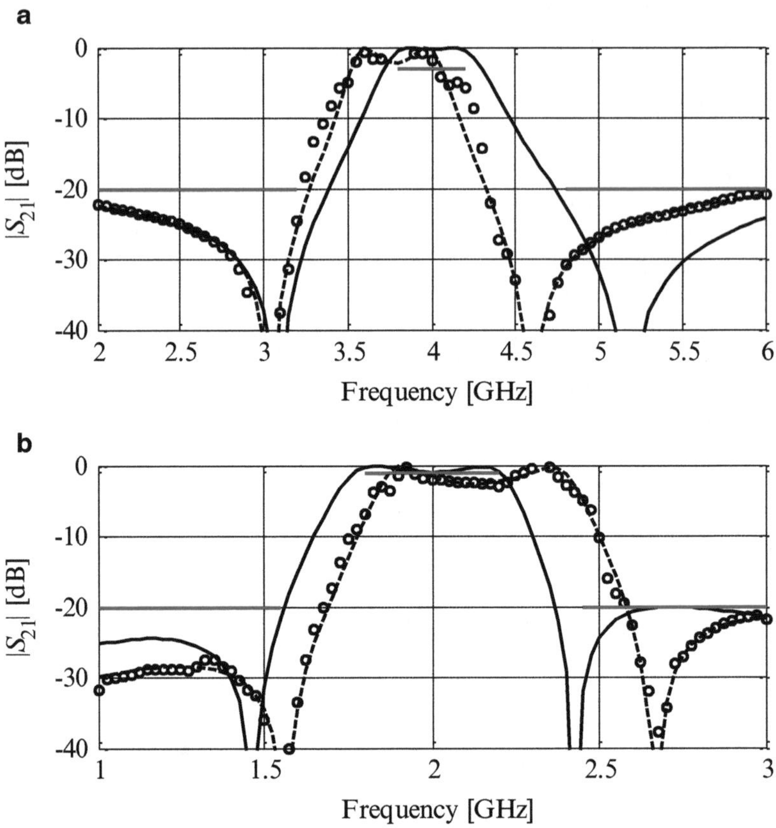

Fig. 10 Optimization results for the three filter structures: responses of $\boldsymbol{R}_s$ (*open circle*) and $\boldsymbol{R}_f$ (*dashed line*) at the initial design, and $\boldsymbol{R}_f$ at the optimized design (*solid line*) for (**a**) Filter 1, (**b**) Filter 2. *Horizontal lines* denote design specifications. The $\boldsymbol{R}_s$ models represented here were trained on $D_{\text{fine,approx}}$ utilizing 20 % data points simulated at high fidelity

6 Conclusions

In the chapter, a two-stage methodology for Gaussian Process modeling of computational electromagnetic (EM) simulation models has been presented. The approach discussed here involves variable-fidelity EM simulations. The key idea is to exploit—in the first modeling stage—the knowledge embedded in the low-fidelity model at hand to set up a mapping between the EM models of different fidelity. This allows us to substantially reduce the number of actual high-fidelity simulations that need to be performed, without compromising the predictive power of the final surrogate. The operation and performance of the two-stage modeling process has been demonstrated using three examples of antenna structures and two examples of microstrip filters. This comprehensive verification confirmed that satisfactory results can be obtained even if the "approximate" high-fidelity training set contains only 10–20 % targets obtained using fine-discretization simulations (the rest being predicted by the auxiliary model in the first stage of the procedure). For the sake of an additional verification, the two-stage GPR model has been shown to be perfectly usable in a design/optimization context.

References

1. Haupt, R.L.: Antenna design with a mixed integer genetic algorithm. IEEE Trans. Antennas Propag. **55**(3), 577–582 (2007)
2. Pantoja, M.F., Meincke, P., Bretones, A.R.: A hybrid genetic algorithm space-mapping tool for the optimization of antennas. IEEE Trans. Antennas Propag. **55**(3), 777–781 (2007)
3. Rasmussen, C.E., Williams, C.K.I.: Gaussian Processes for Machine Learning. MIT Press, Cambridge, MA (2006)
4. De Villiers, J.P., Jacobs, J.P.: Gaussian process modeling of CPW-fed slot antennas. Prog. Electromagn. Res. **98**, 233–249 (2009)
5. Jacobs, J.P., De Villiers, J.P.: Gaussian-process-regression-based design of ultrawide-band and dual-band CPW-fed slot antennas. J. Electromagn. Waves Appl. **24**, 1763–1772 (2010)
6. Angiulli, G., Cacciola, M., Versaci, M.: Microwave devices and antennas modelling by support vector regression machines. IEEE Trans. Magn. **43**, 1589–1592 (2007)
7. Jacobs, J.P.: Bayesian Support Vector Regression with Automatic Relevance Determination Kernel for Modeling of Antenna Input Characteristics. IEEE Trans. Antennas Propag. **60**(4), 2114–2118 (2012)
8. Jacobs, J.P., Koziel, S.: Two-stage framework for efficient Gaussian process modeling of antenna input characteristics. IEEE Trans. Antennas Propag. **62**(2), 706–713 (2014)
9. Jacobs, J.P., Koziel, S.: Reduced-cost microwave filter modeling using a two-stage Gaussian process regression approach. Int. J. RF Microw. Comput. Aided Eng. (2014). doi: 10.1002/mmce.20880
10. Devabhaktuni, V.K., Yagoub, M.C.E., Zhang, Q.J.: A robust algorithm for automatic development of neural network models for microwave applications. IEEE Trans. Microw. Theory Tech. **49**, 2282–2291 (2001)
11. Couckuyt, I., Declercq, F., Dhaene, T., Rogier, H., Knockaert, L.: Surrogate-based infill optimization applied to electromagnetic problems. Int. J. RF Microw. CAE **20**(5), 492–501 (2010)
12. Jacobs, J.P., Koziel, S., Ogurtsov, S.: Computationally efficient multi-fidelity Bayesian support vector regression modeling of planar antenna input characteristics. IEEE Trans. Antennas Propag. **61**(2), 980–984 (2013)
13. CST Microwave Studio, ver. 2011. CST AG, Darmstadt (2011)
14. Jiao, J.-J., Zhao, G., Zhang, F.-S., Yuan, H.-W., Jiao, Y.-C.: A broadband CPW-fed T-shape slot antenna. Prog. Electromagn. Res. **76**, 237–242 (2007)
15. Petosa, A.: Dielectric Resonator Antenna Handbook. Artech House, Norwood (2007)
16. Koziel, S., Bandler, J.W., Madsen, K.: A space mapping framework for engineering optimization: theory and implementation. IEEE Trans. Microw. Theory Tech. **54**(10), 3721–3730 (2006)
17. Alexandrov, N.M., Lewis, R.M.: An overview of first-order model management for engineering optimization. Optim. Eng. **2**(4), 413–430 (2001)
18. Kolda, T.G., Lewis, R.M., Torczon, V.: Optimization by direct search: new perspectives on some classical and modern methods. SIAM Rev. **45**(3), 385–482 (2003)

19. Koziel, S.: Multi-fidelity multi-grid design optimization of planar microwave structures with Sonnet. In: International Review of Progress in Applied Computational Electromagnetics, Tampere, Finland, pp. 719–724, 26–29 April 2010
20. Manchec, A., Quendo, C., Favennec, J.-F., Rius, E., Person, C.: Synthesis of capacitive-coupled dual-behavior resonator (CCDBR) filters. IEEE Trans. Microw. Theory Tech. **54**(6), 2346–2355 (2006)
21. Beachkofski, B., Grandhi, R.: Improved distributed hypercube sampling. In: American Institute of Aeronautics and Astronautics, Paper AIAA 2002–1274, 2002
22. FEKO: Suite 5.3. EM Software & Systems-S.A. (Pty) Ltd, Stellenbosch (2008)
23. Salleh, M.H.M., Prigent, G., Pigaglio, O., Crampagne, R.: Quarter-wavelength side-coupled ring resonator for bandpass filters. IEEE Trans. Microw. Theory Tech. **56**(1), 156–162 (2008)
24. Yelten, M.B., Zhu, T., Koziel, S., Franzon, P.D., Steer, M.B.: Demystifying surrogate modeling for circuits and systems. IEEE Circuits Syst. Mag. **12**(1), 45–63 (2012)
25. Cheng, Q.S., Bandler, J.W., Koziel, S., Bakr, M.H., Ogurtsov, S.: The state of the art of microwave CAD: EM-based optimization and modeling. Int. J. RF Microw. Comput. Aided Eng. **20**(5), 475–491 (2010)

Efficient Reconfigurable Microstrip Patch Antenna Modeling Exploiting Knowledge Based Artificial Neural Networks

Murat Simsek and Ashrf Aoad

Abstract Artificial neural network (ANN) is widely used for modeling and optimization in antenna design problems. It is a very convenient alternative for using computationally intensive 3D-Electromagnetic (EM) simulation in design. The reconfigurable microstrip patch antennas have been considered to ensure operational frequencies for different kind of purposes. ANN is used for modeling of antenna design problems to obtain a surrogate based model instead of a computationally intensive 3D-EM simulation. Further improvement in modeling, a prior knowledge about the problem such as an empirical formula, an equivalent circuit model, and a semi-analytical equation is directly embedded in ANN structure through a knowledge based modeling technique. Knowledge based techniques are developed to improve some properties of conventional ANN modeling such as accuracy and data requirement. All these improvements ensure better accuracy compared to conventional ANN modeling. The necessary knowledge can be obtained by the coarse model which is a complex 3D-EM simulation in terms of grid size selection. Knowledge based techniques can improve the performance of conventional ANN through the guidance of the coarse model. As long as the coarse model approximates to the computationally intensive 3D-EM simulation, the performance of the knowledge based surrogate model can converge to the design targets. The efficiency of modeling strategies is demonstrated by a reconfigurable 5-fingers microstrip patch antenna. The antenna has four modes of operation, which are controlled by two PIN diode switches with ON/OFF states, and it resonates at multiple frequencies between 1 and 7 GHz. The number of training data is changed in terms of selected parameters from the design space. Three different sets are used to show modeling performance according to the size of training data. The simulation results show that knowledge based neural networks ensure considerable savings in computational costs as compared to the computationally intensive 3D-EM simulation while maintaining the accuracy of the fine model.

M. Simsek (✉) • A. Aoad
Faculty of Aeronautics and Astronautics, Istanbul Technical University, Istanbul, Turkey
e-mail: simsekmu@itu.edu.tr

S. Koziel et al. (eds.), *Simulation-Driven Modeling and Optimization*,
Springer Proceedings in Mathematics & Statistics 153,
DOI 10.1007/978-3-319-27517-8_8

Keywords Artificial neural network modeling • Knowledge based modeling • Antenna design • Reconfigurable microstrip patch antenna

MSC codes: 93A30, 65K05, 74P99, 78M50, 90C31, 65D17, 62M45

1 Introduction

Over the years, several numerical and analytical methods that employ detailed electromagnetic models of active/passive components have been developed for designing antennas. However, these methods come with their own set of limitations such as high computational cost and memory requirements. To overcome these challenges, artificial neural network (ANN) has been used as efficient alternative to conventional methods in RF and microwave modeling [23]. Several studies have been carried out for designing antennas using ANN. In the context of reconfigurable antennas, neural network was recently used as an optimization technique to activate the switches in order to realize a given reconfiguration state (e.g., resonating at certain frequency bands) [5, 10].

ANN has been extensively preferred as a modeling technique to obtain a surrogate model instead of a fine model which has high computational burden. Surrogate based modeling [20] is required to overcome this computational burden of the fine model. Surrogate based models can be fundamentally developed in two ways. First way only requires input or output mapping without any change in the computationally cheap coarse model. Space mapping based modeling [3, 9, 14, 16–18] is developed considering this approach. Second way is based on updating the coarse model during modeling process for the coarse model. ANN is very convenient to obtain this kind of coarse model.

ANN provides an efficient strategy to solve modeling and optimization problems which are essential in engineering design where only input–output data are available instead of mathematical formulations [4, 7, 11, 23, 24]. ANN modeling is generally used to construct a mapping from the input to the output depending on the data obtained from detailed physical/EM simulation models or measurements (fine model) and generate approximate results depending on some tunable parameters such as training set, topological structure, and complexity of the fine model.

Since ANN technique constitutes input–output mapping highly depending on the training set, when the points outside of the training range (extrapolation) are used as inputs for the final model after training process, responses of the model are probably unsatisfactory compared to the points inside of the training set (interpolation). ANN and the existing knowledge about the fine model should be combined in the same modeling process in order to reduce complexity of the fine model, while improving extrapolation performance or lowering data requirements for training process.

In some cases, modeling involves numerous training data to satisfy specific design purposes such as good accuracy, better extrapolation, and less computational burden. However training process takes longer time and modeling accuracy cannot be good enough with respect to design purposes. To overcome this problem, knowledge based ANN (KBANN) techniques emerged to generate an efficient model. Knowledge based modeling techniques have been developed to embed existing knowledge into the conventional ANN modeling [6, 14, 15, 19, 23]. Knowledge based models utilize less training data as compared to the conventional ANN. The knowledge provides coarse information for modeling and ANN completes rest of the information using less training data. This modeling approach provides more accuracy and better extrapolation performance than ANN models and offers less computational burden compared to the detailed physical/EM simulation models.

Knowledge based models are applied to reconfigurable 5-fingers microstrip patch antenna using ANNs in this chapter. Source difference (SD), prior knowledge input (PKI), and prior knowledge input with difference (PKI-D) [12, 14, 15, 19] methods are considered as knowledge based neural networks. Employing fine and coarse models in order to train the networks enables to develop fast and accurate EM-ANN models. The developed antenna has four modes of operation, which are controlled by two PIN diode switches with ON/OFF states, and it resonates at multiple frequencies between 1 and 7 GHz. The antenna has several attractive features such as reconfigurability, small size, and low cost. This example handles the increasing requests for the continuing application of ANN in the reconfigurable microstrip antenna design: reduction of model development cost and improving the accuracy.

Conventional ANN modeling and knowledge based modeling techniques will be presented in Section 2 and 3. Design of reconfigurable 5-fingers microstrip patch antenna will be presented in Section 4. Three different cases such as ON–ON, ON–OFF, and OFF–OFF will be handled with three training sets which have different number of samples in Section 4. Simulation results demonstrate considerable savings in computational costs as compared to the 3D-EM simulation results obtained by CST while maintaining the same level of accuracy as the 3D-EM simulation.

2 Conventional ANN Modeling Concept

ANN has been used as an important technique in engineering modeling and optimization. ANN has been widely preferred for modeling purposes in many disciplines such as function approximation, pattern recognition, signal processing, microwave design, and so on [14, 23]. The main reason for ANN being so popular among other modeling techniques is that ANN needs only input–output information obtained from the detailed physical/EM simulation models. ANN usually involves some necessary steps during training such as scaling, initialization of weight coefficients, calculating error which is used for updating weight coefficients. The main purpose of the training process is to reduce the error value as given in Fig. 1, and to increase the generalization capability of the ANN model. Weight coefficients can be obtained by the optimization process defined as

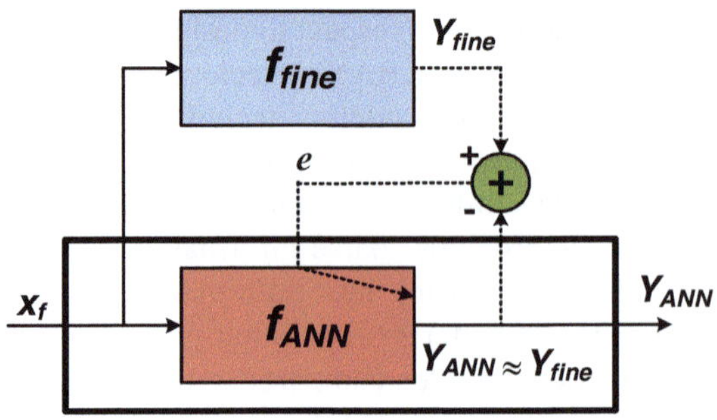

Fig. 1 The training process (updating weight coefficients in terms of error values) and the final model of the conventional ANN technique

$$w^* = \arg\min_{w} \left\| \cdots e^{(i)T} \cdots \right\| \qquad i = 1, 2, \ldots, N \tag{1}$$

where w indicates weight coefficient of the ANN model and N indicates the number of training data. i represents which training data is evaluated by the training process. The error term in (1) can be defined as

$$e^{(i)} = f_{fine}\left(x_f^{(i)}\right) - f_{ANN}\left(x_f^{(i)}\right) \tag{2}$$

where f_{fine} and f_{ANN} indicate the fine model and the ANN model responses, respectively. x_f indicates input of the problem. After the training process of the ANN model, the final response of the ANN model can be given by

$$Y_{ANN} = f_{ANN}\left(x_f\right). \tag{3}$$

Since generalization of ANN is mostly determined by training data set, after the training process the ANN model can generate response in terms of this data set. Extrapolation data are selected differently than training data that's why ANN response will not be highly accurate as interpolation data. The problem specific knowledge based on experience with respect to the engineering problem is required to reduce the data dependency of the conventional ANN.

3 Fundamentals of Knowledge Based Modeling Technique

In engineering design problems, an accurate model for a wide application interval can be obtained by a detailed physical/EM simulation model but it is highly nonlinear and complex, so it is called fine model that has computationally intensive mathematical expressions. In contrast, a less accurate and less computationally intensive model can be utilized instead of the fine model for modeling and optimization purposes, so it is called coarse model that has computationally less complex mathematical expressions than the fine model.

Surrogate based modeling and optimization has been developed to dispose the computational burden of the fine model exploiting a coarse model. In design optimization, the coarse model is used for the optimization process to find optimum design parameters satisfying the design purpose. But the convergence of the optimization process is directly effected by the accuracy of the coarse model. When the coarse model generates quite similar response compared to the fine model response, convergence is probably ensured during surrogate based optimization process.

The knowledge based ANN (KBANN) has emerged to fulfill the requirement for a more accurate model generation than the conventional ANN. The KBANN techniques can create new model exploiting coarse model and this new model can perform better accuracy and improve the generalization capability for interpolation and extrapolation data. The key idea behind the success of the KBANN techniques is that the more accuracy that is needed the more knowledge from the problem space has to be obtained by the coarse model. Another way to overcome the need for more knowledge instead of using the coarse model is to have more training data which requires more effort for data generation.

3.1 Source Difference Method

The source difference method [21, 22] is one of the earliest methods utilizing the knowledge based concept. The target response of the source difference method is the difference between the fine and coarse models (existing approximate model) responses. The coarse model imposes general knowledge behavior of the fine model, thus extrapolation performance and generalization capability of the difference method increase while the number of training data set decreases. In Fig. 2, training phase and final model of SD method are denoted as the dotted line and the bold box, respectively. The training process of ANN during SD modeling can be defined as

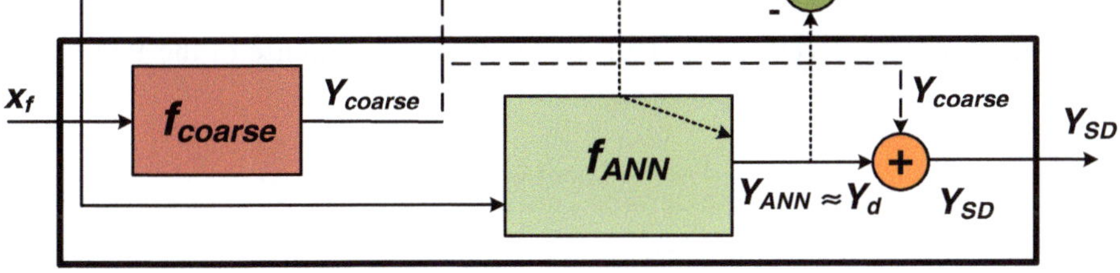

Fig. 2 The training process (updating weight coefficients in terms of error values) and the final model of SD technique embedded the coarse model as the difference between fine and coarse outputs

$$w^* = \arg\min_{w} \left\| \cdots e^{(i)T} \cdots \right\| \qquad i = 1, 2, \ldots, N. \tag{4}$$

The error term in (4) can be defined as

$$e^{(i)} = \left(\underbrace{f_{fine}\left(x_f^{(i)}\right) - Y_{coarse}^{(i)}}_{Y_d} \right) - f_{ANN}\left(x_f^{(i)}\right) \tag{5}$$

where Y_{coarse} indicates the coarse model response and Y_d indicates the difference between the fine model and the coarse model responses. After the training process of the SD model, the final response of the SD model can be given by

$$Y_{SD} = f_{ANN}\left(x_f\right) + Y_{coarse}. \tag{6}$$

The complexity of ANN can be reduced by the coarse model due to Y_d. Therefore the SD model which is trained by less training data can provide similar accuracy obtained by the ANN model.

3.2 Prior Knowledge Input Method

One of the knowledge based techniques is PKI which requires coarse model response as an extra input besides other inputs that belong to the modeling problem [19, 22]. Since extra inputs which contain extra knowledge other than model inputs enables complexity reduction for the modeling problem. ANN can be formed easily to generate a more accurate response. The training process of ANN during PKI modeling can be defined as

$$w^* = \arg\min_{w} \left\| \cdots e^{(i)T} \cdots \right\| \qquad i = 1, 2, \ldots, N. \tag{7}$$

The error term in (7) can be defined as

$$e^{(i)} = f_{fine}\left(x_f^{(i)}\right) - f_{ANN}\left(x_f^{(i)}, Y_{coarse}^{(i)}\right). \tag{8}$$

After the training process of the PKI model, the final response of the PKI model Y_{PKI} can be given by

$$Y_{PKI} = f_{ANN}\left(x_f, Y_{coarse}\right) \tag{9}$$

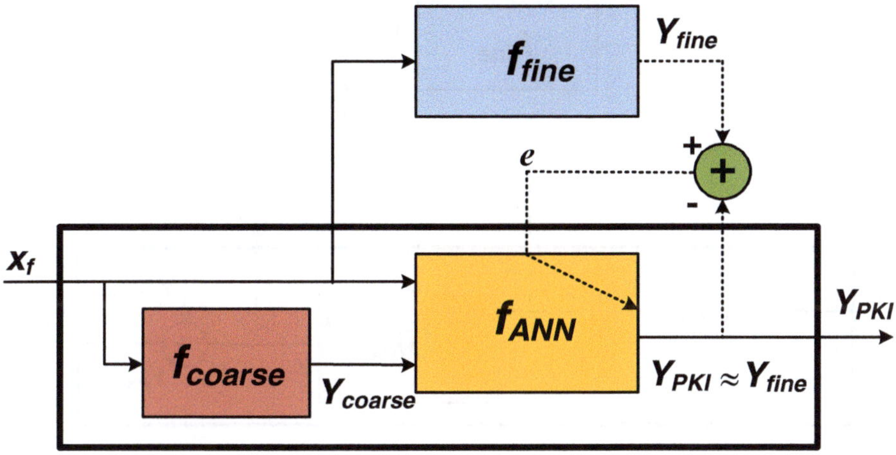

Fig. 3 The training process (updating weight coefficients in terms of error values) and the final model of PKI technique embedded the coarse model as an extra inputs

where Y_{coarse} is used for extra input to the ANN model, hence the accuracy of the PKI model can increase higher than conventional ANN modeling. The training phase and the final model of the PKI are denoted as the dotted line and the bold box in Fig. 3, respectively.

3.3 *Prior Knowledge Input with Difference Method*

PKI-D as shown in Fig. 4 is developed [13, 14, 19] to exploit the advantage of utilizing the coarse model twice. PKI-D combines extra input property of PKI and learning the output difference Y_d calculated as the difference of fine Y_{fine} and coarse Y_{coarse} models in difference method [19]. ANN forms nonlinear mapping from extended input space with coarse model response to difference between fine and coarse model responses. During the training process, weight coefficients are updated by

$$w^* = \arg\min_{w} \left\| \cdots e^{(i)T} \cdots \right\| \qquad i = 1, 2, \ldots, N \tag{10}$$

considering the error term defined as

$$e^{(i)} = \left(\underbrace{f_f\left(x_f^{(i)}\right) - Y_{coarse}^{(i)}}_{Y_d} \right) - f_{ANN}\left(x_f^{(i)}, Y_{coarse}^{(i)}\right). \tag{11}$$

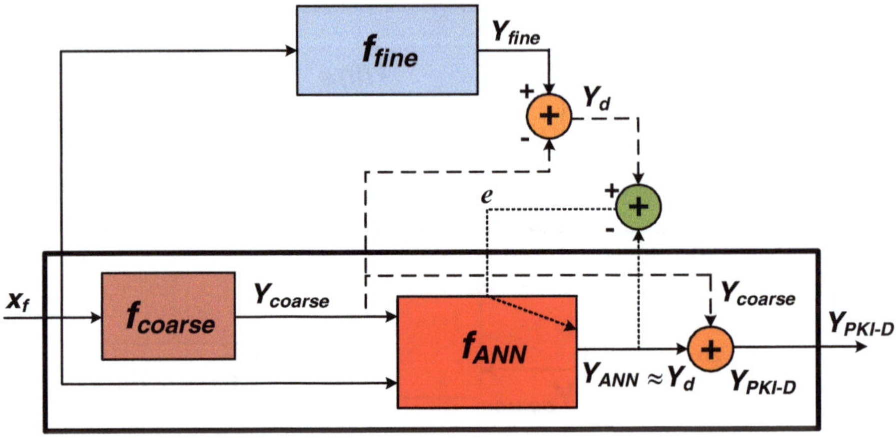

Fig. 4 The training process (updating weight coefficients in terms of error values) and the final model of PKI-D technique embedded the coarse model two times as the extra input and as the difference between fine and coarse outputs

After training is completed, the final model response is ready for the test purpose as follows

$$Y_{PKI-D} = f_{ANN}\left(x_f, Y_{coarse}\right) + Y_{coarse} \tag{12}$$

where Y_{coarse} is used for extra input to the ANN model and also used for obtaining the difference Y_d, hence the accuracy of the PKI-D model can increase higher than conventional ANN modeling due to using this knowledge twice and PKI-D generally provides better accuracy than even other KBANN methods. The training phase and the final model of PKI-D are denoted as the dotted line and the bold box in Fig. 4, respectively.

4 Reconfigurable 5-Fingers Shaped Microstrip Patch Antenna

The Reconfigurable 5-Fingers Shaped Microstrip Patch Antenna (R5FSMPA) [2] is used to perform efficiency of the knowledge based modeling through its three configurations such as ON–ON, ON–OFF, and OFF–OFF states. Since ON–OFF and OFF–ON generate same result, only ON–OFF state is considered. Design parameters of R5FSMPA which are indicated in Fig. 5 are L_1, L_2, and L_3 which represent the length of the radiating patches and W_1, W_2 which represent the width of the radiating patches and W_3 which represents the unfilled space that includes the two PIN diodes (D_1 and D_2) [1]. The feeding coaxial conductor is centered in the middle of L_3 with a radius of 0.065 cm. Two different resistors (R_{D1} and R_{D2}) are

Fig. 5 Physical parameters of R5FSMPA

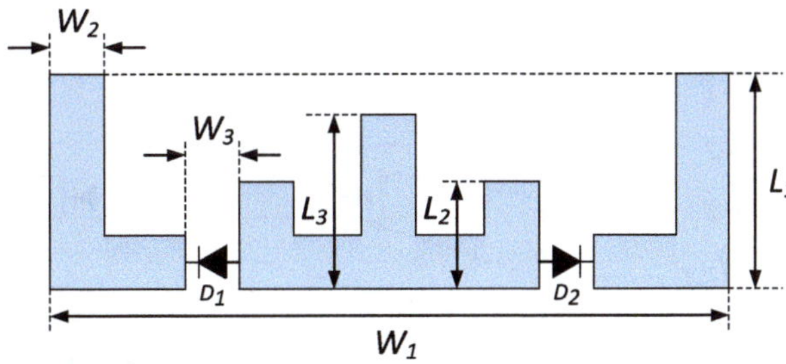

Table 1 Parameters of reconfigurable 5-fingers shaped microstrip patch antenna (r5fsmpa) and data sets in terms of number of samples

		Training data set		Number of samples		
Type of parameters	Input parameters	Minimum	Maximum	Set-1	Set-2	Set-3
Physical geometry	L_1 (cm)	1.2825	1.4175	3	4	5
	L_2 (cm)	0.7125	0.7875	3	4	5
	L_3 (cm)	0.9975	1.1025	3	4	5
Diode states (ON or OFF)	R_{D1}	5 (ON)	1000 (OFF)	3	3	3
	R_{D2}	5 (ON)	1000 (OFF)			
Frequency sweep	f	1 GHz	7 GHz	200	200	200

utilized with 1000 ohm and 5 ohm values for ON and OFF states of the PIN diodes [8]. Right and left patches of R5FSMPA can be activated through ON and OFF states hence three different combinations can be obtained by two diodes. This section is divided into three parts in terms of the training data set. Each training set has three geometrical parameters, two resistors of diode (ON and OFF states)and frequency as input parameters. Return loss S_{11} as output response is obtained by CST 3D-EM simulations. Physical dimensions of R5FSMPA are given in Table 1.

Input–output relationships of R5FSMPA are shown in Fig. 6 and S_{11} (return loss) is obtained by 3D-EM simulation of CST in terms of 200 number of frequency points between 1 GHz and 7 GHz. The relationship between frequency and S_{11} is indicated by Fig. 7. Three different states of R5FSMPA are modeled via one ANN structure while three states were modeled by three ANN structures in the previous study [2].

In order to demonstrate the efficiency of knowledge based methods, three different data sets can be considered. Selection of data sets is summarized in Table 2 including three data sets. Each data set is utilized as training samples for two different number of neurons in ANN hidden layers. Therefore, all methods can be analyzed in terms of the fundamental ANN properties such as the number of data and the number of neurons to reveal the correlation between accuracy and other ANN parameters.

ANN structure for the conventional ANN is realized by feed-forward multi-layer perceptron (MLP) function in MATLAB Toolbox which utilizes Levenberg-Marquard algorithm and such optimization parameters are: two hidden layer

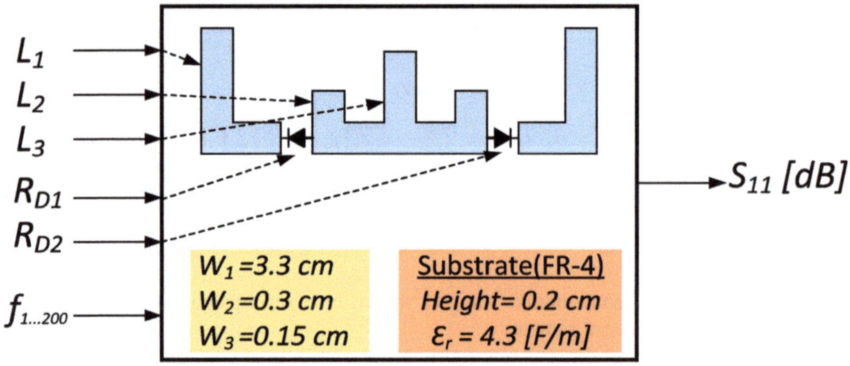

Fig. 6 Input–output relationship of the fine model for R5FSMPA

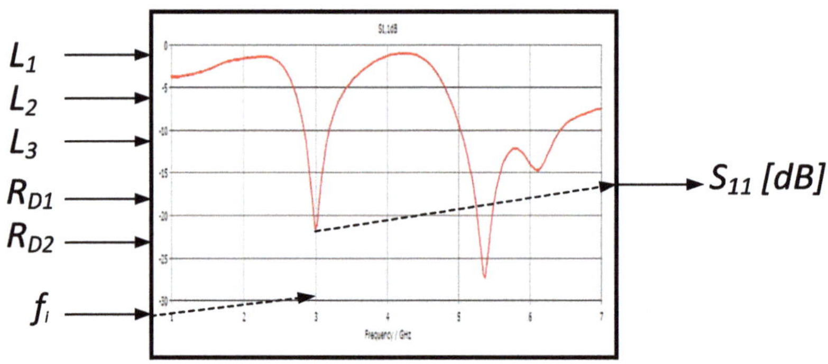

Fig. 7 Frequency-S_{11} relationship of the fine model for R5FSMPA

Table 2 Number of samples for three training data sets and test data

Data type		Geometry L_1 L_2 L_3	Antenna switching states	Frequency 1–7 [GHz]	Total samples
Training	Set-1	3*3*3 = 27	3	200	3*3*3*3*200 = 16,200
	Set-2	4*4*4 = 64	3	200	4*4*4*3*200 = 38,400
	Set-3	5*5*5 = 125	3	200	5*5*5*3*200 = 75,000
Test		3	3	200	3*3*200 = 1800

with different number of neurons, learning rate = 0.1, momentum = 0.2, and regularization = 0.2. Two hidden layer is so suitable for highly nonlinear engineering problem hence it is preferred to form required ANN structure for the knowledge based ANN and conventional ANN methods.

Error calculation is an important part of the comparison. Normalized test error can be formulated by

$$Normalized\ Error = \frac{|Y_{Fine} - Y_{Model}|}{Y_{Fine}} \tag{13}$$

where Y_{Fine} and Y_{Model} indicate the fine model response and the model response which is compared with the fine model response. Normalized mean error can be formulated by

$$Normalized\ Mean\ Error = \frac{1}{N} \times \sum_{i=1}^{N} \frac{|Y_{Fine,i} - Y_{Model,i}|}{Y_{Fine,i}} \tag{14}$$

where i indicates the number of test samples. Normalized max error can be formulated by

$$Normalized\ Max\ Error = \max_i \left\{ \frac{|Y_{Fine,i} - Y_{Model,i}|}{Y_{Fine,i}} \right\}. \tag{15}$$

After 20 runs are completed, normalized mean value of S_{11} is calculated for each test sample. Normalized mean and maximum errors are calculated using (14) and (15) in terms of S_{11} obtained from 20 runs.

4.1 *Data Set* – 1*:* 16,200 *Samples*

In this part, three states of the reconfigurable patch antenna are considered in terms of the accuracy and time consumption for data $set - 1$ which consists of three parameters, three states (ON–ON, ON–OFF, and OFF–OFF) and 200 frequency points. The total number of data samples is $16,200$ obtained by three samples selected from the training data interval for three physical geometries which are multiplied by three states and 200 frequencies. Test data which includes nine different geometry is selected from the training interval but each test geometry is different than the training geometry. The test samples consist of three different geometries for three states. Test performance can be demonstrated by one geometry for each states of reconfigurable antenna. Conventional ANN and knowledge based ANN methods run 20 times and average responses of test samples for EM, ANN, and PKI-D are given in Fig. 8 for three different geometries. In addition, normalized test errors of PKI-D and the conventional ANN are given in Fig. 9 for three different geometries.

Accuracy of all methods are summarized in Table 3 for two different ANN structure such as (30–30) and (30–20). Time consumptions of generating data set and the training phase for all methods are given in Table 4 for ANN structure with (30–20) neurons. Since the fine model is computationally complex, it requires more computational time than the coarse model. The coarse model improves the accuracy of all knowledge based methods compared to conventional ANN. The coarse model is used for twice during training of PKI-D, which reduces the complexity of modeling problem. Therefore, time consumption of PKI-D can be less than other knowledge based methods such as SD and PKI.

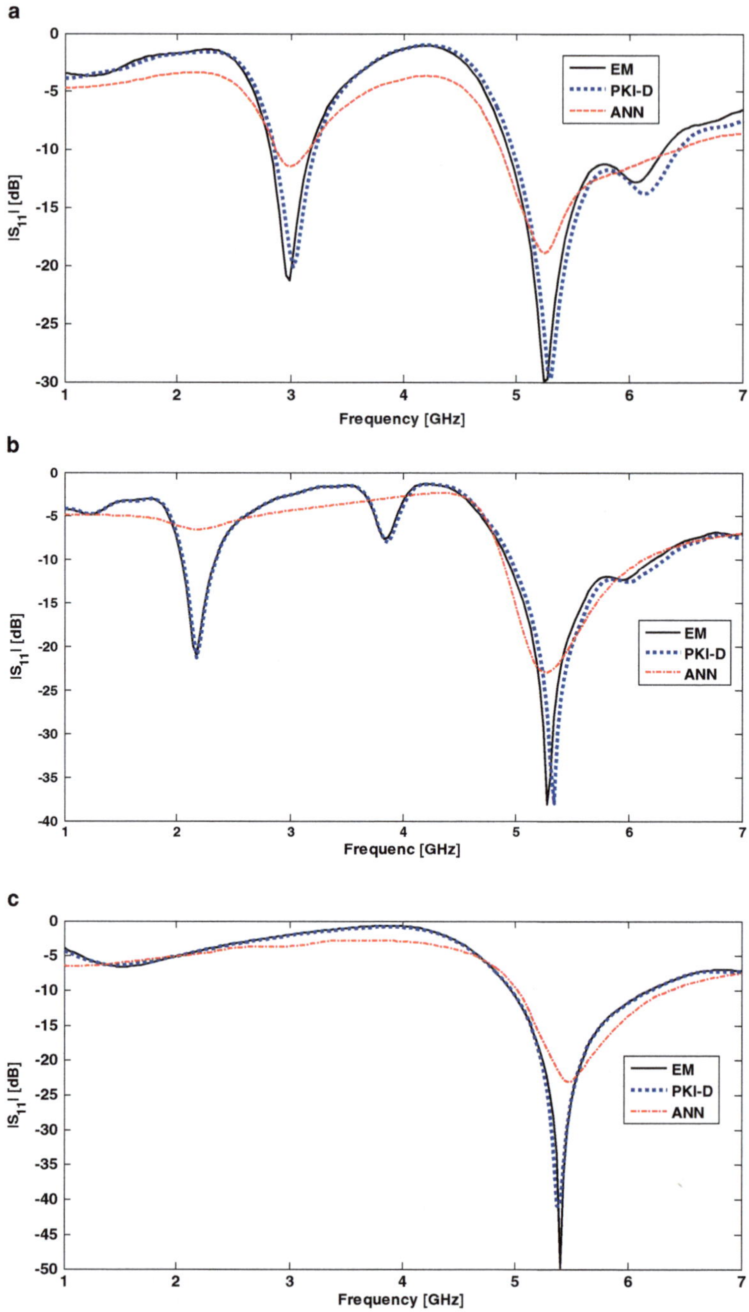

Fig. 8 MLP with two hidden layers (30–30 neurons) trained by 16,200 samples to show EM, PKI-D, and conventional ANN results. (**a**) Magnitude of S_{11} for *Geometry* − 3 (ON–ON case) (**b**) Magnitude of S_{11} for *Geometry* − 6 (ON–OFF case) (**c**) Magnitude of S_{11} for *Geometry* − 9 (OFF–OFF case)

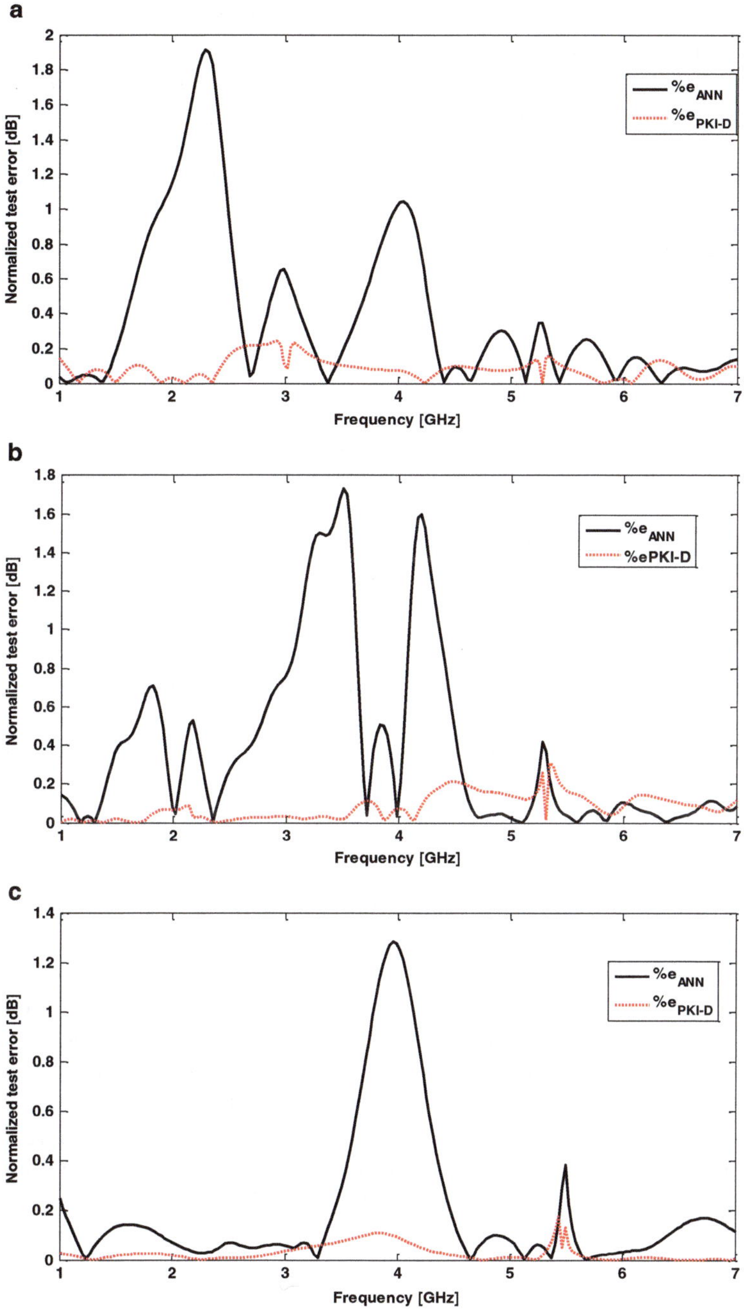

Fig. 9 MLP with two hidden layers (30–30 neurons) trained by 16,200 samples to show EM, PKI-D, and conventional ANN results. (**a**) Normalized test error for *Geometry* − 3 (ON–ON case) (**b**) Normalized test error for *Geometry* − 6 (ON–OFF case) (**c**) Normalized test error for *Geometry* − 9 (OFF–OFF case)

Table 3 Normalized mean errors at 16,200 data samples for all switching states

Tow hidden layers	Error	Coarse (Training)	Coarse (Test)	ANN	SD	PKI	PKI-D
30–30	Mean	0.0485	0.0479	0.3739	0.0628	0.0748	0.0453
	Max	0.4954	0.4719	3.1766	0.4647	1.1451	0.3056
30–20	Mean	0.0485	0.0479	0.3086	0.0630	0.0514	0.0433
	Max	0.4954	0.4719	1.7161	0.4683	0.3423	0.3008

Table 4 Time consumption results of all methods trained by 16,200 data samples for all switching states

	ANN	SD	PKI	PKI-D
Fine	1 h, 11 m	1 h, 11 m	1 h, 11 m	1 h, 11 m
Coarse	–	0 h, 47 m	0 h, 47 m	0 h, 47 m
Max Training	0.462 m	0.217 m	0.220 m	0.203 m
Total	1 h, 11.462 m	1 h, 58.217 m	1 h, 58.220 m	1 h, 58.203 m

4.2 *Data Set* – 2: 38,400 *Samples*

In this part, three states of the reconfigurable patch antenna are considered in terms of the accuracy and time consumption for data $set - 2$ which consists of three parameters, three states (ON–ON, ON–OFF, and OFF–OFF) and 200 frequency points. The total number of data samples is $34,800$ obtained by four samples selected from the training data interval for three physical geometries which are multiplied by 3 states and 200 frequencies. The same test samples are used for comparing $set - 1$ with $set - 2$. PKI is utilized instead of PKI-D to demonstrate the general performance of knowledge based methods. Conventional ANN and knowledge based ANN methods run 20 times and average responses of test samples for EM, ANN, and PKI are given in Fig. 10 for three different geometries. In addition, normalized test errors of PKI and conventional ANN are given in Fig. 11 for three different geometries.

Accuracy of all methods are summarized in Table 5 for two different ANN structure such as (30–30) and (40–30). Time consumptions of generating data set and the training phase for all methods are given in Table 6 for ANN structure with (40–30) neurons. Since extra knowledge obtained by the coarse model reduces the complexity of modeling problem, knowledge based methods require less time for the training process of ANN structure. Time efficiency in training process of knowledge based methods can be realized in Table 6.

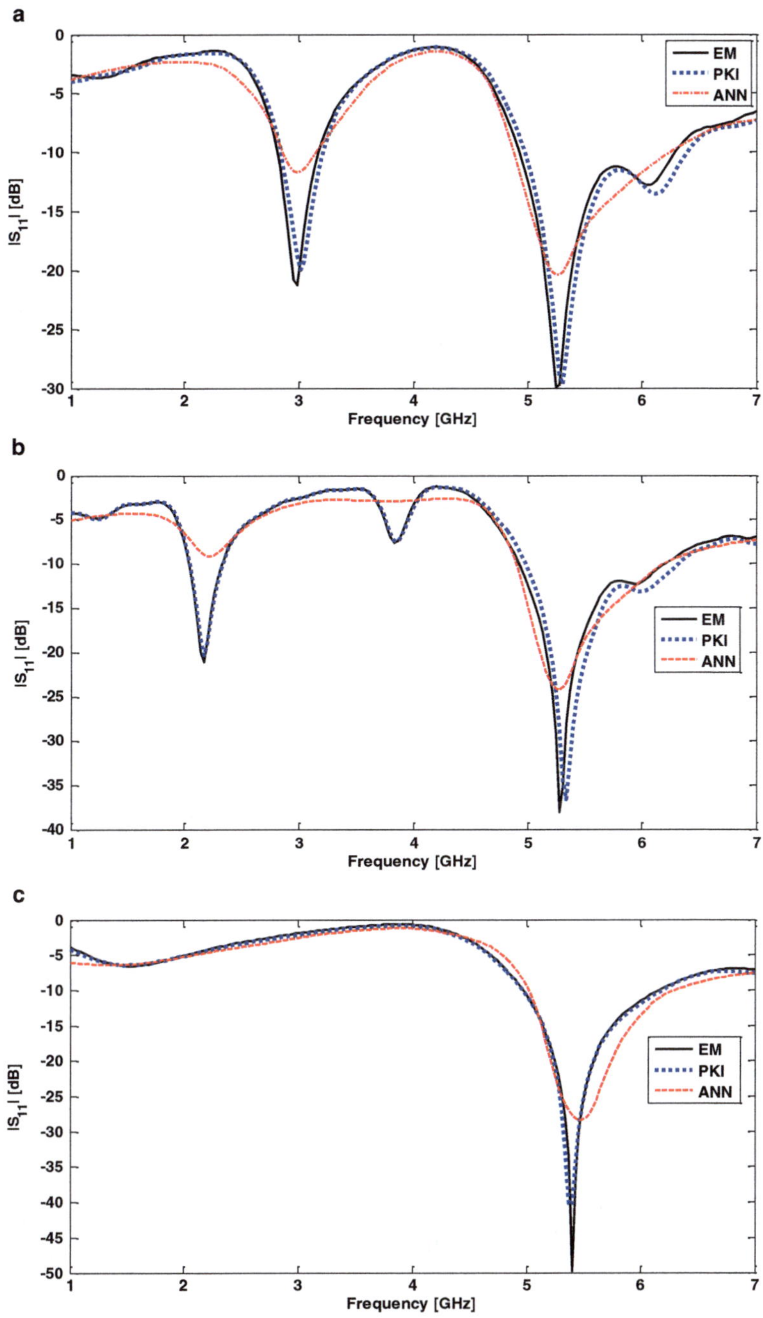

Fig. 10 MLP with two hidden layers (40–30 neurons) trained by 38,400 samples to show EM, PKI, and conventional ANN results. (**a**) Magnitude of S_{11} for *Geometry* − 3 (ON–ON case) (**b**) Magnitude of S_{11} for *Geometry* − 6 (ON–OFF case) (**c**) Magnitude of S_{11} for *Geometry* − 9 (OFF–OFF case)

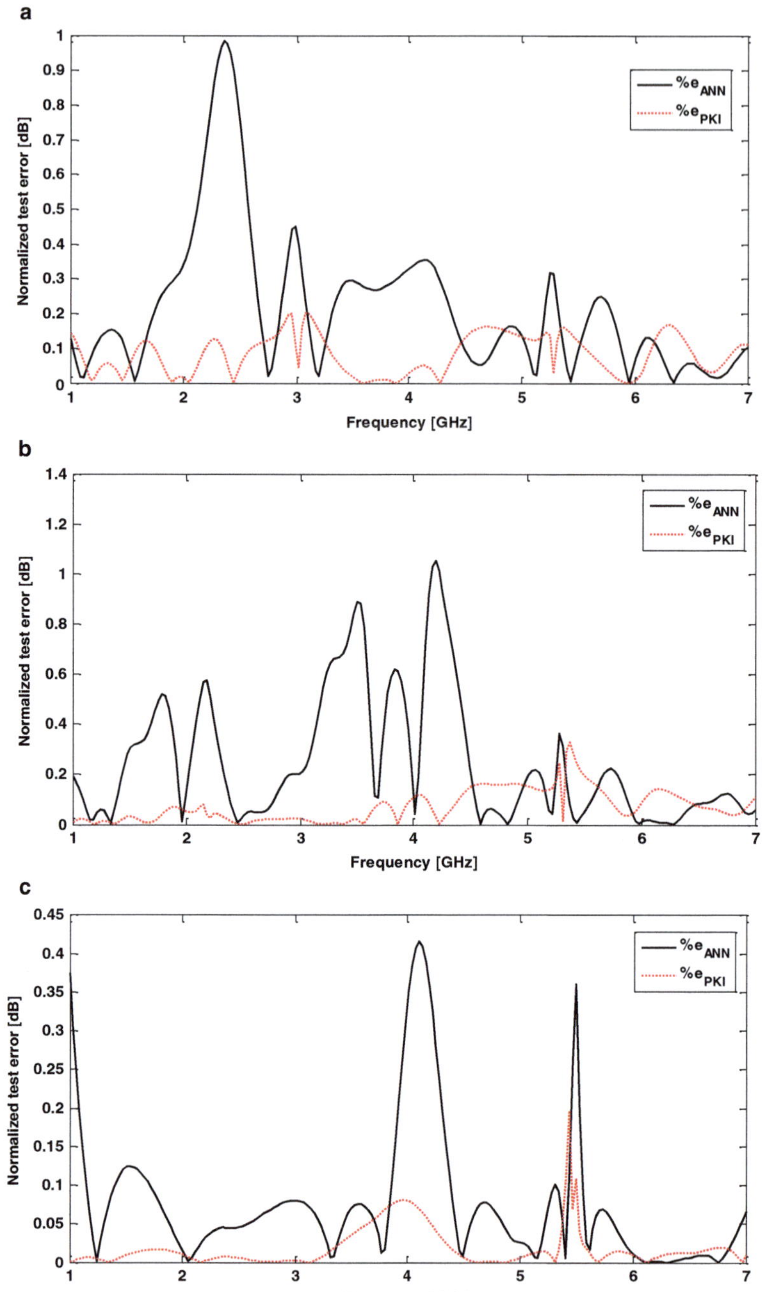

Fig. 11 MLP with two hidden layers (40–30 neurons) trained by 38,400 samples to show EM, PKI, and conventional ANN results. (**a**) Normalized test error for *Geometry*−3 (ON–ON case) (**b**) Normalized test error for *Geometry*−6 (ON–OFF case) (**c**) Normalized test error for *Geometry*−9 (OFF–OFF case)

Table 5 Normalized mean errors at 38,400 data samples for all switching states

Tow hidden layers	Error	Coarse (Training)	Coarse (Test)	ANN	SD	PKI	PKI-D
30–30	Mean	0.0480	0.0479	0.2338	0.0606	0.0475	0.0409
	Max	0.5288	0.4719	1.3846	0.4685	0.3787	0.2645
40–30	Mean	0.0480	0.0479	0.2126	0.0635	0.0428	0.0403
	Max	0.5288	0.4719	1.4164	0.4676	0.3269	0.2747

Table 6 Time consumption results of all methods trained by 38,400 data samples for all switching states

	ANN	SD	PKI	PKI-D
Fine	2 h, 44 m	2 h, 44 m	2 h, 44 m	2 h, 44 m
Coarse	–	2 h, 34 m	2 h, 34 m	2 h, 34 m
Max training	2.679 m	1.028 m	1.116 m	1.016 m
Total	2 h, 46.679 m	5 h, 19.028 m	5 h, 19.116 m	5 h, 19.016 m

4.3 Data Set – 3: 75,000 Samples

In this part, three states of the reconfigurable patch antenna are considered in terms of the accuracy and time consumption for data $set-3$ which consists of three parameters, three states (ON–ON, ON–OFF, and OFF–OFF) and 200 frequency points. The total number of data samples is $75,000$ obtained by five samples selected from training data interval for three physical geometries which are multiplied by 3 states and 200 frequencies. The same test data is used for comparing $set-1$ and $set-2$ with $set-3$. SD is utilized instead of PKI to demonstrate the general performance of knowledge based methods. Conventional ANN and knowledge based ANN methods run 20 times and average responses of test samples for EM, ANN, and SD are given in Fig. 12 for three different geometries. In addition, normalized test errors of SD and conventional ANN are given in Fig. 13 for three different geometries.

Accuracy of all methods are summarized in Table 7 for two different ANN structure such as (45–45) and (50–40). Time consumptions of generating data set and the training phase for all methods are given in Table 8 for ANN structure with (50–40) neurons. Time efficiency in training process of knowledge based methods can be realized in Table 8.

Knowledge based methods generally improve the accuracy of ANN model using even less training data. This improvement is based on extra knowledge about input–output relationship of the modeling problem. This extra knowledge enables to reduce the complexity of the problem. Thus, more accurate results can be obtained by knowledge based methods which utilize less data and fast modeling process. Knowledge based methods provide more accurate results for $16,200$ samples compared to $38,400$ samples for conventional ANN. The performance of knowledge based methods with less training data can be realized in Table 9. Knowledge based

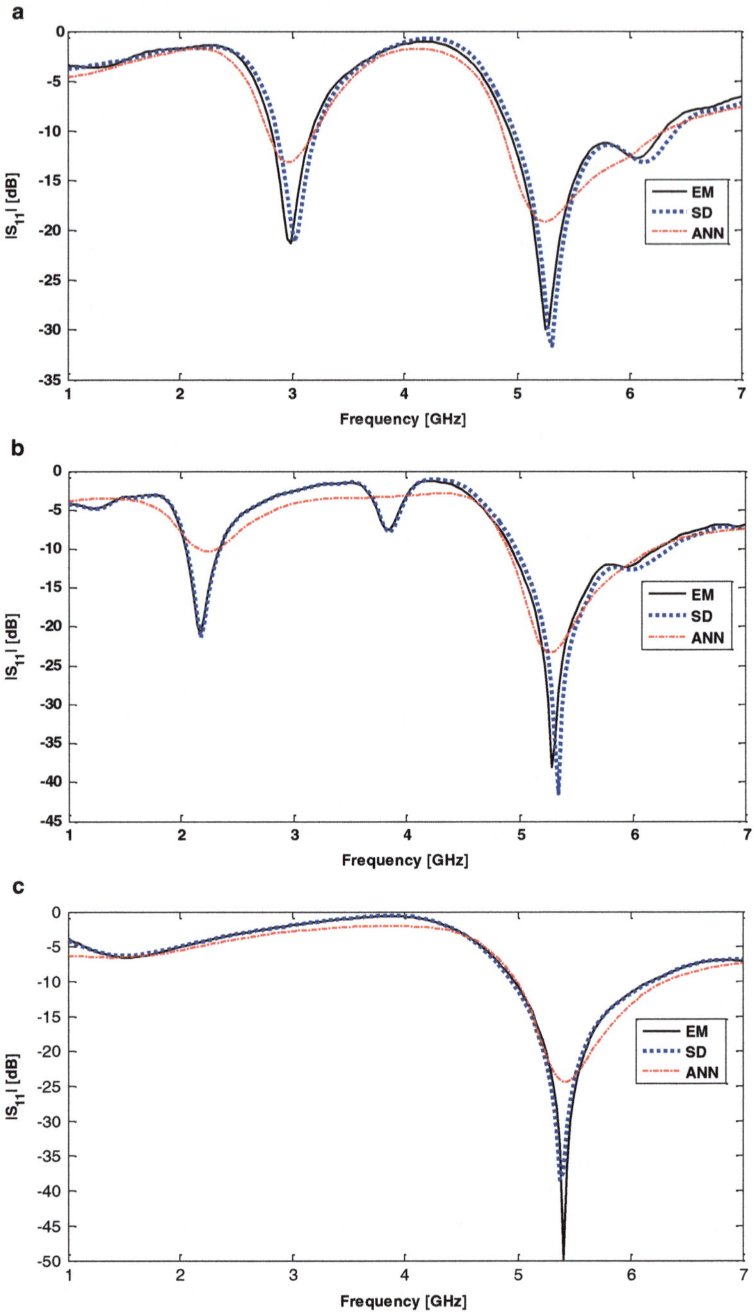

Fig. 12 MLP with two hidden layers (50–40 neurons) trained by 75,000 samples to show EM, SD, and conventional ANN results. (**a**) Magnitude of S_{11} for *Geometry* − 3 (ON–ON case) (**b**) Magnitude of S_{11} for *Geometry* − 6 (ON–OFF case) (**c**) Magnitude of S_{11} for *Geometry* − 9 (OFF–OFF case)

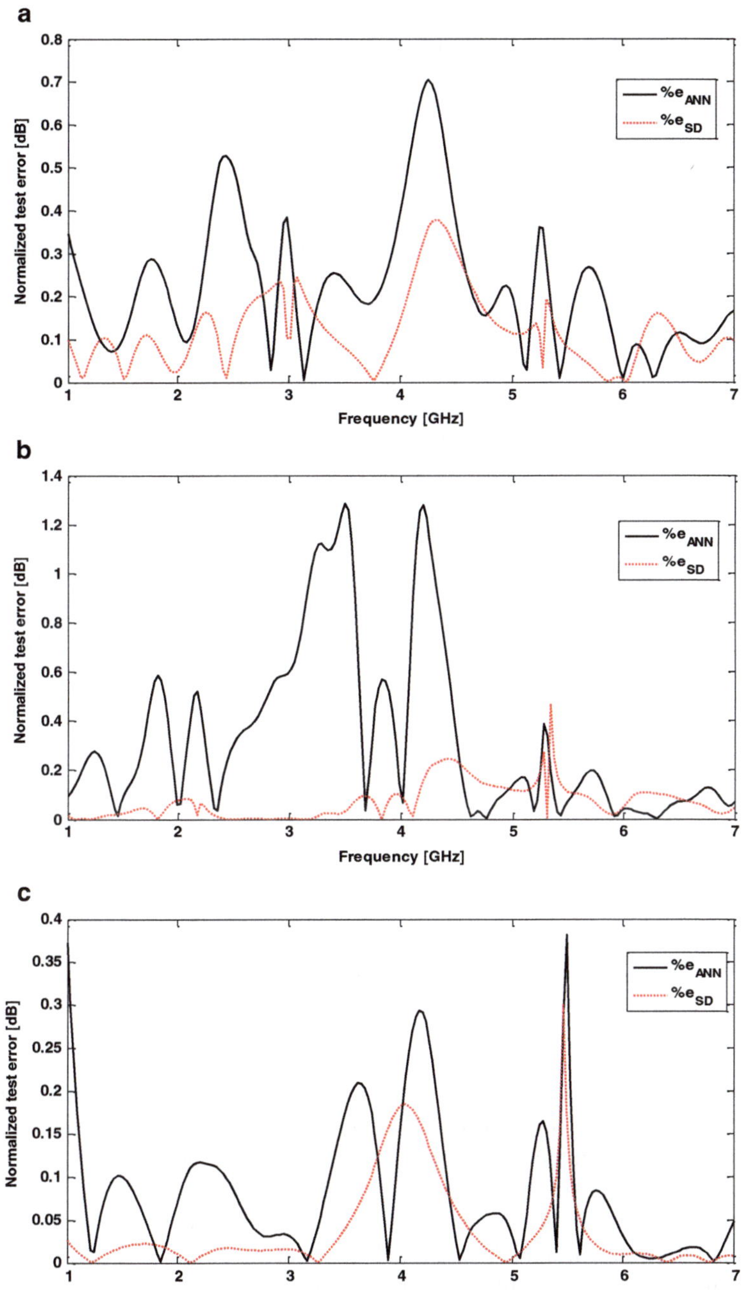

Fig. 13 MLP with two hidden layers (50–40 neurons) trained by 75,000 samples to show EM, SD, and conventional ANN results. (**a**) Normalized test error for *Geometry* − 3 (ON–ON case) (**b**) Normalized test error for *Geometry* − 6 (ON–OFF case) (**c**) Normalized test error for *Geometry* − 9 (OFF–OFF case)

Table 7 Normalized mean errors at 75,000 data samples for all switching states

Tow hidden layers	Error	Coarse (Training)	Coarse (Test)	ANN	SD	PKI	PKI-D
45–45	Mean	0.0475	0.0479	0.1903	0.0592	0.0409	0.0382
	Max	0.6593	0.4719	1.4485	0.4690	0.3682	0.3271
50–40	Mean	0.0475	0.0479	0.2076	0.0583	0.0478	0.0397
	Max	0.6593	0.4719	2.0250	0.4684	0.3265	0.3180

Table 8 Time consumption results of all methods trained by 75,000 data samples for all switching states

	ANN	SD	PKI	PKI-D
Fine	4 h, 22 m	4 h, 22 m	4 h, 22 m	4 h, 22 m
Coarse	–	3 h, 49 m	3 h, 49 m	3 h, 49 m
Max training	3.718 m	1.620 m	1.508 m	2.229 m
Total	4 h, 25.718 m	8 h, 12.620 m	8 h, 12.508 m	8 h, 13.229 m

Table 9 The accuracy comparison of all methods with different data samples and time consumption results for all switching states

Methods	Data samples	Tow hidden layers	Mean error	Max error	Time consumption
SD	16,200	30–30	0.0628	0.4647	1 h, 58.217 m
PKI			0.0748	1.1451	1 h, 58.220 m
PKI-D			0.0453	0.3056	1 h, 58.203 m
ANN	38,400	30–30	0.2338	1.3846	2 h, 46.679 m
ANN	75,000	50–40	0.2076	2.0250	4 h, 25.718 m

methods provide more accurate result for less training data, hence they are so suitable to embed existing knowledge into modeling step of the engineering design process.

5 Conclusion

Knowledge based modeling is applied to engineering modeling relevant to reconfigurable 5-fingers shaped microstrip patch antenna. The aim of this modeling problem is to obtain S_{11} of antenna design parameters corresponding to the frequency. Number of data and number of neurons directly effect ANN performance hence both of them are utilized for the analysis and comparison between knowledge based models and the conventional ANN model. Knowledge based methods with less data are used in order to obtain more accurate results compared to conventional ANN with more data. In addition, knowledge based methods require less time consumption and even less training data through the coarse model efficiency. Knowledge based methods should be selected for the engineering design problem to embed the existing knowledge into the design process. Reconfigurable 5-fingers shaped microstrip patch antenna is selected to demonstrate the efficiency of knowledge based methods which are easily applied to the modeling problem in the engineering design process.

References

1. Aoad, A., Aydin, Z., Korkmaz, E.: Design of a tri band 5-fingers shaped microstrip patch antenna with an adjustable resistor. In: Antenna Measurements Applications (CAMA), pp. 1–4 (2014)
2. Aoad, A., Simsek, M., Aydin, Z.: Design of a reconfigurable 5-fingers shaped microstrip patch antenna by artificial neural networks. Int. J. Adv. Res. Comput. Sci. Softw. Eng. (IJARCSSE) **4**(10), 61–70 (2014)
3. Bandler, J.W., Cheng, Q.S., Dakroury, S.A., Mohamed, A.S., Bakr, M.H., Madsen, K., Sondergaard, J.: Space mapping: The state of the art. IEEE Trans. Microwave Theory Tech. **52**(1), 337–361 (2004)
4. Burrascano, P., Fiori, S., Mongiardo, M.: A review artificial neural networks applications in microwave computer-aided design. Int. J. RF Microwave Comput.-Aided Eng. **9**(3), 158–174 (1999)
5. Costantine, J.: Design, optimization and analysis of reconfigurable antennas. Ph.D. thesis, Electrical and Electronics Engineering, The University of New Mexico, Albuquerque, New Mexico (2009)
6. Devabhaktuni, V.K., Chattaraj, B., Yagoub, M.C.E., Zhang, Q.J.: Advanced microwave modeling framework exploiting automatic model generation, knowledge neural networks, and space mapping. IEEE Trans. Microw. Theory Tech. **51**(7), 1822–1833 (2003)
7. Haykin, S.: Neural Network - A Comprehensive Foundation, 2nd edn. Printice Hall Inc., New Jersey (1999)
8. Huff, G.H., Bernhard, J.T.: Reconfigurable antennas. In: Modern Antenna Handbook, pp. 369–398. Wiley, Inc., New York (2007)
9. Koziel, S., Bandler, J.W.: Modeling of microwave devices with space mapping and radial basis functions. Int. J. Numer. Model: Electron. Networks, Devices Fields **21**(1-2), 187–203 (2008)
10. Patnaik, A., Anagnostou, D., Christodoulou, C.G., Lyke, J.C.: A frequency reconfigurable antenna design using neural networks. In: Antennas and Propagation Society International Symposium, vol. 2A, pp. 409–412 (2005)
11. Rayas-Sanchez, J.E.: Em-based optimization of microwave circuits using artificial neural networks: the state-of-the-art. IEEE Trans. Microwave Theory Tech. **52**(1), 420–435 (2004)
12. Simsek, M.: Developing 3-step modeling strategy exploiting knowledge based techniques. In: The 20th European Conference on Circuit theory and Design, Linkoping, Sweden, Aug 29-31 (2011)
13. Simsek, M.: Knowledge based three-step modeling strategy exploiting artificial neural network. In: Koziel, S., Leifsson, L., X.-S. Yang, (eds.) Solving Computationally Expensive Engineering Problems, volume 97 of Springer Proceedings in Mathematics Statistics, pp. 219–239. Springer International Publishing, New York (2014)
14. Simsek, M., Sengor, N.S.: A knowledge-based neuromodeling using space mapping technique: compound space mapping-based neuromodeling. Int. J. Numer. Model: Electron. Networks, Devices Fields **21**(1-2), 133–149 (2008)
15. Simsek, M., Sengor, N.S.: An efficient inverse ANN modeling approach using prior knowledge input with difference method. In: The European Conference on Circuit theory and Design, Antalya, Turkey, Aug 23 to 27 (2009)
16. Simsek, M., Sengor, N.S.: The efficiency of difference mapping in space mapping-based optimization. In: Koziel, S., Leifsson, L. (eds.) Surrogate-Based Modeling and Optimization, pp. 99–120. Springer, New York (2013)

17. Simsek, M., Serap Sengor, N.: Solving inverse problems by space mapping with inverse difference method. In: Roos, J., Costa, L.R.J. (eds.) Scientific Computing in Electrical Engineering SCEE 2008, Mathematics in Industry, pp. 453–460. Springer, Berlin/ Heidelberg (2010)
18. Simsek, M., Tezel, N.S.: The reconstruction of shape and impedance exploiting space mapping with inverse difference method. IEEE Trans. Antennas Propag. **60**(4), 1868–1877 (2012)
19. Simsek, M., Zhang, Q.J., Kabir, H., Cao, Y., Sengor, N.S.: The recent developments in microwave design. Int. J. Math. Modelling and Numer. Optim. **2**(2), 213–228 (2011)
20. Sondergard, J.: Optimization using surrogate models by the space mapping technique. Ph.D. thesis, Informatics and Mathematical Modelling, Technical University of Denmark, Lyngby, Denmark, Jan (2003)
21. Watson, P.M., Gupta, K.C.: EM-ANN models for microstrip vias and interconnects in multilayer circuits. IEEE Trans. Microw. Theory Tech. **44**, 2495–2503 (1996)
22. Watson, P.M., Gupta, K.C., Mahajan, R.L.: Development of knowledge based artificial neural network models for microwave components. In: International Microwave Symposium Digest, pp. 9–12. IEEE, New York (1998)
23. Zhang, Q.J., Gupta, K.C.: Neural Networks for RF and Microwave Design. Artech House, Boston (2000)
24. Zhang, Q.J., Gupta, K.C., Devabhaktuni, V.K.: Artificial neural networks for RF and microwave design - from theory to practice. IEEE Trans. Microwave Theory Tech. **51**(4), 1339–1350 (2003)

Expedited Simulation-Driven Multi-Objective Design Optimization of Quasi-Isotropic Dielectric Resonator Antenna

Adrian Bekasiewicz, Slawomir Koziel, Wlodzimierz Zieniutycz, and Leifur Leifsson

Abstract Majority of practical engineering design problems require simultaneous handling of several criteria. Although many of design tasks can be turned into single-objective problems using sufficient formulations, in some situations, acquiring comprehensive knowledge about possible trade-offs between conflicting objectives may be necessary. This calls for multi-objective optimization that aims at identifying a set of alternative, Pareto-optimal designs. The most popular solution approaches to genuine multi-objective optimization include population-based metaheuristics. Unfortunately, such methods are not practical for problems involving expensive computational models, particularly for antenna engineering, where reliable design requires CPU-intensive electromagnetic (EM) analysis. In this work, we discuss two methodologies for expedited multi-objective design optimization of a six-parameter dielectric resonator antenna (DRA) with respect to three design criteria. The considered solution approaches rely on surrogate-based optimization (SBO) paradigm, where the design speedup is obtained by shifting the optimization burden into a cheap replacement model referred to as a surrogate. The latter is utilized for generating the initial approximation of the Pareto front representation as well as further refinement of the initially obtained Pareto-optimal solutions.

A. Bekasiewicz (✉) • S. Koziel
Engineering Optimization & Modeling Center, School of Science and Engineering, Reykjavik University, Menntavegur 1, 101 Reykjavik, Iceland
e-mail: bekasiewicz@ru.is; koziel@ru.is

W. Zieniutycz
Faculty of Electronics, Telecommunications and Informatics, Gdansk University of Technology, 80-233 Gdansk, Poland
e-mail: wlz@pg.gda.pl

L. Leifsson
Department of Aerospace Engineering, Iowa State University, Ames, IA 50011, USA
e-mail: leifur@iastate.edu

S. Koziel et al. (eds.), *Simulation-Driven Modeling and Optimization*, Springer Proceedings in Mathematics & Statistics 153,
DOI 10.1007/978-3-319-27517-8_9

Keywords Computer-aided design (CAD) • Multi-objective optimization • Antenna design • Evolutionary algorithm • Response surface approximation • Kriging interpolation • Design space reduction • Dielectric resonator antenna

MSC codes: 78M50, 62P30, 93A30, 65D17

1 Introduction

Design closure of contemporary antennas, especially in terms of automated adjustment of geometry parameters with the aim of improving electrical performance, is a difficult task. A realistic antenna setup comprises not only the radiator together with the feeding structure but also other components such as connectors, housing, or nearest environment of the structure [1–4]. This creates a considerable challenge in terms of automated determination of the antenna geometry using numerical optimization techniques: accurate performance evaluation can only be realized using computationally expensive electromagnetic (EM) analysis and (conventional) optimization algorithms require large number of analyses. Consequently, one of the most popular approaches to adjustment of antenna geometrical parameters is based on repetitive parameter sweeps (usually, one parameter at a time)—utilization of engineering experience allows educated guesses about necessary dimension changes so that satisfactory designs can be obtained in reasonable time. Nevertheless, automation of parameter adjustment process is highly desirable. However, as mentioned above, it may be impractical when conventional numerical optimization methods (particularly, gradient-based [5], or derivative-free [6]) are utilized. Apart from a large number of objective function evaluations (and, consequently, EM simulations) necessary for the algorithm to converge to an optimum solution, there are other issues, e.g., the presence of numerical noise [7].

Real-world design problems are even more challenging because of the necessity of handling multiple and often conflicting criteria. Multi-objective optimization typically aims at yielding the entire set of alternative designs corresponding to the best possible trade-offs between such conflicting objectives [8]. The typical performance criteria for antenna design include: minimization of the return loss response within a defined frequency band [9, 10], maximization of the antenna gain [11, 12], reduction of the side-lobe level [11, 13], etc. There are also additional requirements related to antenna geometry, i.e., structure footprint or volume [7, 9, 14]. Conventional design optimization is based on aggregation of many objectives—for the sake of simplicity—into a scalar merit function. Alternatively, the primary objective can be optimized directly with the remaining ones handled through design constraints. Identification of a set of alternative designs representing possible trade-offs between conflicting objectives usually requires genuine multi-objective optimization [7, 8]. In many practical situations, a priori preference articulation regarding design objectives is either not possible or not desirable, e.g., when the best possible design trade-offs between conflicting objectives (so-called Pareto front) are to be found. The most popular solution approaches for multi-objective optimization involve population-based metaheuristics with particular emphasis on evolutionary algorithms [9, 11, 15, 16] and particle swarm optimizers [17–20]. The main advantage of such algorithms is their ability to process and outcome the entire set of solutions in a single algorithm run. Unfortunately population-based routines require thousands or even tens of thousands of objective function evaluations to converge [9, 11]. Thus, in case of antenna design, metaheuristic-based optimization is only practical when design evaluation time is not of a major concern [20, 21].

The problem related to high computational cost of multi-objective antenna optimization can be alleviated to some extent by exploiting adjoint sensitivity techniques, however, their availability in commercial simulation software packages is still limited [22]. Improved computational efficiency can also be obtained by means of surrogate-based optimization (SBO) approaches [22–24], particularly space mapping [25], manifold mapping [26], or shape preserving response prediction [27]. In SBO, the computational cost of the design process is reduced by replacing the high-fidelity antenna model by an auxiliary low-fidelity representation (usually in the form of coarsely discretized EM model). The former is iteratively optimized and corrected to elevate it to the high-fidelity model level. The design cost may be further reduced by a combination of SBO methods with response surface approximation (RSA) models [28, 29]. Although SBO is mostly utilized in the context of single-objective antenna design [30–32], a few successful applications to multi-objective problems have been reported in the literature [7, 14].

Recently, an efficient procedure for multi-criteria antenna design that combines SBO, low-fidelity EM models, and multi-objective evolutionary algorithm (MOEA) has been proposed in [7]. The technique also features a procedure that permits refinement of the Pareto-optimal solutions obtained by MOEA to elevate them to the high-fidelity EM model level. An interesting variation of the approach exploiting co-Kriging surrogates have been proposed in [14]. Unfortunately applicability of these methods is limited to low-dimensional cases, because computational cost of training data acquisition for RSA model generation grows exponentially with the dimensionality of the search space. Thus, it is only practical for cases with a few adjustable parameters. This challenge has been addressed by utilization of appropriate design space reduction routines that allows for identification or the reliable RSA model using reasonably small number of samples even in higher-dimensional spaces. Design cases with more than a dozen of adjustable parameters have been successfully solved [33–35] using this approach. In most test cases demonstrated in the literature only two design objectives are explicitly considered (with additional criteria handled through appropriately defined constraints) [34].

In this work, we discuss two design approaches for fast multi-objective optimization of a contemporary antenna with respect to three non-commensurable design requirements. Both techniques exploit variable-fidelity electromagnetic (EM) models, fast RSA models, as well as model correction mechanisms which allow for precise allocation of the high-fidelity Pareto-optimal design solutions. The methods are demonstrated and compared using the example of a compact quasi-isotropic dielectric resonator antenna (DRA). Additionally, we investigate the influence of possible imperfections and statistical variability of multi-objective evolutionary optimization of the approximation surrogate (an intermediate step of the design process leading to the initial approximation of the Pareto front) on the quality of the final Pareto set.

The chapter is organized as follows. In Section 2, we introduce the considered DRA and describe the design objectives. The multi-objective optimization problem, variable-fidelity EM-models, design space reduction techniques, as well as methods of constructing the surrogate model are discussed in Section 3. In the same section, we also describe the two alternative algorithmic frameworks for multi-objective antenna design. The multi-objective optimization algorithms are applied to our test case in Section 4, where we also discuss comparative experiments and investigate the influence of evolutionary algorithm imperfections on the accuracy and repeatability of the results. Section 5 concludes the chapter.

2 Design Case

The design example considered in this chapter is a compact, quasi-isotropic DRA shown in Fig. 1 [34, 36, 37]. The antenna consists of a cuboid shape Taconic CER-10 dielectric resonator ($\varepsilon_r = 10$, $\tan\delta = 0.0035$) which is driven through a cylindrical probe, fed from the bottom by a coaxial 50 ohm transmission line. The design is based on a reference structure of [36] and slightly modified to introduce additional degrees of freedom for the probe location, so that a better control of the structure behavior during optimization process is ensured. The antenna geometry is parameterized by a variable vector $\boldsymbol{x} = [a\ b\ c\ o_1\ o_2\ l]^T$, whereas dimensions $d = 1.26$ and $g = 0.82$ remain fixed to ensure 50 ohm input impedance. The conductor thickness is fixed to $T = 0.05$ (all dimensions are in mm).

The design objectives are: F_1—minimization of the antenna return loss $|S_{11}| = |S_{11}(\boldsymbol{x},f)|$ in the frequency band from $f_L = 2.4$ GHz to $f_H = 2.5$ GHz, F_2—reduction of the difference between minimal and maximal E-field strength $\Delta G = \Delta G(\boldsymbol{x},f)$ in x–z plane (see Fig. 1c) at the center frequency $f_C = 2.45$ GHz, and F_3—minimization of antenna volume defined as a cuboid $V = a \times b \times c$. The requirements can be rigorously formulated as follows:

$$F_1(\mathrm{x}) = \max\{|S_{11}(\mathrm{x},f)| : f_L \le f \le f_H\} \tag{1}$$

$$F_2(\mathrm{x}) = \Delta G(\mathrm{x},f) = \{|\max\{G(\mathrm{x},f)\} - \min\{G(\mathrm{x},f)\}| : f = f_C\} \tag{2}$$

$$F_3(\mathrm{x}) = V \tag{3}$$

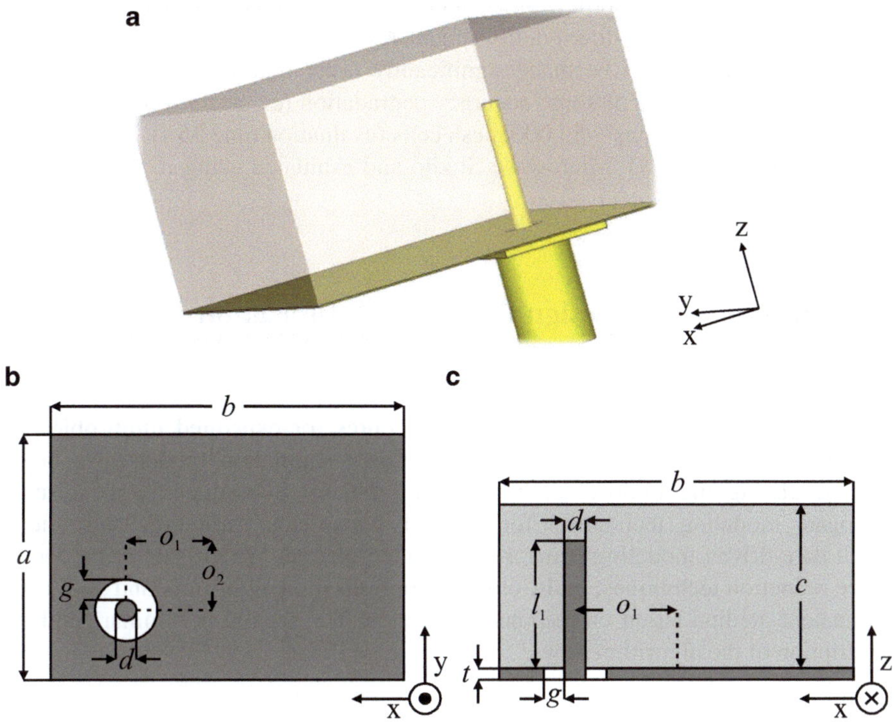

Fig. 1 A compact DRA: (**a**) structure visualization, as well as geometry of the antenna with highlighted dimensions: (**b**) bottom-view and (**c**) cross-section view [35]. The *light-* and *dark-shade gray* represent metallization (copper) and TLC-10, respectively; the *white color* represents vacuum

It should be noted that only the designs for which the in-band return loss is below −10 dB are considered acceptable. Consequently, design objectives (1) and (3) are reformulated by introducing the penalty function components

$$U_i(\mathrm{x}) = F_i(\mathrm{x}) + \beta_i \left(\max \left\{ \left| \frac{-10 - F_i(\mathrm{x})}{-10} \right|, 0 \right\} \right) \quad (4)$$

where i is objective index (1 or 3) for (1) and (3), respectively, and β_i are experimentally selected penalty factors (here, $\beta_1 = 10^5$ and $\beta_3 = 100$).

The initial solution space is defined using the following bounds: $\boldsymbol{l} = [3\ 3\ 3\ -0.45 \bullet a\ -0.45 \bullet b\ 0]^T$ and $\boldsymbol{u} = [30\ 30\ 30\ 0.45 \bullet a\ 0.45 \bullet b\ 0.9 \bullet c]^T$. The linear constraints are necessary to ensure that the probe is located within the resonator. The reference design is $\boldsymbol{x}_0 = [27\ 27\ 14.5\ 1\ 0\ 0.9]^T$.

The high-fidelity antenna model $\boldsymbol{R}_f$ consists of about ~1,000,000 hexahedral mesh cells and its average evaluation time on a dual Intel Xeon E5540 machine with 6 GB RAM is 21 min. Due to high computational cost of evaluating the $\boldsymbol{R}_f$ model its direct multi-objective optimization is impractical. The process can be accelerated by utilizing an auxiliary low-fidelity model $\boldsymbol{R}_c$, which is a coarse-discretization counterpart of $\boldsymbol{R}_f$. It can be made significantly faster than $\boldsymbol{R}_f$ [7], which can be achieved at the expense of some accuracy degradation (cf. Section 3.2). The low-fidelity model $\boldsymbol{R}_c$ contains ~55,000 mesh cells (evaluation time 35 s). Both models are implemented in CST Microwave Studio and evaluated using its time domain solver [38].

3 Multi-Objective Design Problem: Optimization Methodology

In this section, we briefly outline the procedures for expedited multi-objective optimization of expensive electromagnetic (EM) simulation models. We begin by formulating the multi-objective design problem. Subsequently, we discuss surrogate modeling techniques for narrowband antenna structures and briefly recall data-driven modeling using Kriging interpolation. We also describe design space reduction techniques, multi-objective optimization algorithm, and Pareto set refinement method based on response correction. The section is summarized by a description of the algorithm flows.

3.1 Problem Formulation

Let $\boldsymbol{R}_f(\boldsymbol{x})$ denote a computational model of the antenna structure under design. This original (or high-fidelity) model is normally obtained through accurate but computationally expensive electromagnetic (EM) simulation. The response vector $\boldsymbol{R}_f(\boldsymbol{x})$ represent relevant figures of interest such as antenna return loss versus frequency, gain, directivity, etc. Designable parameters (i.e., antenna dimensions) are represented by a vector $\boldsymbol{x}$.

Let $F_k(\boldsymbol{x})$, $k = 1, \ldots, N_{obj}$, be a kth design objective. Normally, the primary design goal is to minimize the antenna return loss response (in particular, to ensure $|S_{11}| \leq -10$ dB in a predefined frequency band [9]). However, other objectives, related to minimization of geometry (e.g., maximal lateral size, height, the maximal area of the footprint, volume, etc.), maximization of gain, or minimization/maximization of other performance measures may also be of interest [8, 21].

If $N_{obj} > 1$ then any two designs $\boldsymbol{x}^{(1)}$ and $\boldsymbol{x}^{(2)}$ for which $F_k(\boldsymbol{x}^{(1)}) < F_k(\boldsymbol{x}^{(2)})$ and $F_l(\boldsymbol{x}^{(2)}) < F_l(\boldsymbol{x}^{(1)})$ for at least one pair $k \neq l$, are not commensurable, i.e., none is better than the other in the multi-objective sense. In this case, a Pareto dominance relation $\prec$ is utilized [8]. For any two designs $\boldsymbol{x}$ and $\boldsymbol{y}$, one says that $\boldsymbol{x}$ dominates over $\boldsymbol{y}$ ($\boldsymbol{x} \prec \boldsymbol{y}$) if $F_k(\boldsymbol{x}) \leq F_k(\boldsymbol{y})$ for all $k = 1, \ldots, N_{obj}$, and $F_k(\boldsymbol{x}) < F_k(\boldsymbol{y})$ for all at least one k. A goal of multi-objective optimization is to find a representation of a so-called Pareto-optimal set X_P composed of the non-dominated designs from the solution space X, such that for any $\boldsymbol{x} \in X_P$, there is no $\boldsymbol{y} \in X$ for which $\boldsymbol{y} \prec \boldsymbol{x}$ [8].

3.2 Surrogate Models for Narrowband Antenna Design

Considerable numerical cost of evaluating the high-fidelity model $\boldsymbol{R}_f$ makes its direct multi-objective optimization prohibitive. Computational speedup can be obtained using surrogate-assisted techniques [1, 7, 33], where the optimization burden is shifted into the cheaper low-fidelity representation of the structure of interest referred to as $\boldsymbol{R}_c$. For the sake of brevity we omit here the detailed description of the low-fidelity model setup. The details can be found in [23, 29].

Evaluation cost of the considered DRA antenna model (cf. Section 2) is reduced by means of coarse mesh discretization. Consequently, the $\boldsymbol{R}_c$ is lacking accuracy and it has to be corrected to become a reliable representation of $\boldsymbol{R}_f$, especially in terms of its return loss response (other characteristics, e.g., gain, are normally well aligned for both $\boldsymbol{R}_f$ and $\boldsymbol{R}_c$). In case of narrowband antennas such as the considered DRA, the major type of discrepancy is frequency shift which can be reduced by suitable frequency scaling.

The frequency-scaled $\boldsymbol{R}_c$ model (denoted as $\boldsymbol{R}_{cF}$) is defined as

$$\boldsymbol{R}_{cF}(\boldsymbol{x}) = \boldsymbol{R}_c(\boldsymbol{x}, \boldsymbol{\alpha} F) \tag{5}$$

where $\boldsymbol{R}_c(\boldsymbol{x},F)$ denotes explicit dependency of $\boldsymbol{R}_{cF}$ on frequency F (in particular, it can be evaluation of antenna $|S_{11}|$ for a range of frequencies). Further, we have

$$\boldsymbol{\alpha} F = \overline{\alpha}_0 + \overline{\alpha}_1 F \tag{6}$$

where (6) describes the affine frequency scaling [39]. The frequency scaling parameters are obtained as

$$[\overline{\alpha}_0 \ \overline{\alpha}_1] = \arg \min_{[\alpha_0\, \alpha_1]} \sum_{k=1}^{N_r} \left|\left| \boldsymbol{R}_f\left(\boldsymbol{x}^{ref.k}\right) - \boldsymbol{R}_c\left(\boldsymbol{x}^{ref.k}, \alpha_0 + \alpha_1 F\right) \right|\right| \tag{7}$$

where $\boldsymbol{x}^{ref.k}$, $k = 1, \ldots, N_r$, are certain reference designs (normally, the extreme points of the Pareto set, cf. Section 3.4). The remaining (vertical) misalignment can be reduced using multi-point response correction of the form

$$\boldsymbol{R}_c(\boldsymbol{x}) = \overline{A} \cdot \boldsymbol{R}_{cF}(\boldsymbol{x}) \tag{8}$$

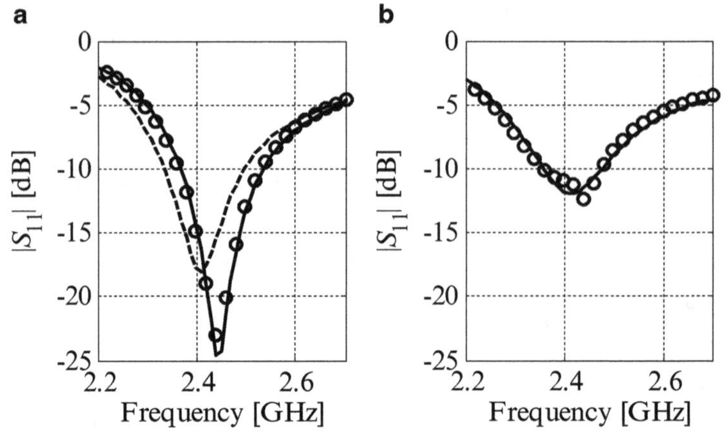

Fig. 2 Low-fidelity model correction of the return loss: (**a**) low- (*dashed line*) and high-fidelity model (*solid line*) responses at one of the reference designs, as well as corrected low-fidelity model response (*open circle*) at the same design; (**b**) corrected low- (*open circle*) and high-fidelity (*solid line*) model response at another design [35]

where $\bar{A} = \text{diag}([a_1 \; \ldots \; a_m])$ is a diagonal correction matrix obtained as

$$\overline{A} = \arg\min_{A} \sum_{k=1}^{N_r} || \boldsymbol{R}_f\left(\boldsymbol{x}^{ref.k}\right) - A \cdot \boldsymbol{R}_c\left(\boldsymbol{x}^{ref.k}, \overline{\alpha}_0 + \overline{\alpha}_1 F\right) ||^2 \tag{9}$$

The problem (9) can be reformulated as a linear regression problem which has an analytical least-square solution.

The return loss characteristics of the low- and the high-fidelity model of the antenna structure considered in Section 2, as well as corrected low-fidelity model $\boldsymbol{R}_c$ at one of the reference designs and some other (verification) design are shown in Fig. 2.

While the coarsely discretized model $\boldsymbol{R}_c$ is usually 10–50 times faster than $\boldsymbol{R}_f$, it is still too expensive to be directly utilized for multi-objective optimization of the considered DRA. Instead, we utilize a fast RSA surrogate model $\boldsymbol{R}_s$, which is constructed from sampled $\boldsymbol{R}_c$ data. Here, the RSA model is created using Kriging interpolation [24]. The model accuracy is determined using cross-validation technique [24], and the number of training samples is adaptively increased to ensure that the average relative root mean square (RMS) error of the model (calculated as $||\boldsymbol{R}_s(\boldsymbol{x}) - \boldsymbol{R}_c(\boldsymbol{x})||/||\boldsymbol{R}_c(\boldsymbol{x})||$ and averaged over the testing set) is at most 3 %. The selected design of experiments technique utilized to allocate the training samples is Latin hypercube sampling (LHS). For more detailed description of the methods for automated construction of the RSA model see [29, 33].

3.3 Kriging Interpolation

Kriging is one of the most popular techniques for approximating sampled data sets [40]. Here it is utilized to construct surrogate model $\boldsymbol{R}_s$ for multi-objective optimization procedure, in particular, to generate the initial approximation of the Pareto-optimal set and its further refinement [7, 33]. In this section, we provide a brief background information about Kriging interpolation. Detailed surveys can be found in [41, 42].

Let $X_B = \{\boldsymbol{x}^1, \boldsymbol{x}^2, \ldots, \boldsymbol{x}^N\}$ denote a base set, such that the responses $\boldsymbol{R}_c(\boldsymbol{x}^j)$ are known for $j = 1, 2, \ldots, N$. Let $\boldsymbol{R}_c(\boldsymbol{x}) = [R_{c.1}(\boldsymbol{x}) \ \ldots \ R_{c.m}(\boldsymbol{x})]^T$ (components of the model response vector may correspond to certain parameters, e.g., return loss or gain evaluated at m selected frequency points). Here, ordinary Kriging (Simpson et al. [40]) is utilized to estimate deterministic function f as $f_p(\boldsymbol{x}) = \mu + \varepsilon(\boldsymbol{x})$, where μ is the mean of the response at base points, and ε is the error with zero expected value, and with a correlation structure being a function of a generalized distance between the base points. Here, we use a Gaussian correlation function of the form

$$R\left(\boldsymbol{x}^i, \boldsymbol{x}^j\right) = \exp\left[\sum\nolimits_{k=1}^{N} \theta_k \left|x_k^i - x_k^j\right|^2\right] \tag{10}$$

where θ_k are unknown correlation parameters used to fit the model, while $x_k{}^i$ and $x_k{}^j$ are the kth components of the base points $\boldsymbol{x}^i$ and $\boldsymbol{x}^j$.

The Kriging-based coarse model $\boldsymbol{R}_s$ is defined as follows:

$$\boldsymbol{R}_s(\boldsymbol{x}) = [R_{s.1}(\boldsymbol{x}) \quad \ldots \quad R_{s.m}(\boldsymbol{x})]^T \tag{11}$$

where

$$R_{s.j}(\boldsymbol{x}) = \overline{\mu}_j + \boldsymbol{r}^T(\boldsymbol{x})\boldsymbol{R}^{-1}\left(\boldsymbol{f}_j - \boldsymbol{1}\overline{\mu}_j\right) \tag{12}$$

with $\boldsymbol{1}$ being an N-vector of ones,

$$\boldsymbol{f}_j = \left[R_{c.j}\left(\boldsymbol{x}^1\right) \quad \ldots \quad \ldots R_{c.j}\left(\boldsymbol{x}^N\right)\right]^T \tag{13}$$

$\boldsymbol{r}$ is the correlation vector between the point $\boldsymbol{x}$ and base points

$$\boldsymbol{r}^T(\boldsymbol{x}) = \left[R\left(\boldsymbol{x}, \boldsymbol{x}^1\right) \quad \ldots \quad \ldots R\left(\boldsymbol{x}, \boldsymbol{x}^N\right)\right]^T \tag{14}$$

whereas $\boldsymbol{R}$ is the correlation matrix between the base points

$$\boldsymbol{R} = \left[R\left(\boldsymbol{x}^j, \boldsymbol{x}^k\right)\right]_{j,k=1,\ldots,N} \tag{15}$$

The mean $\overline{\mu}_j$ is given by $\overline{\mu}_j = \left(\mathbf{1}^T \boldsymbol{R}^{-1} \mathbf{1}\right)^{-1} \mathbf{1}^T \boldsymbol{R}^{-1} \boldsymbol{f}_j$. The corelation parameters θ_k are found by maximizing [43]

$$-\left[N \ln\left(\overline{\sigma}^2\right) + \ln|\boldsymbol{R}|\right]/2 \tag{16}$$

in which the variance $\overline{\sigma}_j^2 = \left(\boldsymbol{f}_j - \mathbf{1}\overline{\mu}_j\right)^T \boldsymbol{R}^{-1}\left(\boldsymbol{f}_j - \mathbf{1}\overline{\mu}_j\right)/N$ and $|\boldsymbol{R}|$ are both functions of θ_k. Here, the Kriging model is implemented using the DACE Toolbox [44].

3.4 Design Space Reduction by Means of Single-Objective Optimizations

Because of its low evaluation cost, the Kriging interpolation surrogate $\boldsymbol{R}_s$ can be directly utilized (cf. Section 3.5) for generating the initial Pareto set using MOEA. Even despite large number of model evaluations required to complete the optimization process, the cost of such optimization is negligible compared to simulation cost of the high-fidelity EM model. On the other hand, the number of training points required to ensure desired accuracy of the interpolation model grows very quickly with the number of design variables. Consequently, the cost of setting up the Kriging model in high-dimensional spaces may quickly surpass the computational savings of the entire surrogate-based multi-objective optimization procedure. Moreover, the initial ranges of antenna geometrical parameters are usually set rather wide to ensure that the Pareto-optimal set can be captured. Clearly, construction of the RSA model in the large initial space may be impractical. Therefore, the reduction of the initially defined design spaces is considered a crucial step for successful EM-driven optimization.

Fortunately, the Pareto-optimal set normally resides in a very small region of the initial design space [33]. Consequently, the ranges of parameters can be limited so that the reduced space is as small as possible yet contains majority of the Pareto front. The reduction procedure can be formulated as follows. Let $\boldsymbol{l}$ and $\boldsymbol{u}$ be the initially defined lower and upper bounds for the design parameters. We define

$$\boldsymbol{x}_c^{*(k)} = \arg\min_{\boldsymbol{l} \le \boldsymbol{x} \le \boldsymbol{u}} F_k\left(\boldsymbol{R}_c\left(\boldsymbol{x}\right)\right) \tag{17}$$

where $k = 1, \ldots, N_{obj}$ be an optimum design of the coarsely discretized antenna model $\boldsymbol{R}_c$ obtained with respect to the kth objective. In the second stage, we obtain the corresponding (approximate) high-fidelity model optima

$$\boldsymbol{x}_f^{*(k)} = \arg\min_{\boldsymbol{l} \le \boldsymbol{x} \le \boldsymbol{u}} F_k\left(\boldsymbol{R}_f\left(\boldsymbol{x}\right)\right) \tag{18}$$

The designs $\boldsymbol{x}_f^{*(k)}$ are found using SBO. The most popular techniques utilized to solve (18) include frequency scaling combined with additive response correction is

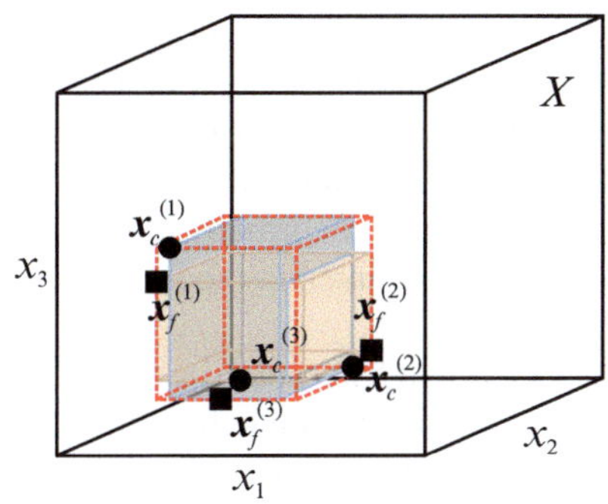

Fig. 3 Reduction of the initial solution space X by means of single-objective optimizations (three-objective case) [33]. The extreme Pareto-optimal points corresponding to the approximate extreme Pareto solutions of the low- (*filled square*) (cf. (8)) and the high-fidelity model (*filled circle*) (cf. (9)) determine the reduced design space X_R (*dashed lines*)

utilized [7]. The typical cost of such a process corresponds to a few evaluations of $\boldsymbol{R}_f$ [33]. The bounds of the reduced design space X_R (see Fig. 3) are then defined as follows:

$$\boldsymbol{l}^* = \min\left\{\boldsymbol{x}_c^{*(1)}, \ldots, \boldsymbol{x}_c^{*(N_{obj})}, \boldsymbol{x}_f^{*(1)}, \ldots, \boldsymbol{x}_f^{*(N_{obj})}\right\} \tag{19}$$

and

$$\boldsymbol{u}^* = \max\left\{\boldsymbol{x}_c^{*(1)}, \ldots, \boldsymbol{x}_c^{*(N_{obj})}, \boldsymbol{x}_f^{*(1)}, \ldots, \boldsymbol{x}_f^{*(N_{obj})}\right\} \tag{20}$$

It is worth mentioning that for typical shapes of the Pareto front the refined solution space contains both the front for the low- and the high-fidelity models. The former is essential because the RSA model created in $[\boldsymbol{l}^*, \boldsymbol{u}^*]$ is a representation of $\boldsymbol{R}_c$. The latter is important to ensure sufficient room for improving the high-fidelity designs if the refinement of the initial Pareto-optimal set is to be performed (cf. Section 3.6).

3.5 Initial Pareto Set Approximation

The initial approximation of the Pareto front is obtained by optimizing the Kriging surrogate model $\boldsymbol{R}_s$ using a MOEA. In this work, a standard MOEA with fitness sharing, Pareto-dominance tournament selection, and mating restrictions is utilized [7, 8]. For the sake of brevity we omit the description of the algorithm components. Interested reader is referred to [8, 45].

It should be noted that because the RSA model is very fast, one can afford execution of MOEA at this stage of the process. In particular, in order to obtain a good representation of the Pareto front for the design problem with three objectives, significant amount of model evaluations is needed. The algorithm setup utilized in the numerical experiments reported in Section 4 is: 2000 individuals per iteration for 50 generations with a total of 10^6 $\boldsymbol{R}_s$ simulations. Nevertheless, the computational cost of $\boldsymbol{R}_s$ optimization using MOEA is normally low and corresponds only to a few evaluations of the high-fidelity EM antenna model, thus, if needed, the number of EM simulations could be even larger.

Another important aspect of MOEA optimization of $\boldsymbol{R}_s$ is repeatability of the results. The variance of the obtained Pareto front representations can be greatly reduced by large population size mentioned in the previous paragraph to the extent of being practically insignificant as indicated by our numerical experiments reported in Section 4.4. On the other hand, for one of the algorithms considered here (see Section 3.8), the initial Pareto set undergoes surrogate-based refinement process (cf. Section 3.7). The latter allows further reducing statistical variations resulting from MOEA optimization.

3.6 Design Space Confinement

While the design space reduction step allows for computationally feasible identification of the Pareto-optimal solutions, considerable part of the reduced space would contain designs that violate the fundamental requirement upon acceptable in-band return loss level (cf. Section 2). This is due to high correlation between antenna dimensions and its electrical properties [35]. Consequently, even a small change of certain parameters may lead to unacceptable modification of antenna response. In order to obtain more precise information about the interesting part of the Pareto front, an additional confinement procedure is executed as described in the remaining part of this section. It utilizes the feasible part of the Pareto set obtained within the initially reduced design space.

Let $X_F = \{\boldsymbol{x}_f^{(k)}\}_{k=1,\ldots,Nf}$, be the feasible subset of the initial Pareto set. The aim of the confinement procedure is to identify a set of vectors $\boldsymbol{v}_k$, $k = 1, \ldots, n$, and dimensions $d_{p.k}$, $d_{n.k}$, $k = 1, \ldots, n$ (see Fig. 4 for symbol explanation) which define the confined space X_C, with respect to the center point $\boldsymbol{x}^c = (1/N_f)\sum_{k=1,\ldots,Nf} \boldsymbol{x}_f^{(k)}$ (X_C center).

Let us assume that the vectors $\boldsymbol{v}_1, \ldots, \boldsymbol{v}_{k-1}$ are already known and let $M_k = R^n \setminus \mathrm{span}(\{\boldsymbol{v}_1, \ldots, \boldsymbol{v}_{k-1}\})$ be the orthogonal complement of the n-dimensional Euclidean space R^n and the subspace spanned by $\boldsymbol{v}_1$ through $\boldsymbol{v}_{k-1}$; we also have $M^k = R^n$ for $k = 1$. The vector $\boldsymbol{v}_k$ is found as the direction at which the diameter of the orthogonal projection of $X_F - \{\boldsymbol{x}^c\}$ onto M_k, $P_k(X_F - \{\boldsymbol{x}^c\})$ reaches minimum, i.e.,

$$\boldsymbol{v}_k = \arg \min_{\boldsymbol{v} \in M_k} D\left(P_k\left(X_F - \{\boldsymbol{x}^c\}\right), \boldsymbol{v}\right) \tag{21}$$

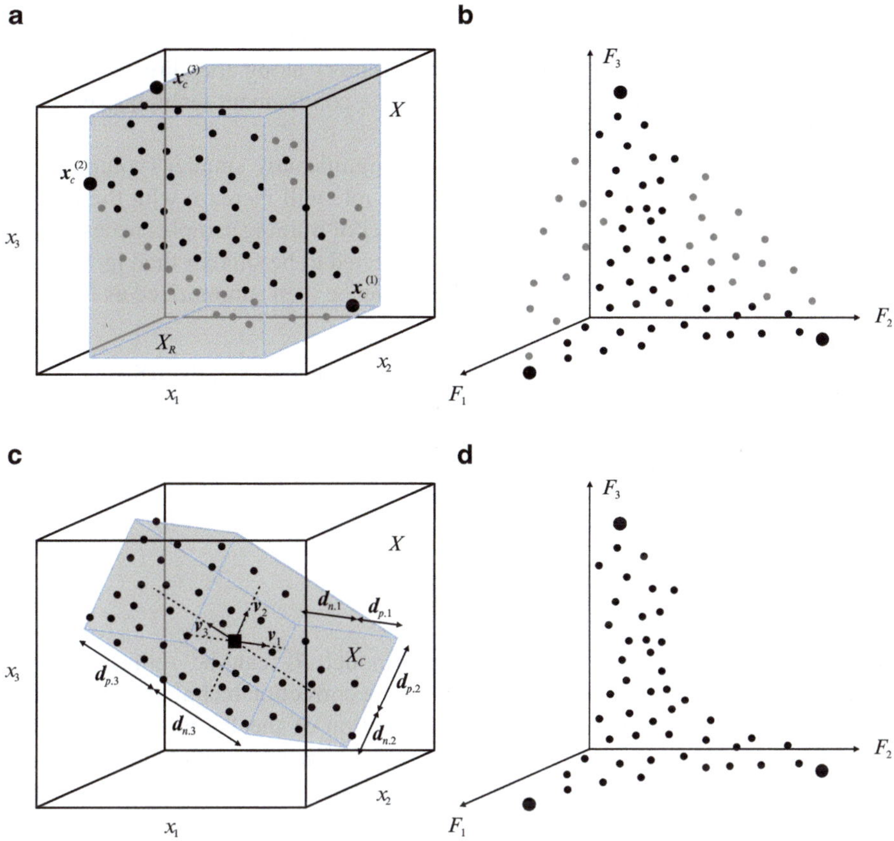

Fig. 4 Conceptual illustration of the design space confinement technique: (**a**) the initial Pareto-optimal set obtained within the initially reduced solution space X_R and (**b**) its feature space representation. The feasible and infeasible solutions (i.e., those with return loss above and below the level of -10 dB) are marked using *black* and *gray dots*, respectively. The confined solution space X_C (**c**) is a box of the smallest possible volume that contains all feasible Pareto-optimal solutions (**d**) obtained by MOEA optimization of the model identified within X_R. The positive and negative dimensions of the X_C are obtained with respect to the center point x_c (*filled square*). The unit vectors are $\boldsymbol{v}_1$, $\boldsymbol{v}_2$, and $\boldsymbol{v}_3$

Here, the diameter of a set Y in the direction of $\boldsymbol{v}$, $D(Y,\boldsymbol{v})$, is defined as

$$D(Y,\boldsymbol{v}) = \max_{y\in Y}\left\{\boldsymbol{v}^T\boldsymbol{y}\right\} - \min_{y\in Y}\left\{\boldsymbol{v}^T\boldsymbol{y}\right\} \tag{22}$$

Having the vectors $\boldsymbol{v}_k$, one can determine the sizes of the confined portion of the space:

$$d_{p.k} = \max_{x\in X_F}\left\{\boldsymbol{v}_k^T\boldsymbol{x}\right\} \qquad d_{n.k} = -\min_{x\in X_F}\left\{\boldsymbol{v}_k^T\boldsymbol{x}\right\} \tag{23}$$

for $k = 1, \ldots, n$. The obtained dimensions determine the maximum distance between the center point $\boldsymbol{x}^c$ and the points from X_F (both "positive" and "negative" as $\boldsymbol{x}^c$ is not necessarily the confined space center) along the directions of the vectors $\boldsymbol{v}_k$. A conceptual illustration of the space confinement technique is shown in Fig. 4.

The confined design space X_C is normally significantly smaller (volume-wise) than the initially reduced one and the reduction rate will increase with the number of the design variables. Consequently, the number of training samples necessary for creating a reliable RSA model within X_C is expected to be smaller than in X_R.

The overall design space reduction scheme can be then summarized as follows:

1. Perform initial design space reduction;
2. Sample the reduced space X_R, acquire $\boldsymbol{R}_c$ data, and construct the RSA;
3. Find initial Pareto set by optimizing RSA using MOEA;
4. Confine design space;
5. Reset the RSA model in the confined space.

3.7 Pareto Set Refinement

As shown in Section 3.2, the low-fidelity model $\boldsymbol{R}_c$ can be refined prior to construction of the RSA model using appropriate correction techniques. While this step allows obtaining more accurate representation of the Pareto front, it is not mandatory. On the other hand, the Pareto set obtained by optimizing $\boldsymbol{R}_s$ model constructed using only the low-fidelity model data is merely an approximation of the true Pareto front, corresponding to the RSA model (which, in turn, is an approximation of the low-fidelity EM simulation model $\boldsymbol{R}_c$). The refinement procedure [7, 33] described below aims at elevating a set of the selected designs $\boldsymbol{x}_s^{(k)}$, $k = 1, \ldots, K$, extracted from the initial Pareto set found by MOEA to the high-fidelity EM model level. The refinement stage exploits the additive response correction (output space mapping, OSM) [23, 25, 33] of the following form:

$$\boldsymbol{x}_f^{(k.i+1)} = \arg\min_{\boldsymbol{x}} F_1\left(\boldsymbol{R}_s(\boldsymbol{x}) + \left[\boldsymbol{R}_f\left(\boldsymbol{x}_s^{(k.i)}\right) - \boldsymbol{R}_s\left(\boldsymbol{x}_s^{(k.i)}\right)\right]\right) \tag{24}$$

subject to

$$F_q(\boldsymbol{x}) \le F_q\left(\boldsymbol{x}_s^{(k.i)}\right), \quad q = 2, \ldots, N_{obj} \tag{25}$$

The optimization process (24) is constrained not to increase the remaining objectives as compared to $\boldsymbol{x}_s^{(k)}$. The surrogate model $\boldsymbol{R}_s$ is corrected using the OSM term $\boldsymbol{R}_f(\boldsymbol{x}_s^{(k.i)}) - \boldsymbol{R}_s(\boldsymbol{x}_s^{(k.i)})$ that ensures zero-order consistency between the models (here, $\boldsymbol{x}_f^{(k.0)} = \boldsymbol{x}_s^{(k)}$) [46]. Consequently, the corrected surrogate coincides with $\boldsymbol{R}_f$ at the beginning of each iteration. In practice, only 2–3 iterations of (10) are sufficient to find a refined high-fidelity model design $\boldsymbol{x}_f^{(k)}$. After completing this stage, a set of Pareto-optimal high-fidelity model designs is created.

3.8 Optimization Flows

The flow of the first multi-objective optimization algorithm considered in this chapter can be summarized as follows [33] (Algorithm I):

1. Reduce design space using sequential single-objective optimizations (cf. Section 3.4);
2. Sample the reduced solution space X_R and acquire $\boldsymbol{R}_c$ model data;
3. Construct the Kriging interpolation model $\boldsymbol{R}_s$ (cf. Section 3.3);
4. Obtain the initial Pareto front by MOEA optimization of the RSA;
5. Refine selected elements of the Pareto set, $\boldsymbol{x}_s^{(k)}$, to obtain the corresponding high-fidelity model designs $\boldsymbol{x}_f^{(k)}$ (cf. Section 3.7).

It should be noted that the algorithm involves high-fidelity model evaluations only at its final stage. Typically, only two to three $\boldsymbol{R}_f$ simulations are required per design to elevate the low-fidelity Pareto-optimal solutions to the high-fidelity model level. Moreover, the number of $\boldsymbol{R}_f$ evaluations during the refinement process is pretty much independent of the dimensionality of the design space. The Pareto set obtained after Stage 5 is the final outcome of the multi-objective optimization process.

The algorithm is numerically efficient and simple to implement. However, in more problematic cases, such as the DRA considered in this chapter (more than two objectives and majority of the Pareto set containing unacceptable designs from the point of view of electrical performance of the antenna structure), it may yield a very sparse representation of the interesting part of the Pareto front.

Better results can be obtained with the algorithm exploiting the design space confinement of Section 3.6 and summarized here [35] (Algorithm II):

1. Perform design space reduction (cf. Section 3.4);
2. Correct low-fidelity model (cf. Section 3.2);
3. Sample the reduced design space X_R and acquire the $\boldsymbol{R}_c$ data;
4. Construct the Kriging interpolation model $\boldsymbol{R}_s$;
5. Obtain the initial Pareto front by optimizing $\boldsymbol{R}_s$ using MOEA;
6. Confine design space (cf. Section 3.6);
7. Reset the RSA model in the confined solution space X_C;
8. Find final Pareto-optimal set by MOEA optimization of the RSA model.

The major difference between the algorithms are Steps 6–8 where the emphasis is put on the part of the design space that contains acceptable Pareto-optimal solutions. It should be noted that low computational cost of the algorithm is ensured by utilization of the high-fidelity model only during $\boldsymbol{R}_c$ model correction step. At the same time, the number of low-fidelity simulations required for construction of the RSA model is typically a few hundred to more than a thousand, so that the computational cost of the optimization process corresponds to a few dozen of high-fidelity model evaluations (depending on the time evaluation ratio between $\boldsymbol{R}_f$ and $\boldsymbol{R}_c$ which is problem dependent). On the other hand, the method requires construction of two RSA models instead of just one as in the first algorithm.

4 Numerical Results

In this section we discuss numerical results of multi-objective design optimization of the compact dielectric antenna of Section 2. Two approaches are considered: design of the antenna within initial solution space and further refinement of the initial Pareto set using response correction technique as well as design of the structure using corrected surrogate model identified within confined solution space. Both considered methods are compared. Moreover, the influence of MOEA operation on the shape of the Pareto front is also investigated.

4.1 Antenna Optimization Using Algorithm I

Here, we present the optimization results of the DRA of Section 2 using Algorithm I (cf. Section 3.8). In the first step, the initial search space X has been refined using the technique of Section 3.4. The ranges of the reduced solution space are: $\boldsymbol{l}_r = [3.3\ 24.5\ 14.5\ {-0.03}{\bullet}a\ 0.37{\bullet}b\ 0.63{\bullet}c]^T$ and $\boldsymbol{u}_r = [29.8\ 30\ 15.8\ 0.16{\bullet}a\ 0.45{\bullet}b\ 0.98{\bullet}c]^T$. The refined space is four orders of magnitude smaller (volume-wise) than the initial one. Subsequently, a Kriging interpolation model $\boldsymbol{R}_s$ has been identified using 576 $\boldsymbol{R}_c$ samples (500 samples obtained using LHS scheme [47] supplemented with 64 corners of the hypercube and a total of 12 samples obtained at the center of each design space face). The average relative RMS error of the $\boldsymbol{R}_s$ model, determined using cross-validation [24], is 3 %. The initial Pareto front has been obtained using MOGA. Finally, a set of 14 designs selected along the initial Pareto front has been refined using response correction technique (cf. Section 3.7). Figure 5 shows the initial front and the refined high-fidelity model designs.

Detailed information on the selected Pareto-optimal designs is provided in Table 1. The high-fidelity design featuring the smallest E-field discrepancy of 4.97 dB has the volume of 9781 mm^3 and return loss of -10 dB. The smallest antenna design listed in Table 1 is characterized by the largest E-field discrepancy of 8.1 dB and return loss of -10.8 dB. Finally, the lowest in-band $|S_{11}| = -13.2$ dB has been obtained for the design with E-field discrepancy of 5.43 dB and volume of 10,164 mm^3. The ranges of variability of the three objectives F_1, F_2, and F_3 along the Pareto front are 3.2 dB, 3.1 dB, and 7308 mm^3, respectively. The results indicate that the influence of antenna miniaturization on the remaining objectives is significant. The return loss characteristics and the radiation patterns of selected high-fidelity antenna designs are shown in Figs. 6 and 7.

The overall cost of DRA optimization corresponds to about 75 $\boldsymbol{R}_f$ simulations ($\sim$26.2 h of CPU-time) and includes: 610 $\boldsymbol{R}_c$ and 567 $\boldsymbol{R}_c$ evaluations for determination of reduced space bounds and identification of the Kriging interpolation model, as well as 42 $\boldsymbol{R}_f$ evaluations required for the refinement of the selected Pareto-optimal designs.

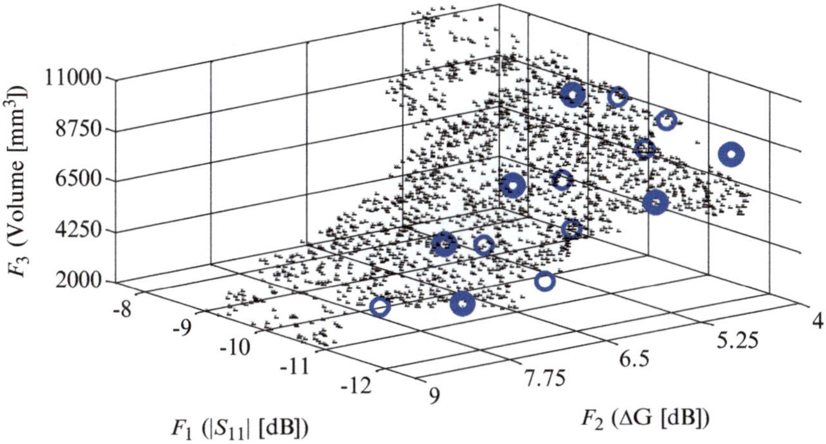

Fig. 5 The initial (*cross*) and refined (*open circle*) Pareto set representation obtained by the first multi-objective optimization algorithm of Section 3.8. *Thick-line circles* denote selected designs listed in Table 1. Note that considerable fraction of the initial Pareto set violates the requirement concerning maximum acceptable in-band return loss $|S_{11}|$

Table 1 Selected Pareto-optimal designs of the DRA antenna

				Design variables [mm]					
No.	F_1 [dB]	F_2 [dB]	F_3 [mm^3]	a	b	c	o_1	o_2	l
1	−10.0	5.0	9780	25.5	25.5	15.1	1.0	11.0	9.9
2	−10.2	5.9	6564	16.1	27.2	15.0	1.0	11.8	9.9
3	−10.4	6.9	4792	11.3	28.5	14.8	0.8	11.6	10.2
4	−11.8	7.8	3962	9.2	29.7	14.6	1.0	12.1	10.0
5	−13.2	5.4	10,163	26.6	26.3	14.5	1.3	11.1	9.2
6	−12.6	6.0	7870	19.9	27.3	14.5	1.6	12.3	9.2

4.2 Antenna Optimization Using Algorithm II

In this section, we present the optimization results of the DRA using the algorithm of Section 3.8. The initially reduced solution space is the same as the one obtained in Section 4.1. Next, the $\boldsymbol{R}_f$ responses of the extreme designs that form the refined space X_R (cf. Section 3.4) have been simulated and utilized for correction of the $\boldsymbol{R}_c$ model responses (cf. Section 3.2). The RSA model has been constructed using the same set of 576 samples and optimized using MOGA to determine the initial Pareto front. In the next step, the acceptable part of the Pareto set (i.e., the designs with in-band return loss below −10 dB) has been utilized to confine the solution space using the procedure of Section 3.6. The confined space is five orders of magnitude smaller (volume-wise) than X. Subsequently, the $\boldsymbol{R}_s$ model has been reset within X_C using only 170 $\boldsymbol{R}_c$ samples generated using LHS scheme [47]. The average relative RMS error of the final RSA model is only 1.5 %. Finally, the Pareto front has

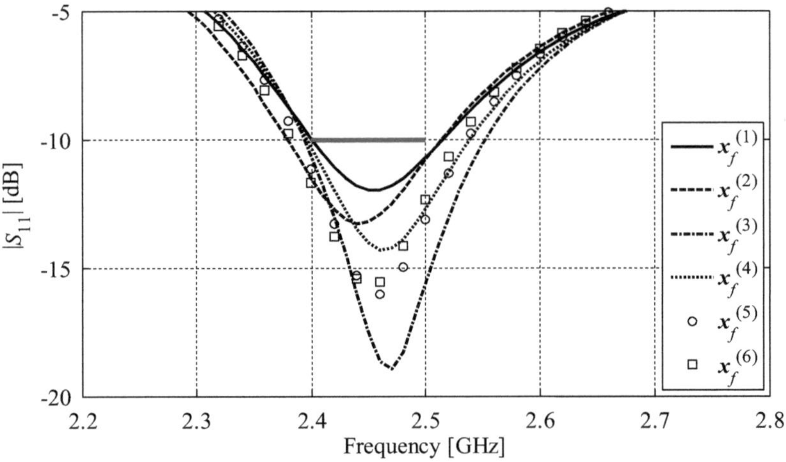

Fig. 6 Return loss characteristics of the DRA designs from Table 1

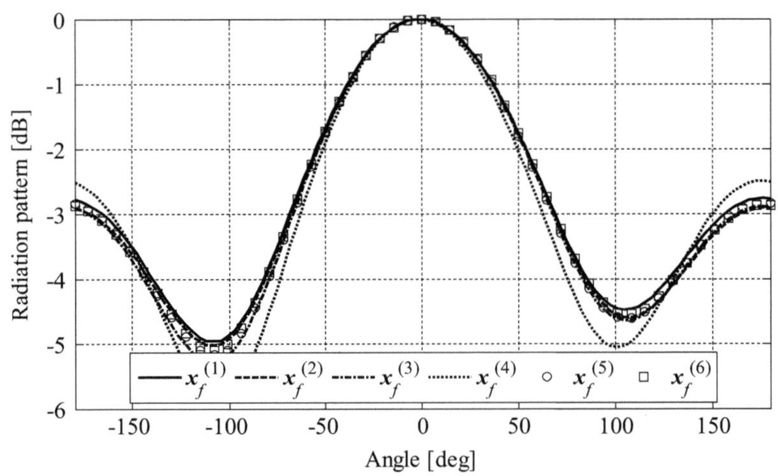

Fig. 7 E-field radiation patterns of the DRA designs from Table 1

been obtained by MOGA optimization of the new $\boldsymbol{R}_s$ model. It should be noted that due to correction of the $\boldsymbol{R}_c$ model responses prior construction of the RSA, the obtained results are the final outcome of the design optimization procedure (no need for further refinement).

The Pareto-optimal set obtained within the confined solution space is shown in Fig. 8, whereas dimensions of the selected high-fidelity model designs evaluated for verification purposes are listed in Table 2. Overall, the results are in good agreement; however, a slight misalignment of $|S_{11}|$ between $\boldsymbol{R}_s$ and $\boldsymbol{R}_f$ model responses can be observed. It is due to residual inaccuracy of the corrected low-fidelity model

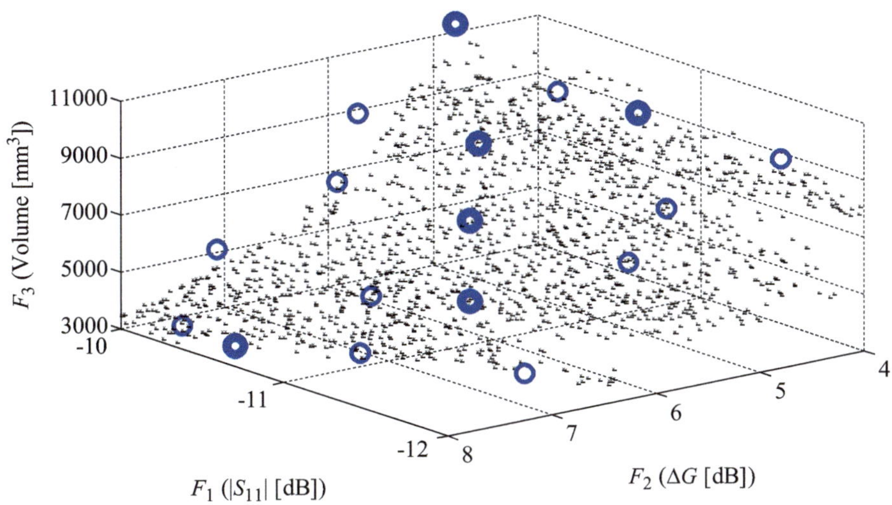

Fig. 8 The initial (*cross*) and refined (*open circle*) Pareto set representation obtained by the first multi-objective optimization algorithm of Section 3.8. The *thin-line circles* denote high-fidelity designs evaluated along the Pareto set for verification, whereas the *thick-line circles* denote designs gathered in Table 2

Table 2 Selected Pareto-optimal designs of the DRA antenna

				Design variables [mm]					
No.	F_1 [dB]	F_2 [dB]	F_3 [mm^3]	a	b	c	o_1	o_2	l
1	−10.0	4.8	11,337	29.2	25.2	15.4	2.5	10.7	10.0
2	−11.3	5.1	10,812	28.9	25.4	14.7	2.0	10.3	9.4
3	−10.8	5.9	6760	16.5	27.2	15.1	0.9	11.5	9.7
4	−11.4	6.8	5699	13.3	28.6	15.0	1.7	11.1	9.7
5	−10.6	7.9	3514	8.0	29.6	14.8	0.8	11.7	10.6
6	−10.5	5.4	8522	21.4	26.2	15.2	1.2	10.9	9.8

(the discrepancies are below 0.5 dB). At the same time, the responses of F_2 and F_3 remain accurate. Among the evaluated high-fidelity designs, the antenna with smallest volume exhibits largest E-field discrepancy of 7.9 dB and return loss of −10.6 dB. The design with lowest E-field variations of 4.81 dB features the largest volume of over 11,000 mm^3 and barely acceptable return loss of −10 dB. Finally, the lowest in-band return loss of −11.4 dB has been obtained for relatively compact design with a volume of almost 5700 mm^3 and E-field discrepancy of 6.84 dB. The ranges of variability of objectives F_1, F_2, and F_3 along the Pareto front are 2.3 dB, 3.1 dB, and 7823 mm^3 (69 %), respectively. Return loss characteristics and radiation patterns of antenna designs from Table 2 are shown in Figs. 9 and 10, respectively.

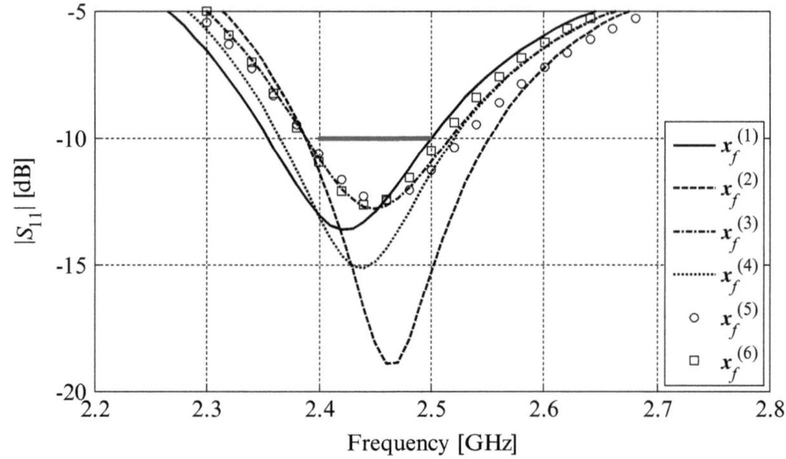

Fig. 9 Return loss characteristics of the DRA designs from Table 2

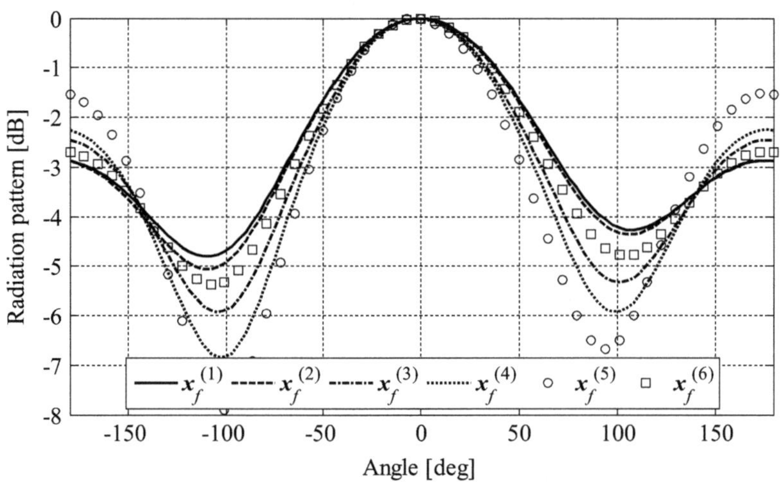

Fig. 10 E-field radiation pattern of the DRA designs from Table 2

The cost of multi-objective DRA optimization using the space confinement technique corresponds to about 42 $\boldsymbol{R}_f$ simulations (~14.5 h of CPU-time) and includes: 610 $\boldsymbol{R}_c$ evaluations for the initial design space reduction, 4 $\boldsymbol{R}_f$ and 567 $\boldsymbol{R}_c$ simulations for correction of the $\boldsymbol{R}_c$ and identification of the initial RSA model, as well as 170 $\boldsymbol{R}_c$ to establish second Kriging model within the confined design space.

4.3 Comparison of the Methods

The obtained results indicate that the cost of space reduction using the confinement technique (cf. Section 3.6) is significantly larger than for the basic method of Section 3.4. On the other hand, correction of the low-fidelity data prior to construction of the RSA model allows reducing the overall cost of multi-objective optimization within the confined space (cf. Section 4.2) by 44 % in comparison to the results of Section 4.1. This is because the cost of model correction for the first algorithm of Section 3.8 grows approximately linearly with the number of samples selected for correction, whereas for the second algorithm it is pretty much independent of the size of the sample set. At the same time, both methods provide similar results in the feasible region of the solution space.

The most important feature of the confinement technique is its ability to accurately capture the region of the search space that contains acceptable designs. Thus, more comprehensive information about the Pareto-optimal designs that are acceptable for the electrical performance standpoint can be obtained. This is not the case for Algorithm I, where most of the Pareto-optimal designs are of no practical value because of violating design specifications with respect to the return loss (see Fig. 5).

4.4 Statistical Analysis of MOEA

The influence of the MOEA operation on the results of the discussed design optimization algorithms has been verified through statistical analysis. It should be noted that the quality of the results obtained by MOEA may be of importance only for the procedure described in Section 4.2, because MOEA optimization is the final stage of the design process. In Algorithm I, the designs are further refined (cf. Section 4.1).

The MOEA has been reset 50 times within refined solution space. Clearly the Pareto fronts for the design problems with three objectives form a landscape, so that the results of statistical analysis and their mean are shown only for selected planes (see Fig. 11 for plots). The calculated standard deviation and average distance from the mean value are 0.04 dB and 0.03 dB, respectively. At the same time the largest discrepancy between the obtained Pareto sets and the mean within defined solution space is 0.32 dB. Such variability of the results has no meaningful influence on the structure behavior from the engineering standpoint. Consequently, the effect of the MOEA operation on the results of the discussed multi-objective optimization scheme is negligible.

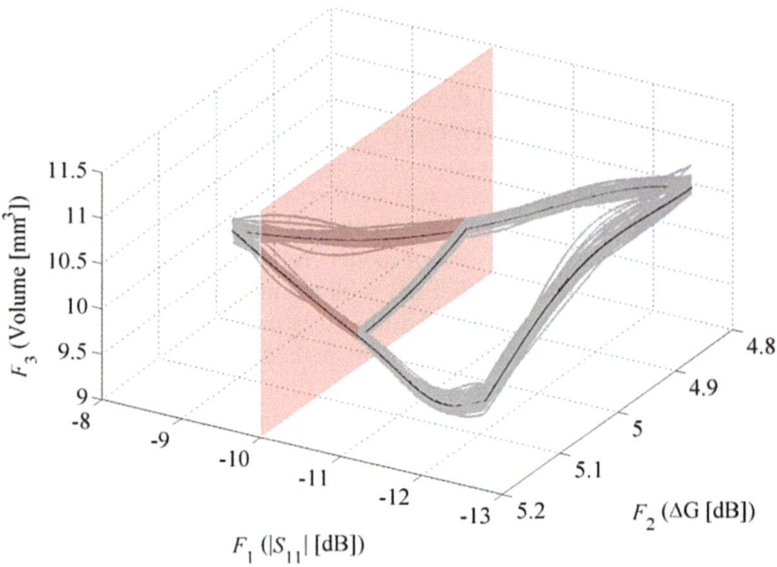

Fig. 11 Projections of the Pareto-optimal sets obtained in 50 runs of MOEA (*gray lines*) and the average (*black line*). Discrepancies between individual runs are insignificant from the engineering point of view, particularly given the range of variability for the design objectives. It should be noted that the results beyond the red plane violate the requirements upon minimal acceptable in-band |S11| (assumed to be below −10 dB), so that they are irrelevant

5 Conclusions

In this chapter, the problem of fast multi-objective design optimization of narrow-band DRA with respect to three non-commensurable objectives is considered. The presented techniques exploit RSA models, variable-fidelity EM simulations, MOEA and surrogate-based correction techniques. Construction of an accurate RSA model is possible by restricting the initial search space to the region that contains Pareto-optimal solutions.

Two reduction techniques based on sequential single-objective optimizations and confinement of the feasible fraction of Pareto set are discussed. The second method, allows for more precise identification of the relevant part of the Pareto front, in particular, identification of those designs are acceptable from the point of view of electrical performance parameters. For both algorithms, the computational cost of obtaining a representative subset of the Pareto-optimal designs corresponds to only a few dozen of evaluations of the high-fidelity EM simulation model of the antenna structure.

The obtained results indicate that changes of the antenna geometry noticeably influence its return loss response and radiation pattern characteristics. Statistical analysis of MOEA revealed that its stochastic nature has negligible influence on accuracy of the optimization results. Overall, the frameworks discussed in this work may be considered as a step towards computationally efficient multi-objective design optimization of antenna structures involving expensive EM simulation models.

Acknowledgements The authors would like to thank Computer Simulation Technology AG, Darmstadt, Germany, for making CST Microwave Studio available. This work is partially supported by National Science Center, Poland under grants 2013/11/B/ST7/04325 and 2014/12/T/ST7/00045.

References

1. Bekasiewicz, A., Koziel, S.: Structure and computationally-efficient simulation-driven design of compact UWB monopole antenna. IEEE Antennas Wirel. Propag. Lett. **14**, 1282–1285 (2015)
2. Koziel, S., Ogurtsov, S.: Rapid optimisation of omnidirectional antennas using adaptively adjusted design specifications and Kriging surrogates. IET Microwaves Antennas Propag. **7**, 1194–1200 (2013)
3. Guha, D., Gupta, B., Kumar, C., Antar, Y.M.M.: Segmented hemispherical DRA: new geometry characterized and investigated in multi-element composite forms for wideband antenna applications. IEEE Trans. Antennas Propag. **60**, 1605–1610 (2012)
4. Koziel, S., Bekasiewicz, A.: A structure and simulation-driven design of compact CPW-Fed UWB antenna. IEEE Antennas Wirel. Propag. Lett. to appear (2015)
5. Nocedal, J., Wright, S.: Numerical Optimization, 2nd edn. Springer, New York (2006)
6. Conn, A., Scheinberg, K., Vincente, L.N.: Introduction to Derivative-Free Optimization. MPS-SIAM Series on Optimization. SIAM, Philadelphia (2009)
7. Koziel, S., Ogurtsov, S.: Multi-objective design of antennas using variable-fidelity simulations and surrogate models. IEEE Trans. Antennas Propag. **61**, 5931–5939 (2013)
8. Deb, K.: Multi-Objective Optimization Using Evolutionary Algorithms. John Wiley & Sons, Chichester (2001)
9. Koziel, S., Bekasiewicz, A.: Strategies for Computationally Feasible Multi-Objective Simulation-Driven Design of Compact RF/Microwave Components. Eng. Comput., to appear (2015)
10. Koziel, S., Bekasiewicz, A., Kurgan, P.: Rapid multi-objective simulation-driven design of compact microwave circuits. IEEE Microwave Wireless Compon. Lett. **25**, 277–279 (2015)
11. Kuwahara, Y.: Multiobjective optimization design of Yagi-Uda antenna. IEEE Trans. Antennas Propag. **53**, 1984–1992 (2005)
12. Cao, W., Zhang, B., Liu, A., Yu, T., Guo, D., Wei, Y.: Broadband high-gain periodic endfire antenna by using I-shaped resonator (ISR) structures. IEEE Antennas Wirel. Propag. Lett. **11**, 1470–1473 (2012)
13. Sharaqa, A., Dib, N.: Position-only side lobe reduction of a uniformly excited elliptical antenna array using evolutionary algorithms. IET Microwaves Antennas Propag. **7**, 452–457 (2013)
14. Koziel, S., Bekasiewicz, A., Couckuyt, I., Dhaene, T.: Efficient multi-objective simulation-driven antenna design using Co-Kriging. IEEE Trans. Antennas Propag. **62**, 5900–5905 (2014)
15. Ramos, R.M., Saldanha, R.R., Takahashi, R.H.C., Moreira, F.J.S.: The real-biased multiobjective genetic algorithm and its application to the design of wire antennas. IEEE Trans. Magn. **39**, 1329–1332 (2003)
16. Yang, X.-S., Ng, K.-T., Yeung, H.S., Man, K.F.: Jumping genes multiobjective optimization scheme for planar monopole ultrawideband antenna. IEEE Trans. Antennas Propag. **56**, 3659–3666 (2008)

17. Robinson, J., Rahmat-Samii, Y.: Particle swarm optimization in electromagnetics. IEEE Trans. Antennas Propag. **52**, 397–407 (2004)
18. Nanbo, J., Rahmat-Samii, Y.: Hybrid real-binary particle swarm optimization (HPSO) in engineering electromagnetics. IEEE Trans. Antennas Propag. **58**, 3786–3794 (2010)
19. Chamaani, S., Mirtaheri, S.A., Abrishamian, M.S.: Improvement of time and frequency domain performance of antipodal Vivaldi antenna using multi-objective particle swarm optimization. IEEE Trans. Antennas Propag. **59**, 1738–1742 (2011)
20. Nanbo, J., Rahmat-Samii, Y.: Advances in particle swarm optimization for antenna designs: real-number, binary single-objective and multiobjective implementations. IEEE Trans. Antennas Propag. **55**, 556–567 (2007)
21. Aljibouri, B., Lim, E.G., Evans, H., Sambell, A.: Multiobjective genetic algorithm approach for a dual-feed circular polarised patch antenna design. Electron. Lett. **36**, 1005–1006 (2000)
22. Forrester, A.I.J., Keane, A.J.: Recent advances in surrogate-based optimization. Prog. Aerosp. Sci. **45**, 50–79 (2009)
23. Koziel, S., Yang, X.-S.: Computational Optimization, Methods and Algorithms. Studies in Computational Intelligence, vol. 356. Springer, Berlin (2011)
24. Queipo, N.V., Haftka, R.T., Shyy, W., Goel, T., Vaidynathan, R., Tucker, P.K.: Surrogate based analysis and optimization. Prog. Aerosp. Sci. **41**, 1–28 (2005)
25. Bandler, J.W., Cheng, Q.S., Dakroury, S.A., Mohamed, A.S., Bakr, M.H., Madsen, K., Søndergaard, J.: Space mapping: the state of the art. IEEE Trans. Microwave Theory Tech. **52**, 337–361 (2004)
26. Koziel, S., Leifsson, L., Ogurtsov, S.: Reliable EM-driven microwave design optimization using manifold mapping and adjoint sensitivity. Microw. Opt. Technol. Lett. **55**, 809–813 (2013)
27. Koziel, S., Ogurtsov, S., Szczepanski, S.: Rapid antenna design optimization using shape-preserving response prediction. Bull. Polish Acad. Sci. Tech. Sci. **60**, 143–149 (2012)
28. Siah, E.S., Sasena, M., Volakis, J.L., Papalambros, P.Y., Wiese, R.W.: Fast parameter optimization of large-scale electromagnetic objects using DIRECT with Kriging metamodeling. IEEE Trans. Microwave Theory Tech. **52**, 276–285 (2004)
29. Bekasiewicz, A., Koziel, S., Zieniutycz, W.: Design space reduction for expedited multi-objective design optimization of antennas in highly-dimensional spaces. In Solving Computationally Expensive Engineering Problems: Methods and Applications, pp. 113–147. Springer (2014)
30. Koziel, S., Ogurtsov, S.: Rapid design optimization of antennas using space mapping and response surface approximation models. Int. J. RF Microwave CAE **21**, 611–621 (2011)
31. Ouyang, J., Yang, F., Zhou, H., Nie, Z., Zhao, Z.: Conformal antenna optimization with space mapping. J. Electromagn. Waves Appl. **24**, 251–260 (2010)
32. Zhu, J., Bandler, J.W., Nikolova, N.K., Koziel, S.: Antenna optimization through space mapping. IEEE Trans. Antennas Propag. **55**, 651–658 (2007)
33. Koziel, S., Bekasiewicz, A., Zieniutycz, W.: Expedite EM-driven multi-objective antenna design in highly-dimensional parameter spaces. IEEE Antennas Wirel. Propag. Lett. **13**, 631–634 (2014)
34. Koziel, S., Bekasiewicz, A., Leifsson, L.: Multi-objective design optimization of Planar Yagi-Uda antenna using physics-based surrogates and rotational design space reduction. Int. Conf. Comp. Sci., in Procedia Comp. Sci., vol. 51, pp. 825–833. Reykjavik (2015)
35. Koziel, S., Bekasiewicz, A.: Fast multi-objective optimization of narrow-band antennas using RSA models and design space reduction. IEEE Antennas Wirel. Propag. Lett. **14**, 450–453 (2015)
36. Pan, Y.-M., Leung, K.W., Lu, K.: Compact quasi-isotropic dielectric resonator antenna with small ground plane. IEEE Trans. Antennas Propag. **62**, 577–585 (2014)
37. Koziel, S., Bekasiewicz, A.: Fast EM-driven size reduction of antenna structures by means of adjoint sensitivities and trust regions. IEEE Antennas Wirel. Propag. Lett., to appear (2015)
38. CST Microwave Studio, ver. 2013 (2013) CST AG, Bad Nauheimer Str. 19, D-64289 Darmstadt, Germany

39. Yelten, M.B., Zhu, T., Koziel, S., Franzon, P.D., Steer, M.B.: Demystifying surrogate modeling for circuits and systems. IEEE Circuits Syst. Mag. **12**, 45–63 (2012)
40. Simpson, T.W., Pelplinski, J.D., Koch, P.N., Allen, J.K.: Metamodels for computer-based engineering design: survey and recommendations. Eng. Comput. **17**, 129–150 (2001)
41. Kleijnen, J.: Design and Analysis of Simulation Experiments. Springer, New York (2008)
42. Sacks, J., Welch, W.J., Mitchell, T., Wynn, H.P.: Design and analysis of computer experiments. Stat. Sci. **4**, 409–435 (1989)
43. Forrester, A.I., Sobester, A., Keane, A.J.: Multi-fidelity optimization via surrogate modelling. Proceedings of the Royal Society, 463, pp. 3251–3269 (2007)
44. Lophaven, S.N., Nielsen, H.B., Søndergaard, J.: DACE: a Matlab Kriging toolbox. Technical University of Denmark (2002)
45. Coello Coello, C.A., Lamont, G.B., Van Veldhuizen, D.A.: Evolutionary Algorithms for Solving Multi-Objective Problems, 2nd edn. Springer, New York (2007)
46. Alexandrov, N.M., Lewis, R.M.: An overview of first-order model management for engineering optimization. Optim. Eng. **2**, 413–430 (2001)
47. Beachkofski, B., Grandhi, R.: Improved distributed hypercube sampling. American Institute of Aeronautics and Astronautics, Paper AIAA 2002–1274 (2002)

Optimal Design of Photonic Crystal Nanostructures

Abdel-Karim S.O. Hassan, Nadia H. Rafat, and Ahmed S.A. Mohamed

Abstract Simulated-driven optimization plays a vital role in the optimal design of engineering systems. The presented work in this chapter considers approaches for obtaining the optimal design of some photonic crystal (PC) nanostructures. PCs are periodic dielectric/dielectric or dielectric/metallic nanostructures manipulating the flow of light. They are one of the most emerging physical systems that have attracted the attention of engineers and scientists, in the last few decades, for their promising applications in many areas. Two optimization approaches are used for achieving the optimal design of one-dimensional (1D) PC nanostructures. The first approach is based on minimax optimization criterion that best fits the design specifications, while the second one is based on design centering criterion, to maximize the probability of satisfying design specifications. The proposed approaches allow considering problems of higher dimensions, in addition, optimizing over the PC layers' thickness and/or its material type. Two practical examples are given to demonstrate the flexibility and efficiency of these approaches. The first is a 1D PC-based optical filter operating in the visible range. The second example is a 1D PC-based spectral control filter, working in the infrared range, and enhances the efficiency of thermophotovoltaic systems.

Keywords Photonic filters design • Photonic crystals • Simulated-driven optimization • Optimal system design • Design centering

MSC code: 78M50

A.S.O. Hassan (✉) • N.H. Rafat • A.S.A. Mohamed
Engineering Mathematics and Physics Department, Faculty of Engineering, Cairo University, Giza, 12613 Egypt
e-mail: asho_hassan@yahoo.com; nhrafat@ieee.org; aashiry@ieee.org

S. Koziel et al. (eds.), *Simulation-Driven Modeling and Optimization*, Springer Proceedings in Mathematics & Statistics 153,
DOI 10.1007/978-3-319-27517-8_10

1 Introduction

Simulated-driven optimization plays a vital role in the design cycle of engineering systems. Besides achieving the optimal system design, it provides an important road map to save time and money before doing the fabrication step. In general, an engineering system is characterized by a set of designable parameters $\boldsymbol{\phi} \in \mathbb{R}^n$, where n is the number of design parameters, and its performance is described in terms of some measurable quantities $f_i(\boldsymbol{\phi})$, $i = 1, 2, \ldots, m$. These performance measures are usually evaluated through numerical system simulations. According to the application, design specifications are suggested by designer through specifying bounds on the performance measures. These design specifications define a region in the design parameter space called feasible region $\mathbf{R}_f$. If a point $\boldsymbol{\phi}$ lies inside the feasible region $\mathbf{R}_f$, then all the corresponding design specifications are satisfied. The objective is to find the optimal design point within $\mathbf{R}_f$ that best fits the predefined design specifications. Generally, the problem of finding optimal design point can be formulated as an optimization problem according to the used criterion [1], i.e., finding the optimal design point of the system necessitates the solution of an optimization problem.

Various optimization criteria can be used to achieve the optimal design that best fits the required specifications of the system. These criteria may be, for example, least-squares criterion, minimax criterion, and design centering criterion. Accordingly, several optimization techniques can be employed in optimizing engineering systems. Each technique has its own advantages and drawbacks. Nevertheless, there is no specific technique that has the absolute superiority over the others. The superiority of a technique over another depends on the type of application and engineering system involved.

Photonic crystal (PC) structures are one of the most emerging physical systems that have attracted the attention of engineers and scientists, in the last few decades, for their promising applications [2]. Generally, the PCs are composed of dielectric–dielectric or metallic–dielectric nanostructures which are repeated regularly in a way that may result in the appearance of what is being called photonic band gap (PBG). The PBG is defined as a range of forbidden frequencies within which transmission of light is blocked, as it is totally reflected and/or absorbed. This PBG exists as a result of the multiple Bragg scattering of the incident electromagnetic waves (EMW). According to the number of directional axes in which dielectric materials exhibit periodicity, PCs can be classified into three types: one-, two-, or three-dimensional PC structures.

The one-dimensional photonic crystal (1D PC) structures consist of a periodically repeated configuration of double/triple layers. Although they are the simplest among these types due to the ease of fabrication and analyzing, they have received the most attention of researchers and engineers because of their numerous promising applications. Among these applications: high reflecting omni-directional mirrors, anti-reflection coatings, low-loss waveguides, low threshold lasers, high quality resonator cavities, and photonic-based active devices [3, 4]. Moreover, these 1D PC

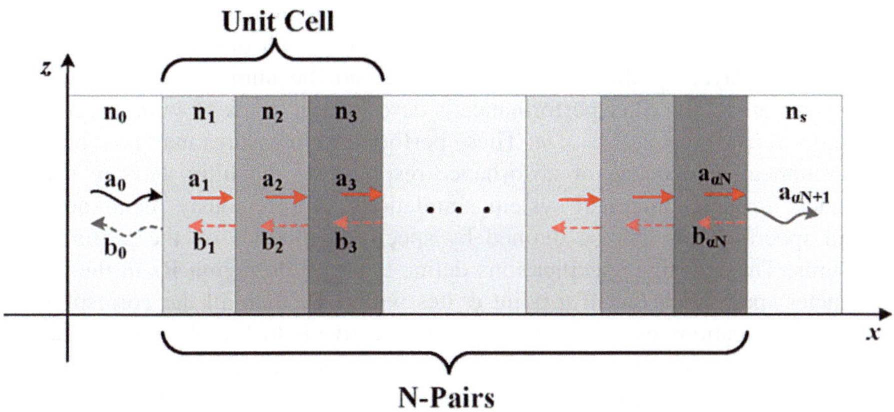

Fig. 1 Basic structure of a 1D PC

structures can form the basis of photonic filters operating over frequency spectrum ranging from radio waves up to optical wavelengths, passing by infrared and ultraviolet ranges.

Nowadays 1D PC filters, consisting of a periodically repeated configuration of double/triple layers as shown in Fig. 1, are playing a vital role in a wide range of applications of many areas of science. For instance, these applications include ultrahigh speed wireless communications, eye protection glasses, and anti-reflecting coating for solar cells. They are also utilized in biological, chemical imaging, and security screening. In addition to this, photonic filters are commonly used in space science and laser applications as well as thermophotovoltaic (TPV) applications [4–20].

According to the application type and the desired specifications, the required filter differs among wide band, narrow band, or selective pass/stop filters at selected wavelength ranges. However, the only common factor among these cases remains in the problem of finding the optimal design which best fits the desired performance. The traditional design of such 1D PC filters is the quarter-wave thick (QWT) design, in which, for a given material type the layer thickness is adjusted to satisfy the condition: $n_j d_j = \frac{\lambda_0}{4}$, where n_j and d_j are the layer refractive index and thickness, respectively. λ_0 may be the central wavelength of band gap or stopband of the corresponding filter.

In order to improve the characteristics and performance of a given 1D photonic filter, it is necessary to change the design approach from the traditional design of QWT layers design to a non-quarter-wave thick design [12]. This leads to an increase of the designable parameters which makes the filter design problem much more complicated. Also, it becomes so hard to predict, by intuitive PBG analysis, the behavior of the filter response due to specific variations of parameters. Therefore, the need for employing a simulated-driven optimization approach, seeking to achieve the optimal filter design, highly increases.

The general structure of 1D PC filter, like any engineering system design, is characterized by a set of designable parameters $\boldsymbol{\phi}$ which might be the refractive index of the layer n_j, the layer thickness d_j, and the number of periods N, as shown in Fig. 1. The filter performance is described in terms of some measurable quantities $f_i(\boldsymbol{\phi})$, $i = 1, 2, \ldots, m$. These performance measures may be the output transmittance, reflectance, or absorbance response of the filter and are usually evaluated through numerical system simulations. As previously mentioned, the design specifications can be defined by specifying bounds on the performance measures. These design specifications define the feasible region $\mathbf{R}_f$ in the design parameter space such that if a point $\boldsymbol{\phi}$ lies within $\mathbf{R}_f$ then all the corresponding design specifications will be satisfied. The objective is to find the optimal design point within the feasible region $\mathbf{R}_f$ that best fits the desired defined design specifications. Such problem of finding optimal filter design point can be formulated as an optimization problem according to the used criterion [21–29]. In this study we will use two different criteria in obtaining the optimal design namely: minimax optimization criterion and design centering optimization criterion.

The optimal design of 1D PC filters has been treated, in literature, through different strategies [12–18]. For example, Asghar et al. obtained the optimal design of a wide band pass optical filter (WBP-OF) by employing an approach based on genetic optimization algorithm [12]. Another design technique based on genetic algorithm was also studied, by Jia et al. [13]. Besides, Celanovic et al. [14] and Xuan et al. [15] applied genetic algorithm to achieve the optimal design of EMW filters. Moreover, Baedi et al. [16] have achieved the optimal design of a narrow band pass filter by using an approach based on particle swarm optimization. Recently, Badaoui and Abri [17] have proposed an optimization technique based on simulated annealing to obtain the optimal design of selective filters.

However, all these presented approaches have some drawbacks, mainly, the excessive number of required function evaluations, some practical cases need more than thousands of function evaluations, and low convergence rate in reaching to an optimal design point. Also, all of these approaches are not guaranteed to converge to an optimal design point, starting from any initial point, since their optimization algorithms are based on uncertainties. Inflexibility to adapt the aforementioned techniques to other filter design problems is another pitfall.

Another strategy, seeking to find the optimal refractive index values of multi-layered structures using convex optimization, was proposed by Swillam et al. [18]. However, this approach fixes the layers thickness to their QWT values, in order to guarantee a convex optimization problem. In addition, the complexity of the corresponding design problem increases in order to realize values of the optimized refractive indices.

In this chapter the optimal design of one-dimensional (1D) PC structures is treated through two different optimization approaches. The first approach is based on minimax optimization criterion [26–29] that belongs to the class of nominal design optimization where nominal values of designable parameters are varied to achieve a feasible design point that best fits the design specifications. This approach is very commonly used in Electromagnetic (EM)-based design [30–33]. The second approach is based on statistical design centering criterion belonging to design centering optimization approaches [1, 21–25, 34, 35], in which the objective is to maximize the probability of satisfying design specifications (yield function). The aim of the design centering approach is to make the optimized system immune against statistical variations, occurring in the system parameters, inherent to the fabrication process and model uncertainties.

The proposed optimization approaches allow considering problems of higher dimension than usually done before. The validity, flexibility, and efficiency of the proposed approaches are demonstrated by applying them to obtain the optimal design of two practical examples. The first is a 1D PC-based optical filter operating in the visible range. Contrarily, the second example is a 1D PC-based spectral control filter, working in the infrared range, which is used for enhancing the efficiency of thermal-photovoltaic systems. The approaches show very good ability to converge to the optimal solution, for different design specifications, regardless of the starting design point. Optimizing over both layers thickness and material types is considered as well. The optimized structures exhibit output responses which go far beyond typical physical intuition on periodic media design. This ensures that the introduced approaches are robust and general enough to be employed for achieving the optimal design of all 1-D photonic crystals promising applications.

2 1-D Photonic Crystal Filters: Structure and Governing Equations

The general structure of photonic filters is a 1-D PC which comprises of a unit cell repeated N times. This unit cell consists of two or three dielectric/metallic layers, as shown in Fig. 1. The filter is surrounded from front and back by incident and substrate media with refractive indices n_0 and n_s, respectively. This periodic layered structure of the filter configuration is defined by the number of periods, N, the layer thicknesses, d_j, and their refractive indices, n_j.

Generally, the propagation of EMW onto periodic photonic crystal structures is governed by the decoupled Maxwell's equation for non-magnetic, charge free, and current free materials [2] which is stated as:

$$\nabla \times \nabla \times \mathbf{E}(\mathbf{r}) = \frac{\omega^2}{c^2}\varepsilon_r(\mathbf{r})\,\mathbf{E}(\mathbf{r}), \tag{1}$$

where $\varepsilon_r(\mathbf{r})$ is the periodic relative permittivity of the structure, $\mathbf{E}$ is the electric field vector, ω is the angular frequency, and c is the speed of EMW in free space. Consider the case of 1D layered structure where the stack layers are normal to the x-axis (see Fig. 1). Also, assume normal incidence of polarized light, that is, the light propagates through layers and falls perpendicularly on each layer's interface. Thus, Eq. (1) is reduced to:

$$\frac{\partial^2 E_z(x)}{\partial x^2} + \frac{\omega^2}{c^2}\varepsilon_r(x)E_z(x) = 0, \tag{2}$$

Solving Eq. (2) leads to a general form of the electric field in the j-th layer:

$$E_j(x) = a_j e^{in_jk(x-x_j)} + b_j e^{-in_jk(x-x_j)}, \tag{3}$$

where a_j and b_j are the forward and backward electric field amplitudes in the j-th layer, respectively (as shown in Fig. 1). While x_j is the coordinate of the j-th interface and k is the free space wave number. By solving Eq. (2) in all regions together with the boundary conditions, we can determine the transmittance, reflectance, and absorbance response of the concerned filter.

The standard transfer matrix method (TMM) [36, 37], which is frequently used in optics, is used to analyze the propagation of the EMW signals through the layered media of the filter. The TMM is based on applying the simple continuity conditions of the electric field, that follow from Maxwell's equations, at each interface. The tangential component of the electric field and its first derivative must be continuous across boundaries from one medium to the next. Hence, by imposing these boundary conditions, the electric field amplitudes between any two successive layers can be related by the following transfer matrix:

$$M_{j,j+1} = \begin{bmatrix} \frac{1}{2}\left(1+\gamma_j\right)e^{in_jkd_j} & \frac{1}{2}\left(1-\gamma_j\right)e^{-in_jkd_j} \\ \frac{1}{2}\left(1-\gamma_j\right)e^{in_jkd_j} & \frac{1}{2}\left(1+\gamma_j\right)e^{-in_jkd_j} \end{bmatrix}, \tag{4}$$

where $\gamma_j = n_j/n_{j+1}$ is the refractive index ratio between the j-th and the $\left(j+1\right)$-th layer. Thus,

$$\begin{bmatrix} a_{j+1} \\ b_{j+1} \end{bmatrix} = M_{j,j+1} \begin{bmatrix} a_j \\ b_j \end{bmatrix}, \tag{5}$$

A total transfer matrix of the system, M_T, arises from the product of all transfer matrices related to all structure layers $M_T = M_{\alpha N,S}\, M_{\alpha N-1,\alpha N} \ldots M_{2,3}\, M_{1,2}\, M_{0,1}$ where αN is the total number of layers of the concerned structure (as declared in Fig. 1), and α is equal to 2 or 3 according to the number of layers per unit cell.

The forward amplitude of the incident medium, a_0, is assigned to a fixed value r_0, and by introducing one more boundary condition on the backward amplitude of the substrate, $b_{\alpha N+1} = b_s$, to be equal to zero; the amplitude of the transmission and reflection coefficients (t and r) can be related by the total transfer matrix M_T as:

$$\begin{bmatrix} t \\ 0 \end{bmatrix} = M_T \begin{bmatrix} r_0 \\ r \end{bmatrix}, \tag{6}$$

Thus, from this relation, the amplitude transmission and reflection coefficients can be derived from the elements of the total transfer matrix, yielding:

$$t = \left(M_{11} - \frac{M_{12}M_{21}}{M_{22}}\right) r_0 \quad \text{and} \quad r = -\frac{M_{21}}{M_{22}} r_0, \tag{7}$$

where M_{11}, M_{12}, M_{21}, and M_{22} are the elements of the total transfer matrix, M_T.

The associated transmittance, T, and reflectance, R, which are often of more practical use, are calculated from:

$$T = \frac{n_s}{n_0}|t|^2 \quad \text{and} \quad R = |r|^2, \tag{8}$$

Finally, the absorbance A is obtained from: $A = 1 - T - R$.

3 Minimax Optimal Design of 1-D PC Filters

In fact, a wide class of engineering system design problems can be formulated as minimax optimization problem. The minimax objective functions result from lower and/or upper specifications imposed on the performance measures of the system. Although the concept of minimax is rather traditional [26], it is still effectively used in many engineering design problems, especially filter design problems [26–33]. Moreover, the minimax optimization criterion is preferable for such filter design problems because it tends to achieve an equal-ripple response of the obtained optimal design [28]. Besides, the minimax approach searches for a better design point even if all the desired specifications were satisfied which allow to have much exceeded satisfactions of the desired specifications.

In the presented work, the design problem of 1D PC filter is formulated as a minimax optimization problem [38]. For solving the resultant minimax optimization design problem, it is formulated as a nonlinear programming problem. This problem can be solved by making use of readily available software tools, e.g., MATLAB [39]. The efficiency, reliability, and flexibility of the proposed approach are demonstrated by applying it to obtain the optimal design of two practical filters, WBP-OF and spectral filter, that are operating in two different spectrum regions and are used in two separate applications.

In general, the photonic crystal filter system parameters (d_j and n_j) are assembled in a vector, $\boldsymbol{\phi} \in \mathbb{R}^{2p}$, where $\boldsymbol{\phi} = \left[d_1\, d_2 \ldots d_p\, n_1\, n_2 \ldots n_p\right]^T$, N is the number of periods, and $p = \alpha N$ is the total number of layers.

A certain photonic filter is required to pass the incident EMW signals related to wavelength values located at passband region(s) and to stop/reject the signals within the stopband region(s). The ranges and locations of both the passband and stopband regions are varying according to the concerned application and needed filter type. The desirable ideal transmittance response of any filter can be expressed as satisfying the following condition:

$$T_I(\lambda) = \begin{cases} 100\%, & \text{for } \lambda \in \Lambda_p \\ 0\%, & \text{for } \lambda \in \Lambda_s \end{cases}, \tag{9}$$

where $T_I(\lambda)$ is the ideal transmittance at certain wavelength value λ, whereas Λ_p and Λ_s are the sets of wavelength values located in the passband and stopband regions, respectively.

In practice, the condition required in Eq. (9) can never be achieved. Therefore, a set of some design specifications $U(\lambda)$ is introduced to describe an acceptable transmittance response:

$$U(\lambda) \triangleq \begin{cases} \tau - \frac{\Gamma}{2} < T < \tau + \frac{\Gamma}{2}, & \text{for } \lambda \in \Lambda_p \\ T < \beta, & \text{for } \lambda \in \Lambda_s \end{cases}, \tag{10}$$

where τ and β are bounds of the desired transmittance or average transmittance at the passband and stopband regions, respectively, whereas Γ represents the acceptable percentage of ripples within the passband region (PBR).

Practically, we consider a finite number of wavelength samples in the spectrum range such that satisfying the specifications at these points implies satisfying them almost everywhere. Let m_p and m_s be the number of sample points in the passband and stopband regions, respectively. In this case the continuous specification function (10) is approximated by the discrete specification function:

$$U_i \triangleq \begin{cases} \tau - \frac{\Gamma}{2} < T(\lambda_i) < \tau + \frac{\Gamma}{2}, & i \in I_1 \\ T(\lambda_i) < \beta, & i \in I_2 \end{cases}, \tag{11}$$

where $I_1 = \{1, 2, \ldots, m_p\}$ and $I_2 = \{1, 2, \ldots, m_s\}$ are finite sets of integers.

On the other hand, it should be emphasized that not all the design parameters, $\boldsymbol{\phi}$, are supposed to be optimized; only a subset of $\boldsymbol{\phi}$, denoted by the design variables $\boldsymbol{x} \subset \boldsymbol{\phi}$, is selected. Fabrication techniques or technologies may also introduce additional restrictions on the possible values of these design variables. Thus, we may have upper bounds, ξ_{uj}, or lower bounds, ξ_{lj}, on some design variables, $\boldsymbol{x}_j$. These constraints can be stated as follow:

$$\xi_{lj} < \boldsymbol{x}_j < \xi_{uj}, \qquad j \in J, \tag{12}$$

where $J = \{1, 2, \ldots, n\}$ is a finite set of integers.

Error functions, e_i, arise from the deviation between each desired specification, U_i, and its corresponding calculated response, $T_i(\mathbf{x})$. However, to have a single type of errors, the upper and lower specifications should be formulated properly. So that, the constraints on design variables can be defined as:

$e_i(\mathbf{x}) < 0,$ *for all* $i \in I_1 \cup I_2$, where

$$e_i(\mathbf{x}) \triangleq \begin{cases} \tau - \frac{\Gamma}{2} - T(\mathbf{x}, \lambda_i), & i \in I_1 \\ T(\mathbf{x}, \lambda_i) - \tau - \frac{\Gamma}{2}, & i \in I_1 \\ T(\mathbf{x}, \lambda_i) - \beta, & i \in I_2 \end{cases} \tag{13}$$

It is clear that a negative error function indicates a satisfaction of the corresponding specification, whereas a non-negative error function indicates a violation of the corresponding specification. The union of all of these error functions forms the feasible region $\mathbf{R}_f \subset \mathbb{R}^n$, which is defined as:

$$\mathbf{R}_f = \{\mathbf{x} \in \mathbb{R}^n : \; e_i(\mathbf{x}) < 0, \quad \text{for all } i \in I_1 \cup I_2\} \tag{14}$$

A certain design point $\mathbf{x}$ is called feasible design point if it corresponds to entirely negative error functions, which in turn implies that it satisfies all the design specifications.

Now, starting from an initial point, $\mathbf{x}^{(0)}$, which may be feasible or infeasible, we seek to for, not only a feasible point, but also to the optimal feasible design point which best fits the desired specifications in a minimax optimization criterion. This point is found by solving the following minimax optimization problem:

$$\begin{aligned} &\min_{\mathbf{x}} \left\{ \max_{i} \{e_i(\mathbf{x})\} \right\} \\ \text{subject to} \\ &\xi_{lj} < x_j < \xi_{uj} \end{aligned} \tag{15}$$

where $i \in I_1 \cup I_2$ and $j \in J$. In (15), if the maximum error obtained at the optimal solution, $\mathbf{x}^*$, is negative, then it implies that all the design specifications and the design variables restrictions are being satisfied. Contrarily, if the maximum error is positive, it means that at least one specification or restriction is violated.

One of the most efficient methods, used for solving a minimax problem, is to convert it to a nonlinear programming problem [26], by introducing an additional variable z and the equivalent nonlinear programming problem becomes:

$$\begin{aligned} &\text{minimize} \quad z \\ &\text{subject to} \begin{cases} f_i(z, \mathbf{x}) > 0, & i \in I \\ x_j - \xi_{lj} > 0, & j \in J \\ -x_j + \xi_{uj} > 0, & j \in J \\ -z > 0 \end{cases} \end{aligned} \tag{16}$$

where $f_i(z, \mathbf{x}) = z - e_i(\mathbf{x})$, which is the i-th constraint of the nonlinear programming problem.

The last restriction in (16) is added to enforce assigning the variable z with negative quantities. Hence, to obtain the optimal minimax solution, we solve problem (16). Actually, such conventional nonlinear programming problem can be solved using any nonlinear programming algorithm. Here, MATLAB [39] is employed.

4 Minimax Optimal Design of a Wide Band Pass Optical Filter

4.1 Brief Description of the Filter

As a practical example, the proposed optimal design approach is applied for achieving the optimal design of a wide band pass optical filter (WBP-OF). The suggested structure of the filter is a 1-D photonic crystal, which comprises of a unit cell repeated *N* times. This unit cell consists of two different dielectric layers with a single metal layer between them. This structure is denoted by (dielectric$_1$/metal/dielectric$_2$)^N. Particularly, the structure of $(SiC/Ag/SiO_2)^N$ is considered, due to its good performance as a WBP-OF working in the visible range [19]. Moreover, such materials configuration is applicable and of ease to be fabricated [40–42].

In this example, we focus on the EMW spectrum range from 300nm to 900nm. The objective is to achieve a WBP-OF that passes EMW in the visible range (from 450nm to 700nm), and rejects both infrared and ultraviolet ranges. The range under study is divided into three main regions: the lower stopband region (LSBR) below 350nm, the PBR from 400nm to 700nm, and the upper stopband region (USBR) above 800nm.

All results of this section are obtained while the refractive indices of the incident and substrate media are assumed to be 1 and 1.52, respectively. A unity value is assigned to the forward amplitude of the incident medium (r_0), and the refractive index of SiO_2 is set to 1.45, whereas the refractive indices of Ag and SiC are assigned to practical measured values obtained from [43], so that the frequency dependency of the layered media can be considered.

The performance of the filter transmittance response is measured with respect to some figures of merit (FOM). First of these is the band factor, BF, which indicates the passband sharpness of the calculated transmittance response and is defined as $BF = \Delta\lambda_{50}/\Delta\lambda_{10}$ [19], where $\Delta\lambda_\delta = \lambda_{u\delta} - \lambda_{l\delta}$; $\delta = 10$ and 50. $\lambda_{l\delta}$ and $\lambda_{u\delta}$ are the wavelengths at which the transmittance equals to $\delta\,\%$. The other merits used are: T_{max}, λ_m, λ_{l10}, λ_{l50}, λ_{u10}, and λ_{u50}, where T_{max} is the maximum calculated transmittance and λ_m is the wavelength at which T_{max} occurs. In addition, we introduce T_{avg_LSBR}, T_{avg_PBR}, and T_{avg_USBR} which are the average transmittance at LSBR, PBR, and USBR, respectively.

Then, we seek to design a filter having the highest possible T_{max}, BF, and T_{avg_PBR}. In contrary, the filter should have the smallest possible T_{avg_LSBR} and T_{avg_USBR}. Also, the filter should have λ_{l50} and λ_{u50} which best match the visible range limits.

4.2 Optimization Procedures and Results

4.2.1 Periodic Structure with Constant Thicknesses Optimization

First, we consider the case of periodic structure with constant thicknesses (PSCT). Hence, the optimization variables are limited only to 3-variables, which are assembled to the design vector, $\mathbf{x} = [d_1\ d_2\ d_3]^T$, where d_1, d_2, and d_3 are the thickness of SiC, Ag, and SiO_2 layers, respectively. The variables are restricted to $\xi_{lj} = 3$ nm and $\xi_{uj} = 300$ nm, for $j = 1,\ 2,$ and 3. Furthermore, the discretization parameters m_p and m_s , the number of sample points in the passband and stopband regions, respectively, are set to 25 and 20, respectively. However, in order to guarantee a minimal transmittance of 75 % for all samples located at the PBR, and a maximal transmittance of 7 % for those sample points located either at the LSBR or at the USBR; the design specifications' parameters (τ and β) are set to 75 % and 7 %, respectively. Here, the ripples constraint is ignored; by neglecting the upper bound on bandpass transmittance, defined in (13), and by setting Γ to 0. We assign the design point suggested in [19], $\mathbf{x} = [20\ 10\ 70]^T$ nm, to be our initial point, $\mathbf{x}^{(0)}$. After applying our optimization approach, we move (only after 13-iterations with 65-TMM system simulations) to an optimal design point, $\mathbf{x}^* = [18.14\ 8.94\ 53.68]^T$ nm, that achieves the desired specifications. Let's denote solution A, to this optimal design point obtained by repeating the optimal design values of the three layers for the five periods to cover the whole structure as indicated in Table 1. Fig. 2 shows how the filter response is improved at solution A. The optimized T_{avg_PB} is incremented from 82.55 % to 83.33 % without violating any of the defined specifications, as declared in Table 1. Besides, the transmittance response is slightly shifted to the

Table 1 Figures of merit of the $(SiC/Ag/SiO_2)^5$ filter (different design points)

Design point	Initial point [19]	Solution A	Solution B	Solution C
T_{avg_LSB} (%)	0.75	2.67	0.48	0.78
T_{avg_PB} (%)	82.55	83.33	84.53	82.04
T_{avg_USB} (%)	4.56	2.67	1.48	1.04
BF (%)	59.68	73.92	84.08	85.47
T_{max} (%)	74.3	89.66	94.03	82.67
λ_m (nm)	543	502	458	455
λ_{l50} (nm)	496	410	442	432
λ_{u50} (nm)	718	736	743	738
$\delta\lambda_{50}$ (nm)	222	326	301	306
λ_{l10} (nm)	403	349	425	418
λ_{u10} (nm)	775	790	783	776
$\delta\lambda_{10}$ (nm)	372	441	358	358

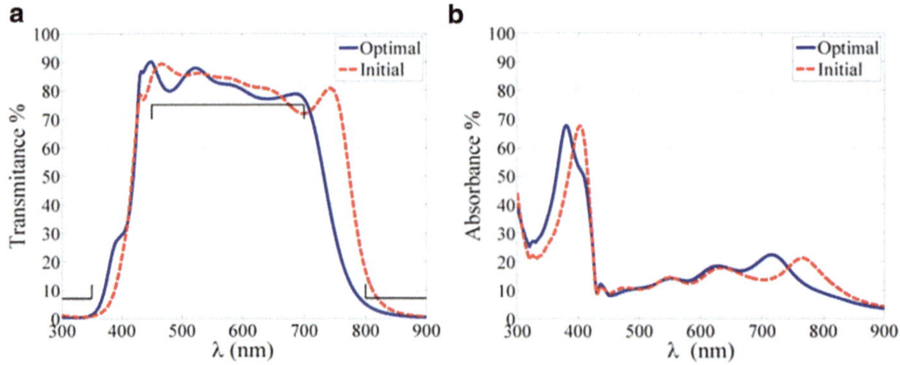

Fig. 2 (**a**) The transmittance and (**b**) The absorbance of $(SiC/Ag/SiO_2)^5$, before and after optimization, starting from an initial point: $d_1 = 20nm, d_2 = 10nm$ and $d_3 = 70nm$. The case of PSCT is considered

left such that it becomes more centered around the visible spectrum. Finally, the achieved enhancement in the FOM of the filter, for solution A, is summarized in Table 1.

4.2.2 Periodic Layered Structure with Variable Thickness Optimization

In order to enhance the performance of the filter response, the dimensionality of the problem is increased; by considering a periodic layered structure with variable thickness (PLSVT) assigned to each layer. Thus, the designable variables become $\mathbf{x} = [d_1\, d_2 \ldots d_{15}]^T$, where d_j is the j-th layer thickness. The new values of the parameters of the optimization approach are set as follow: $m_p = 25$, $m_s = 20$; $\tau = 82\%$, $\beta = 5\%$; $\xi_{lj} = 3\,\text{nm}$ and $\xi_{uj} = 300\,\text{nm}$, for $j = 1, 2, \ldots 15$. We also ignored the ripples constraint for this design problem. Starting from the design point solution A, we apply the proposed design approach with the new considerations. Accordingly, we move (after 108-iterations with 1862-TMM system simulations) into the new optimal design point $\mathbf{x}^* \in \mathbb{R}^{15}$, namely, solution B (see Table 2). Fig. 3 shows the optimal response at solution B, where T_{avg_PB} is incremented to 84.53 % without violating any other specification. In fact we could achieve a transmittance lower than 3 % over the whole stopband spectrum except for wavelength value of 800nm at which 5 % of transmittance is obtained. Moreover the sharpness (as indicated in Fig. 3) of the response's edges is considerably improved. The characteristics of the optimized filter are illustrated in Table 1.

4.2.3 Optimizing for the Least Level of Ripples at the PBR

Now, we try to find the optimal flat response which corresponds to the highest possible PBR transmittance with the least possible level of ripples. Thus, this time

Table 2 Design data of the filter $(SiC/Ag/SiO_2)^5$ (three different design solutions)

Design variable (*nm*)	Solution A	Solution B	Solution C
d_1	18.14	30.05	29.31
d_2	8.94	17.31	17.05
d_3	53.68	18.60	11.66
d_4	18.14	51.44	41.40
d_5	8.94	16.87	10.24
d_6	53.68	13.80	75.43
d_7	18.14	48.32	29.81
d_8	8.94	16.20	21.80
d_9	53.68	18.58	11.66
d_{10}	18.14	52.36	55.98
d_{11}	8.94	17.62	16.03
d_{12}	53.68	3.00	3.00
d_{13}	18.14	37.94	38.22
d_{14}	8.94	3.00	4.80
d_{15}	53.68	65.37	71.00

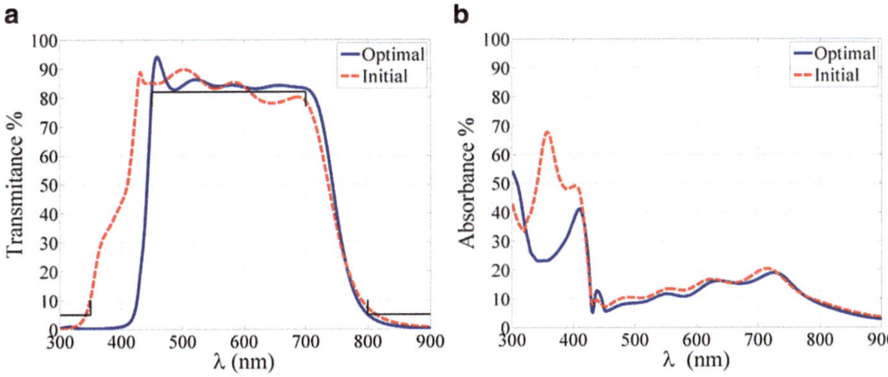

Fig. 3 (**a**) The transmittance and (**b**) The absorbance of $(SiC/Ag/SiO_2)^5$, before and after the optimization. Solution A is assigned to the initial design point. The case of PLSVT is considered

the ripples constraint, defined in (13), is considered. We set the acceptable ripples parameter, Γ, to be 2 % and by adjusting the other parameters as: $m_p = 25$, $m_s = 20$; $\tau = 82\%$, $\beta = 5\%$; $\xi_{lj} = 3$, $\xi_{uj} = 300$, for $j = 1, 2, \ldots 15$. The approach is applied, starting from solution B (see Table 2). We obtain an optimal design point (after 79-iterations with 1375-TMM system simulations) denoted by solution C (see Table 2). The obtained response and the FOM of solution C are illustrated in Fig. 4 and Table 1, respectively. Although the PBR's transmittance is slightly degraded, the transmittance response became almost flat around 82 %. Actually, such flat response is much preferable for many applications, as it prevents phase noise, resulting from the passband ripples.

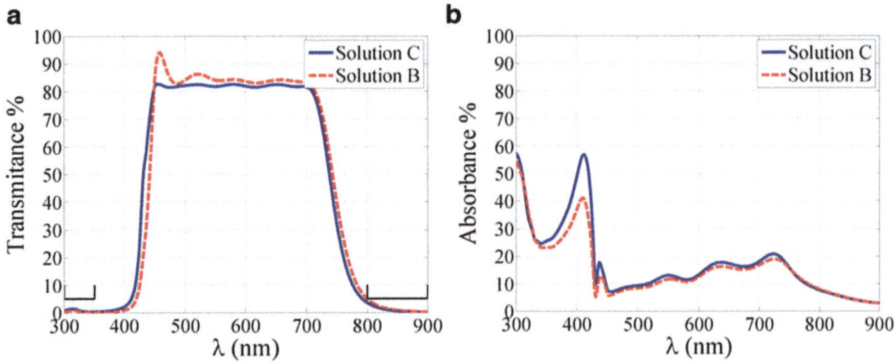

Fig. 4 (**a**) The transmittance and (**b**) The absorbance of $(SiC/Ag/SiO_2)^5$, before and after optimizing; Solution B is assigned to the initial design point. The case of PLSVT is considered and the ripples constraint is taken into consideration

5 Minimax Optimal Design of a Spectral Filter

5.1 *Brief Description of the Filter*

In this section, the proposed minimax optimization approach is applied to obtain the optimal design of spectral control filters, required for enhancing the efficiency of thermophotovoltaic (TPV) systems.

The TPV system is an energy converter that converts thermal heat into electrical energy [11]. It consists of an emitter, a photovoltaic (PV) cell, and a spectral control filter. The emitter is a thermal heater. It emits EMW onto the PV cell. Some of the radiation is reflected at the front surface of the cell (due to difference in refractive indices between cell and incident medium) and returned back to the emitter and the rest is transmitted through the cell. Photons of such transmitted radiation having energy greater or equal to the band gap of the PV cell can be absorbed by the cell and electron–hole pairs are generated. Photons having energy less than the band gap will not be absorbed by the cell and will be lost which limits the overall efficiency of the TPV system. Hence, in order to enhance the TPV efficiency, a spectral filter is located between the emitter and the PV cell to transmit photons that are suitable to the PV cell and reflect the remaining back to the emitter. In other words, an ideal spectral filter can be considered as a low band pass photonic filter that passes all the radiations with wavelength below the PV cell band gap wavelength (λ_g), and reflects all the radiations corresponding to wavelength higher than λ_g, where λ_g is the wavelength of the incident radiation in free space corresponding to photons of energy equals the band gap.

The performance of the spectral filter and the whole TPV system is assessed with respect to three suggested FOM which are: (1) the passband efficiency of the filter (η_p) which is the ratio between the above band gap power density transmitted from the filter to the PV cell (P_{abg}) and that transmitted from an ideal filter ($P_{abg|I}$); (2) the filter stopband efficiency (η_s) which is the ratio between the amount of below band gap power density (P_{bbg}) reflected from the filter back to the emitter and that amount of power reflected in case of an ideal filter ($P_{bbg|I}$); and (3) the spectral efficiency of the TPV system (η_{sp}) which is the ratio of P_{abg} to the net power density (P_{net}) radiated by the emitter, where P_{net} is the total power density radiated from the emitter minus the amount of the power reflected from the filter and returned back to the emitter. These power density quantities are estimated as follow:

$$P_{abg} = \int_0^{\lambda_g} I(\lambda, T_{em})\, T(\lambda)\, d\lambda, \tag{17a}$$

$$P_{bbg} = \int_{\lambda_g}^{\infty} I(\lambda, T_{em})\, R(\lambda)\, d\lambda \tag{17b}$$

where $T(\lambda)$ and $R(\lambda)$ are the transmittance and reflectance responses of the filter, respectively. $I(\lambda, T_{em})$ is the radiant intensity of the blackbody at wavelength λ and temperature T_{em}. It is calculated as [11]:

$$I(\lambda, T_{em}) = \frac{2\pi hc^2}{\lambda^5 \left(e^{hc/\lambda KT_{em}} - 1\right)}, \tag{18}$$

where h, K, and c are Planck constant, Boltzmann constant, and the speed of EMW in free space, respectively.

For an ideal spectral filter η_p, η_s, and η_{sp} equal 100 %. However, a practical filter does not perfectly transmit power in the passband, which negatively affects the above band gap power radiated to the PV cell, i.e., decreases η_p. Besides, in order to obtain a very high passband transmittance response, the filter will not be able to effectively reflect all the below band gap power which decreases η_s—as it is practically impossible to have a sharp edge transition at λ_g. Therefore, we should compromise between the passband efficiency and the stopband efficiency to achieve the highest possible spectral efficiency.

5.2 Optimization Procedures and Results

All results of this subsection are obtained, assuming the emitter radiation as an ideal blackbody radiation with 1500 K temperature. The PV material is assumed to be gallium antimonide (GaSb), which is the most common material used for the fabrication of TPV cells [11, 15]. GaSb has a refractive index of 3.8 and a band gap energy of 0.7 eV, which is equivalent to band gap wavelength $\lambda_g = 1.78\ \mu m$. Thus, the spectral filter is supposed to transmit all photons below 1.78 μm to the PV cell, and to reflect all photons above 1.78 μm. However, due to low energy of the blackbody radiation below 0.85 μm and above 6.5 μm, the filter is just designed to have a transmittance as high as possible in the wave band of 0.85–1.78 μm and becomes as low as possible in the band of 1.79–6.5 μm. In order to guarantee a steep transition at λ_g, the passband and the stopband regions are defined strictly closed to each other, and as an alternative approach the design specifications (τ and β) are defined on the average transmittance of the passband and stopband regions, respectively. Moreover, the ripples constraint is ignored for this design problem due to the minor impact of the passband ripples in this application.

The filter structure is suggested as a 1-D PC comprising of N periodically repeated unit cell. The unit cell consists of two consecutive metallic (M) and dielectric (D) layers. Thus, the filter is denoted by $D(MD)^N$, where the first dielectric layer is added to improve the filter matching with the incident medium. Initially, the metal and dielectric layers are assumed to be Ag and SiO_2 in respective [20]. The number of periods, N, is fixed to 3. The refractive indices of the incident and substrate media are assumed to be 1 and 3.8, respectively, and a unity value is assigned to the forward amplitude of the incident medium (r_0). The refractive index of SiO_2 is set to 1.5. Besides, the absorption and frequency dependency of Ag-layers are considered, by using the Drude model [11] to calculate the refractive index of Ag.

5.2.1 Thickness Optimization

First, the designable parameters are supposed to be the thickness of the layers. Thus, the design vector is considered as $\mathbf{x} = [d_1 \; d_2 \; \ldots \; d_7]^T$, where d_j is the thickness of j-th layer. The variables are restricted to $\xi_{lj} = 3\,\text{nm}$ and $\xi_{uj} = 300\,\text{nm}$, for $j = 1, 2, \ldots 7$, and the discretization parameters (m_p and m_s) are set to 20 and 40, respectively. Initially, the Ag-layers thickness is set to 10nm and the thickness of SiO_2 is assumed to follow the well-known quarter-wave thick design (QWTD), at which: $\mathrm{n_{SiO_2} d_{SiO_2}} = \frac{\lambda_g}{4}$, where n_{SiO2} and d_{SiO2} are the refractive index and thickness of the SiO_2 layers, respectively.

Figures 5 and 6 show optimal responses of two solutions, obtained using the proposed minimax optimization approach, namely solution 1 and solution 2, respectively. In Fig. 5, we optimize for the highest possible transmittance in the passband, whilst we optimize for the least possible transmittance within the stopband in Fig. 6. In solution 1, the design specifications are set as $\tau = 83\%$ and $\beta = 3.5\%$, to have the highest possible transmittance in the passband. Contrarily, to obtain the least possible stopband transmittance in solution 2, the design specifications are adjusted as $\tau = 79\%$ and $\beta = 2\%$. Solution 1 is obtained after 54-iterations and 486-TMM system simulations. In contrast, 62 number of iterations and 558-TMM system simulations are executed, till achieving solution 2. Table 3 shows the optimized values obtained, for the two solutions. The FOM regarding the two solutions and the initial point are compared in Table 4. Starting from initial passband efficiency equal to 15.2 % in the initial point, η_p significantly improved to 83.7 % in solution 1, as opposed to only 77.6 %, achieved, in solution 2. However, solution 2 is practically better than solution 1, because it achieves much reflection of the radiation in the stopband; resulting in higher spectral efficiency than solution 1 (67 % comparing to 59.8 % in solution 2).

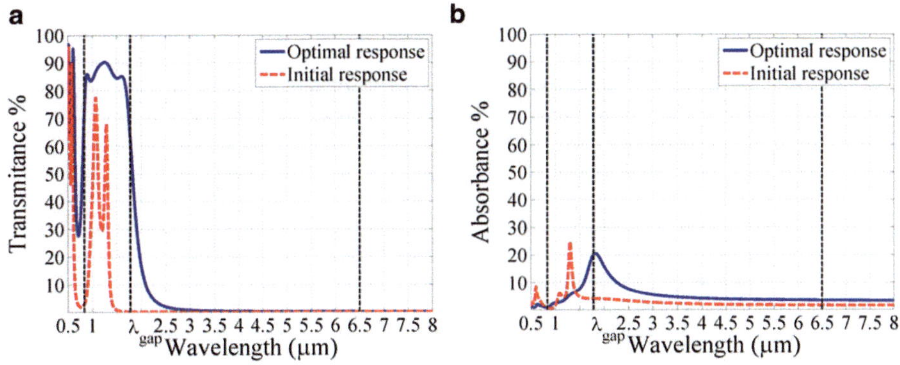

Fig. 5 (**a**) The transmittance and (**b**) The absorbance of the $SiO_2(Ag/SiO_2)^3$ spectral filter, before and after optimizing for the highest passband transmittance. The initial point is assumed as the QWTD for SiO_2-layers, while the Ag-layers are fixed to 10 nm. The obtained optimal solution is referred as solution 1

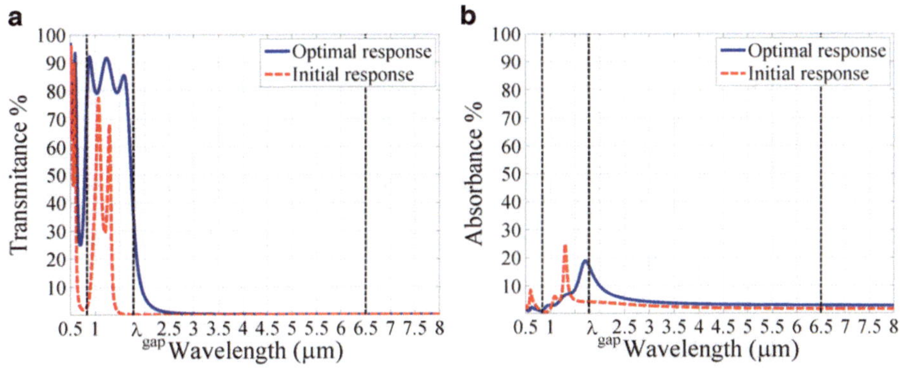

Fig. 6 (**a**) The transmittance and (**b**) The absorbance of the $SiO_2(Ag/SiO_2)^3$ spectral filter, before and after optimizing for the least stopband transmittance. The initial point is assumed as the QWTD for SiO_2-layers, while the Ag-layers are fixed to 10 nm. The obtained optimal solution is referred as solution 2

Table 3 Design data of the $D(Ag/D)^3$ spectral filter, where D refers to a dielectric layer

Design variable	Solution 1	Solution 2	Solution 3
$d_1(nm)$	131.22	120.59	80.87
$d_2(nm)$	3.14	4.15	7.91
$d_3(nm)$	256.74	259.74	164.34
$d_4(nm)$	3.05	3.61	6.28
$d_5(nm)$	245.93	238.82	154.96
$d_6(nm)$	3.00	3.00	5.06
$d_7(nm)$	332.71	327.15	194.60
n_D	1.5	1.5	2.38

Three optimized design points are given

Table 4 Figures of merit of the $D(Ag/D)^3$ spectral filter, where D refers to a dielectric layer

Design point	Initial point	Solution 1	Solution 2	Solution 3
T_{avg_PB} (%)	20.88	83.95	79.42	82.51
T_{avg_SB} (%)	0.01	3.23	1.53	1.06
η_p (%)	15.23	83.69	77.63	79.50
η_s (%)	97.27	87.40	92.44	94.92
η_{sp} (%)	47.90	59.82	66.95	77.46

Different design points are considered

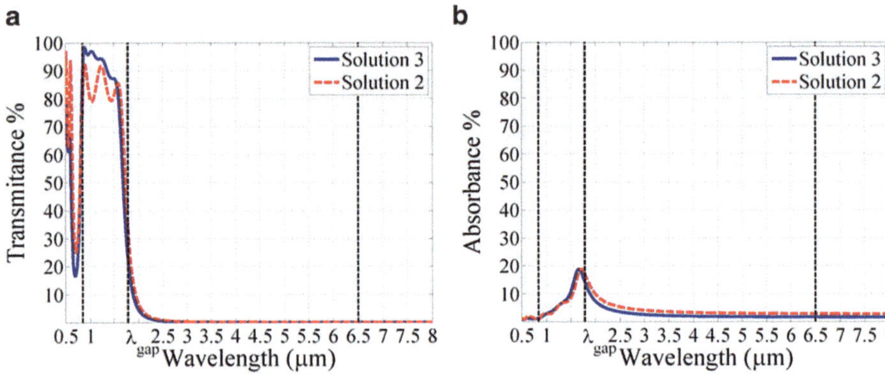

Fig. 7 (**a**) The transmittance and (**b**) the absorbance of the $D(Ag/D)^3$ spectral filter, before and after optimizing for the least possible stopband transmittance, where D refers to a dielectric layer. Solution 2 is considered as an initial point. Both the refractive index of the dielectric layers and the thickness of layers are optimized

5.2.2 Dielectric Material Optimization

In order to improve the performance and efficiency of the filter, the type of dielectric material can also be optimized. Thus, the designable parameters are the thickness of the layers (d_j), as well as the refractive index of the dielectric material, n_D, i.e., the design vector becomes $\mathbf{x} = [d_1\ d_2\ \ldots d_7\ n_D]^T$. Although performing optimization on the refractive index of the dielectric material may seem just a theoretical study as it may result on non-realizable materials, it is still a potential future study as we may be capable of fabricating materials with these exact refractive indices values once day. Another solution to address this problem is to replace the optimized refractive index with its nearest, in refractive index value, realizable (already exact) material.

The optimization parameters are set as: $m_p = 20$, $m_s = 40$; $\tau = 82.5\%$, $\beta = 1.5\%$; $\xi_{lj} = 3$ nm and $\xi_{uj} = 300$ nm, for $j = 1, 2, \ldots 7$. Besides, the dielectric refractive index is bounded as $1 \leq n_D \leq 5$. Starting from solution 2, we apply the proposed design approach under these new considerations. Accordingly, we obtain (after 77-iterations with 798-TMM system simulations) the new optimal design point $\mathbf{x}^* \in \mathbb{R}^8$, namely, solution 3 (see Table 3). Fig. 7 shows the obtained optimized response, where the passband transmittance is significantly enhanced. The FOM of the optimized filter are illustrated in Table 4. The spectral efficiency is noticeably incremented to 77.5 % with considerably high passband efficiency, $\eta_p = 79.5\%$. Fortunately, the optimized refractive index value ($n_D{}^* = 2.38$) is very near to the refractive index of Titanium Dioxide (TiO_2) dielectric material which is commonly used in the fabrication of 1-D PC structures [9]. That makes the achieved spectral efficiency in solution 3, 77.5 %, applicable in reality.

6 Design Centering of 1-D PC Filters

In general, system parameters are subject to known but unavoidable statistical fluctuations inherent to the manufacturing processes used or due to model uncertainties. This may cause some of the manufactured devices to violate the design specifications. The percentage of outcomes that satisfy the design specifications is called the production yield. Production yield is an important factor of the fabrication cost, it is always said "the smaller the yield, the higher the manufacturing cost." A vital goal of optimal system design is to maximize the production yield prior to the fabrication process. Production yield maximization can be achieved through design centering process, which seeks the values of optimal designable system parameters that maximize the probability of satisfying design specifications (yield function). Therefore, we propose a second optimization approach, belonging to class of design centering optimization approaches. The introduced approach is a statistical design centering optimization approach [23, 24, 35, 44–50] in which the objective is to maximize the yield function. The aim of this approach is to achieve a robust optimized system which has immunity against statistical variations that affect the system parameters. In this design centering approach, the design problem is formulated as an unconstrained yield optimization problem. This problem is solved by using derivative-free trust region based algorithm (NEWUOA) coupled with a variance reduction technique for estimating the yield function values. This enables to reduce the large number of required system simulations.

To demonstrate efficiency of the approach, it is employed to obtain optimal design center point of the aforementioned practical example WBP-OF. The approach can be applied starting from either an initial infeasible, feasible design point or the minimax point, for example.

6.1 Design Centering Problem Formulation: Statistical Design Centering

In order to simulate the statistical fluctuations that affect the system parameters, the parameters are assumed to be random variables with a joint probability density function (PDF) $P(\mathbf{x}, \boldsymbol{\nu})$, where $\boldsymbol{\nu}$ the distribution parameters like the mean vector μ

and covariance matrix $\mathbf{\Sigma}$. Thus, the probability that a certain design point $\mathbf{x}$ satisfies the desired design specifications (yield function) can be defined as:

$$Y(\nu) = \int_{R_f} P(\mathbf{x}, \nu)\,\mathrm{dx}., \tag{19}$$

where $\mathbf{R}_f$ is the feasible region defined by (14).

The design centering optimization process assumes that yield function depends only on the nominal values of system design parameter $\mathbf{x}_0$ and seeks for their optimal values that maximize the yield function. Hence, the design centering problem is formulated as:

$$\max_{\mathbf{x}_0} \; Y(\mathbf{x}_0), \tag{20}$$

In general, design centering approaches can be classified as statistical and geometrical. Geometrical approaches optimize the yield function implicitly by approximating the feasible region using a convex body, e.g., a hyperellipsoid. Then the center of this body is considered as the design center. Statistical approaches, on the other hand, optimize the yield function in a straightforward way, regardless the size of the problem or its convexity [1]. Hybrid methods, combining both approaches, may also be used for solving such problems [25].

For statistical design centering the evaluation of the yield values requires computing an n-dimensional integral (19) over a non-explicitly-defined region ($\mathbf{R}_f$). Therefore, the yield value for a given nominal design point $\mathbf{x}_0$ cannot be evaluated analytically; however, it can be estimated. The Monte Carlo method [51] is one of the famous methods used to estimate the yield values (19). The method depends on introducing an acceptance index function, $\mathrm{I_a} : \mathbb{R}^n \to \mathbb{R}$, defined as:

$$\mathrm{I_a}(\mathbf{x}) = \begin{cases} 1 \; if & x \in \mathbf{R}_f \\ 0 \; if & x \notin \mathbf{R}_f \end{cases}, \tag{21}$$

where $\mathbf{R}_f$ is the feasible region (14). By using this acceptance index function the yield integral (19) can be rewritten as:

$$Y(\mathbf{x}_0) = \int_{\mathbb{R}^n} \mathrm{I_a}(\mathrm{x})\, P(\mathbf{x}, \mathbf{x}_0)\, dx = E\{\mathrm{I_a}(\mathbf{x})\}, \tag{22}$$

where $P(\mathbf{x}, \mathbf{x}_0)$ is the PDF of the design parameters and $E\{.\}$ denotes expectation.

The yield value at a nominal vector $\mathbf{x}_0$ can be estimated by generating a set of samples in the designable parameter space using the PDF of designable parameters. Let $\mathbf{x}^{(i)}$, $i = 1, 2, \ldots, K$ be the generated samples around the nominal parameter vector $\mathbf{x}_0$. The system is simulated for each sample point $\mathbf{x}^{(i)}$, and the acceptance index function is evaluated. Hence, the yield function at the nominal parameter value $\mathbf{x}_0$ can be estimated as:

$$Y(\mathbf{x}_0) \approx \frac{1}{K}\sum_{i=1}^{K} \mathrm{I_a}\left(\mathbf{x}^{(i)}\right) = \frac{k}{K}, \tag{23}$$

where $\mathbf{x}^{(i)}$ is the generated i-th sample, k is the number of acceptable sample points, K is the total number of generated samples.

In fact, the error (variance) in estimating the yield value, using (23) is inversely proportional to the number of generated samples K [52]. Therefore, to obtain an accurate yield estimation, a large number of samples should be generated. This means that a large number of system simulations are required which, in turn, necessitates large computational time. However, several variance reduction techniques (like—importance sampling [52], stratified Monte Carlo method [49], and Latin Hypercube Sampling (LHS) [53]) can be used to achieve the same accuracy level with much smaller number of required samples. The notion of such variance reduction techniques is to spread the generated samples as evenly as possible around the interior design space. In this work, LHS technique is used, since it is computationally inexpensive and does not require any a priori knowledge about the simulated system, as well as it provides more accurate estimate of the yield function value than the other techniques. The main idea of this technique is to divide the design space into equiprobable non-overlapping sub-regions. Then K samples are selected such that all sub-regions are sampled.

In general, statistical design centering has some permanent special difficulties. One of these difficulties is the cost of finding a multitude of the yield function evaluations during the optimization process. Another difficulty in statistical yield optimization is the need for a derivative-free optimizer due to the absence of any exact or approximate gradient information about the yield. Any method can be used to approximate the gradient of the yield highly increases the computational overhead. Moreover the estimated yield values are usually contaminated by some numerical noise resulting from the estimation uncertainty.

One of the most reliable derivative-free trust region optimization algorithms is NEWUOA [54, 55]. The NEWUOA algorithm uses a quadratic interpolation scheme and a trust region mechanism to recursively construct and maximize quadratic models of the yield within a trust region. It guarantees global convergence together with fast local convergence. The basic idea of NEWUOA is to approximate the yield function $Y(\mathbf{x}_0)$ using a quadratic model, which is maximized within a trust region (sphere, for example). Then, the quadratic model is updated iteratively through the optimization process. The estimated yield function values submitted to the optimizer via system simulations and employing the LHS sampling technique.

7 Design Centering of a Wide Band Pass Optical Filter

In this section, the proposed design centering approach is applied to achieve the optimal design center point of the WBP-OF described in Section 4. The feasible region of this problem is defined by the following constraint functions:

$$f_i\left(r\left(\mathbf{x}\right)\right) = \begin{cases} r_i\left(\mathbf{x}\right) - LB, & for\ 300nm < \lambda_i < 350nm \\ UB - r_i\left(\mathbf{x}\right), & for\ 450nm < \lambda_i < 700nm \\ r_i\left(\mathbf{x}\right) - \mathrm{UB}, & \mathrm{for}\ 800\mathrm{nm} < \lambda_i < 900nm \end{cases}, \tag{24}$$

where $r_i\left(\mathrm{x}\right) = T\left(\lambda_i\right)$ is the transmittance at wavelength λ_i, whereas LB and UB are lower and upper bounds of the desired transmittance proportion at the passband and stopband regions, respectively.

Two design problems, differ in dimension, are considered and their results are reported separately. The first is a 3-D structure of periodic thickness, whereas the second problem is a 15-D problem with an aperiodic thickness structure. All results of this section are obtained while the designable parameters are assumed to have normal statistical distributions. All yield values are estimated by 100 sample points. For the two problems, both the independent and the correlated parameter cases are studied. Yield values and graphs are given to compare the obtained results either to the initial design points or to the results obtained using the minimax optimization approach.

7.1 *Periodic Thickness*

In this subsection, we consider a periodic filter structure, both from materials and thicknesses points of view, i.e., the unit cell is repeated 5-times with the same materials and thicknesses. Thus, the designable parameter vector $\mathbf{x}$ includes only three variables, namely d_1, d_2, and d_3, which are the periodic thicknesses of SiC, Ag, and SiO_2 layers, respectively.

Initially we consider the case of uncorrelated parameters with maximum deviation of 0.1 nm for each variable. In other words, we start the yield optimization with a diagonal covariance matrix Σ_1, whose all diagonal elements are set to 0.1. The constraints are adjusted as LB $= 75\%$ and UB $= 7\%$. Then, starting from an infeasible initial design point $\mathbf{x}^{(0)} = [20\ 10\ 70]^T$, as in [19], with initial yield equal to 0 %, we can achieve (after 53-TMM system simulations) an optimal solution $\mathbf{x}_1^* = [21.4604\ 10.0851\ 54.0878]^T$ which raises the yield to 54 %. Although the achieved yield is still small, it reflects the large sensitivity of our problem. The transmittance and absorbance response of the filter are compared in Fig. 8 for both the initial and optimal design center points. It is obvious that, the transmittance response is increased and slightly shifted to the left such that it becomes more centered around the visible spectrum. On the other hand, in order to visually declare

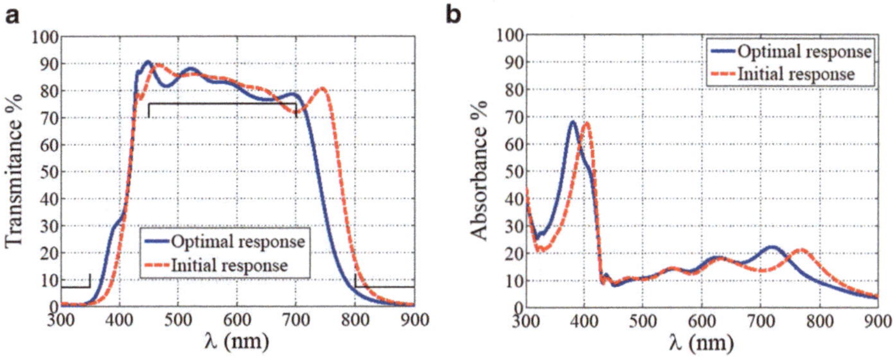

Fig. 8 (**a**) The transmittance and (**b**) the absorbance of $(SiC/Ag/SiO_2)^5$, before and after design centering

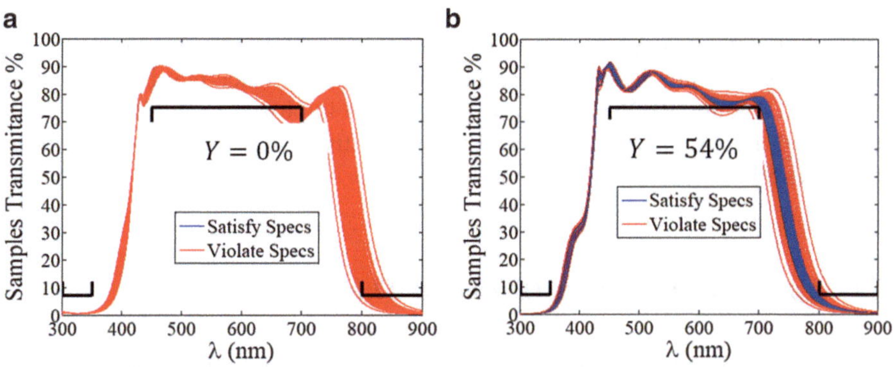

Fig. 9 The transmittance responses of samples generated around (**a**) initial nominal design point $x^{(0)}$ and (**b**) optimal nominal design x_1^*. The covariance matrix Σ_1 is considered

the achieved enhancement of the yield, Fig. 9 compares between the transmittance responses of the samples generated around the nominal design points $\mathbf{x}^{(0)}$ and $\mathbf{x}_1^*$. For comparison, the yield value is estimated at the minimax solution and it was 47 %.

Secondly, the parameters are considered to be correlated with covariance matrix which best describes the feasible region of the concerned problem. We fix the design parameters and optimize over the elements of the covariance matrix leading to the oriented ellipsoidal covariance matrix $\mathbf{\Sigma}_2$ given as:

$$\mathbf{\Sigma}_2 = \begin{pmatrix} 0.1656218 & 0.01169495 & -0.07308409 \\ 0.01169495 & 0.003012833 & -0.01386125 \\ -0.07308409 & -0.01386125 & 0.5322393 \end{pmatrix}$$

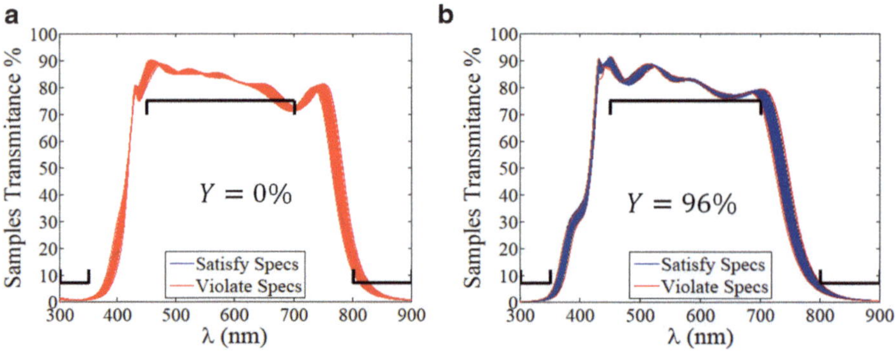

Fig. 10 The transmittance responses of samples generated around (**a**) initial point $\mathbf{x}^{(0)}$ and (**b**) optimal design $\mathbf{x}_2^*$. The covariance matrix $\boldsymbol{\Sigma}_2$ is considered

The optimal design point, $\mathbf{x}_2^* = [20.9968\ 9.9625\ 54.1049]^T$, is obtained (after 64-TMM system simulations) when the correlated covariance matrix $\boldsymbol{\Sigma}_2$ is being considered. Figure 10 illustrates how the yield is improved to 96 % when $\boldsymbol{\Sigma}_2$ is considered. For comparison, the yield value is estimated at the minimax solution and it was 89 %.

7.2 *Aperiodic Thickness (Different Thicknesses)*

Here, we consider the filter periodicity to be on materials only. Thus, the designable parameter vector becomes $\mathbf{x} = [d_1\ d_2 \ldots d_{15}]^T$, where d_j is the j-th layer thickness in nm. The feasible region is defined by the same constraints defined in (24). An uncorrelated covariance matrix, regarding the 15-designable parameters, is considered namely Cov_1, which is the periodic repetition of the uncorrelated covariance matrices, Σ_1, defined in the previous subsection. The constraints to be set as LB = 80% and UB = 10%. Then, starting from the same infeasible initial design point $\mathbf{x}^{(0)}$ [19], we obtain (after 259-TMM system simulations) the optimal center point

$$\begin{pmatrix} 18.623,\ 9.986,\ 69.558,\ 20.198,\ 11.194,\ 68.912,\ 18.854,\ 11.065,\ 61.159, \\ 20.473,\ 10.189,\ 69.325,\ 18.884,\ 7.784,\ 70.856 \end{pmatrix}$$

which is denoted by center point (CP1). A dramatic increase of the yield, from 0 % to 90 %, is achieved at CP1. Also, a very good filter response enhancement is achieved in the new design point, CP1, as shown in Fig. 11. The transmittance responses of the samples generated around the nominal design points $\mathbf{x}^{(0)}$ and CP1 are depicted in Fig. 12. The yield value at the minimax point using the same covariance matrix Cov_1 is 79 %.

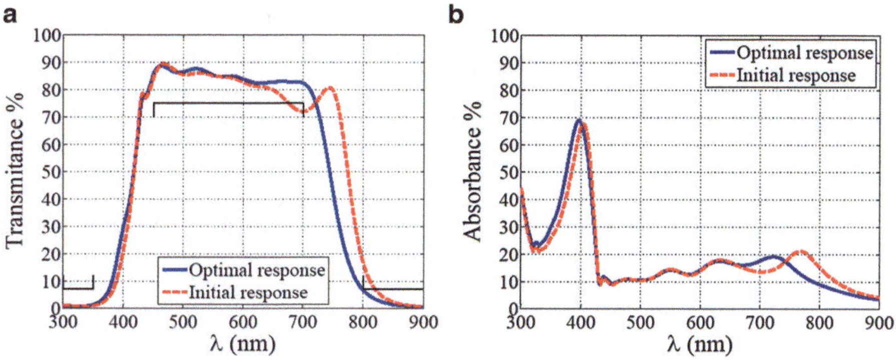

Fig. 11 (**a**) The transmittance and (**b**) the absorbance of $(\mathbf{SiC/Ag/SiO_2})^5$, before and after optimization. The initial point is $\mathbf{x}^{(0)}$ and the optimal point is **CP**1

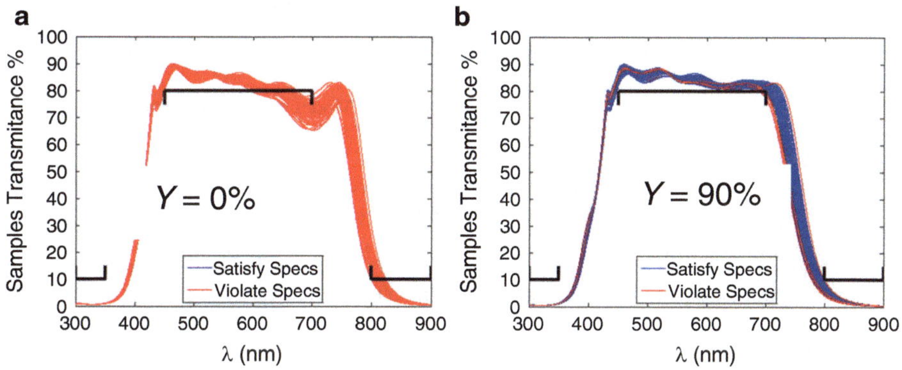

Fig. 12 The transmittance responses of samples generated around (**a**) the initial nominal design point $\mathbf{x}^{(0)}$ and (**b**) the optimal nominal design point **CP**1. The covariance matrix $\mathbf{Cov}_1$ is considered. The constraints **LB** = 80% and **UB** = 10% are considered

Acknowledgments Authors would like to thank Prof. Slawomir Koziel, School of Science and Engineering, Reykjavik University, for his invitation to contribute to this book. Authors also would like to acknowledge the contributions to the original work by, Eng. Mahmoud Taha Maghrabi, Faculty of Engineering, Cairo University, now with Electrical and Computer Engineering Department, McMaster University, which has been reviewed in this chapter.

References

1. Hassan, A.S.O., Mohamed, A.S.A.: Surrogate-based circuit design centering. In: Koziel, S., Leifsson, L. (eds.) Surrogate-Based Modeling and Optimization, pp. 27–49. Springer, New York (2013)
2. Joannopoulos, J.D., Johnson, S.G., Winn, J.N., Meade, R.D.: Photonic Crystals: Molding the Flow of Light. Princeton University Press, Princeton (2011)

3. Prather, D.W.: Photonic Crystals, Theory, Applications and Fabrications. John Wiley & Sons, Hoboken (2009)
4. Srivastava, S.K., Ojha, S.P.: Omnidirectional reflection bands in one-dimensional photonic crystal structure using fullerene films. Prog. Electromagn. Res. **74**, 181–194 (2007)
5. Kumar, A., Suthar, B., Kumar, V., Singh, K.S., Bhargava, A.: Tunable wavelength demultiplexer for DWDM application using 1-D photonic crystal. Prog. Electromag. Res. Lett. **33**, 27–35 (2012)
6. Baldycheva, A., Tolmachev, V.A., Perova, T.S., Zharova, Y.A., Astrova, E.V., Berwick, K.: Silicon photonic crystal filter with ultrawide passband characteristics. Opt. Lett. **36**(10), 1854–1856 (2011)
7. Xu, X.-f., Ding, J.-y.: A wide band-pass filter of broad angle incidence based on one-dimensional metallo-dielectric ternary photonic crystal. Opt. Quant. Electron **41**, 1027–1032 (2009)
8. He, J., Liu, P., He, Y., Hong, Z.: Narrow bandpass tunable terahertz filter based on photonic crystal cavity. Appl. Opt. **51**(6), 776–779 (2012)
9. Wang, Z.-Y., Chen, X.-M., He, X.-Q., Fan, S.-L., Yan, W.-Z.: Photonic crystal narrow filters with negative refractive index structural defects. Prog. Electromagn. Res. **80**, 421–430 (2008)
10. Kurt, H., Citrin, D.S.: Photonic crystals for biochemical sensing in the terahertz region. Appl. Phys. Lett. **87**, 041108 (2005)
11. Chubb, D.: Fundamentals of Thermophotovoltaic Energy Conversion. Elsevier, Amsterdam (2007)
12. Asghar, M.H., Shoaib, M., Placido, M., Naseem, S.: Modeling and preparation of practical optical filters. Curr. Appl. Phys. **9**, 1046–1053 (2009)
13. Jia, W., Deng, J., Reid, B.P.L., Wang, X., Chan, C.C.S., Wua, H., Li, X., Taylor, R.A., Danner, A.J.: Design and fabrication of optical filters with very large stopband ($\approx$500 nm) and small passband (1 nm) in silicon-on-insulator. Photonics Nanostruct. Fundam. Appl. **10**, 447–451 (2012)
14. Celanovic, I., O'Sullivan, F., Ilak, M., Kassakian, J., Perreault, D.: Design and optimization of one-dimensional photonic crystals for thermophotovoltaic applications. Opt. Lett. **29**(8), 863–865 (2004)
15. Xuan, Y., Xue, C., Yuge, H.: Design and analysis of solar thermophotovoltaic systems. Renew. Energy **36**(1), 374–387 (2011)
16. Baedi, J., Arabshahi, H., Armaki, M.G., Hosseini, E.: Optical design of multilayer filter by using PSO Algorithm. Res. J. Appl. Sci. Eng. Technol. **2**, 56–59 (2010)
17. Badaoui, H.A., Abri, M.: One-dimensional photonic crystal selective filters design using simulated annealing optimization technique. Prog. Electromag Res. B **53**, 107–125 (2013)
18. Swillam, M.A., Bakr, M.H., Li, X.: The design of multilayer optical coatings using convex optimization. Lightwave Technol. **25**(4), 1078–1085 (2007)
19. Rafat, N.H., El-Naggar, S.A., Mostafa, S.I.: Modeling of a wide band pass optical filter based on 1D ternary dielectric-metallic-dielectric photonic crystals. J. Opt. **13**, 085101 (2011)
20. Mostafa, S.I., Rafat, N.H., El-Naggar, S.A.: One-dimensional metallic-dielectric (Ag/SiO_2) photonic crystals filter for thermophotovoltaic applications. Renew. Energy **45**, 245–250 (2012)
21. Koziel, S., Leifsson, L.: Surrogate-Based Modeling and Optimization. Springer, New York (2013)
22. Hassan, A.S.O.: Normed distances and their applications in optimal circuit design. Optim. Eng. **4**(3), 197–213 (2003)
23. Hassan, A.S.O., Mohamed, A.S.A., El-Sharabasy, A.Y.: Statistical microwave circuit optimization via a non-derivative trust region approach and space mapping surrogates. In IEEE MTT-S Int. Microw. Symp. Dig., Baltimore, MD, USA, pp. 1–4, (2011)
24. Hassan A.S.O., Mohamed A.S.A., El-Sharabasy A.Y.: EM-based yield optimization exploiting trust-region optimization and space mapping technology. Int. J. RF Microw. CAE, Wiley, (2014, in Press). doi:10.1002/mmce.20878

25. Hassan, A.S.O., Abdel-Naby, A.: A new hybrid method for optimal circuit design using semi-definite programming. Eng. Optm. **44**(6), 725–740 (2012)
26. Waren, A.D., Lasdon, L.S., Suchman, D.F.: Optimization in engineering design. Proc. IEEE **55**, 1885–1897 (1967)
27. Charalambous, C., Conn, A.R.: An efficient method to solve the minimax problem directly. SIAM J. Numer. Anal. **15**(1), 162–187 (1978)
28. Bandler, J.W., Kellermann, W., Madsen, K.: A superlinearly convergent minimax algorithm for microwave circuit design. IEEE Trans. Microw. Theory Tech. **33**, 1519–1530 (1985)
29. Hald, J., Madsen, K.: Combined LP and quasi-Newton methods for minimax optimization. Math. Program. **20**, 49–62 (1981)
30. Chemmangat, K., Ferranti, F., Dhaene, T., Knockaert, L.: Optimization of high-speed electromagnetic systems with accurate parametric macromodels generated using sequential sampling of the design space. In Electromagnetics in Advanced Applications (ICEAA), International Conference on. IEEE, Cape Town, pp. 128–131, (2012)
31. Jen, J., Qian, M., Aliyazicioglu, Z., Hwang, H.K.: Performance studies of antenna pattern design using the minimax algorithm. In Proceedings of the 5th WSEAS international conference on Circuits, systems, signal and telecommunications, World Scientific and Engineering Academy and Society (WSEAS), Wisconsin, USA, pp. 50–55, (2011)
32. Koziel, S., Leifsson, L.: Low-cost parameter extraction and surrogate optimization for space mapping design using Em-based coarse models. Prog. Electromag. Res. B **31**, 117–137 (2011)
33. Kats, B.M., Lvov, A.A., Meschanov, V.P., Shatalov, E.M., Shikova, L.V.: Synthesis of a wideband multiprobe reflectometer. Microw. Theory Tech. IEEE Trans. **56**, 507–514 (2008)
34. Hassan, A.S.O., Abdel-Malek, H.L., Mohamed, A.S.A.: Optimal design of computationally expensive EM-based systems: a Surrogate-based approach. In: Koziel, S., Leifsson, L., Yang, X.-S. (eds.) Solving Computationally Expensive Engineering Problems, pp. 171–194. Springer, New York (2014)
35. Singhal, K., Pinel, J.F.: Statistical design centering and tolerancing using parametric sampling. IEEE Trans. Circuits Syst. **28**, 692–702 (1981)
36. Pendry, J.B.: Photonic band structures. J. Mod. Opt. **41**, 209–229 (1994)
37. Botten, L.C., Nicorovici, N.A., McPhedran, R.C., de Martijn Sterke, C., Asatryan, A.A.: Photonic band structure calculations using scattering matrices. Phys. Rev. E **64**(4), 046603 (2001)
38. Hassan, A.S.O., Mohamed, A.S.A., Maghrabi, M.M.T., Rafat, N.H.: Optimal design of 1D photonic crystal filters using minimax optimization approach. Appl. Opt. **54**(6), 1399–1409 (2015)
39. Matlab, Version 7.10., The MathWorks, Inc., 3 Apple Hill Drive, Natick, MA 01760–2098, USA, (2010)
40. Nehmetallah, G., Aylo, R., Powers, P., Sarangan, A., Gao, J., Li, H., Achari, A., Banerjee, P.P.: Co-sputtered SiC + Ag nanomixtures as visible wavelength negative index metamaterials. Opt. Express **20**, 7095–7100 (2012)
41. Chen, S., Wang, Y., Yao, D., Song, Z.: Absorption enhancement in 1D Ag/SiO_2 metallic-dielectric photonic crystals. Opt. Appl. **39**, 473–479 (2009)
42. Jaksic, Z., Maksimovic, M., Sarajlic, M.: Silver–silica transparent metal structures as bandpass filters for the ultraviolet range. J. Opt. A Pure Appl. Opt. **7**, 51–55 (2005). doi:10.1088/1464-4258/7/1/008
43. Ni, X., Liu, Z., Kildishev, A.V.: PhotonicsDB: Optical constants. [Online]. Available: http://nanohub.org/resources/3692, (2010)
44. Hocevar, D.E., Lightner, M.R., Trick, T.N.: An extrapolated yield approximation for use in yield maximization. IEEE Trans. Comput. Aided Des. **3**, 279–287 (1984)
45. Styblinski, M.A., Oplaski, L.J.: Algorithms and software tools for IC yield optimization based on fundamental fabrication parameters. IEEE Trans. Comput. Aided Des. **5**, 79–89 (1986)
46. Yu, T., Kang, S.M., Hajj, I.N., Trick, T.N.: Statistical performance modeling and parametric yield estimation of MOS VLSI. IEEE Trans. Comput. Aided Des. **6**, 1013–1022 (1987)

47. Elias, N.J.: Acceptance sampling: an efficient accurate method for estimating and optimizing parametric yield. IEEE J. Solid State Circuits **29**, 323–327 (1994)
48. Zaabab, A.H., Zhang, Q.J., Nakhla, M.: A neural network modeling approach to circuit optimization and statistical design. IEEE Trans. Microw. Theory Tech. **43**, 1349–1358 (1995)
49. Keramat, M., Kielbasa, R.: A study of stratified sampling in variance reduction techniques for parametric yield estimations. IEEE Trans. Circuits Syst. II: Analog Digit. Signal Process. **45**(5), 575–583 (1998)
50. Hassan, A.S.O., Abdel-Malek, H.L., Rabie, A.A.: None-derivative design centering algorithm using trust region optimization and variance reduction. Eng. Optim. **38**, 37–51 (2006)
51. Metropolis, N., Ulam, S.: The Monte-Carlo method. J. Am. Stat. Assoc. **44**, 335–341 (1949)
52. Hocevar, D.E., Lightner, M.R., Trick, T.N.: A study of variance reduction techniques for estimating circuit yields. IEEE Trans. Comput. Aided Des. **2**(3), 180–192 (1983)
53. McKay, M.D., Beckman, R.J., Conover, W.J.: A comparison of three methods for selecting values of input variables in analysis of output from a computer code. Technometrics **21**(2), 239–245 (1979)
54. Powell, M.J.D.: The NEWUOA software for unconstrained optimization without derivatives. In: Di Pillo, G., Roma, M. (eds.) Large-Scale Nonlinear Optimization, pp. 225–297. Springer, New York (2006)
55. Powell, M.J.D.: A view of algorithms for optimization without derivatives. Technical Report NA 2007/03, University of Cambridge, Department of Applied Mathematics and Theoretical Physics, Cambridge, England, (2007)

Design Optimization of LNAs and Reflectarray Antennas Using the Full-Wave Simulation-Based Artificial Intelligence Models with the Novel Metaheuristic Algorithms

Filiz Güneş, Salih Demirel, and Selahattin Nesil

Abstract In this chapter, the two primarily important highly nonlinear design problems of the contemporary microwave engineering which are "Low Noise Amplifier (LNA)"s and "Reflect-array Antenna (RA)"s are solved as "Design Optimization problems." For this purpose, firstly the design problem is defined in terms of the feasible design variables (FDVs), the feasible design target space (FDTS), both of which are built up by integrating the artificial intelligence black-box models based upon the measurements or full-wave simulations and a suitable metaheuristic search algorithm. In the second stage, feasible design target (FDT) or objective function of the optimization procedure is determined as a sub-space of the FDTS. Thirdly, the cost function evaluating the objective is minimized employing a suitable metaheuristic search algorithm with respect to the FDVs. Finally the completed designs are verified by the professional Microwave Circuitor3-D EM simulators.

Keywords Microwave engineering • Low-noise amplifier • Reflectarray • Full-wave simulation • Artificial intelligence • Metaheuristic search algorithm • Feasible design target space • Feasible design variables

MSC codes: 78-01, 68-04, 94-01

F. Güneş (✉) • S. Demirel • S. Nesil
Department of Electronics and Communication Engineering, Yildiz Technical University, Esenler, Istanbul 34220, Turkey
e-mail: gunes@yildiz.edu.tr; fgunes51@gmail.com

S. Koziel et al. (eds.), *Simulation-Driven Modeling and Optimization*, Springer Proceedings in Mathematics & Statistics 153,
DOI 10.1007/978-3-319-27517-8_11

1 Design Optimization of LNAs

1.1 LNA Design Problem

As the electronic industry moves towards higher integration and lower cost, RF and wireless design demands increasingly more "concurrent" engineering. Typically, today's most receivers are hand-held or battery-operated devices; one of the major challenges in these receivers is to design a low-noise amplifier (LNA) that has very low power consumption and operates from a very low supply voltage with the provided trade-off of noise measure and mismatch losses. Since the two transistor configurations consume more power from the higher voltage supply than the single transistor configurations, the two transistor configurations are unsuitable for this type of applications. Thus, after selection of a transistor among the available high technology transistors, then a low-noise design approach consists of trading off among the often contrasting goals of low noise, high gain, and input and output match within the device operation domain.

The design optimization method used for a microstrip LNA is given in a flow chart in Fig. 1. However the method can easily be applied to the LNAs using different wave guiding systems.

1.2 Feasible Design Target Space (FDTS)

Since the design optimization of an LNA necessitates the physical limits and compromise relations of the design hexagon consisting of bias voltage V_{DS}, bias current I_{DS}, noise F, gain G_T, input VSWR V_{in}, and output VSWR V_{out} belonging to the employed transistor, in the other words, the "Feasible Design Target Space (FDTS)" must be constructed as an important stage of the design optimization procedure. Certainly, within the optimization process, one can easily embed the desired performance goals without knowing the physical realizability conditions and compromise relations appropriately, but in this case, the device is utilized either under its performance potential or unrealizable requirements that result in failure in the design.

The block diagram of the FDTS is given in Fig. 2 where all the compatible performance quadrates (noise figure F, input VSWR V_{in}, output VSWR V_{out}, transducer gain G_T); the corresponding operation bandwidth B and the source Z_S and load Z_L terminations are obtained as the continuous functions of the device's operation parameters which are bias condition (V, I) and frequency at a chosen configuration type. Let us consider the most commonly used configuration which is common source configuration. Firstly a soft model of the transistor is constructed using either a suitable artificial intelligent network or an equivalent circuit built by a parameter extraction method to obtain the device's scattering S and noise N parameters as the continuous functions of the operation parameters V_{DS}, I_{DS}, *f*. Typical works for the S and N parameters modeling of a microwave transistor can

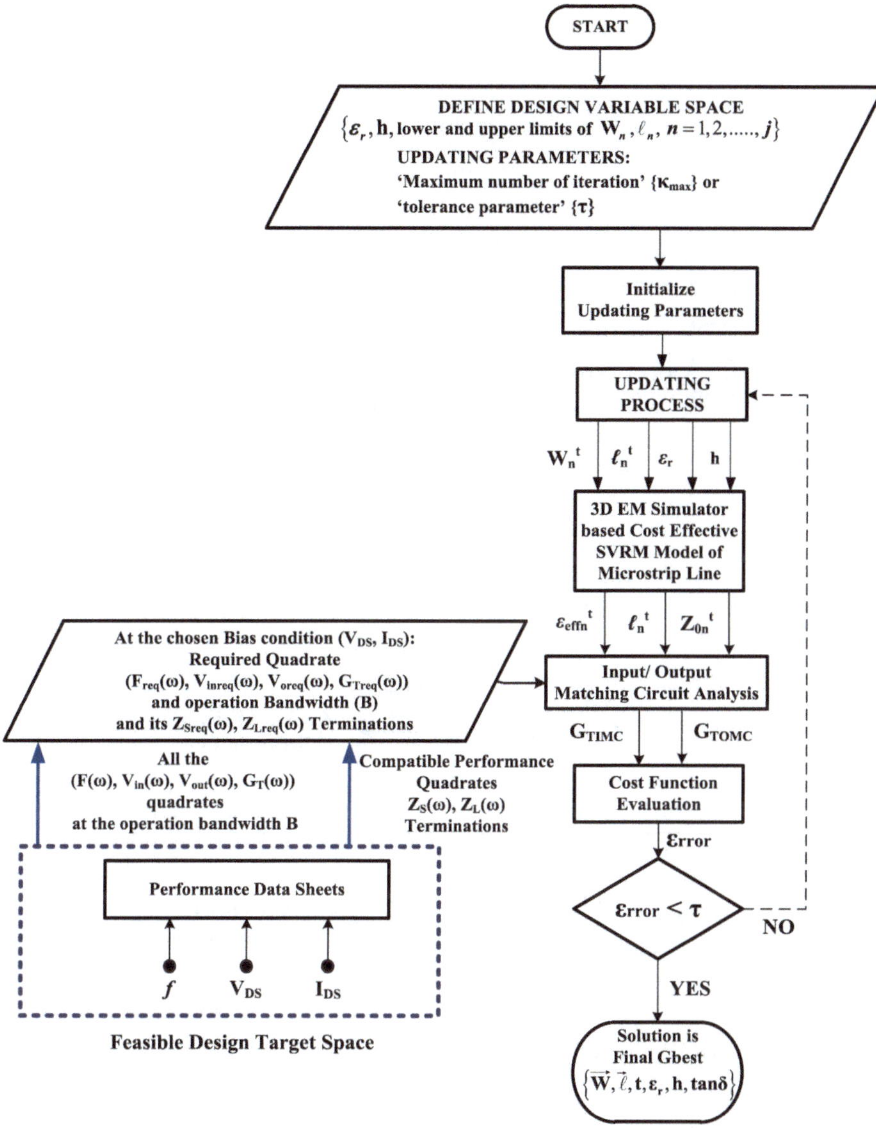

Fig. 1 Design flow chart of the microstrip LNA

be found in [1–7]. Secondly, potential performance of the microwave transistor is analyzed in terms of the S and N parameters at a chosen bias condition. This analysis has been achieved by solving the highly nonlinear performance equations of the transistor using either the analytical approaches based on the constant performance ingredient circles [8–14] or optimization methods without using the complicated microwave theory [15–17].

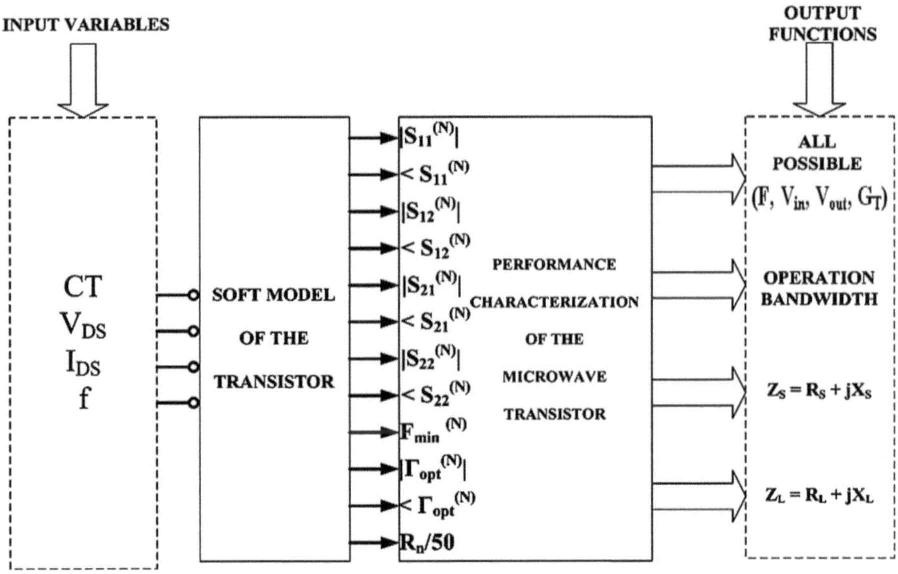

Fig. 2 Block diagramme for performance data sheets

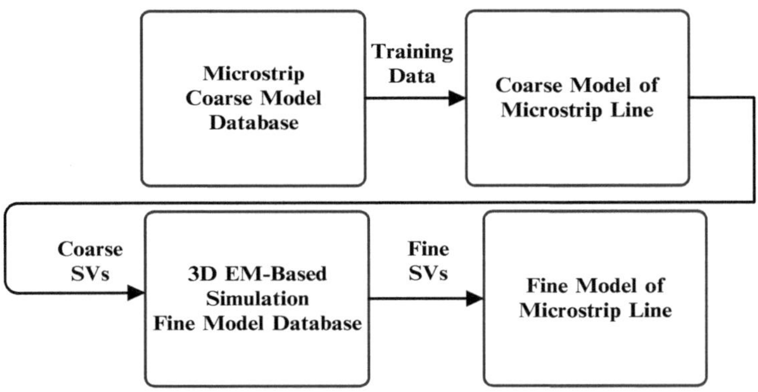

Fig. 3 3D EM simulation-based SVRM model of the microstripline

1.3 Feasible Design Variables (FDVs)

The second stage is modeling of the feasible design variables (FDVs) using the 3-D EM simulation-based support regression vector machine (SVRM). In this modeling, one-to-one mapping is built between the input domain consisting of the microstrip width W, substrate (ε_r, h) parameters, and frequency *f*, and the output domain defined by the equivalent transmission line parameters which are the characteristic impedance Z_0 and effective dielectric constant ε_{eff} (Fig. 3) [18, 19].

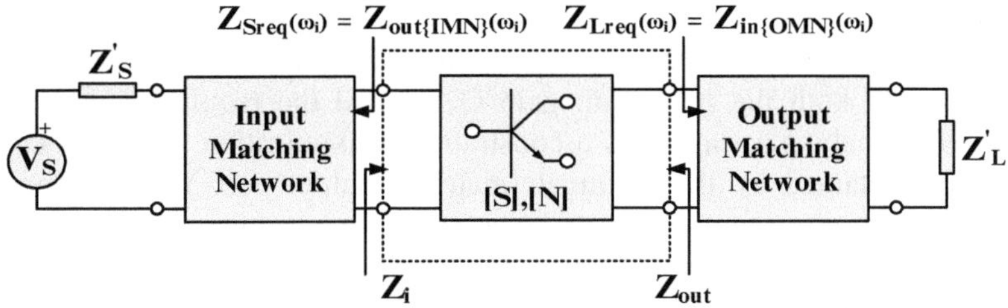

Fig. 4 Transistor with the input and output matching circuits of the compatible performance terminations Z_S, Z_L, respectively

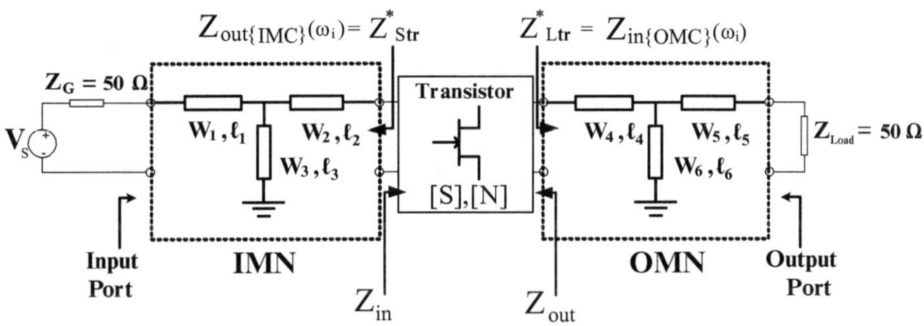

Fig. 5 LNA with the T-type microstrip matching networks

1.4 Design of Input and Output Matching Circuits

Final stage is the design of the input and output matching circuits. Transistor with the Darlington equivalencies of the compatible performance terminations Z_S, Z_L are given in Fig. 4. Input IMC and Output Matching Circuit (OMCs) are designed using either the gain or port impedance optimization of the two independent matching circuits given in Fig. 5 by either a gradient or metaheuristic algorithm. In the next subsection design strategies of LNA are given briefly.

1.5 Design Strategies

Here to fore the two different design strategies can be put forward for the LNAs: In the first strategy, considering $F(Z_S)$ and $V_{in}(Z_S, Z_L)$ as the free variables, $G_{Tmin} \leq G_T \leq G_{Tmax}$ and the corresponding termination Z_S, Z_L couple are determined solving the nonlinear transistor's performance equations with either analytical approaches using the constant performance ingredient circles or a constrained optimization problem. Thus, with the resulted V_{out}, the FDTS can be built consisting of the compatible ($F \geq F_{req}$, $V_{in} \geq 1$, $V_{out} \geq 1$, $G_{Tmin} \leq G_T \leq G_{Tmax}$) and the associated Z_S, Z_L terminations [8–11, 15, 16].

In the second strategy, only noise $F(Z_S)$ is considered as a free variable and the nonlinear performance equations are solved for the input termination Z_S satisfying simultaneously both the maximum gain G_{Tmax} and the required noise F, again either in the analytical way or as a constrained maximization problem. Then the load Z_L is obtained by the conjugate-matched output port $V_{out} = 1$ condition. Mismatching at the input port can be adjusted by degrading either noise and mismatching at the output port. Thus a different FDTS can be built up consisting of the compatible ($F \geq F_{req}$, $V_{in} \geq 1$, $V_{out} \geq 1$, $G_{Tmin} \leq G_T \leq G_{Tmax}$) and the associated Z_S, Z_L terminations [12–14, 17].

Both design strategies are based on the following balance equation:

$$\left(1 - \left|\frac{V_{in}(Z_S, Z_L) - 1}{V_{in}(Z_S, Z_L) + 1}\right|^2\right) .G_{op}(Z_L) = \left(1 - \left|\frac{V_{out}(Z_S, Z_L) - 1}{V_{out}(Z_S, Z_L) + 1}\right|^2\right) .G_{av}(Z_S) \tag{1}$$

Where $G_{op}(Z_L)$ and $G_{av}(Z_S)$ are the operation and available power gains, respectively, which will be taken into account in the study case. Typical LNA designs based on these design strategies using either gradient or metaheuristic algorithms can be found in [20–26]. In the next section, a front-end amplifier design worked out by our research group in [26] will briefly be given as a case study based on the above methodology.

1.6 Case Study: HBMO Design Optimization of an LNA with Support Vector Microstrip Model

In this section, a HBMO design optimization procedure is given in subject to the design flow chart in Fig. 1 for a front-end amplifier so that all the matching microstrip widths, lengths $\{\{\vec{W}, \vec{\ell}\}\}$ can be obtained to provide the (Z_S, Z_L) terminations on a given substrate (ε_r, h, tanδ) for the maximum power delivery and the required noise over the required bandwidth of a selected transistor, respectively [26]. Thus, in the following subsection all the stages of the design procedure will be considered.

1.6.1 Feasible Design Target (FDT)

In this LNA design optimization problem, the design objective is to ensure the maximum output power delivery and the required noise. Thus, hereafter the problem of determination of the source impedance $Z_S = r_S + jx_S$ of a microwave transistor can be described as a mathematically constrained optimization problem so that the transducer gain $G_T(r_S, x_S, r_L, x_L)$ will be maximized simultaneously satisfying the required noise figure $F(r_S, x_S)$ provided that the stability conditions are ensured at each sample frequency throughout the required operation bandwidth. The transistor's load impedance Z_L can be determined using the balanced Eq. (1) by the conjugate-matched output, that is, $V_{out} = 1 \iff Z_L = Z^*_{out}(Z_S)$. Thereby the multi-objective cost function of this constrained optimization process can be expressed as:

$$Cost\,(r_S, x_S,\, f_i) = e^{-\psi_1 G_{AV}(r_S,\, x_S, f_i)} + \psi_2 \left| F\,(r_S,\; x_S, f_i) - F_{req}\,(f_i) \right| \tag{2}$$

with the following constraints for the physical limits and stability of the transistor

$$\Re e\,\{Z_S\} > 0,\ \Re e\,\{Z_L\} > 0,\ F_{req} \geq F_{\min} \tag{3}$$

$$\Re e\,\{Z_{in}\} = \Re e \left\{ z_{11} - \frac{z_{12}\, z_{21}}{z_{22} + Z_L} \right\} > 0,\ \Re e\,\{Z_{out}\} = \Re e \left\{ z_{22} - \frac{z_{12} z_{21}}{z_{11} + Z_S} \right\} > 0 \tag{4}$$

Here the performance measure G_T, G_{AV}, and F functions can be expressed in terms of the transistor's z-parameters and Z_S, Z_L terminations as follows [27]:

$$G_T = \frac{P_L}{P_{avs}} = G_{AV}\,(Z_S)\ .M_{out}\,(Z_S,\ Z_L) \tag{5}$$

where

$$G_{AV}\,(Z_S) = \frac{|z_{21}|^2}{|z_{11} + Z_S|^2} \frac{r_S}{r_{out}},\quad r_{out} \triangleq Re\,\{Z_{out}\}\,,$$

$$M_{out}\,(Z_S, Z_L) = 1 - \left| \frac{Z_{out} - Z^*{}_L}{Z_{out} + Z_L} \right|^2 \leq 1, \tag{6}$$

$$F\,(Z_S) = \frac{\left({}^S/_N\right)_i}{\left({}^S/_N\right)_o} = F_{min} + \frac{R_n\,|Z_S - Z_{opt}|^2}{|Z_{opt}|^2\, r_S} \tag{7}$$

Besides, ψ_1 *and* ψ_2 in the Eq. (2) are the weighting coefficients which can be chosen during the optimization process by trial, which in our case are taken as unity. Thus, the smaller cost is the fitter optimization process we have.

Here, for the ultra-wideband LNA design, the three alternatives are considered for the required noise figure $F_{req}\,f$ of the selected transistor NE3512S02 using the honey bee mating optimization (HBMO): (1) $F_{req}(\omega_i) = F_{min}(\omega_i)$; (2) $F_{req}(\omega_i)$ = constant = 1.0 dB; (3) $F_{req}(\omega_i)$ = constant = 1.5 dB (Fig. 6).

In Fig. 6, the maximum gain variations of the transistor NE3512S02 for the matched output against the input mismatching V_{in} are given as compared with the analytical counterparts [14, 15]. Besides the corresponding terminations of the maximum gain for the matched output and F(*f*) = 1 dB are given in Tab. 1.

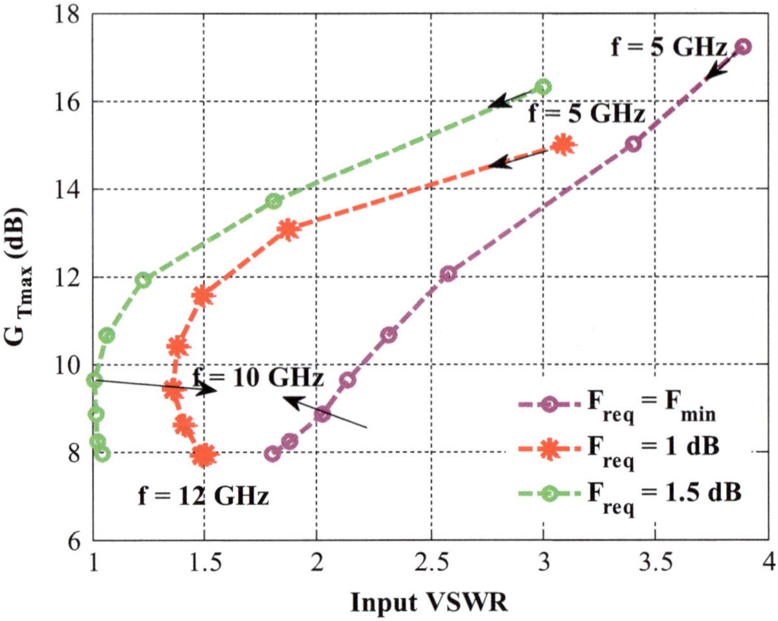

Fig. 6 Maximum gain against input VSWR Vin and for $|\rho_{out}| = 0$ for NE3512S02 at the bias condition $V_{DS} = 2$ V, $I_{DS} = 10$ mA

Table 1 The source Z_S and load Z_L terminations for the maximum gain for $|\rho_{out}| = 0$ and $F(f) = 1$ dB for NE3512S02 at the bias condition $V_{DS} = 2$ V, $I_{DS} = 10$ mA

f(GHz)	V_{in}	G_{TMAX}(dB)	Real(Z_L) Ω	Imag(Z_L) Ω	Real(Z_S) Ω	Imag(Z_S) Ω
5	3.09	15.0	16.32	29.97	14.66	23.23
6	1.87	13.06	19.31	26.14	14.31	12.36
7	1.49	11.55	21.78	22.19	14.78	3.09
8	1.38	10.40	23.52	17.42	15.97	−5.00
9	1.36	9.43	25.35	12.18	18.08	−12.03
10	1.40	8.62	27.06	6.791	20.77	−18.72
11	1.49	7.92	29.34	0.872	24.72	−24.68
12	1.50	7.94	35.67	−5.91	35.37	−28.85

1.6.2 Design Objective for the Matching Networks

Thus, we have the transistor terminations solving the nonlinear performance equations subject to the objective of Eq. (2–7). A novel metaheuristic the HBMO is used in the solution procedure of the equations of Eq. (2–7) that will briefly be given in the following section. In the design optimization procedure, the gain of the input/output matching two-port terminated by the complex conjugate of the $(Z_S(\omega_i)/Z_L(\omega_i))$ as given in Fig. 5 is maximized over the required bandwidth:

$$cost\left(\vec{W},\ \vec{\ell}\right) = Minimum \sum_{i} \left(1 - G_{Ti}\left(f_i, \vec{W},\ \vec{\ell}\right)\right) \tag{8}$$

where $\left\{\vec{W},\ \vec{\ell}\right\}$ is the design variable vector which consists of the microstrip widths and lengths of the problem matching circuit and G_{Ti} is the transducer power gain of the same matching circuit at the sample frequency f_i. In the worked example, T-type matching circuits are considered to be designed. The proposed method can be applied without any difficulty to another different type of matching circuit. In that case, the gain function G_{Ti} given in Eq. (8) should be evaluated for the considered matching circuit.

1.6.3 Design Variables: Microstrip Widths and Lengths $\{W,\ \ell\}$

In this design optimization procedure, the microstrip widths and lengths $\left\{\vec{W},\ \vec{\ell}\right\}$ on a selected substrate $\{\varepsilon_r, h, tan\delta\}$ are directly used by the HBMO optimization of the LNA (Fig. 1) and the cost function (Eq. 8) is evaluated by means of the SVRM microstrip model (Fig. 1). The 3-D SONNET-based SVRM model of the microstrip [18, 19] is employed that provides an accurate, fast, and cost effective generalization from the highly nonlinear discrete mapping from the input domain M (R4) of the microstrip width W, substrate $\{\varepsilon_r, h, tan\delta\}$, and frequency f to the output domain of the characteristic impedance Z_0 and effective dielectric constant ε_{eff}.

Here, the range of input and output domains is given as $\{0.1\ \text{mm} \le W \le 4.6\ \text{mm},\ 2 \le \varepsilon_r \le 10,\ 0.1\ \text{mm} \le h \le 2.2\ \text{mm},\ 2\ \text{GHz} \le f \le 14\ \text{GHz}\}$ and $\{3\,\Omega \le Z_0 \le 240\,\Omega\}$ and $\{1.5 \le \varepsilon_{eff} \le 9.7\}$, respectively.

1.6.4 Building Knowledge-Based Microstrip SVRM Model

Knowledge-based microstrip SVRM is given as block diagram in Fig. 3 where the quasi-TEM microstrip analysis formula is used as a coarse SVRM model database by means of which $n_{freq} \times n_\varepsilon \times n_h \times n_w = 5 \times 5 \times 4 \times 10 = 1000\ \left(\vec{x}_i, \vec{y}_i\right)$ data pairs are obtained to train the coarse SVRM, where n_{freq}, n_ε, n_h, n_w are the number of the samples for the frequency, the dielectric constant, the substrate height and width, respectively. Tab. 2 gives the accuracy of the "Z_0" coarse model with the number of the SVs and the radius of selection tube ϵ. 402 and 367 fine SVs obtained from 3-D SONNET simulator are used to train the fine "Z_0" and "ε_{eff}" SVRMs, respectively, with the accuracy at least 99.4 % (Fig. 7b). Thus the substantial

Table 2 Accuracy of the fine SVRM model

Epsilon (ϵ)	Number of SVs	Accuracy (%)
0.05	583	99.4
0.07	402	98.6
0.1	279	97.9

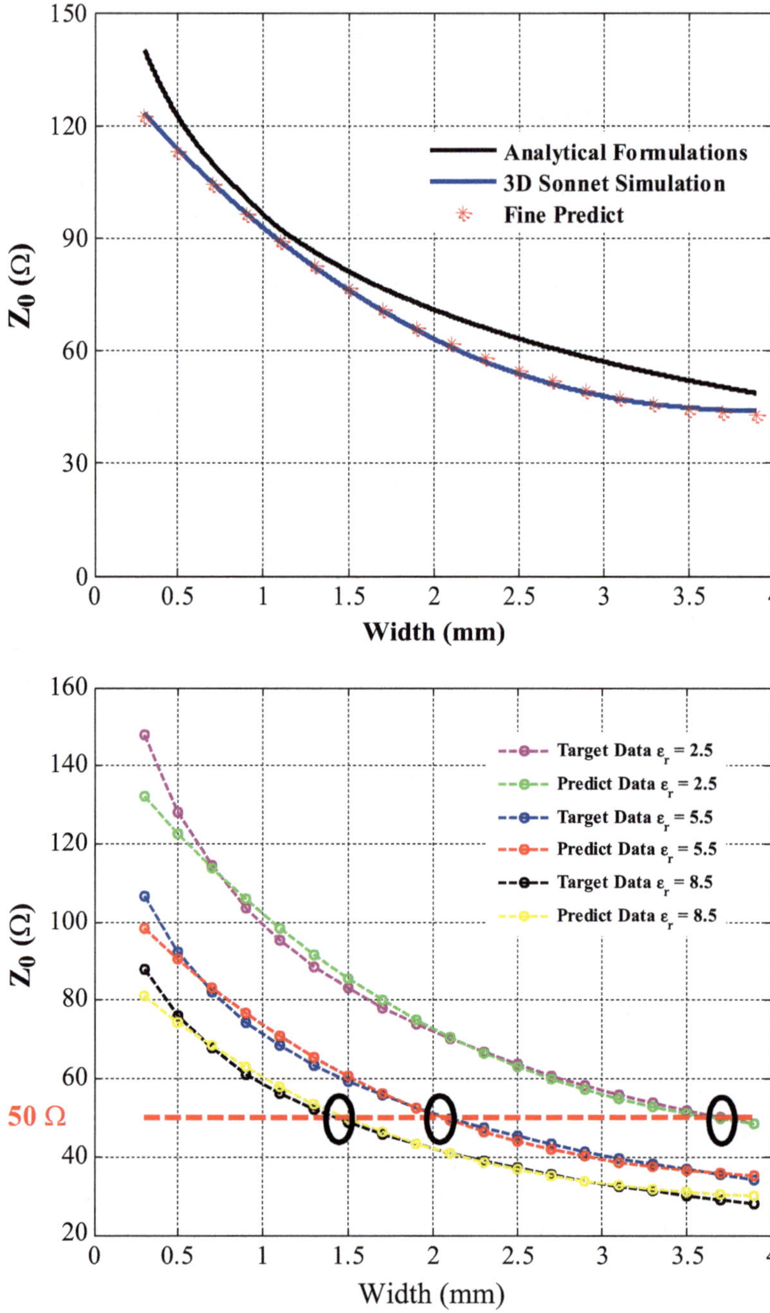

Fig. 7 (continued)

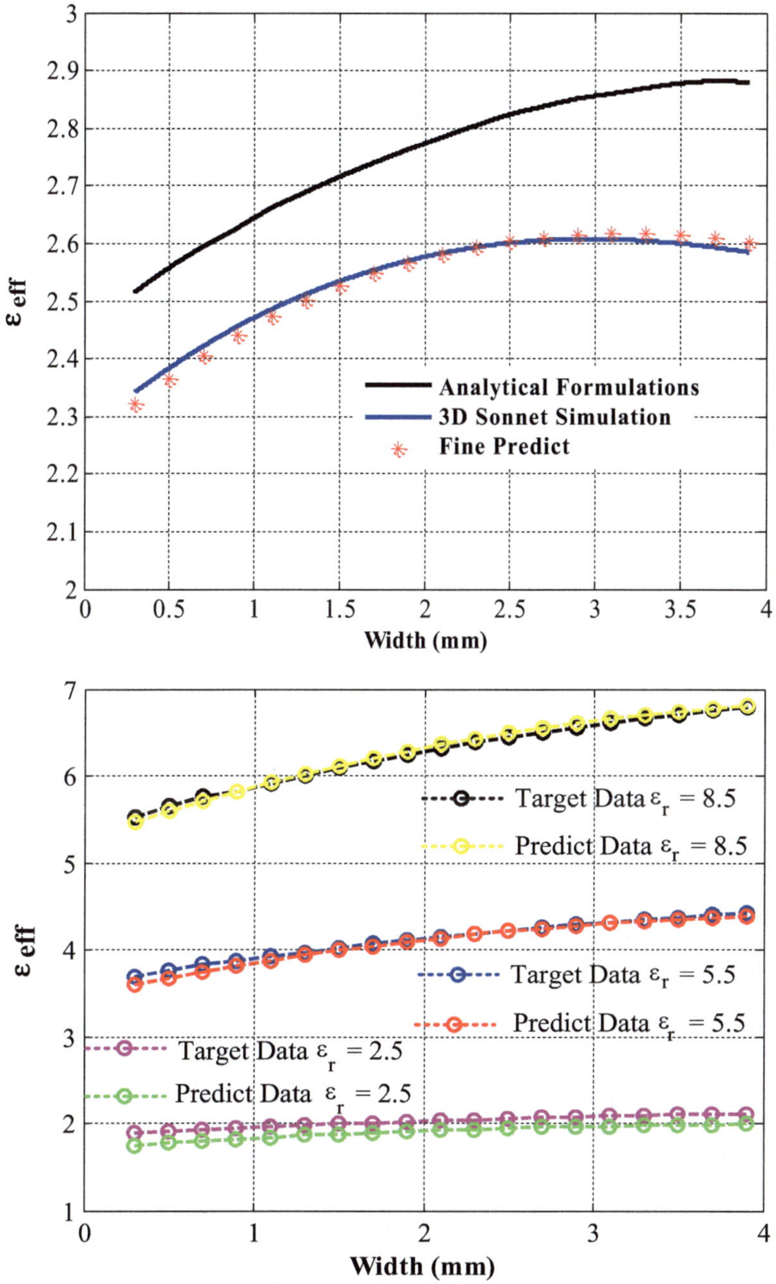

Fig. 7 Comparative variations for characteristic impedance Z_0 and effective dielectric constant $\varepsilon_{\mathrm{eff}}$ vs width of the analytical formulations, fine model and the 3-D SONNET simulation. (**a**) Rogers 435 ($\varepsilon_r = 3.48$, h = 1.524 mm, tanδ = 0.003, t = 0.001 mm) at 4 GHz, (**b**) Z_0 variations for various dielectrics at f = 8 GHz (**c**) Rogers 435($\varepsilon_r = 3.48$, h = 1.524 mm, tanδ = 0.003, t = 0.001 mm) at f = 4 GHz (**d**) ε_{eff} for various dielectrics at f = 8 GHz

reduction (up to 60 %) is obtained utilizing sparseness of the standard SVRM in number of the expensive fine discrete training data with the negligible loss in the predictive accuracy and the resulted fine microstrip SVRM model can be considered as accurate as the 3-D EM simulator and as fast as the analytical formulae. The typical comparative prediction curves of the microstrip SVRM model take place in Fig. 7a–d give Z_0 and ε_{eff} variations with respect to the microstrip width W resulted from the fine SVRM model for the dielectrics at f = 4GHz and 8 GHz, respectively.

1.6.5 HBMO with Royal Jelly for the Amplifier's Matching Network Design Problem

HBMO is a recent swarm-based optimization algorithm to solve highly nonlinear problems, whose based approach combines the powers of simulated annealing, genetic algorithms, and an effective local search heuristic to search for the best possible solution to the problem under investigation within a reasonable computing time.

The flow diagramme of the algorithm is given by Fig. 8. The user-defined parameters of the algorithms are the number of the Drone bees N_{Drone}, maximum iteration number t_{max}, sizes of the genetic inheritance of the Master Queen Q_M, and each Drone bee D_j, m_Q, m_D; maximum number of feeding times of the Master Queen Q_M with Royal Jelly N_{RJ}, maximum E_{max} and minimum E_{min} energies of the Queen at the start and end of the mating flights, respectively, and the required cost $cost_{req}$. In the algorithm, the numbers of the Hive N_{Hive}, Brood N_{Brood}, LarvaN_{Larva}, Fertilization $N_{fertilization}$ are set equal to N_{Gen} which is taken to be equal to t_{max} and the total egg number $N_{Egg} = (N_{Gen})\,5$.

The proposed HBMO algorithm is used effectively and efficiently to design a front-end amplifier. The working mechanism of the HBMO version can briefly be summarized as follows (Fig. 8): In the proposed HBMO algorithm, after initialization, a genetic pool is built by the mating process of a single queen with the drone bees, governed by the probabilistic annealing law, thus a complete solution space between the predefined lower and upper limitations is generated in the form of the queen's and the successful drones's genetic inheritances. Here the entire colony is divided into the N_{hive} hives that facilitates "Sorting" process applied to the sub-colonies step by step, in the other words the search for the new candidates is performed in reduced number of sub-matrices instead of making a search for a single gigantic matrix. This gains the algorithm both simplicity and efficiency. The mating process is also simplified to only energy-based probabilistic decision rule to enable the fittest solutions. Furthermore, a sub-solution space as the "Egg-Population" is built by crossover processing of the entire huge solution space of the genetic pool. Accelerated exploration in the form of the five steps is applied into the egg population to obtain the best solution: 1-Fertilization ($N_{fertilization}$), 2-Larva (N_{Larva}), 3-Brood (N_{brood}), 4-Hive (N_{Hive}), and 5-Generation (N_{Gen}), size of each of these steps is equal to maximum iteration number which is taken to be equal to 20 in our application. The accelerated exploration is based on the "sorting" step

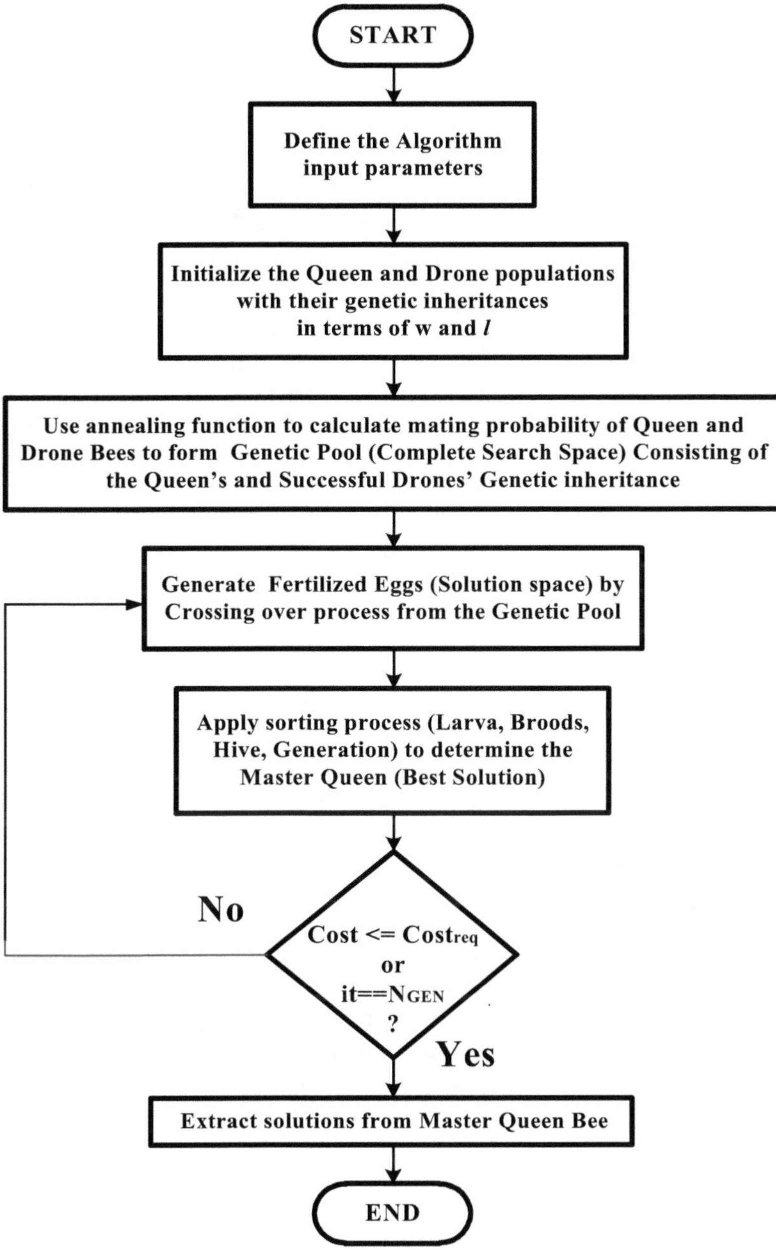

Fig. 8 Flow chart of the HBMO algorithm

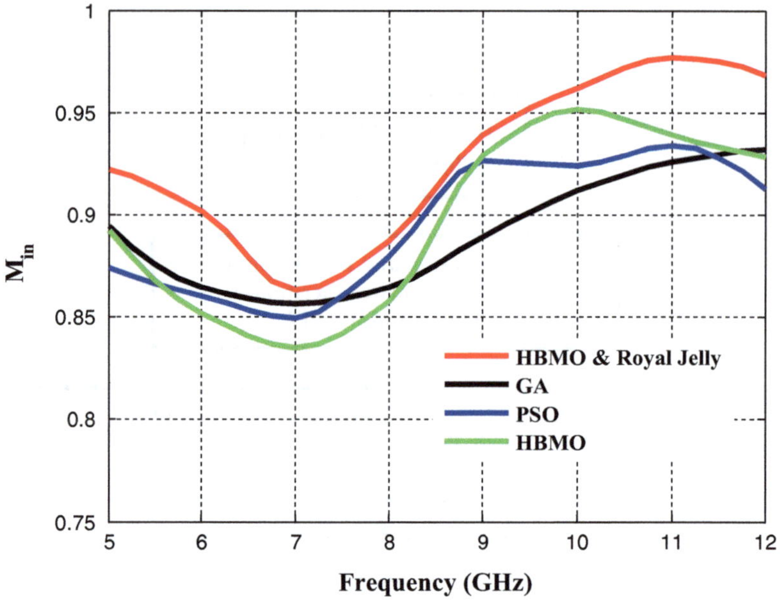

Fig. 9 Mismatching at the input port using standard metaheuristic algorithms

by step and can briefly be summarized as follows: In each step, the current entire population is divided into the subpopulations having (N_{Gen}) members, then the best member with the minimum cost value of each subpopulation is promoted to the next step, and the rest members are discarded. In this final step, only (W_j, ℓ_j) couples having the minimum cost of the competition will be chosen as the new Master Queen bee which will take new mating flights to give born to new members of the next generation of the colony. Besides "Royal Jelly" feed is used in algorithm to make a local search in order to improve the fitness of the Master Queen bee at the end of the each generation or iteration. Thus comparison with the counterpart population-based algorithms (Figs. 9, 10, 11, 13, 14 and 15) verified that a robust and fast convergent algorithm with the minimal problem information is resulted for the most successful design of a front-end amplifier.

1.6.6 Implementation

The user-defined parameters of the HBMO algorithms are set to the following values in the design of the front-end amplifier: $N_{Drone} = 20$, $t_{max} = N_{Gen} = 20$, $m_Q = 1000$, $m_D = 100$, $N_{Rj} = 1000$, $E_{max} = 1$, $E_{min} = 0.2$, $cost_{req} = 0.02$.

In the implementation, NE3512S02 is selected as the microwave transistor and maximum gain $G_{Tmax}(f)$ variations constrained by the minimum noise figures $F_{min}(f)$, $F = 1$ dB and $F = 1.5$ dB are evaluated numerically using the HMO and

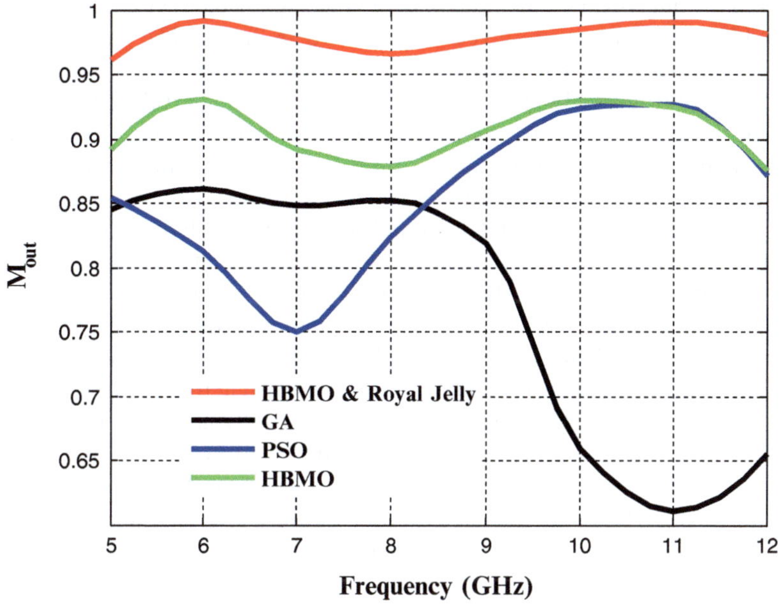

Fig. 10 Mismatching at the output port using standard metaheuristic algorithms

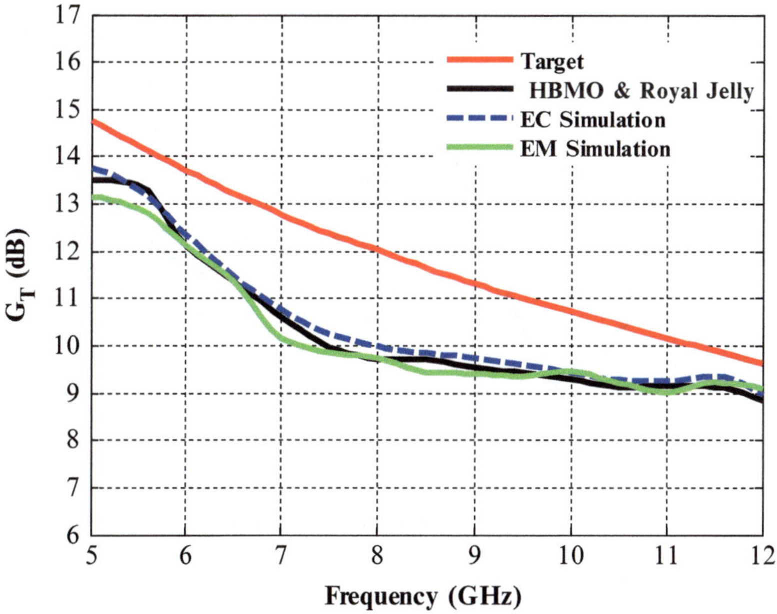

Fig. 11 Gain performance of the amplifier for the maximum power delivery for the noise figure F(*f*) = 1 dB

compared the analytical counterparts [14, 15] in Fig. 6 and the transistor source Z_S and load Z_L terminations are given for F = 1 dB in Tab. 1. The gain performance G_{Tmax} (f) constrained by F = 1 dB at the bias condition (2 V, 10 mA) is designed on the substrate of Rogers 4350 ($\varepsilon_r = 3.48$, h = 1.524 mm, tanδ = 0.003, t = 0.001 mm) along the bandwidth of 5–12 GHz. The solution space of the T-type matching circuits in Fig. 5 is shown in Tab. 2. Impedance mismatching at the input and output ports are given as compared with the genetic algorithm (GA), particle swarm optimization (PSO), and HBMO with and without Royal Jelly in Figs. 9 and 10, respectively. The resulted gain, noise performances, input and output reflections of the amplifier designed by HBMO with Royal Jelly take place by are given in Figs. 11, 12, 13, and 14, respectively, as compared with the targeted performances and obtained by the AWR circuit and 3-D EM simulators. Furthermore the cost and execution time with iteration number of the used counterpart's algorithms which are GA, PSO, and HBMO with and without Royal Jelly are given in Fig. 15. The optimization parameters of the studied algorithms are given in Tab. 5, the parameters of the PSO and GA are taken as their default values of the MATLAB optimization tool, MATLAB 2010b. The cost values and execution times at the 20th iteration of a random run are given in Tab. 3 performed by the Intel Core i7 CPU, 1.60 GHz Processor, 6 GB RAM (Tabs. 4, 5 and 6).

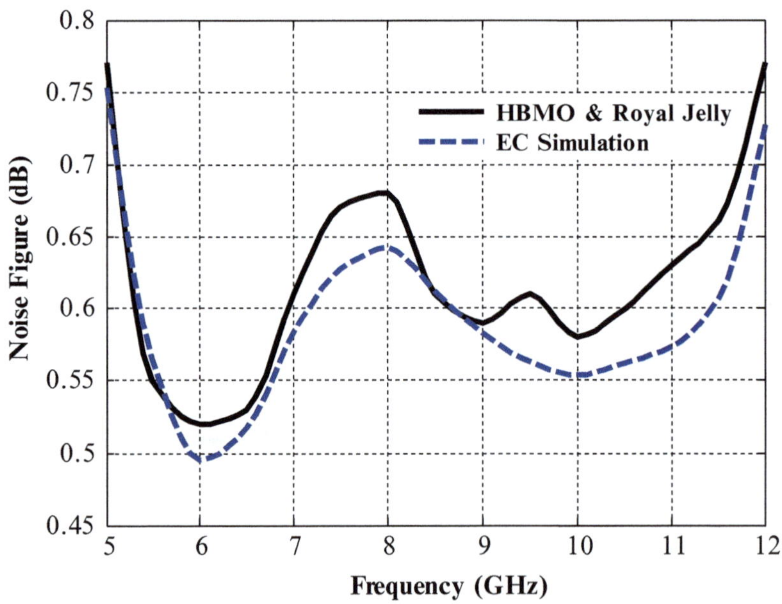

Fig. 12 Synthesized noise performance of the T-type amplifier

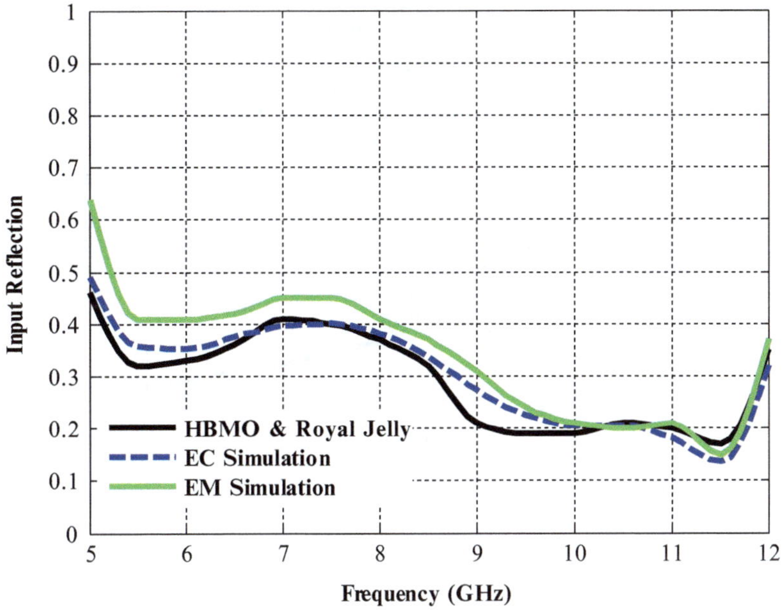

Fig. 13 Input reflection of the T-type amplifier

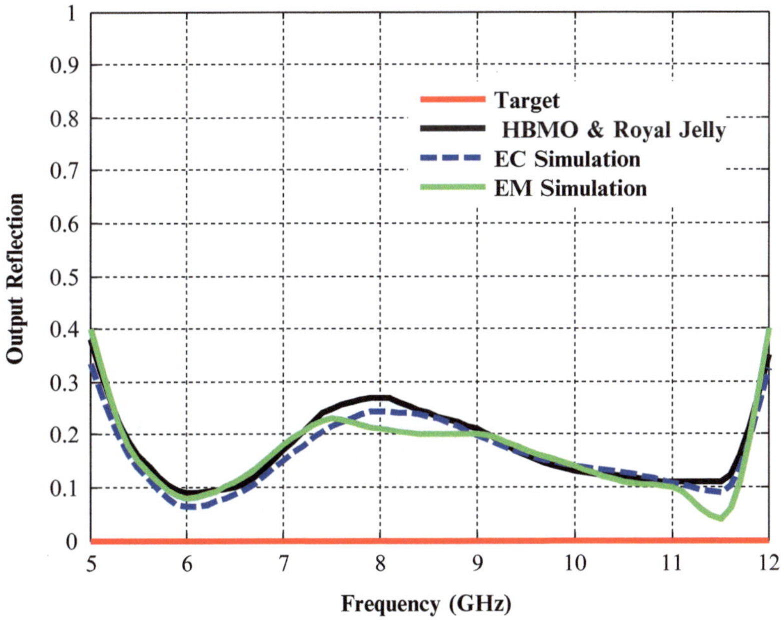

Fig. 14 Output reflection of the T-type amplifier

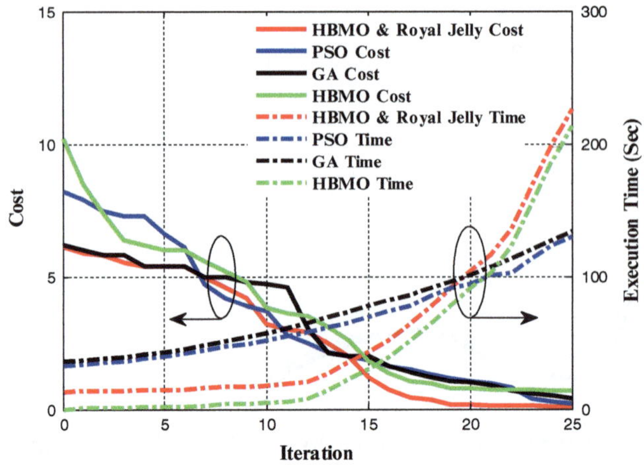

Fig. 15 Cost and execution time variations for PSO, GA, and HBMO and Royal Jelly

Table 3 Benchmarking of cost variation for 10 tries at 20th iteration for all algorithms

Algorithm	Worst	Best	Mean
HBMO and Royal Jelly	0.29	0.12	0.18
HBMO	0.9	0.65	0.74
GA	1.27	0.95	0.99
PSO	1.15	0.9	0.96

Table 4 Solutions of the T-type input and output microstrip matching elements for the maximum output power and the noise figure $F(f) = 1$ dB

W_1(mm)	W_2(mm)	W_3(mm)	W_4(mm)	W_5(mm)	W_6 (mm)
4.58	4.99	4.32	1.28	3.79	4.13
ℓ_1 (mm)	ℓ_2 (mm)	ℓ_3 (mm)	ℓ_4 (mm)	ℓ_5 (mm)	ℓ_6 (mm)
13.93	5.37	0.77	1.73	5.65	14.36

Table 5 Benchmarking at 20th iteration

Algorithm	Cost	Execution time(Sec)
HBMO and Royal Jelly	0.17	84
HBMO	0.77	71
PSO	1.15	84
GA	1.05	89

Table 6 User-defined parameters of the algorithms

Algorithm	Population	Maximum iteration	Special parameters
HBMO and Royal Jelly	Iteration 5	25	NDrone = 20, Emax = 1, Emin = 0.2, NRJ = ±0.01
HBMO	Iteration 5	25	NDrone = 20, Emax = 1, Emin = 0.2
GA	30	25	Gaussian mutation
PSO	30	25	Learning factors c1 = c2 = 2

1.7 Summary

In this part of the chapter a front-end amplifier is formulated as a constrained optimization problem each ingredient of which is carried out rigorously on the mathematical basis. The significance of the work for the microwave circuit theory can mainly be itemized as follows:

(1) First of all, the design needs solely the fundamental microwave circuit knowledge; (2) Design target is feasible based on the potential performance of the used active device that is obtained by solving numerically the nonlinear gain, noise, and input and output mismatching equation using a metaheuristic algorithm subject to the design objective; (3) In the design of the input and output microstrip matching circuits, the cost effective microstrip SVRM model is used as a fast and accurate model so that it facilitates to obtain directly all the matching microstrip widths, lengths $\left\{\vec{W}, \vec{\ell}\right\}$ on a chosen substrate to satisfy the feasible design target (FDT) over the required bandwidth of a selected transistor; (4) Microstrip matching circuit in any configuration can be easily synthesized by either gradient/nongradient optimization.

It can be concluded that the paper presents an attractive design method for a front-end amplifier design based on the transistor potential performance, and it can be adapted to design of the other types of linear amplifiers.

2 Design Optimization of Reflectarray Antennas

Reflectarray antenna (RA) is able to provide equivalent performance of a traditional parabolic reflector, but their simple structures with the low profiles, light weights, and no need any complicated feeding networks. This can be achieved by designing each RA element to reflect the incident wave independently with a phase compensation proportional to the distance from the phase center of the feed-horn to form a pencil beam in a specified direction (θ_0, ϕ_0) as is well-known from the classical array theory. Thus, "Phasing" is very important process in designing reflectarray. In literature different approaches of compensating the phase of each element have been proposed, however, phasing method using the variable size patches is preferable choice in many designs due to its simplicity [28, 29].

Since it is simple to manufacture the microstrip RA on a single layer, in order to satisfy requirements as the capability to radiate a shaped beam or multi-beams, or also to enhance the frequency behavior and bandwidth, the advanced patch configurations are necessary to be worked out in which the structure has a lot of degrees of freedom and all concur to the performances of the whole antenna. The management of different parameters and the need of satisfying requirements that could be also in opposite each other could however make the design of a reflectarray quite complex. Therefore first of all for a computationally efficient optimization process, an accurate and rapid model for the reflection phase of a unit element is needed to establish it as a continuous function in the input domain of the patch geometry and substrate variables, then it could be convenient to carry this model out adopting a hybrid "global + local" search method to find the best solution among all the possible solutions.

Thus, the systematic design optimization procedure for the Minkowski RA is presented in this chapter. It can briefly be summarized in the following steps: The first step is devoted to the discretization of the 5-D Minkowski space of (m, n, ε_r, h, f) to obtain the training and validation data for MLP NN. In the next part, the gain and bandwidth optimization of MLP NN model with respect to the input variables will be presented using the hybrid combination of Genetic and Nelder-Mead algorithms. In addition, the sensitivity and yield analyses are performed for the tolerance analysis in order to specify the tolerance limits of optimized design parameters. Design and performance analysis of the Minkowski RAs with the optimized or non-optimized antenna parameters will be taken place in the fourth and fifth sections, respectively. Finally the paper ends with the conclusions.

2.1 Reflection Phase Characterization of a Minkowski Element

2.1.1 Minkowski Space

In the design of microstrip RA, the shape and geometry selection of the RA element is the crucial part as well as the substrate properties chosen. In this work, the geometry of radiating element has been proposed to be a resonant element shape for a periodic RA structure, which is a first fractal type, as called the Minkowski shape. Fig. 16a shows the geometrical representation of Minkowski shape patch element.

The relationship between the Minkowski parameters is formulated as:

$$n = \frac{s}{m/3}, \qquad 0 \leq n \leq 1 \tag{9}$$

In Eq. (9), s is the indention and m is the width of the patch, respectively, and n refers the indention ratio. The reflection response of unit cell and phase of reflected wave are generated by the 3-D CST MWS-based analysis implemented

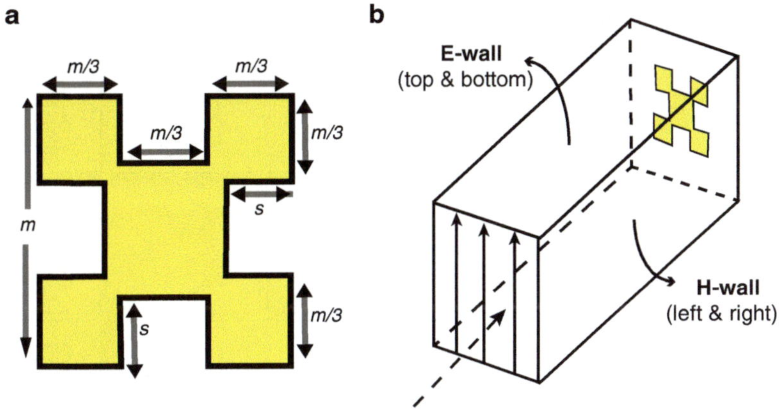

Fig. 16 (**a**) Minkowski patch geometry, (**b**) The H-wall waveguide simulator

to the H-wall waveguide simulator which is shown in Fig. 16b. The top and bottom surfaces of the H-wall waveguide simulator are perfectly electric conducting walls, while the right and left walls are perfectly magnetic field walls [29]. The vertically polarized incoming waves will be incident normally onto the element at the end of the waveguide at the broadside direction and then scattered back also at the broadside direction with a set of amplitude and phase information. The 5-D discretized Minkowski space of (m, n, ε_r, h, f) is constructed by totally 5400 samples to be used in the training and validation of the MLP NN model using the H-wall waveguide simulator analyzed by 3-D CST MWS as follows:

The operation bandwidth of 8–12 GHz is swept as the intervals of 1 GHz and the resulted number of the sample frequencies is fs = 5. Then, Minkowski sampling matrix (Fig. 17) is generated as $n\,\mathrm{s} \times \mathrm{ms}$ for each sampled substrate properties (εr, h) at each sampling frequency where n s = 6 and ms = 5 are the number of samples for the indention factor and patch width within the ranges of $0.15 \le \mathrm{n} \le 0.9$ and $\mathrm{m} \pm (\Delta \mathrm{m}/\mathrm{m})\,\mathrm{max} = \mathrm{m} \pm \%\ 20$ where m is the resonant length at 11 GHz, respectively. Simultaneously the substrate thickness h is sampled as the intervals of 0.5 mm between them $0.5\ \mathrm{mm} \le \mathrm{h} \le 3$ mm and the total number of the thickness sampling is h s = 6. In addition, the dielectric permittivity of substrate $(\varepsilon \mathrm{r})$ is totally sampled εs = 6 times between $1 \le \varepsilon\mathrm{r} \le 6$. Thus, the entire Minkowski space is discretized totally into the $\varepsilon\mathrm{s} \times \mathrm{fs} \times \mathrm{hs} \times \mathrm{ms} \times \mathrm{ns} = 5400$ Minkowski configurations [30–32].

2.1.2 The Modeling of MLP NN

The employed MLP NN model of Minkowski patch, which is generalization process, is depicted in Fig. 18. The MLP NN has the two hidden layers each of which consists of 10 neurons activating by the tangential sigmoid function. The

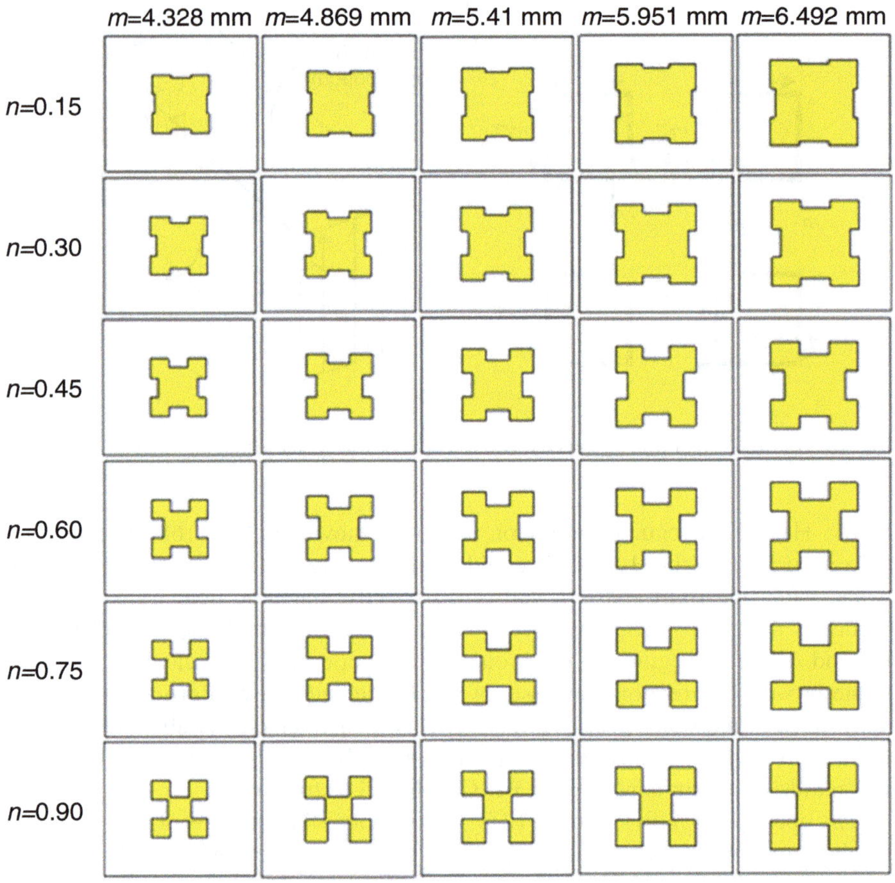

Fig. 17 Sampling Minkowski patch variation matrix (*ns* x *ms* = 6 x 5 = 30)

input and output vectors $\left(\vec{x}, \vec{y}\right)$ are 5- and 1-dimensioned, respectively, and can be expressed as Eq. (10):

$$\vec{x} = [m \;\; n \;\; \varepsilon_r \, h \; f]^t, \quad \vec{y} = [\varphi_{11}]^t = \varphi_{11}\left(\vec{x}, \vec{w}\right) \tag{10}$$

where $\vec{w}$ is the weighting vector of the MLP NN given in Fig. 13. The output function $\varphi_{11}\left(\vec{x}, \vec{w}\right)$ can be built using the MLP NN theory [8]. The weighting vector $\vec{w}$ is determined by the optimization with mean-squared error (Eq. 11) over the training data using the Levenberg-Marquardt algorithm [33, 34]:

$$MSE = \sum_{k \in T_r} \left(\varphi_{11k} - d_k\right)^2 \tag{11}$$

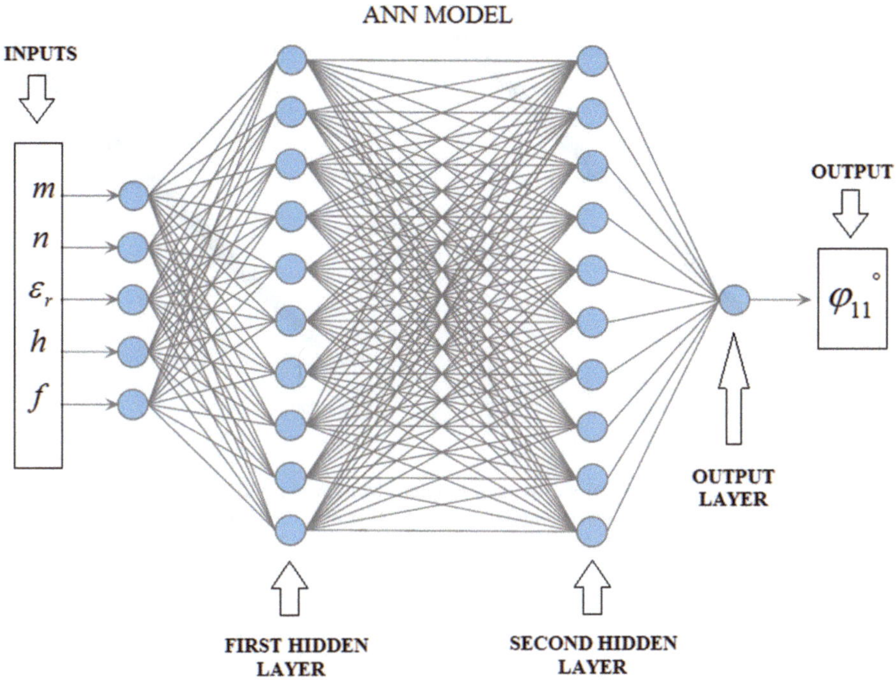

Fig. 18 The MLP NN structure for Minkowski patch

where T_r is an index set of the training data which consists of 3240 $(\vec{x},\ \varphi_{11})$ data pairs corresponding to the patch lengths of 4.328, 5.41, and 6.491, the rest of 2160 $(\vec{x},\ \varphi_{11})$ data pairs are used to validate the MLP NN model. The linear regression scattering plots for the training and the validation process are given with their MSE errors in Fig. 19.

Fig. 20 gives the 3-D view of the reflection phase variations with the patch width m and the relative permittivity of substrate (ε_r) for the constructed and targeted data at the fixed conditions of $h = 1.5$ mm, $n = 0.6$, $f = 11$ GHz. Some examples of modeling performances are depicted in Fig. 21 where the constructed phasing characteristics are compared with their targets. Furthermore thus, it can be inferred that the MLP NN model works very well in generalization of the 5400 $(\vec{x},\ \varphi_{11})$ data pairs to the entire domains to obtain the continuous Minkowski reflection phasing function $\varphi_{11}(\vec{x})$. In the next section, this $\varphi_{11}(\vec{x})$ function will be used directly to determine the phase calibration characteristic and later it will be reversed to synthesize the Minkowski RA in the Memetic optimization procedure.

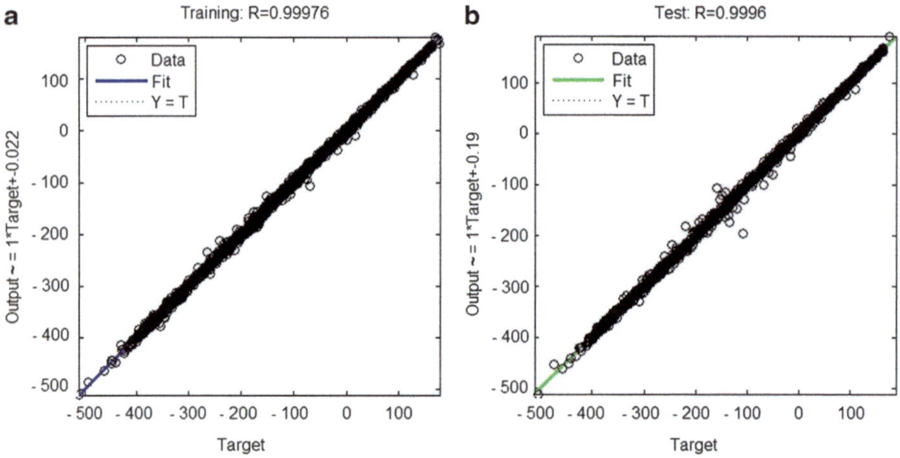

Fig. 19 Regression scattering plots for the complete Minkowski MLP NN model (**a**) training (MSE Error = 9.9564x10^{-5}) and (**b**) validation (MSE error = 1.7264x10^{-4})

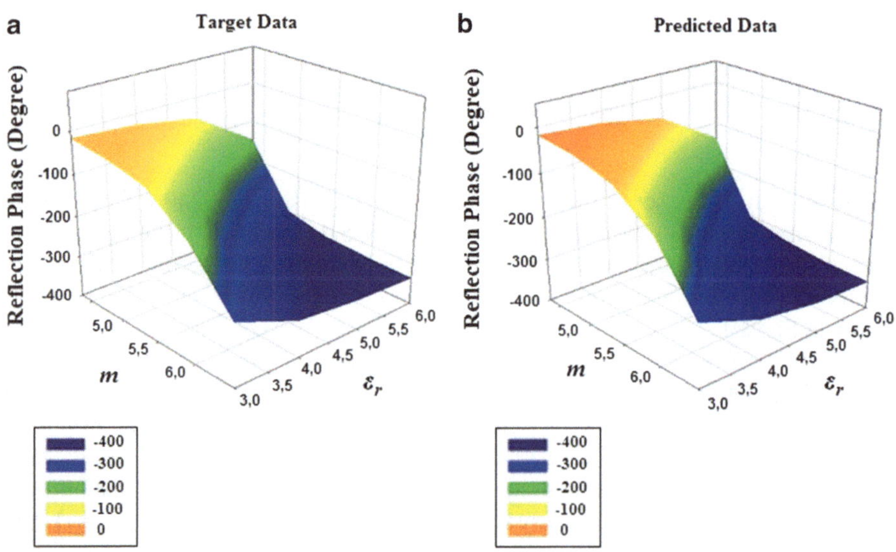

Fig. 20 3-D view of reflection phase variations w.r.t. the patch width m and the relative permittivity ε_r for the fixed conditions of $h = 1.5$ mm, $n = 0.6$, $f = 11$ GHz for (**a**) target and (**b**) constructed data

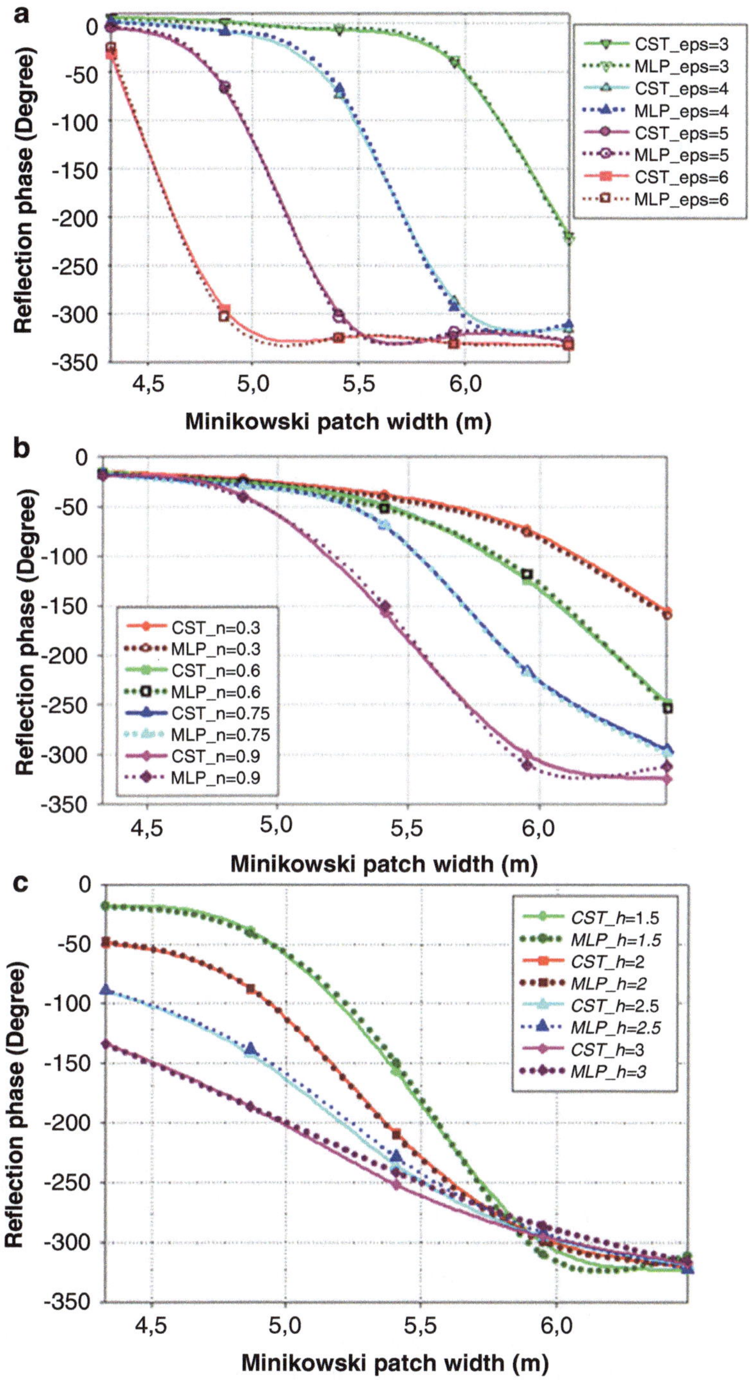

Fig. 21 Reflection phase characteristics for (**a**) $h = 1$ mm, $n = 0.60, f = 11$ GHz; taking dielectric constant ε_r as parameter; (**b**) $\varepsilon_r = 3$, $h = 1.5$ mm, $f = 11$ GHz and indention ratio n is parameter (**c**) $\varepsilon_r = 3, n = 0.90, f = 11$ GHz and substrate thickness h is parameter

2.2 *The Optimization Process*

2.2.1 Objective Function

In the optimization process, a multi-objective procedure is established where the phase calibration characteristic is selected among the phasing characteristics obtained in the previous section as the one having the slower gradient and the wider range with respect to the indention of patch (n) and substrate (ε_r, h) to achieve the wider band and smaller susceptibility to the manufacturing errors. Thus, this objective can be expressed as the sum of the three ingredients as follows:

$$Objective = \underset{n,\, h,\, \varepsilon_r}{Min} \left\{ \sum_{i=l,c,u} \vartheta_i\,(n,\ h,\ \varepsilon_r) \right\} \tag{12}$$

with the following objective ϑ_i at the frequency f_i:

$$\vartheta_i = \left\{ \sum_{\substack{\varepsilon_r=1 \\ \Delta\varepsilon_r=0.01}}^{6} \ \sum_{\substack{h=0.5\text{mm} \\ \Delta h=0.01\text{mm}}}^{3\text{ mm}} \ \sum_{\substack{n=0.15 \\ \Delta n=0.01}}^{0.9} W_1.\in_1(f_i) + W_2.\in_2(f_i) + W_3.\in_3(f_i) \right\} \tag{13}$$

where,

$$\in_1 = e^{-\left(\frac{\varphi_{\max}-\varphi_{\min}}{360}\right)} \tag{14}$$

$$\in_2 = |\varphi_{\max} - \varphi_{center}| - |\varphi_{\min} - \varphi_{center}| \tag{15}$$

$$\in_3 = 1\text{-}\left(\frac{\Delta\varphi_{center}}{\Delta m_{center}}\right) \tag{16}$$

In Eq. (14), $\in_1$ is used to maximize the phase range while $\in_2, \in_3$ provide the centralization of the characteristic with the angle of $\pi/4$. In Eqs. (14), (15), and (16), $\varphi_{\max}$, $\varphi_{\min}$, and φ_{center}, are the reflection phase values at m_{max}, $m_{\min}$, and m_{center} for a certain (n, ε_r, h) set, respectively, at the f_i where l, c, u stand for the lower, center, and the upper frequencies. In the optimization process, the operation frequency range is defined as follows: $f_l = 10$ GHz, $f_c = 11$ GHz, $f_u = 12$ GHz. In Eq. (16), the phase difference between $\varphi_{\max}$ and $\varphi_{\min}$ is normalized by dividing 360 and ($\Delta\varphi_{\text{center}}/\Delta m_{\text{center}}$) is the gradient of the phasing characteristic at the point of ($\varphi_{\text{center}}, m_{\text{center}}$) which is aimed at to be equal to unity corresponding to optimum angle $\pi/4$. All weighting coefficients in the objective function ϑ_i at the frequency f_i

in Eq. (13) have been taken as unity. Optimization process is completed as soon as the iteration number has reached to its maximum value or the predefined cost value. In our case, the optimization ends when the cost value reaches to 0.4353 with the optimized values of all the weighting coefficients.

2.2.2 The Memetic Algorithm: Hybrid Combination of GA-NM Algorithm

A Memetic algorithm (MA) is essentially a combination of a population-based global optimization algorithm with a local search [35]. Recently, Memetic algorithms consisting of the hybrid GA-NM and bacterial swarm optimization BSO-NM algorithms are successfully implemented to designs of the low-noise microwave amplifier and Bow–Tie antennas in [36] and [37], respectively. In this work, a Genetic Algorithm (GA) is used as a population-based global optimizer and a simple local search algorithm called Nelder-Mead (NM) [13] is employed along with the GA to reduce the cost of the solution at each iteration of the optimization procedure.

The GA uses the evolution operations which are the crossover, mutation, and recombination together with the concept of fitness. The population is built by the chromosomes as the solution candidates, binary encoded randomly varied as 0 and 1. The objective function corresponding to each chromosome is evaluated, then chromosomes are ranked according to their fitness's and the least fit ones are discarded and the remaining chromosomes are paired at randomly selected crossover points. In order to prevent the solution from being trapped into the local minima, mutation process is applied by transforming a small percentage of the bits in the chromosome from 0 to 1 or vice versa. The mutation process per iteration is applied for 1 % of the chromosomes.

The MA used in our work can be briefly described through the following abstract description [37]:

Begin
Population initialization
Local search
Evaluation
Repeat
Crossover
Recombination
Mutation
Local Search
Evaluation
Selection
Until termination criterion is satisfied
Return best solution
End

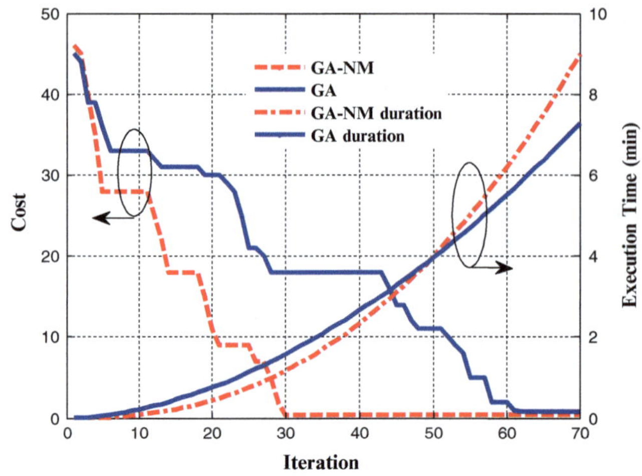

Fig. 22 The convergence curves of the genetic and Memetic optimization

Here, the initial populations are usually generated in a random or controlled manner and then the evolution of these populations is carried out by the genetic operators such as crossover, mutation, and recombination. Local search is utilized to reduce the cost of the resulted solution from the global optimization.

In our GA-NM application, the MATLAB [34] is used for the Memetic algorithm with the selection stochastic uniform operators consisting of a population (chromosome) of 60, number of generation of 900, crossover probability of 0.8 (or crossover fraction for reproduction is 0.8), and mutation probability of 0.001. Mutation function is constraint dependent. Crossover function is scattered. Migration direction is just forward numbered 0.2. The convergence occurs very quickly typically within the 30 iterations shortening 5 times as compared with the 60 iterations GA process, which takes 1 min and 12 s and 5 min and 41 s with Core i7 CPU, 1.60 GHz Processor, 4 GB RAM depending on the initialization values. A typical convergence curve is given in Fig. 22 [38].

2.3 Tolerance Analysis of the Optimized Parameters

The design parameters may usually change in a certain tolerance region during the manufacturing process. Thus it is of interest to which percentage the design specifications are fulfilled. Thus the yield analysis is applied to compute an expected tolerance as percentage. In the implementation of yield analysis, variations in the design parameters are assumed to be small so that the linearization via the sensitivity analysis can be valid. For this reason a yield analysis can only be applied after a successful run of the sensitivity analysis. In the sensitivity analysis, the derivatives of output function with respect to geometric and/or material design parameters can be calculated without re-meshing the example. The first derivative of the network function with respect to a design parameter can be calculated with the information of the nominal value in a small neighborhood of that nominal value. Also the sensitivity information is used for a more efficient optimization.

In this study, sensitivity analysis is applied to the optimum dielectric constant $\varepsilon_{ropt} = 3.164$ by rounding up the other parameters, as $n_{opt} = 0.85$, $h_{opt} = 1.8$. Then the yield analysis is applied to the results of the sensitivity analysis for the three values of the standard deviation belonging to the dielectric constant. The graphics for these results are shown in Fig. 23 [38].

As is seen from Fig. 23, the best tolerance is at the nominal design parameter value with a lower and upper bound (−3*sigma, +3*sigma) of the dielectric permittivity when the sigma is equal to 0.01. The upper and lower bound indicate as the worst case limits of the tolerance for the dielectric property of substrate. The substrate that has closest specifications to the optimized parameters had been searched, and the two commercially available substrates which are Rogers RO3003 and RO4232 have been found. As is seen from Fig. 24, RO4232 is the fittest substrate as commercially available for our optimized parameter result.

2.4 *Design of the Variable–Size RA*

2.4.1 Phase Compensation

In this study, the 15×15 variable sizes Minkowski RA with half-wave spacing at resonant frequency of 11 GHz are designed. The radiation analysis has been generated using available full-wave simulation tool of CST MWS. In the phase compensation unit, a coordinate system has been used to determine the progressive phase distribution on the microstrip reflectarray surface of $M \times N$ arbitrarily spaced patches with a centered focal point that will produce a pencil beam in a direction of normal to the surface [8]. Thus, the required phase to compensate path difference ΔR(x) for a reflectarray element can be given as a function of its radial distance x to the center and the operation frequency f as follows:

$$\varphi\,(x,f) = -\beta\,(\Delta R_{\max} - \Delta R(x)) = -\frac{2\pi f}{c} F\left(\sqrt{1 + (D/F)^2/4} - \sqrt{1 + (x/F)^2}\right) \tag{17}$$

where the minus sign expresses delay, c is the velocity of light. In Eq. (17) D and F are the diameter and the focal length of the feed to the array center, respectively. Quadrature symmetry characteristic of the phase compensation with respect to the element position for the *15* $\times$ *15* reflectarray where frequency is considered as the parameter and F/D is taken as 0.8.

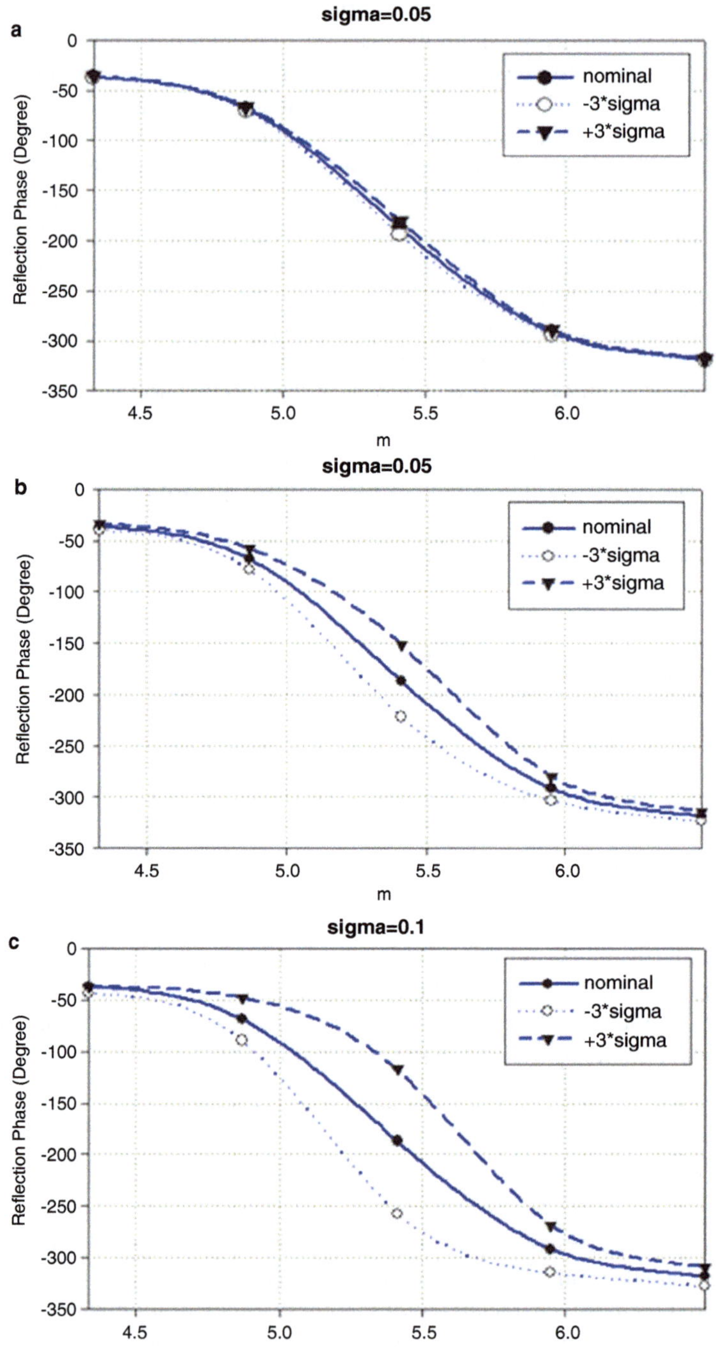

Fig. 23 Sensitivity analysis results for the optimum dielectric constant for the standard deviation (sigma) values: (**a**) $\sigma = 0.01$, (**b**) $\sigma = 0.05$, (**c**) $\sigma = 0.1$ at $f = 11$ GHz

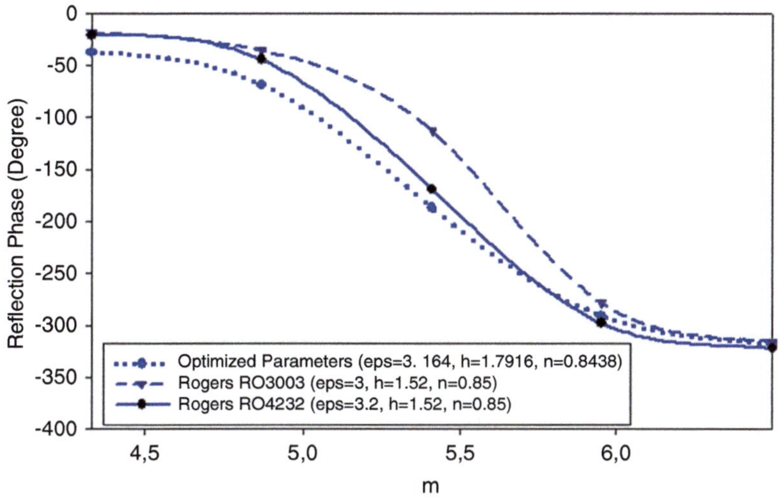

Fig. 24 Comparison of the reflection phase responses for unit cell element designed with optimized parameters and two equivalent commercially available substrates

2.4.2 Determination Size of Each Radiator

Size of each radiator is determined to meet the necessary compensation phase using the phase calibration characteristic. For this purpose, the established ANN model is reversed by inputting optimum values corresponding to the phase calibration characteristic and while input m changes itself using the adaptable size Δm which get exponentially smaller with an adaptation parameter τ as decreasing the squared error as given in Fig. 25 [32].

2.5 Implementation

In the implementation stage, all the radiation performance analyses are made using 3-D CST Microwave Studio. The fully optimized X-band Minkowski reflectarray antenna with the parameters $\varepsilon_{\text{ropt}} = 3.1694$, $h_{opt} = 1.7916$ mm, $n_{opt} = 0.8438$ is designed using the general design procedure (Fig. 25) and its realized gain patterns at the frequencies 10.5, 11, and 11.9 GHz are given in Fig. 26a. Furthermore for the purpose of comparison, the realized gain patterns of an arbitrary non-optimized RA antenna with the parameters of $\varepsilon_r = 2.2$, $h = 1.5$ mm, $n = 0.90$ at the same frequencies are obtained with the same procedure and depicted in Fig. 26b and the compared performance values take place in Tab. 7. In order to examine the influence of dielectric property optimization, the gain variation with respect to the frequency is obtained with the same optimized indention ratio $n_{opt} = 0.8438$ and thickness $h_{opt} = 1.7916$ mm, but on some traditional substrates which are Taconic RF-35 with

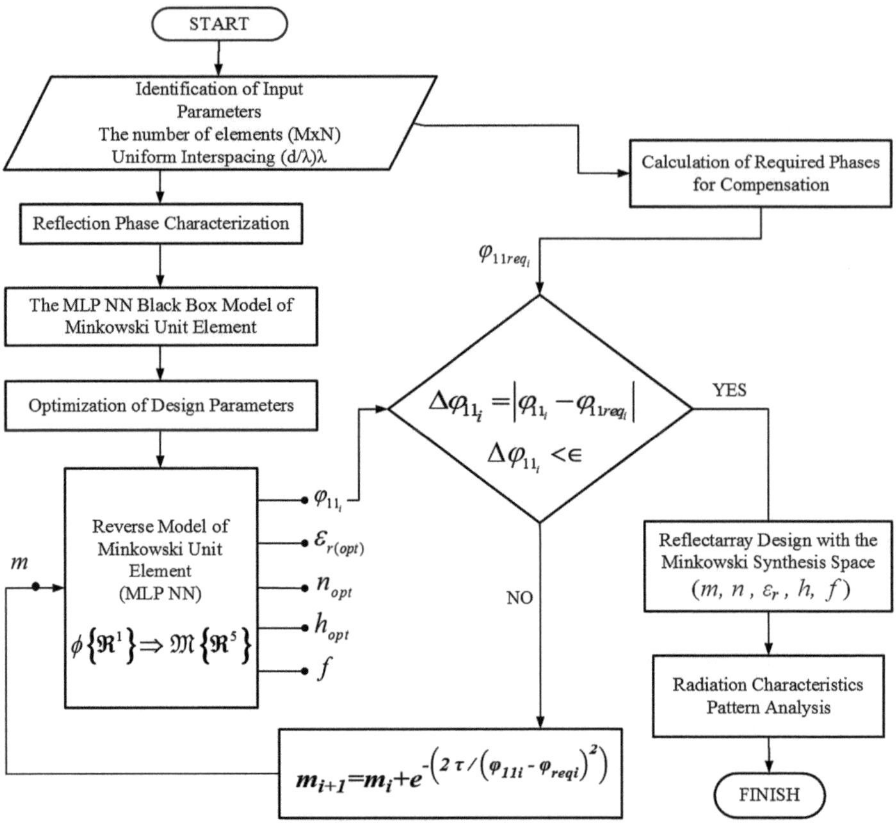

Fig. 25 Design flow chart for the optimum reflectarray antenna

$\varepsilon_r = 3.5$, Taconic TRF41 with $\varepsilon_r = 4.1$, Rogers TMM4 with $\varepsilon_r = 4.5$ and depicted in Fig. 27. The performance values corresponding to Fig. 27 take place in Tab. 8, Fig. 28 depicts the gain versus frequency variations of the optimized RAs designed on the dielectric $\varepsilon_{ropt} = 3.1694$ and the traditional substrates. The performance values belonging to Fig. 28 are given in Tab. 8 (Tab. 9).

2.5.1 Summary

Doubtlessly, microstrip reflectarrays are of prime importance in today's antenna technology, since they combine the advantages of both the printed phased arrays and parabolic reflectors to create a new generation of high gain antennas.

In this part of the chapter, a robust and systematic method is put forward to be used in the design and analysis of a Minkowski reflectarray. The most important and critical stage of a reflectarray design is the design optimization of its element.

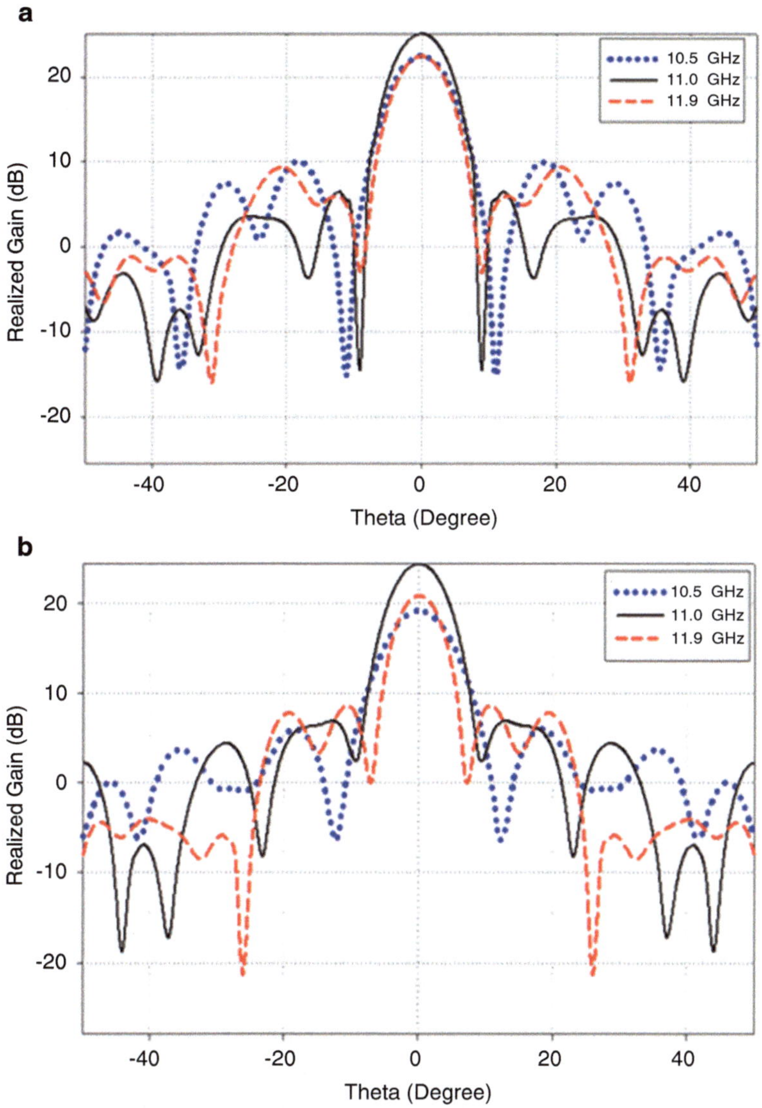

Fig. 26 (**a**) Fully optimized RA with $\varepsilon_{ropt} = 3.1694$, hopt = 1.7916 mm, nopt = 0.8438; (**b**) Non-optimized reflectarray with $\varepsilon_r = 2.2$, h = 1.5 mm, n = 0.90

Therefore, firstly a complete, accurate and fast MLP ANN model of a Minkowski patch radiator is built based on the 3-D CST Microwave Studio MWS that takes into account all the main factors influencing the performance of the Minkowski RA. When the outputs of performed MLP ANN model and 3-D simulations are compared, it is verified that the MLP is very accurate and fast solution method to construct the highly nonlinear phasing characteristics within the continuous domain

Table 7 Performance comparison of the fully optimized reflectarray with a non-optimized reflectarray

Antenna	Frequency (GHz)	Realized gain (dB)	Side lobe level (dB)	Angular width (3 dB) (Deg.)
Optimized RA $\varepsilon_{ropt} = 3.1694$, $h_{opt} = 1.7916$, $n_{opt} = 0.8438$	10.5	22.5	−12.5	7.9
	11	25	−18.6	7.4
	11.9	22.5	−13.2	7.1
Non-optimized RA $\varepsilon_r = 2.2$, $h = 1.5$, $n = 0.90$	10.5	19.2	−13.2	8.8
	11	24.4	−17.5	7.5
	11.9	21	−12.4	6.3

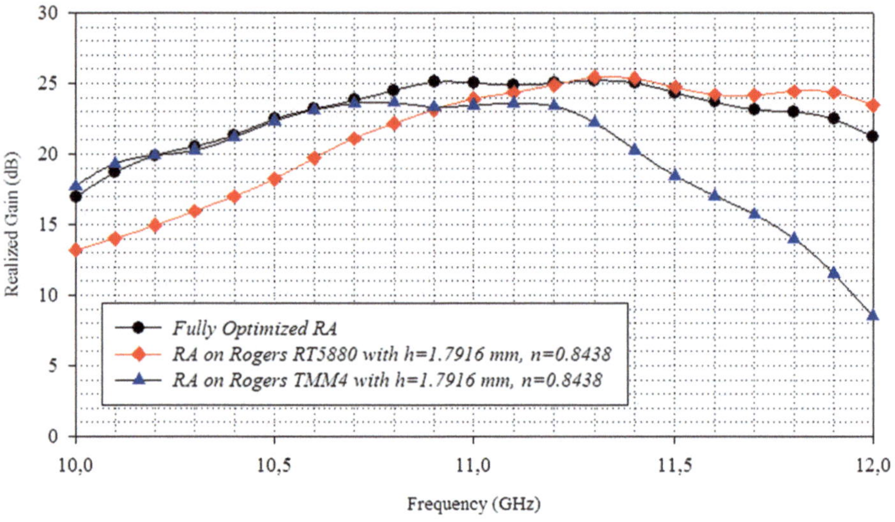

Fig. 27 Realized gain versus frequency graphs for the fully optimized RA and the other RAs on the different substrates with the optimized parameters n_{opt}, h_{opt}

of the geometrical and substrate parameters of the RA element and frequency. All the stages of building the MLP ANN model and its utilization in design of a Minkowski RA are given in details as a general systematic method that can be applied to the differently shaped patch radiators.

Overall parameters of Minkowski RA including dielectric permittivity of the substrate ε_r are optimized for an optimum linear phasing range of an ultra-wideband RA in the X-band by applying a standard novel evolutionary hybrid combination of Global Genetic (GA) and Local Nelder-Mead (NM) algorithms.

In addition to optimization process, the sensitivity and yield analyses are performed as tolerance analysis in order to specify the tolerance limits of optimized design parameters and the commercially available substrate options which are compatible with our optimized design parameters. The optimum dielectric permittivity

Table 8 Comparison of the fully optimized RA and the other RAs designed on the different substrates with same optimized parameters n_{opt}, h_{opt}

	Realized gain (dB)		
Frequency (GHz)	Fully optimized RA $\varepsilon_{ropt} = 3.1694$, $h_{opt} = 1.7916$ mm, $n_{opt} = 0.8438$	Rogers RT5880 $\varepsilon_r = 2.2$, $h_{opt} = 1.7916$ mm, $n_{opt} = 0.8438$	Rogers TMM4 $\varepsilon_r = 4.5$, $h_{opt} = 1.7916$ mm, $n_{opt} = 0.8438$
10	17	13.2	17.7
10.5	22.5	18.2	22.3
11	25	23.9	23.5
11.5	24.3	24.7	18.5
12	21.2	23.5	8.5

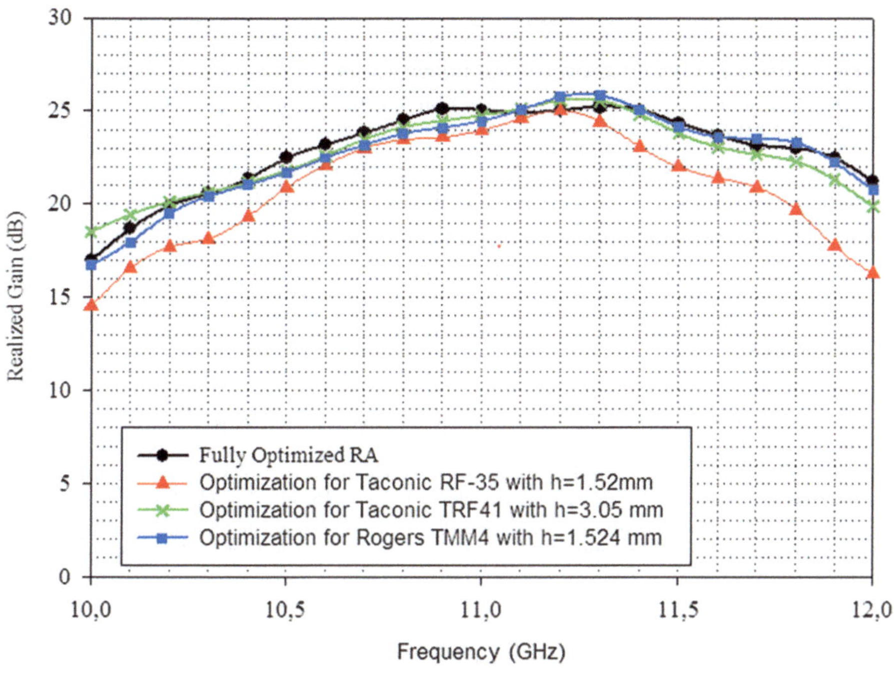

Fig. 28 Gain variations of fully optimized RA with only patch geometry n_{opt} optimized RAs on the given dielectric permittivity ε_r and substrate thickness η

tolerance limits are qualified rounding up the values of the optimum substrate thickness h_{opt} and indention ratio of Minkowski microstrip patch n_{opt} for the three characteristic values of the standard deviation. Thus this tolerance analysis results in the limits of design parameters and the proper commercial available dielectric substrate as Rogers RO4232. Finally, a fully optimized 15×15 Minkowski RA is designed as a worked example. Thus, its radiation characteristics are analyzed

Table 9 Comparison of the fully optimized RA and RAs with the optimized Minkowski shapes on the traditional substrates

Frequency (GHz)	Realized gain (dB)			
	Optimized reflectarray $\varepsilon_{ropt} = 3.1694$, $h_{opt} = 1.7916$, $n_{opt} = 0.8438$	Taconic RF-35 $\varepsilon_r = 3.5$, $h = 1.52$, $n_{opt} = 0.7848$	Taconic TRF41 $\varepsilon_r = 4.1$, $h = 3.05$, $n_{opt} = 0.6212$	Rogers TMM4 $\varepsilon_r = 4.5$, $h = 1.524$, $n_{opt} = 0.3604$
	17	14.5	18.5	16.7
10.5	22.5	20.9	21.8	21.7
11	25	24	24.8	24.5
11.5	24.3	22	23.8	24.1
12	21.2	16.3	19.9	20.7

based on the 3-D CST Microwave Studio MWS and graphically represented, then compared with the performances of the non-optimized and the partially optimized Minkowski RAs.

It may be concluded that the presented method can be considered as a robust and systematic method for the design and analysis of a microstrip reflectarray antenna built by the advanced patches.

References

1. Güneş, F., Gürgen, F., Torpi, H.: Signal - noise neural network model for active microwave device. IEE Proc. Circuits Devices Syst. **143**, 1–8 (1996)
2. Güneş, F., Torpi, H., Gürgen, F.: A multidimensional signal-noise neural network model for microwave transistors. IEE Proc. Circuits Devices Syst. **145**, 111–117 (1998)
3. Giannini, F., Leuzzi, G., Orengo, G., Albertini, M.: Small-signal and large-signal modelling of active Devices using CAD-optimized neural networks. Int. J. RF Microw. Comput. Aided Eng. **12**, 71–78 (2002)
4. Güneş, F., Türker, N., Gürgen, F.: Signal-noise support vector model of a microwave transistor. Int. J. RF Microw. Comput. Aided Eng. **17**, 404–415 (2007)
5. Marinkovic, Z.Z., Pronic -Rancic, O., Markovic, V.: Small-signal and noise modelling of class of HEMTs using knowledge-based artificial neural networks. Int. J. RF Microw. Comput. Aided Eng. **23**, 34–39 (2013)
6. Güneş, F., Özkaya, U., Uluslu, A.: Generalized neural-network based efficient noise modelling of microwave transistors. In: International Symposium on Innovations in Intelligent Systems and Applications (INISTA), pp. 212–216, Istanbul, Turkey, 15–18 June 2011
7. Mahouti, P., Güneş, F., Demirel, S., Uluslu, A., Belen, M.A.: Efficient scattering parameter modeling of a microwave transistor using generalized regression neural network. in Microwaves, Radar, and Wireless Communication (MIKON), 2014, 20th International Conference on, pp. 1–4, 16–18 June 2014
8. Güneş, F., Güneş, M., Fidan, M.: Performance characterisation of a microwave transistor. IEE Proc. Circuits Devices Syst. **141**(5), 337–344 (1994)
9. Güneş, F., Çetiner, B.A.: A Novel Smith chart formulation of performance characterisation for a microwave transistor. IEE Proc. Circuits Devices Syst. **145**(6), 419–428 (1998)

10. Güneş, F., Bilgin, C.: A generalized design procedure for a microwave amplifier: a typical application. Prog. Electromagn. Res. B, **10**, 1–19 (2008)
11. Edwards, M.L., Cheng, S., Sinsky, J.H.: A deterministic approach for designing conditionally stable amplifiers. IEEE Trans. Microw. Theory Tech. **43**(1), 1567–1575 (1995)
12. Demirel, S.: A generalized procedure for design of microwave amplifiers and its applications. PhD Thesis (in Turkish), Yıldız Technical University, Istanbul, Turkey (2009)
13. Demirel, S., Güneş, F.: Performance characterisation of a microwave transistor for maximum output power and the required noise. IET Circuits Devices Syst. **7**(1), 9–20 (2013)
14. Ciccognani, W., Longhi Patrick, E., Colangeli, S., Limiti, E.: Constant mismatch circles and application to low-noise microwave amplifier design. IEEE Trans. Microw. Theory Tech. **61**(12), 4154–4167 (2013)
15. Güneş, F., Özkaya, U., Demirel, S.: Particle swarm intelligence applied to determination of the feasible design target for a low-noise amplifier. Microw. Opt. Technol. Lett. **51**(5), 1214–1218 (2009)
16. Mahouti, P., Güneş, F., Demirel, S.: Honey– bees mating algorithm applied to feasible design target space for a wide– band front–end amplifier. In: ICUWB 2012- IEEE International Conference on Ultra-Wideband, pp. 251–255 (2012). doi: 10.1109/ICUWB.2012.6340406
17. Güneş, F., Demirel, S., Mahouti, P.: Design of a Front–End Amplifier for the Maximum Power Delivery and Required Noise by HBMO with Support Vector Microstrip Model. Radioengineering, **23**(1), (2014)
18. Tokan, N.T., Güneş, F.: Knowledge –based support vector synthesis of the microstrip lines. Prog. Electromagn. Res. **92**, 65–77 (2009)
19. Güneş, F., Tokan, N.T., Gürgen, F.: A knowledge-based support vector synthesis of the transmission lines for use in microwave integrated circuits. Expert Syst. **37**, 3302–3309 (2010)
20. Güneş, F., Cengiz, Y.: Optimization of a microwave amplifier using neural performance data sheets with genetic algorithms. In: International Conference on Artificial Neural Networks (ICANN), Istanbul, pp. 26–29, June 2003
21. Cengiz,Y., Göksu, H., Güneş, F.: Design of a broadband microwave amplifier using neural performance data sheets and very fast simulated reannealing. In: Wang, J. et al. (ed.) Advances in neural networks – ISNN, Pt3, Proceedings, vol. 3973, pp. 815–820. Springer-Verlag, Berlin, Heidelberg (2006)
22. Güneş, F., Demirel, S.: Gain gradients applied to optimization of distributed-parameter matching circuits for a microwave transistor subject to its potential performance. Int. J. RF Microw. CAE **18**, 99–111 (2008)
23. Güneş, F., Demirel, S., Özkaya, U.: A low-noise amplifier design using the performance limitations of a microwave transistor for the ultra-wideband applications. Int. J. RF Microw. Comput. Aided Eng. **20**, 535–545 (2010)
24. Cengiz, Y., Kılıç, U.: Memetic optimization algorithm applied to design microwave amplifier for the specific gain value constrained by the minimum noise over the available bandwidth. Int. J RF Microw. Comput. Aided Eng. **20**, 546–556 (2010)
25. Keskin, A.K.: Design optimization of ultrawide band microstrip amplifier using 3D Sonnet-based SVRM with particle swarm intelligence. MSc thesis, Yıldız Technical University, Istanbul, Turkey (2012)
26. Güneş, F., Demirel, S., Mahouti, P.: Design of a front– end amplifier for the maximum power delivery and required noise by HBMO with support vector microstrip model. Radioengineering **23**(1), 134–142 (2014)
27. Pozar, D.M.: Microwave Engineering. Wiley, New York (2012)
28. Pozar, D.M., Metzler, T.A.: Analysis of a reflectarray antenna using microstrip patches of variable size. Electron. Lett. **27**, 657–658 (1993)
29. Huang, J., Encinar, J.A.: Reflectarray Antennas. Wiley-IEEE Press, Hoboken. New Jersey. ISBN: 978–0470–08491–4, 2007

30. Nesil, S., Güneş, F., Özkaya, U.: Phase characterization of a reflectarray unit cell with Minkowski shape radiating element using multilayer perceptron neural network. In: 7th International Conference on Electrical and Electronics Engineering (ELECO), pp. 219–222, 1–4 December 2011
31. Nesil, S., Güneş, F., Kaya, G.: Analysis and design of X-band Reflectarray antenna using 3-D EM-based artificial neural network model. In: ICUWB, IEEE International Conference on Ultra-Wideband, pp. 532–536, 17–20 September 2012
32. Güneş, F., Nesil, S., Demirel, S.: Design and analysis of Minkowski reflectarray antenna using 3-D CST microwave studio-based neural network model with particle swarm optimization. Int. J. RF Microw. Comput. Aided Eng. **23**, 272–284 (2013)
33. Zhang, Q.J., Gupta, K.C.: Models for RF and Microwave Components. Neural Networks for RF and Microwave Design. Artech House, Norwood, MA (2000)
34. MATLAB and Neural Networks Toolbox Release: The Math Works, Inc., Natick, Massachusetts, United States (2012b)
35. Konstantinidis, A., Yang, K., Chen, H.-H., Zhang, Q.: Energy-aware topology control for wireless sensor networks using Memetic algorithms. Elsevier Comput. Commun. **30**, 2753–2764 (2007)
36. Mahmoud, K.R.: Design optimization of a bow-tie antenna for 2.45GHz RFID readers using a hybrid BSO- NM algorithm. Prog. Electromagn. Res. **17**, 100–105 (2010)
37. Nelder, J.A., Mead, R.: A simplex method for function minimization. Comput. J. **7**, 308–313 (1965)
38. Güneş, F., Demirel, S., Nesil, S.: Novel design approach to X-Band Minkowski reflectarray antennas using the full-wave EM simulation-based complete neural model with a hybrid GA-NM Algorithm. Radioengineering **23**(1), 144–153 (2014)

Stochastic Decision-Making in Waste Management Using a Firefly Algorithm-Driven Simulation-Optimization Approach for Generating Alternatives

Raha Imanirad, Xin-She Yang, and Julian Scott Yeomans

Abstract In solving municipal solid waste (MSW) planning problems, it is generally preferable to formulate several quantifiably good alternatives that provide multiple, disparate perspectives. This is because MSW decision-making typically involves complex problems that are riddled with incompatible performance objectives and possess competing design requirements which are very difficult—if not impossible—to quantify and capture at the time when supporting decision models must be constructed. By generating a set of maximally different solutions, it is hoped that some of the dissimilar alternatives can provide very different perspectives that may serve to satisfy the unmodelled objectives. This maximally different solution creation approach is referred to as modelling-to-generate-alternatives (MGA). Furthermore, many MSW decision-making problems contain considerable elements of stochastic uncertainty. This chapter provides a firefly algorithm-driven simulation-optimization approach for MGA that can efficiently create multiple solution alternatives to problems containing significant stochastic uncertainties that satisfy required system performance criteria and yet are maximally different in their decision spaces. It is shown that this new computationally efficient algorithmic approach can simultaneously produce the desired number of maximally different solution alternatives in a single computational run of the procedure. The efficacy of this stochastic MGA approach for "real world," environmental policy formulation is demonstrated using an MSW case study.

R. Imanirad
Technology and Operations Management, Harvard Business School, Boston, MA 02163, USA
e-mail: rimanirad@hbs.edu

X.-S. Yang
Department of Design Engineering and Mathematics, School of Science and Technology, Middlesex University, Hendon Campus, London NW4 4BT, UK
e-mail: x.yang@mdx.ac.uk

J.S. Yeomans (✉)
OMIS Area, Schulich School of Business, York University, Toronto, ON, Canada M3J 1P3
e-mail: syeomans@schulich.yorku.ca

S. Koziel et al. (eds.), *Simulation-Driven Modeling and Optimization*, Springer Proceedings in Mathematics & Statistics 153,
DOI 10.1007/978-3-319-27517-8_12

Keywords Modelling-to-generate-alternatives • Firefly algorithm • Municipal solid waste

MSC codes: 90B50, 90C59, 90-08, 90C90, 90B99

1 Introduction

The processing of municipal solid waste (MSW) is a multibillion-dollar industry in North America [1, 2]. Since MSW systems generally possess all of the characteristics associated with environmental planning, problems of MSW management have provided an ideal setting for testing a wide variety of modelling techniques used in support of environmental decision-making [3–5]. MSW decision-making frequently involves complex problems that possess design requirements which are very difficult to incorporate into any supporting modelling formulations and tend to be plagued by numerous unquantifiable components [6–13]. Numerous objectives and system requirements always exist that can never be explicitly captured during the problem formulation stage [14, 15]. This commonly occurs in "real world" situations where final decisions must be constructed based not only upon clearly articulated specifications, but also upon environmental, political, and socio-economic objectives that are either fundamentally subjective or not articulated [16–18].

Moreover, in public MSW policy formulation, it may never be possible to explicitly convey many of the subjective considerations because there are numerous competing, adversarial stakeholder groups holding diametrically opposed perspectives. Therefore many of the subjective aspects remain unknown, unquantified, and unmodelled in the construction of any corresponding decision models. MSW policy formulation can prove even more complicated when the various system components also contain considerable stochastic uncertainties [19, 20]. Consequently, MSW policy determination proves to be an extremely challenging and complicated undertaking [10, 21, 22].

Numerous ancillary mathematical modelling approaches have been introduced to support environmental policy formulation (see, for example: [4, 7, 11, 14, 23–25]. However, while mathematically optimal solutions may provide the best answers to these modelled formulations, they generally do not supply the best solutions to the underlying real problems as there are invariably unmodelled aspects not apparent during the model construction phase [6, 10, 11, 21, 26–29]. Furthermore, although deterministic optimization-based techniques are designed to create single best solutions, the presence of the unmodelled issues coupled with the system uncertainties and opposition from powerful stakeholders can actually lead to the outright elimination of any single (even an optimal) solution from further consideration [8, 9, 15, 18–20, 30–33]. Under such conflicting circumstances where no universally optimal exists, it has been stated that "there are no ideal solutions, only trade-offs" [34] and some behavioral aspects taken by decision-makers when faced with such difficulties are described in [26].

In the MSW decision-making domain, there are frequently numerous stakeholder groups holding completely incongruent standpoints, essentially dictating that policy-makers have to establish decision frameworks that must somehow consider numerous irreconcilable points of view simultaneously [8, 9, 14, 20, 33, 35, 36]. Hence, it is generally considered desirable to generate a reasonable number of very different alternatives that provide multiple, contrasting perspectives to the specified problem [1, 13, 33, 37, 38]. These alternatives should preferably all possess near-optimal objective measures with respect to all of the modelled objective(s) that are known to exist, but be as fundamentally different from each as possible other in terms of the system structures characterized by their decision variables. By generating such a diverse set of solutions, it is hoped that at least some of the dissimilar alternatives can be used to address the requirements of the unknown or unmodelled criteria to varying degrees of stakeholder acceptability. Several approaches collectively referred to as modelling-to-generate-alternatives (MGA) have been developed in response to this multi-solution creation requirement [17, 18, 21, 24, 25, 29, 37–42].

The MGA approach was established to implement a much more systematic exploration of a solution space in order to generate a set of alternatives that are good within the modelled objective space while being maximally different from each other in the decision space. The resulting alternatives provide a set of diverse approaches that all perform similarly with respect to the known modelled objectives, yet very differently with respect to any unmodelled issues [13, 43]. Obviously the policy-makers must conduct subsequent comprehensive comparisons of these alternatives to determine which options most closely fulfill their very specific circumstances. Thus, a good MGA process should enable a thorough exploration of the decision space for good solutions while simultaneously allowing for unmodelled objectives to be considered when making final decisions. Consequently, unlike the more customary practice of explicit solution determination inherent in most "hard" optimization methods of Operations Research, MGA approaches are necessarily classified as decision support processes.

As mentioned earlier, the components of most MSW systems contain considerable stochastic uncertainty. Hence, deterministic MGA methods are rendered unsuitable for most MSW policy formulation [2, 11, 14, 19, 28, 30, 31, 35, 44, 45]. Yeomans et al. [46] incorporated stochastic uncertainty directly into MSW planning using an approach referred to as simulation-optimization (SO). SO is a family of optimization techniques that incorporates inherent stochastic uncertainties expressed as probability distributions directly into its computational procedure [47–49]. To address the deficiencies of the deterministic MGA methods, Yeomans [36] demonstrated that SO could be used to generate multiple alternatives which simultaneously integrated stochastic uncertainties directly into each generated option. Since computational aspects can negatively impact SO's optimization capabilities, these difficulties clearly also extend into its use as an MGA procedure [7, 20]. Linton et al. [4] and Yeomans [20] have shown that SO can be considered an effective, though very computationally intensive, MGA technique for MSW policy formulation. However, none of these SO-based approaches could ensure that the created alternatives were sufficiently different in decision variable structure from one another to be considered an effective MGA procedure.

In this chapter, a new stochastic MGA procedure is described that efficiently generates sets of maximally different solution alternatives by implementing a modified version of the nature-inspired Firefly Algorithm (FA) [5, 50, 51] combined with a concurrent, co-evolutionary MGA approach [3, 52–55]. For calculation and optimization purposes, Yang [51] has demonstrated that the FA is more computationally efficient than such commonly used metaheuristic procedures as genetic algorithms, simulated annealing, and enhanced particle swarm optimization (PSO) [56, 57]. The new FA-driven stochastic MGA procedure extends the earlier deterministic approach of [3, 52–55] by extending FA into SO for stochastic optimization and by exploiting the concept of co-evolution within the FA's solution approach to simultaneously generate the desired number of solution alternatives (see [3]). Remarkably, this novel algorithm can concurrently produce the overall best solution together with n locally optimal, maximally different alternatives in a single computational run. Hence, this stochastic FA-driven procedure is extremely computationally efficient for MGA purposes. The efficacy of this approach for environmental decision-making purposes is demonstrated using the MSW case study taken from [46]. More significantly, the practicality of this new stochastic MGA FA-driven approach can easily be adapted to many other stochastic systems and, therefore, can be readily modified to satisfy numerous other planning applications.

2 Modelling-to-Generate-Alternatives

Most mathematical programming methods appearing in the optimization literature have concentrated almost exclusively upon producing single optimal solutions to single-objective problem instances or, equivalently, generating noninferior solution sets to multi-objective formulations [10, 13, 17, 43]. While such algorithms may efficiently generate solutions to the derived complex mathematical models, whether these outputs actually establish "best" approaches to the underlying real problems is certainly questionable [6, 10, 17, 21]. In most "real world" decision environments, there are innumerable system objectives and requirements that are never explicitly apparent or included in the decision formulation stage [6, 13]. Furthermore, it may never be possible to explicitly express all of the subjective components because there are frequently numerous incompatible, competing, design requirements and, perhaps, adversarial stakeholder groups involved [1, 9, 14]. Therefore most subjective aspects of a problem necessarily remain unquantified and unmodelled in the resultant decision models. This is a common occurrence in situations where final decisions are constructed based not only upon clearly stated and modelled objectives, but also upon more fundamentally subjective socio-political-economic goals and stakeholder preferences [1, 37, 38]. Numerous "real world" examples describing these types of incongruent modelling dualities in environmental decision-making appear in [17, 18, 21].

When unquantified issues and unmodelled objectives exist, non-conventional approaches are required that not only search the decision space for noninferior sets of solutions, but must also explore the decision space for discernibly *inferior* alternatives to the modelled problem. In particular, any search for good alternatives to problems known or suspected to contain unmodelled objectives must focus not only on the noninferior solution set, but also necessarily on an explicit exploration of the formulation's entire inferior feasible region.

To illustrate the implications of an unmodelled objective on a decision search, assume that the optimal solution for a quantified, single-objective, maximization decision problem is $\boldsymbol{X}^*$ with corresponding objective value $Z1^*$. Now suppose that there exists a second, unmodelled, maximization objective $Z2$ that subjectively reflects some unquantifiable component such as "political acceptability." Let the solution $\boldsymbol{X}^c$, belonging to the noninferior, 2-objective set, represent a potential best compromise solution if both objectives could somehow have been simultaneously evaluated by the decision-maker. While $\boldsymbol{X}^c$ might be viewed as the best compromise solution to the real problem, it would appear inferior to the solution $\boldsymbol{X}^*$ in the quantified mathematical model, since it must be the case that $Z1^c \leq Z1^*$. Consequently, when unmodelled objectives are factored into the decision-making process, mathematically inferior solutions for the modelled problem can prove optimal to the underlying real problem.

Therefore, when unmodelled objectives and unquantified issues might exist, different solution approaches are needed in order to not only search the decision space for the noninferior set of solutions, but also to simultaneously explore the decision space for inferior alternative solutions to the modelled problem. Population-based solution methods such as the FA permit concurrent searches throughout a decision space and thus prove to be particularly adept procedures for searching through a problem's feasible region.

The primary motivation behind MGA is to produce a manageably small set of alternatives that are quantifiably good with respect to the known modelled objective(s) yet are as different as possible from each other in the decision space. In doing this, the resulting alternative solution set is likely to provide truly different choices that all perform somewhat similarly with respect to the modelled objective(s) yet very differently with respect to any unknown unmodelled issues. By generating a set of good-but-different solutions, the decision-makers can explore desirable qualities within the alternatives that may prove to satisfactorily address the various unmodelled objectives to varying degrees of stakeholder acceptability.

In order to properly motivate an MGA search procedure, it is necessary to apply a more mathematically formal definition to the goals of the MGA process [1, 21, 38]. Suppose the optimal solution to an original mathematical model is $\boldsymbol{X}^*$ with objective value $\boldsymbol{Z}^* = F(\boldsymbol{X}^*)$. The following maximal difference model, subsequently referred to in the chapter as problem [P1], can then be solved to generate an alternative solution that is maximally different from $\boldsymbol{X}^*$:

$$\text{Maximize } \Delta = \sum_i \left| X_i - X_i^* \right|$$
$$\text{Subject to :} \quad X \in D$$
$$\left. F(X) - Z^* \right| \leq T$$

where Δ represents some difference function (for clarity, shown as an absolute difference in this instance), D is the original mathematical model's feasible domain, and T is a targeted tolerance value specified relative to the problem's original optimal objective $\boldsymbol{Z}$*. T is a user-supplied value that determines how much of the inferior region is to be explored in the search for acceptable alternative solutions.

3 Firefly Algorithm for Function Optimization

While this section supplies only a relatively brief synopsis of the FA procedure, more detailed explanations can be accessed in [3, 50–55, 57]. The FA is a nature-inspired, population-based metaheuristic. Each firefly in the population represents one potential solution to a problem and the population of fireflies should initially be distributed uniformly and randomly throughout the solution space. The solution approach employs three idealized rules. (1) The brightness of a firefly is determined by the overall landscape of the objective function. Namely, for a maximization problem, the brightness is simply considered to be proportional to the value of the objective function. (2) The relative attractiveness between any two fireflies is directly proportional to their respective brightness. This implies that for any two flashing fireflies, the less bright firefly will always be inclined to move toward the brighter one. However, attractiveness and brightness both decrease as the relative distance between the fireflies increases. If there is no brighter firefly within its visible neighborhood, then the particular firefly will move about randomly. (3) All fireflies within the population are considered unisex, so that any one firefly could potentially be attracted to any other firefly irrespective of their sex. Based upon these three rules, the basic operational steps of the FA can be summarized within the pseudo code of Figure 1 [51].

In the FA, there are two important issues to resolve: the formulation of attractiveness and the variation of light intensity. For simplicity, it can always be assumed that the attractiveness of a firefly is determined by its brightness which in turn is associated with its encoded objective function value. In the simplest case, the brightness of a firefly at a particular location $\boldsymbol{X}$ would be its calculated objective value $F(\boldsymbol{X})$. However, the attractiveness, β, between fireflies is relative and will vary with the distance r_{ij} between firefly i and firefly j. In addition, light intensity decreases with the distance from its source, and light is also absorbed in the media, so the attractiveness needs to vary with the degree of absorption. Consequently, the overall attractiveness of a firefly can be defined as

```
Objective Function F(X), X = (x_1, x_2,... x_d)
        Generate the initial population of n fireflies, X_i, i = 1, 2,..., n
        Light intensity I_i at X_i is determined by F(X_i)
        Define the light absorption coefficient γ
        while (t < MaxGeneration)
              for i = 1: n , all n fireflies
                      for j = 1: n ,all n fireflies (inner loop)
                              if (I_i < I_j), Move firefly i towards j; end if
                              Vary attractiveness with distance r via e^(-γr)
                      end for j
              end for i
              Rank the fireflies and find the current global best solution G*
        end while
      Postprocess the results
```

Fig. 1 Pseudo code of the firefly algorithm

$$\beta = \beta_0 \, exp\left(-\gamma r^2\right)$$

where β_0 is the attractiveness at distance $r = 0$ and γ is the fixed light absorption coefficient for the specific medium. If the distance r_{ij} between any two fireflies *i* and *j* located at $\boldsymbol{X}_i$ and $\boldsymbol{X}_j$, respectively, is calculated using the Euclidean norm, then the movement of a firefly *i* that is attracted to another more attractive (i.e., brighter) firefly *j* is determined by

$$\boldsymbol{X}_i = \boldsymbol{X}_i + \beta_0 exp\left(-\gamma\left(r_{ij}\right)^2\right)\left(\boldsymbol{X}_i\text{–}\boldsymbol{X}_j\right) + \alpha\boldsymbol{\varepsilon}_i.$$

In this expression of movement, the second term is due to the relative attraction and the third term is a randomization component. Yang [51] indicates that α is a randomization parameter normally selected within the range [0,1] and $\boldsymbol{\varepsilon}_i$ is a vector of random numbers drawn from either a Gaussian or uniform (generally [−0.5,0.5]) distribution. It should be explicitly noted that this expression represents a random walk biased toward brighter fireflies and if $\beta_0 = 0$, it becomes a simple random walk. The parameter γ characterizes the variation of the attractiveness and its value determines the speed of the algorithm's convergence. For most applications, γ is typically set between 0.1 and 10 [51, 57].

In any given optimization problem, for a very large number of fireflies $n \gg k$, where *k* is the number of local optima, the initial locations of the *n* fireflies should be distributed relatively uniformly throughout the entire search space. As the FA proceeds, the fireflies begin to converge into all of the local optima (including the global ones). Hence, by comparing the best solutions among all these optima, the global optima can easily be determined. Yang [51] proves that the FA will approach the global optima when $n \to \infty$ and the number of iterations t is set so that $t \gg 1$. In reality, the FA has been found to converge extremely quickly with n set in the range 20–50 [50, 57].

Two important limiting or asymptotic cases occur when $\gamma \to 0$ and when $\gamma \to \infty$. For $\gamma \to 0$, the attractiveness is constant $\beta = \beta_0$, which is equivalent to having a light intensity that does not decrease. Thus, a firefly would be visible to every other firefly anywhere within the solution domain. Hence, a single (usually global) optima can easily be reached. If the inner loop for j in Figure 1 is removed and $\boldsymbol{X}_j$ is replaced by the current global best $\boldsymbol{G}^*$, then this implies that the FA reverts to a special case of the accelerated PSO algorithm. Subsequently, the computational efficiency of this special FA case is equivalent to that of enhanced PSO. Conversely, when $\gamma \to \infty$, the attractiveness is essentially zero along the sightline of all other fireflies. This is equivalent to the case where the fireflies randomly roam throughout a very thick foggy region with no other fireflies visible and each firefly roams in a completely random fashion. This case corresponds to a completely random search method. As the FA operates between these two asymptotic extremes, it is possible to adjust the parameters α and γ so that the FA can outperform both the random search and the enhanced PSO algorithms [57].

The computational efficiencies of the FA will be exploited in the subsequent MGA solution approach. As noted, between the two asymptotic extremes, the population in the FA can determine both the global optima as well as the local optima concurrently. The concurrency of population-based solution procedures holds huge computational and efficiency advantages for MGA [37, 38]. An additional advantage of the FA for MGA implementation is that the different fireflies essentially work independently of each other, implying that FA procedures are better than genetic algorithms and PSO for MGA because the fireflies will tend to aggregate more closely around each local optimum [51, 57]. Consequently, with a judicious selection of parameter settings, the FA can be made to simultaneously converge extremely quickly into both local and global optima [50, 51, 57].

4 A Simulation-Optimization Approach for Stochastic Optimization

The optimization of large stochastic problems proves to be very complicated when numerous system uncertainties have to be incorporated directly into the solution procedures [47–49]. SO is a broadly defined family of stochastic solution approaches that combines simulation with an underlying optimization component for optimization [47]. In SO, all unknown objective functions, constraints, and parameters are replaced by discrete event simulation models in which the decision variables provide the settings under which the simulation is performed. While SO holds considerable potential for solving a wide range of difficult stochastic problems, it cannot be considered a "procedural panacea" because of its accompanying processing time requirements [47, 48].

The general process of SO can be summarized in the following way. Suppose the mathematical representation of the optimization problem possesses n decision variables, X_i, expressed in vector format as $\boldsymbol{X} = [X_1, X_2, \ldots, X_n]$. If the problem's objective function is designated by F and its feasible region is represented by D, then the related mathematical programming problem is to optimize $F(\boldsymbol{X})$ subject to $\boldsymbol{X} \in D$. When stochastic conditions exist, values for the constraints and objective are determined by simulation. Thus, any direct solution evaluation between two distinct solutions $\boldsymbol{X1}$ and $\boldsymbol{X2}$ requires the comparison of some statistic of F modelled with $\boldsymbol{X1}$ to the same statistic modelled with $\boldsymbol{X2}$ [20, 47]. These statistics are calculated by a simulation performed on the solutions, in which each candidate solution provides the decision variable settings in the simulation. While simulation presents a mechanism for comparing results, it does not provide the means for determining optimal solutions to problems. Hence, simulation, by itself, cannot be used as a stochastic optimization procedure.

Since all measures of system performance in SO are stochastic, every potential solution, $\boldsymbol{X}$, must be determined through simulation. Because simulation is computationally intensive, an optimization algorithm is employed to guide the search for solutions through the problem's feasible domain in as few simulation runs as possible [20, 49]. As stochastic system problems frequently contain numerous potential solutions, the quality of the final solution could be highly variable unless an extensive search has been performed throughout the problem's entire feasible region. Population-based metaheuristic such as the FA are conducive to these extensive searches because the complete set of candidate solutions maintained in their populations permit searches to be undertaken throughout multiple sections of the feasible region, concurrently.

An FA-directed SO approach contains two alternating computational phases; (1) an "evolutionary phase" directed by the FA module and (2) a simulation module [5]. As described earlier, the FA maintains a population of candidate solutions throughout its execution. The evolutionary phase evaluates the entire current population of solutions during each generation of the search and evolves from the current population to a subsequent one. Because of the system's stochastic components, all performance measures are necessarily statistics calculated from the responses generated in the simulation module. The quality of each solution in the population is found by having its performance criterion, F, evaluated in the simulation module. After simulating each candidate solution, their respective objective values are returned to the evolutionary FA module to be utilized in the creation of the ensuing population of candidate solutions.

A primary characteristic of FA procedures is that better solutions in a current population possess a greater likelihood for survival and progression into the subsequent population. Thus, the FA module advances the system toward improved solutions in subsequent generations and ensures that the solution search does not become trapped in some local optima. After generating a new candidate population in the FA module, the new solution set is returned to the simulation module for comparative evaluation. This alternating, two-phase search process terminates when an appropriately stable system state (i.e., an optimal solution) has been attained. The optimal solution produced by the procedure is the single best solution found over the course of the entire search [5].

5 FA-Driven SO Algorithm for Stochastic MGA

Linton et al. [4] and Yeomans [20] have shown that SO can be used as a computationally intensive, stochastic MGA technique. Yeomans [58] examined several approaches to accelerate the search times and solution quality of SO. This section introduces an FA-driven MGA method that incorporates stochastic uncertainty using SO [5] to efficiently generate sets of maximally different solution alternatives.

The FA-driven stochastic MGA approach is designed to generate a pre-determined small number of close-to-optimal, but maximally different alternatives, by essentially adjusting the value of T in [P1] and using the FA to solve each corresponding, maximal difference problem instance. This algorithm provides a stochastic extension to the deterministic approach of [3, 54, 55]. By exploiting the co-evolutionary solution structure within the population of the FA, stratified subpopulations within the algorithm's overall population are established as the fireflies collectively evolve toward different local optima within the solution space. In this process, each desired solution alternative undergoes the common search procedure driven by the FA. However, the survival of solutions depends not only upon how well the solutions perform with respect to the modelled objective(s), but also by how far away they are from all of the other alternatives generated in the decision space.

A direct process for generating these alternatives with the FA would be to iteratively solve the maximum difference model [P1] by incrementally updating the target T whenever a new alternative needs to be produced and then re-running the algorithm. Such an iterative approach would parallel the seminal Hop, Skip, and Jump (HSJ) MGA algorithm of [17] in which, once an initial problem formulation has been optimized, supplementary alternatives are created one-by-one through a systematic, incremental adjustment of the target constraint to force the sequential generation of the suboptimal solutions. While this direct approach is straightforward, it is relatively computationally expensive as it requires a repeated execution of the specific optimization algorithm employed [1, 37, 38, 52, 53].

In contrast, the concurrent FA-driven MGA approach is designed to generate the pre-determined number of maximally different alternatives within the entire population in a single run of the FA procedure (i.e., the same number of runs as if FA were used solely for function optimization purposes) and its efficiency is based upon the concept of co-evolution [52–55]. In this FA-driven co-evolutionary approach, pre-specified stratified subpopulation ranges within the FA's overall population are established that collectively evolve the search toward the creation of the stipulated number of maximally different alternatives. Each desired solution alternative is represented by each respective subpopulation and each subpopulation undergoes the common processing operations of the FA.

The FA-driven approach can be structured upon any standard FA solution procedure containing the appropriate encodings and operators that best correspond to the problem. The survival of solutions in each subpopulation depends simultaneously upon how well the solutions perform with respect to the modelled objective(s) and by how far away they are from all of the other alternatives. Consequently, the evolution of solutions in each subpopulation toward local optima is directly influenced by those solutions currently existing in all of the other subpopulations, which necessarily forces the concurrent co-evolution of each subpopulation toward good but maximally distant regions of the decision space. This co-evolutionary concept enables the simultaneous search for, and production of, the set of quantifiably good solutions that are maximally different from each other according to [P1] [38].

By employing this co-evolutionary concept, it becomes possible to implement an FA-driven MGA procedure that concurrently produces alternatives which possess objective function bounds that are analogous, but inherently superior, to those created by a sequential HSJ-styled solution generation approach. While each alternative produced by an HSJ procedure is maximally different only from the single, overall optimal solution together with a bound on the objective value which is at least x% different from the best objective (i.e., x = 1 %, 2 %, etc.), the concurrent co-evolutionary FA procedure is able to generate alternatives that are no more than x% different from the overall optimal solution but with each one of these solutions being as maximally different as possible from every other generated alternative that is produced. Co-evolution is also much more efficient than a sequential HSJ-styled approach in that it exploits the inherent population-based searches of FA procedures to concurrently generate the entire set of maximally different solutions using only a single population. Specifically, while an HSJ-styled approach would need to run n different times in order to generate n different alternatives, the concurrent algorithm need run only once to produce its entire set of maximally different alternatives irrespective of the value of n. Hence, it is a much more computationally efficient solution generation process.

The steps involved in the stochastic FA-driven co-evolutionary MGA algorithm are as follows:

1. Create the initial population stratified into P equally sized subpopulations. P represents the desired number of maximally different alternative solutions within a prescribed target deviation from the optimal to be generated and must be set a priori by the decision-maker. S_p represents the pth subpopulation set of solutions, $p = 1,\ldots,P$ and there are K solutions contained within each S_p. Note that the target for each S_p could be a common deviation value (e.g., all P alternatives need to be within 10 % of optimal) or the targets for each S_p could represent different selected increments (e.g., one alternative would need to be within 1 % of optimal, another alternative would need to be within 2 %, etc.).
2. Evaluate each solutions in S_1 using the simulation module and identify the best solution with respect to the modelled objective. S_1 is the subpopulation dedicated to the search for the overall optimal solution to the modelled problem. The

best solution residing in S_1 is employed in establishing the benchmarks for the relaxation constraints used to create the maximally different solutions as in P1.

3. Evaluate all solutions in S_p, $p = 2, \ldots, P$, with respect to the modelled objective using the simulation module. Solutions meeting the target constraint and all other problem constraints are designated as *feasible*, while all other solutions are designated as *infeasible*.
4. Apply an appropriate elitism operator to each S_p to preserve the best individual in each subpopulation. In S_1, this is the best solution evaluated with respect to the modelled objective. In S_p, $p = 2, \ldots, P$, the best solution is the feasible solution most distant in decision space from all of the other subpopulations (the distance measure is defined in Step 7). Note: Because the best solution to date is always placed into each subpopulation, at least one solution in S_p will always be feasible. This step simultaneously selects a set of alternatives that, respectively, satisfy different values of the target T while being as far apart as possible (i.e., maximally different in the sense of [P1]) from the solutions generated in each of the other subpopulations. By the co-evolutionary nature of this algorithm, the alternatives are simultaneously generated in one pass of the procedure rather than the P implementations suggested by the necessary HSJ-styled increments to T in problem [P1].
5. Stop the algorithm if the termination criteria (such as maximum number of iterations or some measure of solution convergence) are met. Otherwise, proceed to Step 6.
6. Identify the decision space centroid, C_{ip}, for each of the $K' \leq K$ feasible solutions within $k = 1, \ldots, K$ of S_p, for each of the N decision variables X_{ikp}, $i = 1, \ldots, N$. Each centroid represents the N-dimensional center of mass for the solutions in each of the respective subpopulations, p. As an illustrative example for determining a centroid, calculate $C_{ip} = (1/K') * \sum_k X_{ikp}$. In this calculation, each dimension of each centroid is computed as the straightforward average value of that decision variable over all of the values for that variable within the feasible solutions of the respective subpopulation. Alternatively, a centroid could be calculated as some fitness-weighted average or by some other appropriately defined measure.
7. For each solution $k = 1, \ldots, K$, in each S_q, calculate D_{kq}, a distance measure between that solution and all other subpopulations. As an illustrative example for determining a distance measure, calculate $D_{kq} = \text{Min}\ \{\sum_i \mid X_{ikp} - C_{ip} \mid;\ p = 1, \ldots, P,\ p \neq q\}$. This distance represents the minimum distance between solution k in subpopulation q and the centroids of all other subpopulations. Alternatively, the distance measure could be calculated by some other appropriately defined function.
8. Rank the solutions within each S_p according to the distance measure D_{kq} objective—appropriately adjusted to incorporate any constraint violation penalties. The goal of maximal difference is to force solutions from one subpopulation to be as far apart as possible in the decision space from the solutions of each of the other subpopulations. This step orders the specific solutions in each

subpopulation by those solutions which are most distant from the solutions in all of the other subpopulations.
9. In each S_p, apply the appropriate FA "change operations" to the solutions and return to Step 2.

6 Case Study of Stochastic MGA for Municipal Solid Waste Management Planning

As described in the previous sections, MSW decision-makers generally prefer to be able to select from a set of "near-optimal" alternatives that significantly differ from each other in terms of the system structures characterized by their decision variables. The efficacy of this new FA-driven SO MGA procedure will be illustrated using the MSW management planning study of Hamilton-Wentworth taken from [46]. While this section briefly outlines the case, more extensive details and descriptions can be found in both [1] and [46].

Located at the Western-most edge of Lake Ontario, the Municipality of Hamilton-Wentworth covers an area of 1100 km^2 and includes six towns and cities; Hamilton, Dundas, Ancaster, Flamborough, Stoney Creek, and Glanbrook. The Municipality is considered the industrial center of Canada, although it simultaneously incorporates diverse areas of not only heavy industrial production, but also densely populated urban space, regions of significant suburban development, and large proportions of rural/agricultural environments. Prior to the study of [46], the municipality had not been able to effectively incorporate inherent uncertainties into their planning processes and, therefore, had not performed effective systematic planning for the flow of wastes within the region. The MSW management system within the region is a very complicated process which is impacted by economic, technical, environmental, legislational, and political factors.

The MSW system within Hamilton-Wentworth needed to satisfy the waste disposal requirements of its half-million residents who, collectively, produced more than 300,000 tons of waste per year, with a budget of $22 million. The region had constructed a system to manage these wastes composed of: a waste-to-energy incinerator referred to as the Solid Waste Reduction Unit (or SWARU); a 550 acre landfill site at Glanbrook; three waste transfer stations located in Dundas (DTS), in East Hamilton at Kenora (KTS), and on Hamilton Mountain (MTS); a household recycling program contracted to and operated by the Third Sector Employment Enterprises; a household/hazardous waste depot, and a backyard composting program.

The three transfer stations have been strategically located to receive wastes from the disparate municipal (and individual) sources and to subsequently transfer them to the waste management facilities for final disposal; either to SWARU for incineration or to Glanbrook for landfilling. Wastes received at the transfer stations are compacted into large trucks prior to being hauled to the landfill site. These transfer stations provide many advantages in waste transportation and management; these include reducing traffic going to and from the landfill, providing an effective control mechanism for dumping at the landfill, offering an inspection area where wastes can be viewed and unacceptable materials removed, and contributing to a reduction of waste volume because of the compaction process. The SWARU incinerator burns up to 450 tons of waste per day and, by doing so, generates about 14 million kilowatt hours per year of electricity which can be either used within the plant itself or sold to the provincial electrical utility. SWARU also produces a residual waste ash which must subsequently be transported to the landfill for disposal.

Within this MSW system, decisions have to be made regarding whether waste materials would be recycled, landfilled, or incinerated and additional determinations have to be made as to which specific facilities would process the discarded materials. Included within these decisions is a determination of which one of the multiple possible pathways that the waste would flow through in reaching the facilities. Conversely, specific pathways selected for waste material flows determine which facilities process the waste. It was possible to subdivide the various waste streams with each resulting substream sent to a different facility. Since cost differences from operating the facilities at different capacity levels produced economies of scale, decisions have to be made to determine how much waste should be sent along each flow pathway to each facility. Therefore, any single MSW policy option is composed of a combination of many decisions regarding which facilities received waste material and what quantities of waste are sent to each facility. All of these decisions are compounded by overriding system uncertainties.

The complete mathematical model used for MSW planning appears in the subsequent section. This mathematical formulation was used not only to examine the existing municipal MSW system, but also to prepare the municipality for several potentially enforced future changes to its operating conditions. Yeomans et al. [46] examined three likely future scenarios, with each scenario involving potential incinerator operations. Scenario 1 considered the existing MSW management system and corresponded to a *status quo* case. Scenario 2 examined what would occur should the incinerator operate at its upper design capacity; corresponding to a situation in which the municipality would landfill as little waste as possible. Scenario 3 permitted the incinerator to operate anywhere in its design capacity range; from being closed completely to operating up to its maximum capacity.

6.1 Mathematical Model for MSW Planning in Hamilton-Wentworth

This section provides the complete mathematical model for MSW planning in Hamilton-Wentworth. Extensive details and descriptions of it can be found in [46]. In the model, any uncertain parameter A is represented by $\overset{\leftrightarrow}{A}$. In the model, the various districts from which waste originates will be identified using subscript i;

where $i = 1, 2, \ldots, 17$ denotes the originating district. The transfer stations will be denoted by subscript j, in which $j = 1, 2, 3$ represents the number assigned to each transfer station, where DTS = 1, KTS = 2, and MTS = 3. Subscript k, $k = 1, 2, 3$, identifies the specific waste management facility, with landfill = 1, SWARU = 2, and Third Sector = 3. The decision variables for the problem will be designated by x_{ij}, y_{jk}, and z_{ik} where x_{ij} represents the proportion of solid waste sent from district i to transfer station j; y_{jk} corresponds to the proportion of waste sent from transfer station j to waste management facility k, and z_{ik} corresponds to the proportion of waste sent from district i to waste management facility k. For notational brevity, and also to reflect the fact that no district is permitted to deliver their waste directly to the landfill, define $z_{i1} = 0$, for $i = 1, 2, \ldots, 17$.

The cost for transporting one ton of waste from district i to transfer station j is denoted by $\overset{\leftrightarrow}{t}\, x_{ij}$, that from transfer station j to waste management facility k is represented by $\overset{\leftrightarrow}{t}\, y_{jk}$, and that from district i to waste management facility k is $\overset{\leftrightarrow}{t}\, z_{ik}$. The per ton cost for processing waste at transfer station j is $\overset{\leftrightarrow}{\delta}_j$ and that at waste management facility k is $\overset{\leftrightarrow}{\rho}_k$. Two of the waste management facilities can produce revenues from processing wastes. The revenue generated per ton of waste is $\overset{\leftrightarrow}{r}_2$ at SWARU and $\overset{\leftrightarrow}{r}_3$ at the Third Sector recycling facility. The minimum and maximum processing capacities at transfer station j are $\overset{\leftrightarrow}{K}_j$ and $\overset{\leftrightarrow}{M}_j$, respectively. Similarly, the minimum and maximum capacities at waste management facility k are $\overset{\leftrightarrow}{L}_k$ and $\overset{\leftrightarrow}{N}_k$, respectively. The quantity of waste, in tons, generated by district i is $\overset{\leftrightarrow}{W}_i$, and the proportion of this waste that is recyclable is $\overset{\leftrightarrow}{a}_i$. The proportion of recyclable waste flowing into transfer station j is $\overset{\leftrightarrow}{R}\, W_j$. The proportion of residue (residual wastes such as the incinerated ash at SWARU) generated by waste management facility j is $\overset{\leftrightarrow}{Q}_j$, where $\overset{\leftrightarrow}{Q}_1 = 0$ by definition. This waste residue must be shipped to the landfill for final disposal.

Formulating any single MSW policy corresponds to finding a decision variable solution satisfying constraints (2) through (31), with cost determined by objective (1).

$$\text{Minimize Cost} = \sum_{p=1}^{5} T_p + \sum_{q=1}^{6} P_q - \sum_{r=2}^{3} R_r \tag{1}$$

Subject to:

$$T_1 = \sum_{i=1}^{17} \sum_{j=1}^{3} \overset{\leftrightarrow}{t}\, x_{ij}\, x_{ij}\, \overset{\leftrightarrow}{W}_i \tag{2}$$

$$T_2 = \sum_{i=1}^{17} \sum_{k=1}^{3} \overset{\leftrightarrow}{t}\, z_{ik}\, z_{ik}\, \overset{\leftrightarrow}{W}_i \tag{3}$$

$$T_3 = \sum_{i=1}^{17} \sum_{j=1}^{3} \sum_{k=1}^{3} \overset{\leftrightarrow}{t}\, y_{jk}\, y_{jk}\, x_{ij}\, \overset{\leftrightarrow}{W}_i \tag{4}$$

$$T_4 = \left(\overset{\leftrightarrow}{t}\, sl\right) \overset{\leftrightarrow}{Q}_2 \sum_{i=1}^{17} \overset{\leftrightarrow}{W}_i \left[z_{i2} + \sum_{j=1}^{3} y_{j2} x_{ij} \right] \tag{5}$$

$$T_5 = \left(\overset{\leftrightarrow}{t}\, tl\right) \overset{\leftrightarrow}{Q}_3 \sum_{i=1}^{17} \overset{\leftrightarrow}{W}_i \left[z_{i3} + \sum_{j=1}^{3} y_{j3} x_{ij} \right] \tag{6}$$

$$P_1 = \overset{\leftrightarrow}{\rho}_1 \sum_{i=1}^{17} \overset{\leftrightarrow}{W}_i \sum_{k=1}^{3} \left[\overset{\leftrightarrow}{Q}_k z_{ik} + \sum_{j=1}^{3} x_{ij}\, y_{jk} \right] \tag{7}$$

$$P_2 = \overset{\leftrightarrow}{\rho}_2 \sum_{i=1}^{17} \overset{\leftrightarrow}{W}_i \left[z_{i2} + \sum_{j=1}^{3} x_{ij}\, y_{j2} \right] \tag{8}$$

$$P_3 = \overset{\leftrightarrow}{\rho}_3 \sum_{i=1}^{17} \overset{\leftrightarrow}{W}_i \left[z_{i3} + \sum_{j=1}^{3} x_{ij}\, y_{j3} \right] \tag{9}$$

$$P_4 = \overset{\leftrightarrow}{\delta}_1 \sum_{i=1}^{17} x_{i1}\, \overset{\leftrightarrow}{W}_i \tag{10}$$

$$P_5 = \overset{\leftrightarrow}{\delta}_2 \sum_{i=1}^{17} \overset{\leftrightarrow}{W}_i \left[x_{i2} + \overset{\leftrightarrow}{Q}_3 \left\{ z_{i3} + \sum_{j=1}^{3} x_{ij} y_{j3} \right\} \right] \tag{11}$$

$$P_6 = \overset{\leftrightarrow}{\delta}_3 \sum_{i=1}^{17} x_{i3}\, \overset{\leftrightarrow}{W}_i \tag{12}$$

$$R_2 = \overset{\leftrightarrow}{r}_2 \sum_{i=1}^{17} \overset{\leftrightarrow}{W}_i \left[z_{i2} + \sum_{j=1}^{3} x_{ij}\, y_{j2} \right] \tag{13}$$

$$R_3 = \overset{\leftrightarrow}{r}_3 \sum_{i=1}^{17} \overset{\leftrightarrow}{W}_i \left[z_{i3} + \sum_{j=1}^{3} x_{ij}\, y_{j3} \right] \tag{14}$$

$$\sum_{i=1}^{17} x_{i1}\, \overset{\leftrightarrow}{W}_i \le \overset{\leftrightarrow}{M}_1 \tag{15}$$

$$\sum_{i=1}^{17} \overset{\leftrightarrow}{W}_i \left[x_{i2} + \overset{\leftrightarrow}{Q}_3 \left\{ z_{i3} + \sum_{j=1}^{3} x_{ij}\, y_{j3} \right\} \right] \leq \overset{\leftrightarrow}{M}_2 \tag{16}$$

$$\sum_{i=1}^{17} x_{i3} \overset{\leftrightarrow}{W}_i \leq \overset{\leftrightarrow}{M}_3 \tag{17}$$

$$\sum_{i=1}^{17} x_{i1} \overset{\leftrightarrow}{W}_i \geq \overset{\leftrightarrow}{K}_1 \tag{18}$$

$$\sum_{i=1}^{17} \overset{\leftrightarrow}{W}_i \left[x_{i2} + \overset{\leftrightarrow}{Q}_3 \left\{ z_{i3} + \sum_{j=1}^{3} x_{ij}\, y_{j3} \right\} \right] \geq \overset{\leftrightarrow}{K}_2 \tag{19}$$

$$\sum_{i=1}^{17} x_{i3} \overset{\leftrightarrow}{W}_i \geq \overset{\leftrightarrow}{K}_3 \tag{20}$$

$$\sum_{i=1}^{17} \overset{\leftrightarrow}{W}_i \sum_{k=1}^{3} \left[\overset{\leftrightarrow}{Q}_k\, z_{ik} + \sum_{j=1}^{3} x_{ij}\, y_{jk} \right] \leq \overset{\leftrightarrow}{N}_1 \tag{21}$$

$$\sum_{i=1}^{17} \overset{\leftrightarrow}{W}_i \left[z_{i2} + \sum_{j=1}^{3} x_{ij}\, y_{j2} \right] \leq \overset{\leftrightarrow}{N}_2 \tag{22}$$

$$\sum_{i=1}^{17} \overset{\leftrightarrow}{W}_i \left[z_{i3} + \sum_{j=1}^{3} x_{ij}\, y_{j3} \right] \leq \overset{\leftrightarrow}{N}_3 \tag{23}$$

$$\sum_{i=1}^{17} \overset{\leftrightarrow}{W}_i \left[z_{i2} + \sum_{j=1}^{3} x_{ij}\, y_{j2} \right] \geq \overset{\leftrightarrow}{L}_2 \tag{24}$$

$$\sum_{i=1}^{17} \overset{\leftrightarrow}{W}_i \left[z_{i3} + \sum_{j=1}^{3} x_{ij}\, y_{j3} \right] \geq \overset{\leftrightarrow}{L}_3 \tag{25}$$

$$\sum_{j=1}^{3} x_{ij} + \sum_{k=1}^{3} z_{ik} = 1 \qquad i = 1, 2, \ldots, 17 \tag{26}$$

$$\sum_{j=1}^{3} x_{ij} \overset{\leftrightarrow}{R} W_j + z_{i3} \leq \overset{\leftrightarrow}{a}_i \quad i = 1, 2, \ldots, 17 \tag{27}$$

$$\sum_{k=1}^{3} y_{jk} = 1 \; j = 1, 2, 3 \tag{28}$$

$$\sum_{i=1}^{17} \overset{\leftrightarrow}{W}_i \left[x_{i2} + \overset{\leftrightarrow}{Q}_3 \left\{ z_{i3} + \sum_{j=1}^{3} x_{ij} y_{j3} \right\} \right] = \sum_{i=1}^{17} \sum_{k=1}^{3} x_{i2} \overset{\leftrightarrow}{W}_i \, y_{2k} \tag{29}$$

$$\sum_{i=1}^{17} x_{ij} \overset{\leftrightarrow}{W}_i \, y_{j3} = \overset{\leftrightarrow}{R} \;\; W_j \sum_{i=1}^{17} \;\; x_{ij} \;\overset{\leftrightarrow}{W}_i \quad j = 1, 2, 3 \tag{30}$$

$$x_{ij} \geq 0, \; y_{jk} \geq 0, \; z_{ik} \geq 0 \qquad i = 1, 2, \ldots, 17, \; j = 1, 2, 3, \; k = 1, 2, 3 \tag{31}$$

In the objective function (1), the total transportation costs for wastes generated are provided by equations (2)–(6). Equation (2) calculates the transportation costs for waste flows from the districts (i.e., the cities and towns) to the transfer stations, while equation (3) provides the costs for transporting the waste from the districts directly to the waste management facilities. The total cost for transporting waste from the transfer facilities to the waste management facilities is determined in equation (4). The transportation costs for residue disposal created at SWARU and the Third Sector are given by equations (5) and (6), respectively. The total processing costs for the transfer stations and waste management facilities are expressed in (7) through (12). Here, P_k represents the processing costs at waste management facility k, $k = 1,2,3$, and $P_{(j+3)}$ represents the processing costs at transfer station j, $j = 1,2,3$. The processing cost, P_1, in (7) indicates that the landfill receives wastes from both SWARU and the Third Sector in addition to the waste sent from the transfer stations. The relationship specifying the processing costs at KTS, P_5 in (11), is more complicated than for DTS and MTS, since KTS must also process the Third Sector's unrecyclable residue (this waste processing pattern can also be observed in equations (16) and (19)) and this residue may have been sent there directly from the districts or from the other transfer stations. The revenue generated by SWARU, R_2, and by the Third Sector, R_3, are determined by expressions (13) and (14). All of these cost and revenue elements are amalgamated into objective function (1).

Upper and lower capacity limits placed upon the transfer stations DTS, KTS, and MTS, are provided by constraints (15) through (20), while capacity limits established for the landfill, SWARU, and the Third Sector are given by (21) to (25). The waste processing relationship for the landfill is more complicated than for the other waste management facilities, since the landfill receives residue from both SWARU and the Third Sector. Furthermore, while there is no lower operating requirement placed upon the use of the landfill, both SWARU and the Third Sector require minimum levels of activity in order for their ongoing operations to remain economically viable. Mass balance constraints must also be included to ensure that all generated waste is disposed and that the amount of waste flowing into a transfer facility matches the amount flowing out of it. Constraint (26) ensures the disposal of all waste produced by each district. Recyclable waste disposal is established by constraint (27). In (27), it is recognized that not all recyclable waste produced at a district is initially sent to the Third Sector recycling facility (i.e., some recyclable waste may initially be discarded as "regular" garbage) and that some, but not all, recyclable waste received at a transfer station is subsequently sent for recycling. The expression in (28) ensures that all waste received by each transfer station must be sent to a waste management facility. Equation (29) provides the mass balance constraint for the wastes entering and leaving KTS (which handles more complicated waste patterns than the other two transfer stations). Constraint (30) describes the mass balance requirement for recyclable wastes received by the transfer stations that are then forwarded to the Third Sector. Finally, (31) establishes non-negativity requirements for the decision variables. Hence, any specific MSW policy formulated for Hamilton-Wentworth would require the determination of a decision variable solution that satisfies constraints (2) to (31) and would be evaluated by its resulting cost found using objective (1).

Yeomans et al. [46] ran SO for a 24-hour period to determine best solutions for each scenario. For the existing system (Scenario 1), a solution that would never cost more than $20.6 million was constructed. For Scenarios 2 and 3, Yeomans et al. [46] produced optimal solutions costing $22.1 million and $18.7 million, respectively. In all of these scenarios, SO was used exclusively as a function optimizer with the goal being to produce only single best solutions.

6.2 Using the Co-Evolutionary MGA Method for MSW Planning in Hamilton-Wentworth

As outlined earlier, when public policy planners are faced with difficult and controversial choices, they generally prefer to be able to select from a set of near-optimal alternatives that differ significantly from each other in terms of the system structures characterized by their decision variables. In order to create these alternative planning options for the three MSW system scenarios, it would be possible to place extra target constraints into the original model which would force the generation of solutions that were different from their respective, initial optimal solutions. Suppose, for example, that ten additional planning alternative options were created through the inclusion of a technical constraint on the objective function that increased the total system cost of the original model from 1 % up to 10 % in increments of 1 %. By adding these incremental target constraints to the original SO model and sequentially resolving the problem 10 times for each scenario (i.e., 30 additional runs of the SO procedure), it would be possible to create a specific number of alternative policies for MSW planning.

However, to improve upon the process of running 30 separate instances of the computationally intensive SO algorithm to generate these solutions, the FA-driven MGA procedure described in the previous section was run only once for each scenario, thereby producing the 30 additional alternatives shown in Table 1.

Table 1 Annual MSW costs ($ millions) for 11 maximally different alternatives for Scenario 1, Scenario 2, and Scenario 3

Annual MSW system costs	Scenario 1	Scenario 2	Scenario 3
Best solution overall	20.611	22.102	18.712
Best solution within 1 %	20.744	22.221	18.847
Best solution within 2 %	20.899	22.439	18.896
Best solution within 3 %	21.093	22.620	19.119
Best solution within 4 %	21.388	22.685	19.368
Best solution within 5 %	21.432	23.076	19.540
Best solution within 6 %	21.762	23.348	19.745
Best solution within 7 %	21.997	23.635	19.884
Best solution within 8 %	22.189	23.826	20.019
Best solution within 9 %	22.303	23.943	20.122
Best solution within 10 %	22.464	24.079	20.325

Each column of the table shows the overall system costs for the ten maximally different options generated for each of the three scenarios. Given the performance bounds established for the objective in each problem instance, the decision-makers can feel reassured by the stated performance for each of these options while also being aware that the perspectives provided by the set of dissimilar decision variable structures are as different from each other as is feasibly possible. Hence, if there are stakeholders with incompatible standpoints holding diametrically opposing viewpoints, the policy-makers can perform an assessment of these different options without being myopically constrained by a single overriding perspective based solely upon the objective value.

Furthermore, it should also be explicitly noted that the alternatives created do not differ from the lowest cost solution by *at least* the stated 1 %, 2 %, 3 %, ..., 10 %, respectively, but, in general, actually differ by less than these pre-specified upper deviation limits. This is because each of the best alternatives produced in $S_2, S_3, \ldots, S_{11}$ has solutions whose structural variables differ maximally from those of all of the other alternatives generated while simultaneously guaranteeing that their objective values deviate from the overall best objective by *no more* than 1 %, 2 %, ..., 10 %, respectively. Thus, the goal of the alternatives generated in this MGA procedure is very different from those produced in the more straightforward HSJ-style approach, while simultaneously establishing much more robust guarantees of solution quality.

Although a mathematically optimal solution may not provide the best approach to the real problem, it can be demonstrated that the co-evolutionary procedure does indeed produce very good solution values for the originally modelled problem, itself. Table 2 clearly highlights how the alternatives generated in S_1 by the new MGA procedure are all "good" with respect to their best overall cost measurements relative to the optimal solutions found in [46]. It should be explicitly noted that the cost of the overall best solution produced by the MGA procedure (i.e., the solution in S_1) is actually identical to the one found in the function optimization of [46] for

Table 2 Best annual MSW performance costs (in millions of $) found for (a) existing system structure (Scenario 1), (b) incinerator at maximum operating (Scenario 2), and (c) incinerator at any operating level (Scenario 3)

	Scenario 1	Scenario 2	Scenario 3
Yeomans et al. [46] using SO	20.6	22.1	18.7
Best solution found using co-evolutionary algorithm	20.6	22.1	18.7

each scenario—which is clearly not a coincidence. Expanding the population size in the SO procedure to include the subpopulations S_2, S_3, ... ,S_{11} does not detract from its evolutionary capabilities to find the best, function optimization solution in subpopulation S_1. Hence, in addition to its alternative generating capabilities, the MGA procedure simultaneously performs exceedingly well with respect to its role in function optimization.

This example has demonstrated how the FA-driven SO MGA modelling can be used to efficiently generate multiple, good policy alternatives that satisfy required system performance criteria according to pre-specified bounds within highly uncertain environments and yet remain maximally different in the decision space. As described earlier, public sector, environmental policy problems are typically riddled with incongruent performance requirements that contain significant stochastic uncertainty that is also very difficult to quantify. Consequently, it is preferable to create several quantifiably good alternatives that concurrently provide very different perspectives to the potentially unmodelled performance design issues during the policy formulation stage. The unique performance features captured within these dissimilar alternatives can result in very different system performance with respect to the unmodelled issues, thereby incorporating the unmodelled issues into the actual solution process.

In summary, the computational example underscores several important findings with respect to the concurrent FA-driven MGA method: (1) The FA can be employed as the underlying optimization search routine for SO purposes; (2) Because of the evolving nature of its population-based solution searches, the co-evolutionary capabilities within the FA can be exploited to simultaneously generate more good alternatives than planners would be able to create using other MGA approaches; (3) By the design of the MGA algorithm, the alternatives generated are good for planning purposes since all of their structures are guaranteed to be as mutually and maximally different from each other as possible (i.e., these differences are not just simply different from the overall optimal solution as in the HSJ-style approach to MGA); (4) The approach is very computationally efficient since it need only be run once to generate its entire set of multiple, good solution alternatives (i.e., to generate n maximally different solution alternatives, the MGA algorithm would need to be run exactly the same number of times that the FA would need to be run for function optimization purposes alone—namely once—irrespective of the value of n); and, (5) The best overall solutions produced by the MGA procedure will be identical to the best overall solutions that would be produced by the FA for function optimization purposes alone.

7 Conclusions

MSW decision-making problems provide multidimensional performance specifications which are invariably complicated by unquantifiable performance objectives and incongruent modelling features. These problems often possess incompatible design specifications which are difficult—if not impossible—to capture when the supporting decision models are formulated. Consequently, there are invariably unmodelled problem components, not apparent during model construction, that can greatly impact the acceptability of the model's solutions. These ambiguous and competing components force MSW decision-makers to incorporate many conflicting requirements into their decision process prior to the final solution determination.

Because of this, ancillary modelling techniques used to support the decision formulation process must somehow consider all of these aspects while remaining flexible enough to simultaneously capture the impacts from the inherent stochastic and planning uncertainties. In such situations, instead of determining a single, mathematically optimal solution to the problem formulation, it is more desirable to produce a set of quantifiably good alternatives that provide distinct perspectives to the potentially unmodelled issues. The unique performance features captured within these dissimilar alternatives result in the consideration of very different system performance features, thereby addressing some of the unmodelled issues during the policy formulation stage.

In this chapter, a stochastic FA-driven MGA approach was introduced that demonstrated how the co-evolutionary solution aspects of the computationally efficient, population-based FA could be used to guide a stochastic SO algorithm's search process in order to concurrently generate multiple, maximally different, near-best alternatives. In this stochastic MGA capacity, the co-evolutionary approach produces numerous solutions possessing the required problem characteristics, with each generated alternative guaranteeing a very different perspective. Since FA techniques can be adapted to solve a wide variety of problem types, the practicality of this new FA-driven stochastic MGA approach could clearly be extended into numerous disparate "real world" environmental applications and can be readily modified to many other planning situations. These extensions will be considered in future research studies.

References

1. Gunalay, Y., Yeomans, J.S., Huang, G.H.: Modelling to generate alternative policies in highly uncertain environments: an application to municipal solid waste management planning. J Environ. Inform. **19**(2), 58–69 (2012)
2. Tchobanoglous, G., Thiesen, H., Vigil, S.: Integrated solid waste management: engineering principles and management issues. McGraw-Hill, New York (1993)

3. Imanirad, R., Yang, X., Yeomans, J.S.: Environmental Decision-Making Under Uncertainty Using a Biologically-Inspired Simulation-Optimization Algorithm for Generating Alternative Perspectives. Int. J. Bus. Innov. Res., In Press (2014)
4. Linton, J.D., Yeomans, J.S., Yoogalingam, R.: Policy planning using genetic algorithms combined with simulation: the case of municipal solid waste. Environ. Plann. B: Plann Des. **29**(5), 757–778 (2002)
5. Yeomans, J.S., Yang, X.S.: Municipal waste management optimization using a firefly algorithm-driven simulation-optimization approach. Int. J. Process Manage. Benchmarking **4**(4), 363–375 (2014)
6. Brugnach, M., Tagg, A., Keil, F., De Lange, W.J.: Uncertainty matters: computer models at the science-policy interface. Water Resour. Manag. **21**, 1075–1090 (2007)
7. Castelletti, A., Galelli, S., Restelli, M., Soncini-Sessa, R.: Data-driven dynamic emulation modelling for the optimal management of environmental systems. Environ. Model Softw. **34**(3), 30–43 (2012)
8. De Kok, J.L., Wind, H.G.: Design and application of decision support systems for integrated water management; lessons to be learnt. Phys. Chem. Earth **28**(14–15), 571–578 (2003)
9. Hipel, K.W., Walker, S.G.B.: Conflict analysis in environmental management. Environmetrics **22**(3), 279–293 (2011)
10. Janssen, J.A.E.B., Krol, M.S., Schielen, R.M.J., Hoekstra, A.Y.: The effect of modelling quantified expert knowledge and uncertainty information on model based decision making. Environ. Sci. Policy **13**(3), 229–238 (2010)
11. Lund, J.: Provoking more productive discussion of wicked problems. J. Water Resour. Plan. Manag. **138**(3), 193–195 (2012)
12. Mowrer, H.T.: Uncertainty in natural resource decision support systems: sources, interpretation and importance. Comput. Electron. Agric. **27**(1–3), 139–154 (2000)
13. Walker, W.E., Harremoes, P., Rotmans, J., Van der Sluis, J.P., Van Asselt, M.B.A., Janssen, P., Krayer von Krauss, M.P.: Defining uncertainty – a conceptual basis for uncertainty management in model-based decision support. Integr. Assess. **4**(1), 5–17 (2003)
14. Fuerst, C., Volk, M., Makeschin, F.: Squaring the circle? combining models, indicators, experts and end-users in integrated land-use management support tools. Environ. Manag. **46**(6), 829–833 (2010)
15. Wang, L., Fang, L., Hipel, K.W.: On achieving fairness in the allocation of scarce resources: measurable principles and multiple objective optimization approaches. IEEE Syst. J. **1**(1), 17–28 (2007)
16. Baugh, J.W., Caldwell, S.C., Brill, E.D.: A mathematical programming approach for generating alternatives in discrete structural optimization. Eng. Optim. **28**(1), 1–31 (1997)
17. Brill, E.D., Chang, S.Y., Hopkins, L.D.: Modelling to generate alternatives: the HSJ approach and an illustration using a problem in land use planning. Manag. Sci. **28**(3), 221–235 (1982)
18. Zechman, E.M., Ranjithan, S.R.: Generating alternatives using evolutionary algorithms for water resources and environmental management problems. J. Water Resour. Plan. Manag. **133**(2), 156–165 (2007)
19. Kasprzyk, J.R., Reed, P.M., Characklis, G.W.: Many-objective de Novo water supply portfolio planning under deep uncertainty. Environ. Model Softw. **34**, 87–104 (2012)
20. Yeomans, J.S.: Applications of Simulation-Optimization Methods in Environmental Policy Planning Under Uncertainty. J Environ. Inform. **12**(2), 174–186 (2008)
21. Loughlin, D.H., Ranjithan, S.R., Brill, E.D., Baugh, J.W.: Genetic algorithm approaches for addressing unmodeled objectives in optimization problems. Eng. Optim. **33**(5), 549–569 (2001)
22. van Delden, H., Seppelt, R., White, R., Jakeman, A.J.: A methodology for the design and development of integrated models for policy support. Environ. Model Softw. **26**(3), 266–279 (2012)
23. Lund, J.R., Tchobanoglous, G., Anex, R.P., Lawver, R.A.: Linear programming for analysis of material recovery facilities. ASCE J. Environ. Eng. **120**, 1082–1094 (1994)

24. Rubenstein-Montano, B., Zandi, I.: Application of a genetic algorithm to policy planning: the case of solid waste. Environ. Plann. B: Plann Des. **26**(6), 791–907 (1999)
25. Rubenstein-Montano, B., Anandalingam, G., Zandi, I.: A genetic algorithm approach to policy design for consequence minimization. Eur. J. Oper. Res. **124**, 43–54 (2000)
26. Hamalainen, R.P., Luoma, J., Saarinen, E.: On the importance of behavioral operational research: the case of understanding and communicating about dynamic systems. Eur. J. Oper. Res. **228**(3), 623–634 (2013)
27. Martinez, L.J., Joshi, N.N., Lambert, J.H.: Diagramming qualitative goals for multiobjective project selection in large-scale systems. Syst. Eng. **14**(1), 73–86 (2011)
28. Reed, P.M., Kasprzyk, J.R.: Water resources management: the myth, the wicked, and the future. J. Water Resour. Plan. Manag. **135**(6), 411–413 (2009)
29. Trutnevyte, E., Stauffacher, M., Schlegel, M.: Context-specific energy strategies: coupling energy system visions with feasible implementation scenarios. Environ. Sci. Technol. **46**(17), 9240–9248 (2012)
30. Caicedo, J.M., Zarate, B.A.: Reducing epistemic uncertainty using a model updating cognitive system. Adv. Struct. Eng. **14**(1), 55–65 (2011)
31. He, L., Huang, G.H., Zeng, G.-M.: Identifying optimal regional solid waste management strategies through an inexact integer programming model containing infinite objectives and constraints. Waste Manag. **29**(1), 21–31 (2009)
32. Kassab, M., Hipel, K.W., Hegazy, T.: Multi-criteria decision analysis for infrastructure privatisation using conflict resolution. Struct. Infrastruct. Eng. **7**(9), 661–671 (2011)
33. Matthies, M., Giupponi, C., Ostendorf, B.: Environmental decision support systems: current issues, methods and tools. Environ. Model. Softw. **22**(2), 123–127 (2007)
34. Sowell, T.: A conflict of visions. William Morrow & Co., New York (1987)
35. McIntosh, B.S., Ascough, J.C., Twery, M.: Environmental decision support systems (EDSS) development - challenges and best practices. Environ. Model Softw. **26**(12), 1389–1402 (2011)
36. Yeomans, J.S.: Automatic generation of efficient policy alternatives via simulation-optimization. J. Oper. Res. Soc. **53**(11), 1256–1267 (2002)
37. Yeomans, J.S.: Efficient generation of alternative perspectives in public environmental policy formulation: applying Co-evolutionary simulation-optimization to municipal solid waste management. CEJOR **19**(4), 391–413 (2011)
38. Yeomans, J.S., Gunalay, Y.: Simulation-optimization techniques for modelling to generate alternatives in waste management planning. J.Appl. Oper. Res. **3**(1), 23–35 (2011)
39. Caicedo, J.M., Yun, G.J.: A novel evolutionary algorithm for identifying multiple alternative solutions in model updating. Struct. Health Monit. Int. J. **10**(5), 491–501 (2011)
40. DeCaroli, J.F.: Using modeling to generate alternatives (MGA) to expand our thinking on energy futures. Energy Econ. **33**(2), 145–152 (2011)
41. Ursem, R.K., Justesen, P.D.: Multi-objective distinct candidates optimization: locating a Few highly different solutions in a circuit component sizing problem. Appl. Soft Comput. **12**(1), 255–265 (2012)
42. Zarate, B.A., Caicedo, J.M.: Finite element model updating: multiple alternatives. Eng. Struct. **30**(12), 3724–3730 (2008)
43. Walker, S.G.B., Hipel, K.W., Inohara, T.: Attitudes and preferences: approaches to representing decision maker desires. Appl. Math. Comput. **218**(12), 6637–6647 (2012)
44. Sun, W., Huang, G.H.: Inexact piecewise quadratic programming for waste flow allocation under uncertainty and nonlinearity. J Environ. Inform. **16**(2), 80–93 (2010)
45. Thekdi, S.A., Lambert, J.H.: Decision analysis and risk models for land development affecting infrastructure systems. Risk Anal. **32**(7), 1253–1269 (2012)
46. Yeomans, J.S., Huang, G.H., Yoogalingam, R.: Combining simulation with evolutionary algorithms for optimal planning under uncertainty: an application to municipal solid waste management planning in the regional municipality of Hamilton-Wentworth. J Environ. Inform. **2**(1), 11–30 (2003)
47. Fu, M.C.: Optimization for simulation: theory vs. practice. INFORMS J. Comput. **14**(3), 192–215 (2002)

48. Kelly, P.: Simulation optimization is evolving. INFORMS J. Comput. **14**(3), 223–225 (2002)
49. Zou, R., Liu, Y., Riverson, J., Parker, A., Carter, S.: A nonlinearity interval mapping scheme for efficient waste allocation simulation-optimization analysis. Water Resour. Res. **46**(8), 1–14 (2010)
50. Yang, X.S.: Firefly algorithms for multimodal optimization. Lect. Notes Comput. Sci **5792**, 169–178 (2009)
51. Yang, X.S.: Nature-inspired metaheuristic algorithms, 2nd edn. Luniver Press, Frome (2010)
52. Imanirad, R., Yang, X., Yeomans, J.S.: A computationally efficient, biologically-inspired modelling-to-generate-alternatives method. J. Comput. **2**(2), 43–47 (2012)
53. Imanirad, R., Yang, X., Yeomans, J.S.: A co-evolutionary, nature-inspired algorithm for the concurrent generation of alternatives. J. Comput. **2**(3), 101–106 (2012)
54. Imanirad, R., Yang, X., Yeomans, J.S.: A biologically-inspired metaheuristic procedure for modelling-to-generate-alternatives. Int. J. Eng. Res. Appl. **3**(2), 1677–1686 (2013)
55. Imanirad, R., Yang, X.S., Yeomans, J.S.: Modelling-to-generate-alternatives via the firefly algorithm. J.Appl. Oper. Res. **5**(1), 14–21 (2013)
56. Cagnina, L.C., Esquivel, C.A., Coello, C.A.: Solving engineering optimization problems with the simple constrained particle swarm optimizer. Informatica **32**, 319–326 (2008)
57. Gandomi, A.H., Yang, X.S., Alavi, A.H.: Mixed variable structural optimization using firefly algorithm. Comput. Struct. **89**(23–24), 2325–2336 (2011)
58. Yeomans, J.S.: Waste management facility expansion planning using simulation-optimization with grey programming and penalty functions'. Int. J. Environ. Waste Manage. **10**(2/3), 269–283 (2012)

Linear and Nonlinear System Identification Using Evolutionary Optimisation

K. Worden, I. Antoniadou, O.D. Tiboaca, G. Manson, and R.J. Barthorpe

Abstract While system identification of linear systems is largely an established body of work encoded in a number of key references (including textbooks), nonlinear system identification remains a difficult problem and tends to rely on a "toolbox" of methods with no generally accepted canonical approach. Fairly recently, methods of parameter estimation using evolutionary optimisation have emerged as a powerful means of identifying whole classes of systems with nonlinearities which previously proved to be very difficult, e.g. systems with unmeasured states or with equations of motion nonlinear in the parameters. This paper describes and illustrates the use of evolutionary optimisation methods (specifically the self-adaptive differential evolution (SADE) algorithm) on a class of single degree-of-freedom (SDOF) dynamical systems with hysteretic nonlinearities. The paper shows that evolutionary identification also has some desirable properties for linear system identification and illustrates this using data from an experimental multi-degree-of-freedom (MDOF) system.

Keywords System Identification • Nonlinear systems • Evolutionary optimisation • SADE

MSC code: 93E12

1 Introduction

System Identification (SI) is a technique of considerable importance within the discipline of structural dynamics. In the absence of a complete physics-based description of a system or structure, SI can provide the missing pieces of information that allow the formulation of a descriptive or predictive model. When the structure of interest has linear dynamical behaviour, the problem of SI is well established, to the

K. Worden (✉) • I. Antoniadou • O.D. Tiboaca • G. Manson • R.J. Barthorpe
Dynamics Research Group, Department of Mechanical Engineering, University of Sheffield, Mappin Street, Sheffield S1 3JD, UK
e-mail: k.worden@sheffield.ac.uk

S. Koziel et al. (eds.), *Simulation-Driven Modeling and Optimization*, Springer Proceedings in Mathematics & Statistics 153,
DOI 10.1007/978-3-319-27517-8_13

extent that authoritative textbooks and monographs exist [1, 2]. In the case of linear dynamical systems, it is usually sufficient to consider sets of linear second-order differential equations (*modal* models) or first-order differential equations (*state-space* models) as the appropriate mathematical model structure. In that case, the SI problem is largely reduced to determining the correct number of equations and the numerical parameters in the model. Unfortunately, most structures will in reality display nonlinear characteristics to some extent and the SI problem for nonlinear structures and systems is by no means solved. One of the main problems in nonlinear SI is the number and variety of possible model structures once the variety of possible nonlinearities is taken into account [3, 4].

It is not necessary here to provide a detailed classification of nonlinear SI models and approaches; however, it will prove useful to give a higher-level breakdown of model structures based on their motivation. Predictive models can be divided into three classes: *white*, *grey* and *black-box* models.

White-box models are taken here to be those whose equations of motion have been derived completely from the underlying physics of the problem of interest and in which the model parameters have direct physical meanings. Finite element models constitute one sub-class of such models.

Black-box models are, by contrast, usually formed by adopting a parametrised class of models with some universal approximation property and learning the parameters from measured data; in such a model, like a neural network, the parameters will not generally carry any physical meaning.

Grey-box models, as the name suggests, are usually a hybrid of the first two types above. They are usually formed by taking a basic core motivated by known physics and then adding a black-box component with approximation properties suited to the problem of interest. A good example of a grey-box model is the Bouc–Wen model of hysteresis which will be discussed in more detail later. In the Bouc–Wen model, a mass-spring-damper core is supplemented by an extra state-space equation which allows versatile approximation of a class of hysteresis loops [5, 6].

One can immediately see that black and grey box models are essentially *surrogate* models; they can represent a (perhaps drastically) simplified view of the physics in order to make predictions consistent with measured data. Even if the detailed physics is known, SI allows the formation of surrogates which can be run much faster than full physics-based models. Apart from the hysteretic systems which will be discussed in much more detail later, another good example of the use of SI-based surrogates is provided by the need to model friction. While physics-based models for friction are available, they can be computationally expensive to use and simpler grey-box models like the LuGre model can prove very useful [7].

Once a model structure has been chosen, the problem of SI reduces to estimating any free parameters in the model to bring its predictions into close correspondence with measured data.[1] For linear SI, fairly straightforward linear least-squares

[1]Although it is usually the basis for system identification, fidelity-to-data is only one means of assessing the validity of a model. An essential part of validation is to assess how well an identified

algorithms are often sufficient (in the absence of substantial coloured noise) [1, 2]. The situation is usually more complicated for nonlinear SI and no general approach is sufficient to deal with all classes of nonlinearity. The aim of this paper is to discuss and demonstrate one powerful approach to parameter estimation, based on evolutionary optimisation, which can prove effective in a wide range of circumstances. There is no intention here to provide a survey of nonlinear SI methods (the reference [4] is a fairly recent step in that direction) or even evolutionary SI methods; the discussion will be very much focussed on previous work by the current authors in order to provide illustrations via simulated and experimental data.

The layout of the paper is as follows. The next section discusses how evolutionary SI can be implemented using a specific algorithm—Differential Evolution—and shows how a self-adaptive version of DE offers advantages. It is also shown that confidence intervals for parameters can be obtained within the methodology. The implementation is demonstrated on simulated data from a Bouc–Wen hysteretic system. Section 3 shows how the self-adaptive DE algorithm (SADE) also proves effective in linear SI and is demonstrated on a multi-degree-of-freedom (MDOF) experimental structure. The final section of the paper is concerned with brief conclusions.

2 Identification of a Nonlinear System: The Bouc–Wen Hysteresis Model

2.1 The Bouc–Wen Model

Hysteretic systems are a useful and challenging test of nonlinear system identification algorithms. In the first case, they are of significant importance as they appear in many different engineering contexts. Such systems show significant memory-dependance and the type of phenomena this can cause are observed in many areas of physics and engineering such as electromagnetism, phase transitions and elastoplasticity of solids [8]. This paper will focus attention on one particular parametric model for hysteresis—the Bouc–Wen (BW) model [5, 6]. The second reason why the BW model is interesting is because it shows many of the properties which have traditionally caused technical problems for system identification.

The general BW model is a nonlinear single-degree-of-freedom (SDOF) model where the total internal restoring force is composed of a polynomial non-hysteretic and a hysteretic component based on the displacement $y(t)$ and velocity $\dot{y}(t)$ time-histories. The general representation described in the terms of Wen [6] is represented

model works in a context independent of the training/identification data. This validation process might be as simple as checking fidelity-to-data on a test data set completely independent of the identification process; this would actually be regarded as the minimum in the machine learning community.

below where $g(y, \dot{y})$ is the polynomial part of the restoring force and $z(y, \dot{y})$ the hysteretic,

$$m\ddot{y} + g(y, \dot{y}) + z(y, \dot{y}) = x(t) \tag{1}$$

where m is the mass of the system and $x(t)$ is the excitation force. For the purposes of this paper, the polynomial g will be assumed linear, so that $g(y, \dot{y}) = c\dot{y} + ky$. The overall system equation of motion is then

$$m\ddot{y} + c\dot{y} + ky + z(y, \dot{y}) = x(t) \tag{2}$$

The hysteretic component z is then defined by Wen [6] via an additional equation of motion,

$$\dot{z} = A\dot{y} - \alpha|\dot{y}|z^n - \beta\dot{y}|z^n| \tag{3}$$

for n odd, or,

$$\dot{z} = A\dot{y} - \alpha|\dot{y}|z^{n-1}|z| - \beta\dot{y}z^n \tag{4}$$

for n even.

The parameters α, β and n govern the shape and smoothness of the hysteresis loop. To simplify matters from the point of view of parameter estimation, the stiffness term in equation (2) can be combined with the term $A\dot{y}$ in the state equation for z. As a system identification problem, this set of equations presents a number of difficulties, foremost are:

- The variables available from measurement will generally be the input x and some form of response: displacement, velocity or acceleration. In this paper the response variable will be assumed to be displacement y, although the identification problem can just as easily be formulated in terms of velocity or acceleration. Even if all the response variables mentioned are available, the state z is not measurable and therefore it is not possible to use equation (3) or equation (4) directly in a least-squares formulation.
- The parameter n enters the state equations (3) and (4) in a nonlinear way; this means that a linear least-squares approach is not applicable to the estimation of the full parameter set, some iterative nonlinear least-squares approach is needed as in [9] or an evolutionary scheme can be used as will be shown here.

How these problems are addressed in the context of evolutionary optimisation is discussed in the following section.

2.2 Hysteretic System Identification Using Differential Evolution

Evolutionary computation began with the basic Genetic Algorithm (GA) and even the simplest form of that algorithm proved useful in the identification of hysteretic systems [10]. However, once real-parameter evolutionary schemes like Differential Evolution (DE) emerged [11], it quickly became clear that they offered major advantages for SI. The first application of DE for the BW model appeared in [12]. Later in this paper, an advanced, more adaptive form of the DE algorithm will be demonstrated; for now, the basic algorithm will be introduced and illustrated. As in all evolutionary optimisation procedures, a population of possible solutions (here, the vector of parameter estimates) is iterated in such a way that succeeding generations of the population contain better solutions to the problem in accordance with the Darwinian principle of "survival of the fittest". The problem is framed here as a minimisation problem with the cost function defined as a normalised mean-square error (NMSE) between the "measured" data and that predicted using a given parameter estimate, i.e.,

$$J(m, c, k, \alpha, \beta) = \frac{100}{N\sigma_y^2}\sum_{i=1}^{N}(y_i - \hat{y}_i(m, c, k, \alpha, \beta))^2 \tag{5}$$

where σ_y^2 is the variance of the "measured" sequence of displacements y_i and the caret denotes a predicted quantity; N is the total number of samples. With the normalisation chosen in equation (5), previous experience has shown that a cost value of less than 5.0 represents a good model or parameter estimate, while one with less than 1.0 can usually be considered excellent. Note that this definition of cost function could quite easily be used with velocity or acceleration data; this means that whatever data is sampled, there will be no need to apply numerical differentiation or integration procedures. A further advantage of this approach is that the optimisation does not need measurements of z; the correct prediction for the state is implicit in the approach. This overcomes the first of the problems discussed in the last subsection.

The standard DE algorithm of reference [11] attempts to transform a randomly generated initial population of parameter vectors into an optimal solution through repeated cycles of evolutionary operations, in this case: *mutation*, *crossover* and *selection*. In order to assess the suitability of a certain solution, a cost or fitness function is needed; the cost function in equation (5) is the one used here. Figure 1 shows a schematic for the DE procedure for evolving between populations. The following process is repeated with each vector within the current population being taken as a *target vector*; each of these vectors has an associated cost taken from equation (5). Each target vector is pitted against a *trial vector* in a competition which results in the vector with lowest cost advancing to the next generation.

The mutation procedure used in basic DE proceeds as follows. Two vectors A and B are randomly chosen from the current population to form a vector differential $A - B$. A *mutated* vector is then obtained by adding this differential, multiplied by a scaling factor F, to a further randomly chosen vector C to give the overall expression

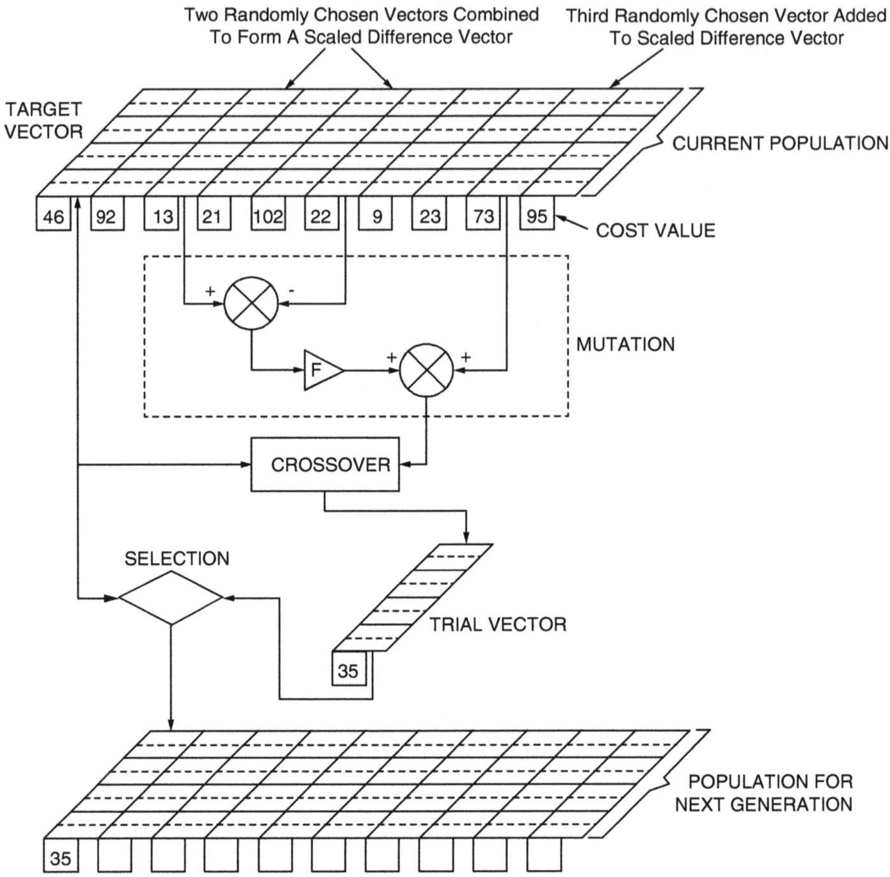

Fig. 1 Schematic for the standard DE algorithm

for the mutated vector: $C + F(A - B)$. The scaling factor, F, is often found to have an optimal value between 0.4 and 1.0. The quantity F would be referred to in the machine learning literature as a *hyperparameter*; this is a parameter of the *algorithm* which must be specified before the algorithm can be applied and is thus distinct from the parameters of the proposed model which are estimated by the algorithm. The hyperparameters can affect the effectiveness of the algorithm to a considerable extent and usually require careful choice. Best practice in a machine learning context would be to *optimise* the hyperparameters in some sense; one of the simplest means of doing this is via cross-validation on an independent validation set [13].

The *trial vector* is the child of two vectors: the target vector and the mutated vector, and is obtained via a crossover process; in this work, uniform crossover is used. Uniform crossover decides which of the two parent vectors contributes to each chromosome of the trial vector by a series of $D - 1$ binomial experiments. Each experiment is mediated by a crossover parameter C_r (where $0 \leq C_r \leq 1$). If a random number generated from the uniform distribution on [0,1] is greater than C_r, the trial vector takes its parameter from the target vector, otherwise the parameter comes from the mutated vector. The crossover parameter C_r is another example of a hyperparameter.

This process of evolving through the generations is repeated until the population becomes dominated by only a few low cost solutions, any of which would be suitable. The algorithm is designed so that the cost function must stay constant or decrease with each generation. Because the cost is monotonically non-increasing and bounded below (by zero) it must clearly converge. Like the vast majority of optimisation algorithms, convergence to the global minimum is not guaranteed; however, one of the benefits of the evolutionary approach is that it is more resistant to finding a local minimum.

The illustration presented here is from computer simulation. The coupled equations (2) and (4) were integrated forward in time using the Matlab function *ode45* for initial value problems. The function in question implements an adaptive $(4,5)$th-order Runge–Kutta method. The parameters for the baseline BW system chosen were: $m = 1$, $c = 20$, $\alpha = 1.5$, $\beta = -1.5$, $A = 6680.0$ and $n = 2$. (The parameters are chosen relative to SI unit choices of kg for the mass, N for the force, etc.) The excitation was a Gaussian white noise sequence with mean zero and standard deviation 9.92. The step-size (or sampling interval) was taken as 0.004 s, corresponding to a sampling frequency of 250 Hz. The data was not corrupted by any artificial noise. The "training set" or identification set used here was composed of 1000 points corresponding to a duration of 4 s. The response variable used in the identification algorithm here was the displacement.

For the identification, DE was implemented using a parameter vector (m, c, k, α, β). (The parameter n was not included here; the reasons for this are discussed in [14], this does not affect the validity of this data as an illustration of the method.) The DE algorithm was initialised with a population of randomly selected parameter vectors or individuals.

The parameters for the initial population were generated using uniform distributions on ranges covering one order of magnitude above and below the true values as in [12]. A population of 30 individuals was chosen for the DE runs with a maximum number of generations of 200. In order to sample different random initial conditions for the DE algorithm, 10 independent runs were made. The initial conditions of the algorithm are essentially another set of hyperparameters; the other hyperparameters chosen here for DE were an F-value of 0.9 and a crossover probability of 0.5; these were chosen on the basis of experience, they have proved effective in a range of applications. This completes the specification of the DE.

Each of the 10 runs of the DE algorithm converged to an acceptable solution to the problem in the sense that cost function values of less than 0.1 were obtained in all cases; the summary results are given in Table 1. The best solution gave a cost function value of 0.059. A comparison between the "true" and predicted responses for the best parameter set is given in Figure 2. The parameter estimates for m, c and A are all very accurate; only α and β show significant deviations from the

Table 1 Summary results for 10 DE identification runs on the simulation data

Parameter	True value	Best model	% error	Minimum	Maximum	Mean	Standard deviation
m	1.0	1.0017	0.17	0.9885	1.0104	1.0002	0.0064
c	20.0	20.014	0.07	19.940	20.355	20.073	0.139
α	1.5	1.555	3.67	1.144	1.609	1.411	0.170
β	−1.5	−1.225	16.6	−2.554	−0.150	−1.141	0.711
A	6680.0	6719.3	0.59	6577.5	6772.3	6704.0	55.1

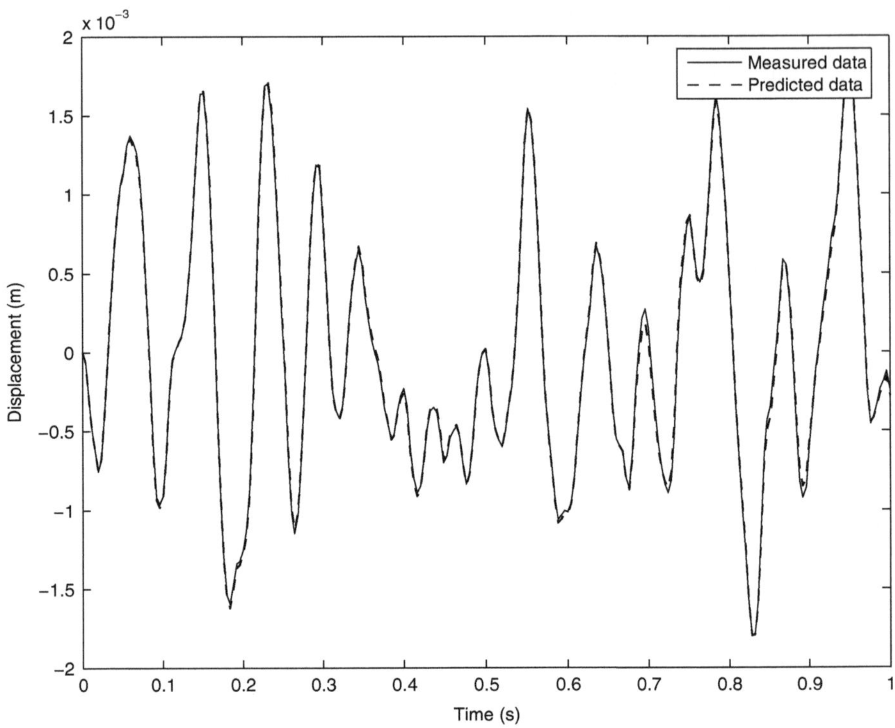

Fig. 2 Comparison of measured and predicted responses for DE algorithm

true parameters. Despite errors in the parameter estimates, the predicted response (Figure 2) is excellent; this is simply because the response is more sensitive to some parameters than others.

In fact, for a rather subtle reason which is discussed in [14], the identification procedure was impeded by the use of the adaptive ODE solver; the effect being to mask the global minimum of the NMSE in artificial "noise". The solution to this was simply to use a fixed-step 4th-order Runge–Kutta algorithm [15]. When a set of 10 DE runs were made with the same hyperparameters as before, the results were as shown in Table 2. The results are markedly better than those obtained using the adaptive solver; the best solution gave a vastly improved cost function value

Table 2 Summary results for 10 DE identification runs on the simulation data: fixed-step solver used

Parameter	True value	Best model	% error	Minimum	Maximum	Mean	Standard deviation
m	1.0	0.9995	0.05	0.996	1.002	1.000	0.001
c	20.0	20.011	0.06	19.950	20.043	20.011	0.029
α	1.5	1.496	0.27	1.419	1.535	1.494	0.032
β	−1.5	−1.423	5.13	−1.683	−1.243	−1.501	0.143
A	6680.0	6678.6	0.02	6657.8	6695.2	6676.6	11.3

of 7.36×10^{-5}. A comparison between the "true" and predicted responses for the best parameter set produced traces which were completely indistinguishable, so the comparison is not shown here. It is important to note here that this model has not been properly validated in the sense discussed earlier; the fidelity of the model is not demonstrated here on an independent test data set. This is simply justified here by the fact that the "true" parameters are known, so fidelity of the model is assured by the very close correspondence shown here between "true" and estimated parameters. In the general case, the "true" parameters would not be known and a principled approach to validation would be required. This is illustrated later in this paper when an experimental data set is used for identification.

2.3 System Identification Using SADE

A potential weakness of the standard implementation of the DE algorithm as described above is that it requires the prior specification of a number of hyper-parameters. Apart from the population size, maximum number of iterations, etc., the algorithm needs a priori specification of the scaling factor F and crossover probability C_r. The values above were chosen on the basis of previous studies where they gave good results; however, they are not guaranteed to work as well in all situations and an algorithm which establishes "optimum" values for these parameters during the course of the evolution is clearly desirable. Such an algorithm is available in the form of the SADE algorithm [16, 17]; the description and implementation of the algorithm here largely follows [17] (the local search option in the latter reference is not implemented here).

The development of the SADE algorithm begins with the observation that Storn and Price, the originators of DE, arrived at five possible strategies for the mutation operation [18]:

1. *rand1*: $M = A + F(B - C)$
2. *best1*: $M = X^* + F(B - C)$
3. *current-to-best*: $M = T + F(X^* - T) + F(B - C)$
4. *best2*: $M = X^* + F(A - B) + F(C - D)$
5. *rand2*: $M = A + F(B - C) + F(D - E)$

where T is the current trial vector, X^* is the vector with (currently) best cost and (A, B, C, D, E) are randomly chosen vectors in the population distinct from T. F is a standard (positive) scaling factor. The SADE algorithm also uses multiple variants of the mutation algorithm as above; however, these are restricted to the following four:

1. *rand1*
2. *current-to-best2*: $M = T + F(X^* - T) + F(A - B) + F(C - D)$
3. *rand2*
4. *current-to-rand*: $M = T + K(A - T) + F(B - C)$

In the strategy *current-to-rand*, K is defined as a coefficient of combination and would generally be assumed in the range $[-0.5, 1.5]$; however, in the implementation of [17] and the one used here, the prescription $K = F$ is used to essentially restrict the number of tunable parameters. The SADE algorithm uses the standard crossover approach, except that at least one crossover is forced in each operation on the vectors. If mutation moves a parameter outside its allowed (predefined) bounds, it is pinned to the boundary. Selection is performed exactly as in DE; if the trial vector has smaller (or equal) cost to the target, it replaces the target in the next generation.

The adaption strategy must now be defined. First, a set of probabilities are defined: $\{p_1, p_2, p_3, p_4\}$, which are the probabilities that a given mutation strategy will be used in forming a trial vector. These probabilities are initialised to be all equal to 0.25. When a trial vector is formed during SADE, a roulette wheel selection is used to choose the mutation strategy on the basis of the probabilities (initially, all equal). At the end of a given generation, the numbers of trial vectors successfully surviving to the next generation from each strategy are recorded as: $\{s_1, s_2, s_3, s_4\}$; the numbers of trial vectors from each strategy which are discarded are recorded as: $\{d_1, d_2, d_3, d_4\}$. At the beginning of a SADE run, the survival and discard numbers are established over the first generations, this interval is called the *learning period* (and is another example of a hyperparameter). At the end of the learning period, the strategy probabilities are updated by

$$p_i = \frac{s_i}{s_i + d_i} \tag{6}$$

After the learning period, the probabilities are updated every generation but using survival and discard numbers established over a moving window of the last N_L generations. The algorithm thus adapts the preferred mutation strategies. SADE also incorporates adaption or variation on the hyperparameters F and C_r. The scaling factor F mediates the convergence speed of the algorithm, with large values being appropriate to global search early in a run and small values being consistent with local search later in the run. The implementation of SADE used here largely follows [16] and differs only in one major aspect, concerning the adaption of F. Adaption of the parameter C_r is based on accumulated experience of the successful values for the parameter over the run. It is assumed that the crossover probability for a trial is normally distributed about a mean $\overline{C}_r$ with standard deviation 0.1. At initiation, the parameter C_r is set to 0.5 to give equal likelihood of each parent contributing a chromosome. The crossover probabilities are then held fixed for each population index for a certain number of generations and then resampled. In a rather similar manner to the adaption of the strategy probabilities, the C_r values for trial vectors successfully passing to the next generation are recorded over a certain greater number of generations and their mean value is adopted as the next $\overline{C}_r$. The record of successful trials is cleared at this point in order to avoid long-term memory effects. The version of the algorithm here adapts F in essentially the same manner as C_r but uses the Gaussian $N(0.5, 0.3)$ for the initial distribution. At this point, the reader might legitimately argue that SADE has simply replaced one set of hyperparameters (F, C_r) with another (duration of the learning period, etc.). In fact, because DE and SADE are heuristic algorithms, there is no analytical counter to this argument. However, the transition to SADE is justified by the fact that the algorithm appears to be very robust with respect to the new hyperparameters.

The SADE algorithm is now illustrated on the same identification problem as considered earlier. To show how robust the algorithm is, the results are presented for the *first* attempt with the algorithm, where the learning period was simply taken as a plausible ten generations; updates were subsequently applied every ten generations. The insensitivity of SADE to these hyperparameters is shown by the fact that a simple first *guess* at their values leads to good control of the mutation strategy and rapid convergence to an excellent minimum of the cost function. As before, the algorithm used a population of 30 individuals and was allowed to run for 200 generations for 10 independent runs with different initial populations, this means that the same number of cost function evaluations took place as in the standard DE run. As before, the solver made use of a fixed step. The results of the SADE run are given in Table 3.

The results from SADE show a radical improvement on the results from the standard DE given in Table 2. The cost function value for the best run is 1.51×10^{-9}, an improvement over DE of over four orders of magnitude. There is little point here in showing a comparison between the model predictions and the original data as the curves are indistinguishable.

Table 3 Summary results for 10 SADE identification runs on the simulation data: fixed-step solver used

Parameter	True value	Best model	% error	Minimum	Maximum	Mean	Standard deviation
m	1.0	1.000	0.0005	1.000	1.000	1.000	0.00008
c	20.0	20.00	0.0006	19.999	20.001	20.000	0.0008
α	1.5	1.500	0.0004	1.498	1.501	1.499	0.0008
β	−1.5	−1.500	0.005	−1.502	−1.495	−1.499	0.0022
A	6680.0	6680.0	0.003	6679.7	6682.0	6680.2	0.67

2.4 Confidence Intervals for Parameter Estimates

A drawback of the optimisation-based approach to SI discussed so far is that it does not immediately provide a measure of confidence in those estimates. In contrast, it is well known that when linear-algebraic least-squares methods are applicable to the identification problem, the methods provide an estimate of the covariance matrix for the parameters [1, 2]. From the covariance matrix, one can extract the standard deviations of the parameter estimates and thus form appropriate confidence intervals. Some means of deriving confidence intervals for the optimisation-based approach is clearly desirable. In fact, a means is available and has been reported in the literature as far back as the classic work by Box and Jenkins on time-series analysis [19]. The discussion here follows that in [20].

The basic principle is straightforward; if the parameter estimates are obtained by minimising a cost function, one would expect the accuracy/precision of the estimates to be related to the curvature of the cost function in the vicinity of the minimum. If the cost function has a very shallow minimum, the algorithm will be likely to converge over a larger range of estimates. If the minimum is very narrow, one would expect a correspondingly narrow range of possible estimates. According to basic principles of differential geometry, the curvature of the cost function is approximated by the second derivatives with respect to the parameters. The relevant equation for the covariance matrix is [19],

$$\Sigma(\underline{w}) \approx 2\sigma_\xi^2 [S]^{-1} \tag{7}$$

where $\underline{w}$ is the vector of parameters and the matrix $[S]$ is defined by its elements,

$$S_{ij} = \left. \frac{\partial^2 J'(\underline{w})}{\partial w_i \partial w_j} \right|_{\underline{w}=\underline{w}^*} \tag{8}$$

where $\underline{w}^*$ is the optimum derived for the parameter estimate and σ_ξ^2 is the residual variance unexplained by the model. The function $J'(\underline{w})$ is the sum-of-squares function and is here directly related to the cost function in equation (5) by

$$J'(\underline{w}) = \frac{N\sigma_y^2}{100} J(\underline{w}) = \sum_{i=1}^{N} (y_i - \hat{y}_i(\underline{w}))^2 \tag{9}$$

One also sees that the residual variance is estimated by

$$\sigma_\xi^2 = \frac{J'(\underline{w})}{N} \tag{10}$$

Equation (7) will be used to estimate the parameter covariance. However, it has been observed that the definition (7) can present problems due to the presence of the second derivatives [21]. In practice, these are evaluated numerically from

Table 4 Summary results for 10 SADE identification runs on the simulation data with 5 % noise: fixed-step solver used

Parameter	True value	Best model	Minimum	Maximum	Mean	Standard deviation
m	1.0	0.9975	0.9975	0.9978	0.9976	0.00007
c	20.0	20.088	20.083	20.089	20.088	0.002
α	1.5	1.440	1.440	1.444	1.440	0.001
β	−1.5	−1.539	−1.567	−1.537	−1.542	0.009
A	6680.0	6663.0	6661.8	6664.7	6663.0	0.7

Table 5 Parameter confidences from estimated covariance matrix (7)

Parameter	True value	Standard deviation	95 % confidence interval
m	1.0	0.002	[0.993, 1.001]
c	20.0	0.06	[19.97, 20.21]
α	1.5	0.06	[1.33, 1.55]
β	−1.5	0.29	[−2.07, −1.01]
A	6680.0	17.0	[6629.7, 6696.3]

evaluations of the cost function at points close to the minimum and can sometimes be inaccurate. As an alternative, the book [22] provides an estimator for the covariance matrix which only requires first derivatives.

The first stage is of course to obtain parameter estimates using the optimisation scheme. Exactly the same procedure as above was followed, except that Gaussian white noise with a standard deviation 5 % of the response standard deviation was added to the simulated response (in order that sensible confidence intervals were obtained from the simulated data). The results of the consequent 10 SADE runs are given in Table 4.

Note that the standard deviation in the final column of Table 5 is *not* an estimate derived from a covariance matrix, it is simply the standard deviation of the estimates over the 10 runs of the SADE algorithm. The NMSE for the best set of parameters was 0.225. The set of parameters from the best run (column 3 in Table 4) will be used as the vector $\underline{w}^*$ in the covariance matrix estimates. In order to estimate the covariance matrices using equation (7) one needs to evaluate the cost function at points near the optimum $\underline{w}^*$ and numerically estimate first and second derivatives. Centred differences were used to estimate the derivatives here. One issue that arises is choice of the step-size for the differences. This was defined here in terms of a fraction of the parameter size by $h = w_i/d_f$ with a user-specified parameter d_f. A simple convergence study showed that a value of $d_f = 50$ was appropriate, and this value was used. The estimator of the covariance matrix based on equation (7) gave the results shown in Table 5.

The estimator for the covariance matrix gives sensible results; the 95 % confidence intervals bracket the true parameters as one would expect for unbiased estimates. Furthermore, the estimator based on first-derivatives [22] was shown to agree with these results in [23] as was an estimator based on Markov Chain Monte Carlo.

3 Linear System Identification of an Experimental Structure

Although the main advantages of the evolutionary SI approach are felt for nonlinear structures, there is no reason why it cannot be applied to linear systems and structures. In fact, there is a reason, alluded to above, why the evolutionary approach may be desirable. The reason is that the identification method can work with a single type of measured response from the system of interest, e.g. a displacement corresponding to each degree of freedom. In contrast, a direct parameter estimation based on linear least-squares analysis would require measurements of acceleration, velocity and displacement for all degrees of freedom of interest. This means that either extensive instrumentation is required (three times the number of sensors and measurement channels) or troublesome numerical integration or differentiation must be employed [24]. For MDOF systems, there is thus a clear advantage of adopting an explicit optimisation-based approach over a "classical" least-squares strategy. In this section, the use of SADE on an experimental MDOF structure is demonstrated.

3.1 The Experimental Rig and Data Capture

For the purposes of the present illustration, a small-scale simulated shear building model was used. This was designed to correspond closely with a structure previously designed and built at Los Alamos National Laboratories (LANL). Within LANL, the experimental rig was referred to informally as the "bookshelf" rig and this informal nomenclature is also adopted here. The bookshelf structure, illustrated in Figure 3 has four levels, floors or shelves, with the lower level being considered the base. Each shelf/floor is composed of a substantial rectilinear aluminium block with a mass of 5.2 kg and dimensions $35 \times 25.5 \times 0.5$ cm ($L \times w \times h$). The shelves are joined by upright beams at each corner; each beam having a mass of 238 g and dimensions $55.5 \times 2.5 \times 0.6$ cm. The blocks used to connect the main plates and the upright beams have a mass of 18 g and dimensions $2.5 \times 2.5 \times 1.3$ cm. For each block, four bolts were used, each of Viraj A2-70 grade and with a mass of 10 g. The structure was mounted on a rail system which was securely clamped onto a substantial testing table; linear bearings were used in order to minimise friction in the rail system. In order to introduce the excitation into the structure, an electrodynamic shaker with a force transducer was connected to the base.

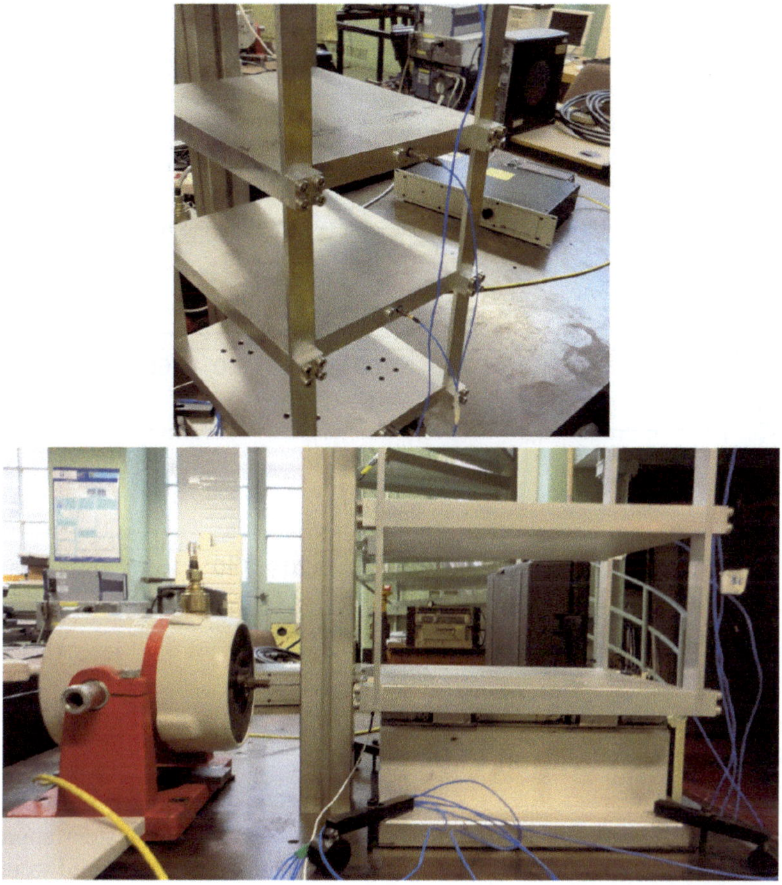

Fig. 3 The "bookshelf" experimental rig showing accelerometer positions, shaker attachment and guide rail system

The experimental data were acquired using an LMS CADA system connected to a SCADAS-3 interface. A total of 93,184 points per channel were recorded at a sampling frequency of 1024 Hz. Lateral accelerations were recorded for each shelf from piezoelectric accelerometers fixed to the edges (as shown in the figures). Transmissibilities between the relative accelerations of the floors and the base acceleration were produced by applying the Welch method to the raw measurement data and are given in Figure 4; the structures shown indicate that a three-DOF model of the rig is likely to capture the main dynamics. The use of relative accelerations allows the expression of the four-DOF dynamics in terms of the three-DOF model.

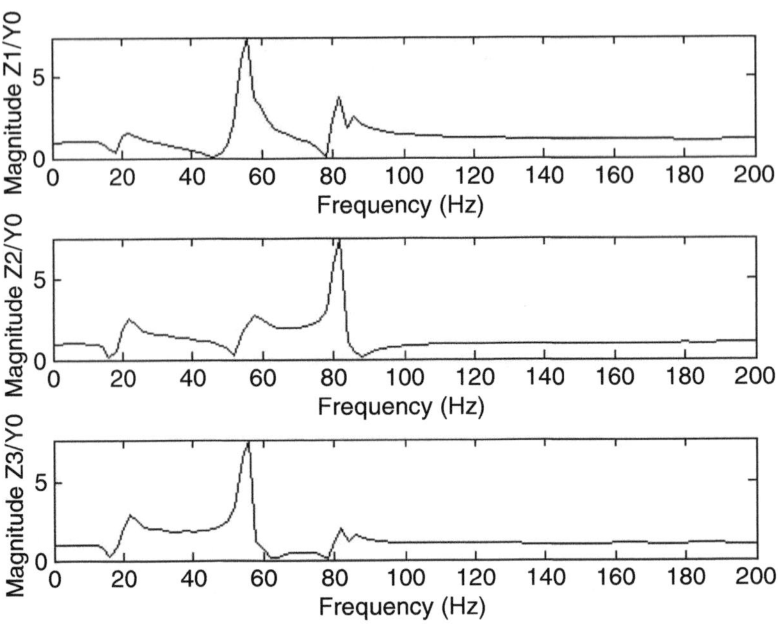

Fig. 4 "FRFs" between the base acceleration of the rig and the relative accelerations of the upper floors

4 System Identification Using SADE

Having established by the FRF analysis that the base-excited system appears to correspond well to a three-DOF system, the model equations considered were

$$\begin{aligned} m_1\ddot{z}_1 + c_1\dot{z}_1 + c_2(\dot{z}_1 - \dot{z}_2) + k_1 z_1 + k_2(z_1 - z_2) &= -m_*\ddot{y}_0 \\ m_2\ddot{z}_2 + c_2(\dot{z}_2 - \dot{z}_1) + c_3(\dot{z}_2 - \dot{z}_3) + k_2(z_2 - z_1) + k_3(z_2 - z_3) &= -m_*\ddot{y}_0 \\ m_3\ddot{z}_3 + c_3(\dot{z}_3 - \dot{z}_3) + k_3(z_3 - z_2) &= -m_*\ddot{y}_0 \end{aligned} \quad (11)$$

where the $\{z_i = y_i - y_0 \ : \ i = 1, \ldots, 3\}$ are displacement coordinates relative to the base displacement. As it is not clear what the actual masses are prior to the identification, an estimate m_* is used for the RHS of the equations. The estimate is based on the physical masses of the shelves and associated fixings. Including m_1, m_2 and m_3 in the parameter vector $\underline{w} = (m_1, m_2, m_3, c_1, c_2, c_3, k_1, k_2, k_3)$ allows the identification algorithm to correct for the contribution of the vertical beams, etc. Based on the design geometry and materials, m_* was taken here as 5.475 kg.

The cost function referred to above was defined again in terms of the prediction errors associated with each DOF. A set of NMSEs J_i were defined by

$$J_i(\underline{w}) = \frac{100}{N\sigma_{\ddot{z}_i}^2} \sum_{i=1}^{N} (\ddot{z}_i - \hat{\ddot{z}}_i(\underline{w}))^2 \tag{12}$$

where $\sigma_{\ddot{z}_i}^2$ is the variance of the measured sequence of relative accelerations and the caret denotes the predicted quantity; N is the number of "training" points used for identification. The total cost function J was then taken as the average of the J_i. In order to generate the predictions $\hat{\ddot{z}}_i$, the coupled equations (11) were integrated forward in time using a fixed-step fourth-order Runge–Kutta scheme as before. The excitations for the predictions were established by the measured base accelerations $\ddot{y}_0$ and the initial estimate m_*. Although a great deal of data were measured in the experiments, the SADE identification scheme is computationally expensive, with the main overhead associated with integrating trial equations forward in time. For this reason, the "training set" or identification set used here was composed of only $N = 5000$ points. To avoid problems associated with transients, the cost function was only evaluated after the first 200 points of each predicted record.

Once the data were generated, the SADE algorithm was applied to the identification problem using a parameter vector $\underline{w}$. Initial ranges for the parameters were required as usual; the initial parameters were generated using uniform distributions on those ranges. Estimates based on engineering judgement were used. The masses in the model were not considered as a problem as the inertia of the system was considered very likely to be dominated by the shelves and fixings, it was therefore expected that the true values would be close to the estimate m_* given above. For this reason, a short range [4.5, 6.0] was taken for the initial population. The situation with the stiffness parameters is a little more complicated as it is not clear what the appropriate boundary conditions are for the upright beams connecting the floors. An approximate value of $k_* = 5.2 \times 10^5$ N/m can be obtained by assuming encastre conditions; however, the true value may vary substantially from k_* if the bolts do not impose a true fixed condition, for example. Because of the uncertainty in the initial physical estimates of the stiffness, the initial ranges for SADE were taken on roughly an order of magnitude below and above k_* i.e. $[5 \times 10^4, 5 \times 10^6]$. Taking into account the values of m_* and k_* and assuming damping ratios in the vicinity of 0.1 % for aluminium, the initial ranges for the damping parameters were estimated at [0.1, 10.0], again an order or magnitude below and above a nominal value of $c_* = 1.0$ Ns/m. It was anticipated that the damping would be at the higher end of the ranges as a result of damping from joints etc.

A population of 100 individuals was chosen for the SADE runs with a maximum number of generations of 200. In order to sample different random initial conditions for the DE algorithm, 5 independent runs were made. The other parameters chosen

Table 6 Parameter estimates from 5 independent SADE runs

Parameter	Best	Maximum	Minimum	Mean	Standard deviation	Coefficient of variation
m_1	5.037	5.044	4.902	5.000	0.062	0.012
m_2	6.000	6.000	5.536	5.753	0.166	0.029
m_3	5.424	5.535	5.190	5,359	0.129	0.024
c_1	9.612	9.683	6.044	8.701	1.522	0.175
c_2	0.100	9.632	0.100	2.167	4.185	1.931
c_3	1.445	9.022	1.445	4.000	3.090	0.772
k_1 ($\times 10^{-5}$)	0.644	2.936	0.644	1.292	0.953	0.738
k_2 ($\times 10^{-5}$)	6.946	6.946	3.362	6.031	1.521	0.252
k_3 ($\times 10^{-5}$)	0.736	5.941	0.500	1.877	2.326	1.239
J	3.383	3.466	3.383	3.418	0.042	0.012

used for SADE were a starting value for F of 0.9 and a starting value for C_p of 0.5 (these values proved to be effective in a number of previous studies); this completes the specification of SADE for the problem.

Each of the 5 runs of the DE algorithm converged to a good solution to the problem in the sense that cost function values of around 2 % or below were obtained in all cases; the summary results are given in Table 6. The best solution gave a cost function value of 1.591. A visual comparison of the experimental responses and predicted responses for the best parameter set is given in Figure 5. As the true parameters are not known in this case, a more rigourous approach to validation is required, so the comparison is based upon a set of 5000 points of testing data that was distinct from the training data used to fit the parameters. If the cost function is deconstructed into the individual errors for degrees of freedom, the result is the set (3.370, 2.952, 3.827) for the training data and (4.882, 5.885, 7.772) for the testing data. The errors on the training data (all less than 5 %) indicate a good model; however, there is some degradation on the testing set. The increased error on the testing set is an indication that the model is not generalising perfectly to independent data sets. There could be a number of explanations for this, a likely one here is that the experimental rig is not perfectly linear; if the linear bearings are at all misaligned, there may be a contribution from contacts or friction. This issue will require further investigation and tuning of the rig.

The results are interesting. Although there are only small variations in the prediction errors, the coefficients of variation (standard deviation of estimate/mean of estimate) are quite high for a number of parameters. This indicates that the errors are rather insensitive to some of the parameters. This would be confirmed by a full sensitivity analysis (in fact, sensitivity analysis is a potential ingredient in a rigorous programme of model validation).

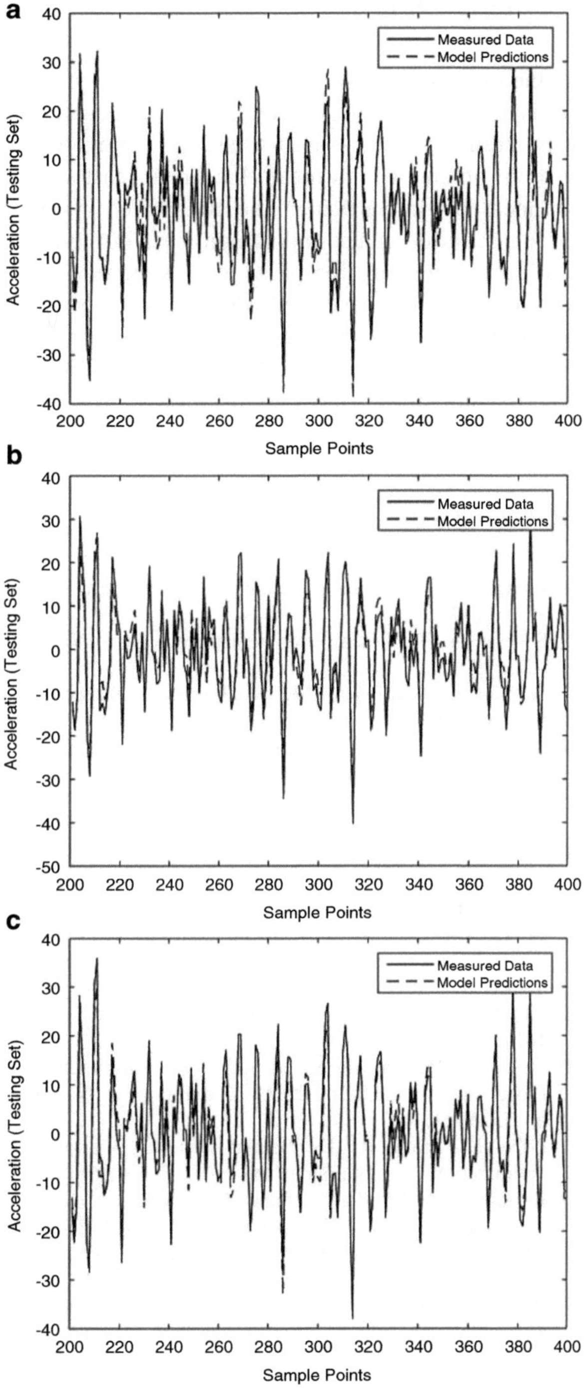

Fig. 5 SADE model predictions on testing data: (**a**) $\ddot{z}_1$, (**b**) $\ddot{z}_2$, (**c**) $\ddot{z}_3$

5 Conclusions

As the purpose of this paper is to illustrate the use of evolutionary SI rather than to display new results, extensive conclusions are not warranted. The example of a hysteretic system is used to show how evolutionary methods overcome some of the problems which mean that linear least-squares methods are ineffective for nonlinear SI. In particular, the Bouc–Wen system is singled out as one which has unmeasured states and is nonlinear-in-the-parameters. It is also shown how confidence intervals for parameter estimates can be obtained straightforwardly within the evolutionary methodology. Another strength of the optimisation-based approach to SI is that one does not need simultaneous measurements of acceleration, velocity and displacement data and this is illustrated via a case study of an experimental three-storey building model.

References

1. Ljung, L.: System Identification: Theory for the User, 2nd edn. Prentice Hall, Upper Saddle River (1999)
2. Soderstrom, T., Stoica, P.: System Identification (New Edition). Prentice Hall, Upper Saddle River (1994)
3. Worden, K., Tomlinson, G.R.: Nonlinearity in Structural Dynamics. Institute of Physics Press, Bristol (2001)
4. Kerschen, G., Worden, K., Vakakis, A.F., Golinval, J.-C.: Past, present and future of nonlinear system identification in structural dynamics. Mech. Syst. Signal Process. **20**, 505–592 (2006)
5. Bouc, R.: Forced vibration of mechanical system with hysteresis. In: Proceedings of 4th Conference on Nonlinear Oscillation, Prague (1967)
6. Wen, Y.: Method for random vibration of hysteretic systems. ASCE J. Eng. Mech. Div. **XX**, 249–263 (1976)
7. Worden, K., Wong, C.X., Parlitz, U., Hornstein, A., Engster, D., Tjahjiwidodo, T., Al-Bender, F., Rizos, D.D., Fassois, S.: Identification of pre-sliding and sliding friction dynamics: grey box and black box models. Mech. Syst. Signal Process. **21**, 514–534 (2007)
8. Visitin, A.: Differential Models of Hysteresis. Springer, Berlin (1994)
9. Yar, M., Hammond, J.K.: Parameter estimation for hysteretic systems. J. Sound Vib. **117**, 161–172 (1987)
10. Deacon, B.P., Worden, K.: Identification of hysteretic systems using genetic algorithms. In: Proceedings of EUROMECH - 2nd European Nonlinear Oscillation Conference, Prague, pp. 55–58 (1996)
11. Price, K., Storn, R.: Differential evolution - a simple and efficient heuristic for global optimization over continuous spaces. J. Glob. Optim. **11**, 341–359 (1997)
12. Kyprianou, A., Worden, K., Panet, M.: Identification of hysteretic systems using the differential evolution algorithm. J. Sound Vib. **248**, 289–314 (2001)
13. Bishop, C.M.: Neural Networks for Pattern Recognition. Oxford University Press, Oxford (1998)
14. Worden, K., Manson, G.: On the identification of hysteretic systems. Part I: fitness landscapes and evolutionary identification. Mech. Syst. Signal Process. **29**, 201–212 (2012)
15. Press, W.H., Teukolsky, S.A., Vetterling, W.T., Flannery, B.P.: Numerical Recipes: The Art of Scientific Computing, 3rd edn. Cambridge University Press, Cambridge (2007)

16. Qin, A.K., Suganthan, P.N.: Self-adaptive differential evolution algorithm for numerical optimization. In: Proceedings of IEEE Congress on Evolutionary Computation (CEC 2005), Edinburgh (2005)
17. Huang, V.L., Qin, A.K., Suganthan, P.N.: Self-adaptive differential evolution algorithm for constrained real-parameter optimization. In: Proceedings of IEEE Congress on Evolutionary Computation (CEC 2006), Vancouver, pp. 17–24 (2006)
18. http://www.icsi.berkeley.edu/~storn/code.html: (Accessed 27th October 2009)
19. Box, G.E.P., Jenkins, G.M.: Time Series Analysis, Forecasting and Control. Holden-Day, San Francisco (1976)
20. Abdulla, F.A., Lettenmaier, D.P., Liang, X.: Estimation of the ARNO model baseflow parameters using daily streamflow data. J. Hydrol. **222**, 37–54 (1999)
21. Bates, D.M., Watts, D.G.: Nonlinear Regression Analysis and Its Applications. Wiley, New York (1988)
22. Bard, Y.: Nonlinear Parameter Estimation. Academic, New York (1974)
23. Worden, K., Becker, W.E.: On the identification of hysteretic systems. Part II: Bayesian sensitivity analysis and parameter confidence. Mech. Syst. Signal Process. **29**, 213–227 (2012)
24. Mohammad, K.S., Worden, K., Tomlinson, G.R.: Direct parameter estimation for linear and nonlinear structures. J. Sound Vib. **152**, 471–499 (1992)

A Surrogate-Model-Assisted Evolutionary Algorithm for Computationally Expensive Design Optimization Problems with Inequality Constraints

Bo Liu, Qingfu Zhang, and Georges Gielen

Abstract The surrogate model-aware evolutionary search (SMAS) framework is a newly emerged model management method for surrogate-model-assisted evolutionary algorithms (SAEAs), which shows clear advantages on necessary number of exact evaluations. However, SMAS aims to solve unconstrained or bound constrained computationally expensive optimization problems. In this chapter, an SMAS-based efficient constrained optimization method is presented. Its major components include: (1) an SMAS-based SAEA framework for handling inequality constraints, (2) a ranking and diversity maintenance method for addressing complicated constraints, and (3) an adaptive surrogate model updating (ASU) method to address many constraints, which considerably reduces the computational overhead of surrogate modeling. Empirical studies on complex benchmark problems and a real-world mm-wave integrated circuit design optimization problem are reported in this chapter. The results show that to obtain comparable results, the presented method only needs 1–10 % of the exact function evaluations typically used by the standard evolutionary algorithms with popular constraint handling techniques.

Keywords Surrogate model assisted evolutionary computation • Constrained optimization • Constraint handling • Expensive optimization • Gaussian Process • Surrogate modeling • mm-wave IC synthesis

MSC codes: 00A06, 41A30, 60G15, 62L05, 68T20

B. Liu (✉)
Department of Computing, Glyndwr University, Wrexham, UK
e-mail: b.liu@glyndwr.ac.uk; Bo.Liu@esat.kuleuven.be

Q. Zhang
City University of Hong Kong, Kowloon, Hong Kong SAR
e-mail: qingfu.zhang@cityu.edu.hk

G. Gielen
ESAT-MICAS, Katholieke Universiteit Leuven, Leuven, Belgium
e-mail: Georges.Gielen@esat.kuleuven.be

S. Koziel et al. (eds.), *Simulation-Driven Modeling and Optimization*,
Springer Proceedings in Mathematics & Statistics 153,
DOI 10.1007/978-3-319-27517-8_14

1 Introduction

Many industrial design optimization problems require expensive simulations for evaluating their candidate solutions. Employing surrogate models to replace computationally expensive exact function evaluations is a routine approach to address these problems [22]. Because many real-world problems have constraints, constrained expensive optimization is receiving increasing attention. However, most research works focus on computationally expensive local optimization [1, 12]. This chapter focuses on handling global optimization problems with both expensive simulations (i.e., only very few exact evaluations are allowed) and complex inequality constraints, which can be found in many real-world applications such as mm-wave integrated circuit design (e.g., [19]). Constraints in many industrial design optimization problems come from design specifications (e.g., $power \leq 1.5\,\mathrm{mW}$) which are inequality constraints and equality constraints are often self-contained in the simulator (e.g., Maxwell's equations). For complex constraints, this chapter mainly aims at a number of constraints, active constraints (tight design specifications), disconnected feasible region, and complex (sophisticated landscape) constraint functions.

Surrogate-model-assisted evolutionary algorithms (SAEAs) have been accepted as an effective approach to deal with expensive optimization. SAEAs take advantages of both evolutionary algorithms (EAs) and surrogate modeling techniques. To develop an SAEA for constrained expensive optimization problems, one must consider three highly related issues:

- Which surrogate modeling method should be used to approximate the objective function and the constraints?
- Which SAEA framework should be used?
- How should the constraints be handled?

The Gaussian process (GP) modeling is one of the most popular surrogate modeling methods used in SAEAs. Some principled expensive optimization approaches with the GP model and with prescreening, such as the efficient global optimization (EGO) method [10], have been well investigated and documented. Moreover, very few empirical parameters are necessary in a GP model, making the surrogate modeling more controllable. Due to these, the GP modeling is adopted for approximating the objective function and the constraints.

To deal with the second issue, several SAEA frameworks have been proposed for accommodating surrogate models. Successful examples include the surrogate-model-assisted memetic evolutionary search (SMMS) framework [16, 34], the meta-model-assisted EA (MAEA) framework [5], and the surrogate model-aware evolutionary search (SMAS) framework [20]. These frameworks balance the surrogate model quality and the optimization efficiency in different manners and have been tested mainly on unconstrained optimization problems. The SMAS framework considers EA-driven function optimization and high-quality surrogate model construction at the same time by controlling the locations of the generated candidate solutions. It has shown clear advantages (up to eight times fewer exact evaluations) over the SMMS and the MAEA frameworks on a set of widely used unconstrained test instances with 20–50 variables in terms of solution quality with a limited number of exact function evaluations [20]. Therefore, SMAS is especially suitable for quite expensive industrial design optimization problems. For this reason, the SMAS framework is selected.

With regard to constraint handling, a number of techniques have been suggested and used in EAs for general (often *inexpensive*) constrained optimization. Besides static penalty function-based methods and the superiority of the feasibility (SF) method [3], some advanced constraint handling methods have been developed for handling complex constraints [21, 25, 31, 32], such as the self-adaptive penalty function-based methods [31], stochastic ranking-based methods [25], and multi-objective ranking-based methods [32]. All these techniques aim at maintaining the population diversity while driving their populations from infeasible region to feasible one by adaptively trading off the objective function optimization and the total constraint violation minimization. In the context of expensive optimization, some general constraint handling methods have successfully been applied to constrained expensive optimization (e.g., [5, 7]). Several prescreening methods have been generalized from unconstrained expensive optimization to constrained expensive optimization [5, 6, 28]. Some modeling methods and model updating methods for constraint function satisfaction and objective function optimization have been developed [2, 14, 26]. In addition, surrogate-model-assisted expensive integer nonlinear programming has been investigated [11].

One focus of this chapter is to handle complex constraints in an efficient manner (i.e., using the SMAS framework). It is not straightforward to combine the above advanced constraint handling methods for inexpensive and expensive constrained optimization with SMAS. The diversity maintenance methods in most advanced constraint handling methods rely on the population updating of a standard EA [21], while the population updating of SMAS is completely different and is critical for its efficiency.

Another focus is to reduce the computational overhead of surrogate modeling. Independent modeling of the constraint functions is needed for constrained expensive optimization problems [5]. When the number of decision variables is large (e.g., 20–50 variables), surrogate model construction itself may cost a few minutes for a single function in some cases (e.g., [20]), and it should be conducted at each iteration for both the objective function and all the constraints. Thus, the computational cost of surrogate modeling can be tremendous, especially for problems with many constraints.

To address these challenges, an improved SMAS framework for efficient constrained expensive optimization, a diversity maintenance method for the SMAS framework to handle complex constraints, and an adaptive surrogate model updating (ASU) method for adaptively saving the computational overhead of surrogate modelling are introduced. Using these three techniques, a Gaussian Process SAEA for computationally expensive inequality constrained optimization problems (GPEEC) is constructed. Empirical studies on 8 benchmark problems that are challenging in terms of constraint handling, a self-developed 20-dimensional benchmark problem whose objective function is highly multimodal and whose constraint function is very complex, and a real-world mm-wave integrated circuit design optimization problem are used as examples. Results show that comparable solution quality is obtained compared to the state-of-the-art constrained optimization methods (without surrogate models), and that only 1–10 % of the number of exact function evaluations are needed compared to the standard EA with the popular SF method.

The remainder of this chapter is organized as follows. Section 2 introduces the basic techniques. The general framework of GPEEC is then presented in Section 3. Sections 4–6 provide details of the algorithm. Section 7 presents the experimental results of GPEEC. The parameter settings of GPEEC are also discussed. The summary is presented in Section 8.

2 Problem Definition and Basic Techniques

2.1 Problem Definition

The following constrained optimization problem is considered in this chapter:

$$\begin{aligned} &\text{minimize} \;\; f(\mathbf{x}) \\ &\text{subject to} \;\; g_i(\mathbf{x}) \leq 0, \; i = 1, \ldots, m. \\ &\qquad\qquad\;\; \mathbf{x} \in [a, b]^d, \end{aligned} \tag{1}$$

where $f(\mathbf{x})$ is the objective function, $g_i(\mathbf{x}) \leq 0$ $(i = 1, \ldots, m)$ are the constraints, and $[a, b]^d$ is the search region. We assume that some constraints $g_i(\mathbf{x}) \leq 0$ can be active. In other words, these constraints become almost equalities at the globally optimal solution. The problem can have disconnected feasible regions, the function of $f(\mathbf{x})$ can be highly multimodal and the function landscape of $g_i(\mathbf{x})$ can be quite complex. We further assume that the calculations of $f(\mathbf{x})$ and the different $g_i(\mathbf{x})$ can be done in a single simulation, which is the case for many real-world expensive optimization problems (e.g., [18]), or can be done in parallel considering the rapid development of parallel computation techniques.

2.2 GP Modeling

To model an unknown function $y = f(\mathbf{x}), \mathbf{x} \in R^d$, the GP modeling assumes that $f(\mathbf{x})$ at any point $\mathbf{x}$ is a Gaussian random variable $N(\mu, \sigma^2)$, where μ and σ are two constants independent of $\mathbf{x}$. For any $\mathbf{x}$, $f(\mathbf{x})$ is a sample of $\mu + \epsilon(\mathbf{x})$, where $\epsilon(\mathbf{x}) \sim N(0, \sigma^2)$. For any $\mathbf{x}, \mathbf{x}' \in R^d$, $c(\mathbf{x}, \mathbf{x}')$, the correlation between $\epsilon(\mathbf{x})$ and $\epsilon(\mathbf{x}')$, depends on $\mathbf{x} - \mathbf{x}'$. More precisely,

$$c(\mathbf{x}, \mathbf{x}') = \exp(-\sum_{i=1}^{d} \theta_i |x_i - x_i'|^{p_i}), \tag{2}$$

where parameter $1 \leq p_i \leq 2$ is related to the smoothness of $f(\mathbf{x})$ with respect to x_i, and parameter $\theta_i > 0$ indicates the importance of x_i on $f(\mathbf{x})$. More details of the GP modeling can be found in [24].

2.2.1 Hyper Parameter Estimation

Given K points $\mathbf{x}^1, \ldots, \mathbf{x}^K \in R^d$ and their f-function values $y^1, \ldots, y^K$, then the hyper parameters μ, σ, $\theta_1, \ldots, \theta_d$, and $p_1, \ldots, p_d$ can be estimated by maximizing the likelihood that $f(\mathbf{x}) = y^i$ at $\mathbf{x} = \mathbf{x}^i$ $(i = 1, \ldots, K)$ [10]:

$$\frac{1}{(2\pi\sigma^2)^{K/2}\sqrt{det(C)}} \exp\left[-\frac{(\mathbf{y} - \mu\mathbf{1})^T C^{-1}(\mathbf{y} - \mu\mathbf{1})}{2\sigma^2}\right] \tag{3}$$

where C is a $K \times K$ matrix whose (i,j)-element is $c(\mathbf{x}^i, \mathbf{x}^j)$, $\mathbf{y} = (y^1, \ldots, y^K)^T$ and $\mathbf{1}$ is a K-dimensional column vector of ones.

To maximize (3), the values of μ and σ^2 must be:

$$\hat{\mu} = \frac{\mathbf{1}^T C^{-1}\mathbf{y}}{\mathbf{1}^T C^{-1}\mathbf{1}} \tag{4}$$

and

$$\hat{\sigma}^2 = \frac{(\mathbf{y} - \mathbf{1}\hat{\mu})^T C^{-1}(\mathbf{y} - \mathbf{1}\hat{\mu})}{K}. \tag{5}$$

Substituting (4) and (5) into (3) eliminates the unknown parameters μ and σ from (3). As a result, the likelihood function depends only on θ_i and p_i for $i = 1, \ldots, d$. Equation (3) can then be maximized to obtain estimates of $\hat{\theta}_i$ and $\hat{p}_i$. The estimates $\hat{\mu}$ and $\hat{\sigma}^2$ can then readily be obtained from (4) and (5).

2.2.2 The Best Linear Unbiased Prediction and Predictive Distribution

Given the hyperparameter estimates $\hat{\theta}_i$, $\hat{p}_i$, $\hat{\mu}$, and $\hat{\sigma}^2$, one can predict $y = f(\mathbf{x})$ at any untested point $\mathbf{x}$ based on the f-function values y^i at $\mathbf{x}^i$ for $i = 1, \ldots, K$. The best linear unbiased predictor of $f(x)$ is [10, 27]:

$$\hat{f}(\mathbf{x}) = \hat{\mu} + \mathbf{r}^T C^{-1}(\mathbf{y} - \mathbf{1}\hat{\mu}) \tag{6}$$

and its mean squared error is:

$$s^2(\mathbf{x}) = \hat{\sigma}^2[1 - \mathbf{r}^T C^{-1}\mathbf{r} + \frac{(1 - \mathbf{1}^T C^{-1}\mathbf{r})^2}{\mathbf{1}^T C^{-1}\mathbf{r}}] \tag{7}$$

where $\mathbf{r} = (c(\mathbf{x}, \mathbf{x}^1), \ldots, c(\mathbf{x}, \mathbf{x}^K))^T$. $N(\hat{f}(\mathbf{x}), s^2(\mathbf{x}))$ can be regarded as a predictive distribution for $f(\mathbf{x})$ given the function values y^i at $\mathbf{x}^i$ for $i = 1, \ldots, K$.

More detailed derivations about the hyperparameter estimation and prediction can be found in [9].

2.2.3 Lower Confidence Bound

We consider minimization of $f(\mathbf{x})$ in this chapter. Given the predictive distribution $N(\hat{f}(\mathbf{x}), s^2(\mathbf{x}))$ for $f(\mathbf{x})$, a lower confidence bound (LCB) of $f(\mathbf{x})$ can be defined as [4]:

$$f_{lcb}(\mathbf{x}) = \hat{f}(\mathbf{x}) - \omega s(\mathbf{x}) \tag{8}$$

where ω is a predefined constant. In the GPEEC algorithm, $f_{lcb}(\mathbf{x})$ instead of $\hat{f}(\mathbf{x})$ itself is used to measure the quality of $\mathbf{x}$. The use of LCB can balance the search between promising areas (i.e., with low $\hat{f}(\mathbf{x})$ values) and less explored areas (i.e., with high $s(\mathbf{x})$ values). Following the suggestion in [4, 5], $\omega = 2$ is used for balancing the exploration and exploitation of LCB.

2.3 *Differential Evolution*

The differential evolution (DE) algorithm is used as the search engine in the GPEEC algorithm. DE is an effective and popular global optimization algorithm. It uses a differential operator to create new candidate solutions [23]. There are quite a few different DE variants and DE/best/1 is used here to generate new solutions for prescreening. The DE/best/1 mutation uses the current best solution as the base vector, so as to increase the speed of generating promising candidates.

Suppose that P is a population and the best individual in P is $\mathbf{x}^{best}$. Let $\mathbf{x} = (x_1, \ldots, x_d) \in R^d$ be an individual solution in P. To generate a child solution $\mathbf{u} = (u_1, \ldots, u_d)$ for $\mathbf{x}$, DE/best/1 works as follows.

A donor vector is first produced by mutation:

$$\mathbf{v} = \mathbf{x}^{best} + F \cdot (\mathbf{x}^{r_1} - \mathbf{x}^{r_2}) \tag{9}$$

where $\mathbf{x}^{r_1}$ and $\mathbf{x}^{r_2}$ are two different solutions randomly selected from P and also different from $\mathbf{x}^{best}$. $F \in (0, 2]$ is a control parameter, often called the scaling factor [23]. Then the following crossover operator is applied to produce $\mathbf{u}$:

1. Randomly select a variable index $j_{rand} \in \{1, \ldots, d\}$,
2. For each $j = 1$ to d, generate a uniformly distributed random number *rand* from (0, 1) and set:

$$u_j = \begin{cases} v_j, \text{ if } (rand \leq CR) | j = j_{rand} \\ x_j, \text{ otherwise} \end{cases} \tag{10}$$

where $CR \in [0, 1]$ is a predefined constant called the crossover rate.

3 Algorithm Framework

As described above, GPEEC adopts and improves the SMAS framework originally proposed for unconstrained expensive optimization [20]. GPEEC maintains a database and iteratively updates surrogate models for the objective function and the constraints until a stopping criterion is met.

- The database is composed of all the evaluated solutions and their exact function values. At the first step, α solutions from $[a, b]^d$ are sampled by an experimental design method and are evaluated (through exact function evaluations) to form the initial database.
- Surrogate models for the objective function and each constraint are constructed at the first step and are then updated at the consecutive iterations.

In each iteration, GPEEC works as follows:

Step 1: Selecting working population: Select the λ best solutions from the current database to form a population P.

Step 2: Diversity maintenance: Check the diversity of P. When necessary, conduct diversity enhancement operations on P.

Step 3: Generating child population: Apply evolutionary operators on P to generate λ child solutions.

Step 4: Prescreening of child solutions: Adaptively update the surrogate models for the objective function and for the constraint functions using information extracted from the database and the available surrogate models. Estimate the quality of the λ child solutions generated in Step 3 based on the updated surrogate models and prescreening methods.

Step 5: Function evaluation: Perform exact function evaluation on the estimated best candidate solution $\mathbf{x}^b$ from Step 4 and then add $\mathbf{x}^b$ and its exact function values to the database.

Since the working population P consists of the best solutions in the current database, the search concentrates on the current promising subregion, which is moving in the search space for exploration. This is necessary because the computational budget for exact function evaluations is very limited. In surrogate modeling, training data points that are close to the child solutions can be obtained so as to construct high-quality surrogate models. This will further be illustrated in Section 5.

4 Constraint Handling and Diversity Maintenance

This section explains and discusses the implementation of Step 1 and Step 2 of GPEEC for handling constraints.

4.1 Basic Constraint Handling Method

It is natural to use the information of constraint satisfaction to rank the candidate solutions. For simplicity and efficiency, a revised SF method proposed in [3] is adopted in Step 1 for selecting the λ best candidate solutions from the database. The ranking rules based on SF are as follows:

1. Feasible solutions rank higher than infeasible ones.
2. Feasible solutions are ranked solely based on their objective function values in ascending order.
3. Infeasible solutions are ranked solely based on their total constraint violation values ($\sum_{i=1}^{m} max\{0, g_i(\mathbf{x})\}$) in ascending order.

4.2 The Diversity Maintenance Method

Many constrained optimization procedures have the following phases [21]: (1) The population moves towards the feasible region and the main driving force is the minimization of the total amount of constraint violations. (2) A part of the population is in the feasible region and the other part is in the infeasible region, and the main driving forces are both the minimization of the total constraint violations and the optimization of the objective function. (3) Most candidate solutions of the population are in the feasible region and the main driving force is optimization of the objective function. An early stage and a late stage are used, which are separated by T, the number of feasible solutions generated so far, which should be set to several multiples of λ (the reason is explained in Section 7). This indicates that at the end of the early stage, most candidates are feasible while a substantial effort is used for objective function optimization; the late stage, on the other hand, mainly focuses on optimizing the objective function.

There are d decision variables x_i ($i = 1, \ldots, d$) in Problem (1). Let $\mathbf{x}^*$ be its globally optimal solution and let P^i contain the x_i values of all the solutions in the current population P. Ideally, each P^i will converge to the x_i value in $\mathbf{x}^*$. However, due to complex constraints and other reasons, some P^i may get trapped at some wrong position and thus lose its diversity at some search stages. If P^i is trapped at a value, x_i is called a trapped variable. Figure 1 provides an example with two variables, illustrating why the trapping of some P^i may happen. In this example, Ω_1 and Ω_2 are two parts of the feasible region. For each value of x_2, the feasible range

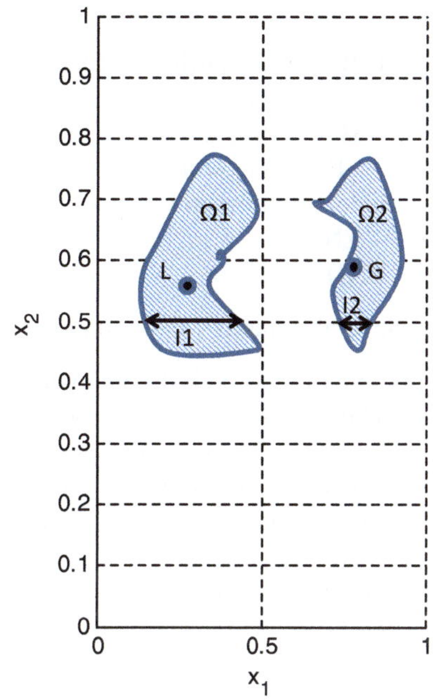

Fig. 1 An illustrative figure of the trapping of variables

of x_1 is different. In Figure 1, the feasible range of x_1 for each fixed x_2 value consists of two disconnected intervals since the feasible region is disconnected. When the early stage begins, the major driving force is the minimization of the total amount of constraint violations. When there are many constraints in Problem (1), it is very likely that the total amount of constraint violations drops significantly when some elements in P^1 enter the interval I_1 instead of I_2. Therefore, once some elements in P^1 fall in I_1, most elements in P^1 will enter I_1 very soon. Since I_2 is not connected with I_1, it is very difficult for x_1 to get out of I_1 just by reproduction operators such as crossovers and small mutations, and P^1 may lose its diversity and then get stuck at a value in I_1 because of the objective function optimization near the end of the early stage. To deal with this issue, the following method is used to improve the diversity.

The variables $x_1, \ldots, x_d$ in $\mathbf{x}$ are treated separately in the DM procedure.[1] For each x_k, its diversity in P, $D_k(P)$, is:

$$D_k(P) = max_{\tilde{\mathbf{x}}=(\tilde{x}_1,\ldots,\tilde{x}_d)\in TP}\{|\tilde{x}_k - x_k^{best}|\}$$

where TP contains the top ν solutions in P based on the SF ranking, $\mathbf{x}^{best} = (x_1^{best}, \ldots, x_d^{best})$ is the best solution in P, and ν is a control parameter.

[1] We assume that all the decision variables are at least related to one of the constraints; otherwise, they can be easily eliminated from the DM method.

The DM method (Step 2 in GPEEC in Section 3) works as follows:

If the total number of feasible solutions in the current database $\leq T$
FOR $k = 1, \dots, d$
IF $D_k(P) \leq \varepsilon$.
For every solution $\widetilde{x} = (\widetilde{x}_1, \dots, \widetilde{x}_d) \in P$, reset $\widetilde{x}_k$ to a uniformly distributed random number in $[a, b]$.
END
END

Several remarks can be made on the above method:

- The DM method is applied when the total number of feasible solutions generated so far is less than a predefined number T. In other words, this method will not be used at the late search stage. A major consideration is that the algorithm should focus on optimizing the objective function starting from a diversity maintained (if necessary) population at the late stage.
- When $D_k(P) \leq \varepsilon$, the x_k values of the top ν solutions in P are very close. It indicates that the current population P does not have a good diversity in x_k. Note that *TP* is the best subset of P and thus the current database. To prevent the search from being trapped in a locally optimal area, the value of x_k is randomly reset for each solution in P.
- At the early search stage, some $x_i(s)$ (often trapped variables) converge much faster than others (see definition and explanation of the trapped variables). Therefore, in the for loop of the DM method, not all the variables will be re-sampled from $[a, b]$. The newly generated solutions still inherit variables with good diversity in the current population. The re-sampling of trapped variables after the detection, in contrast to the random sampling in the beginning of the early stage, is effective for jumping out of the premature convergence. As said above, substantial effort must be spent on objective function optimization when the trapped variable can be detected (i.e., it converges to a very narrow range ε). Hence, many feasible solutions should have been generated and P should be around the feasible region, i.e., most constraints are satisfied. This implies that after re-sampling, each interval of a trapped variable has nearly equal chance to be selected considering the feasibility and the value of the objective function.
- The DM method can easily be used in the SMAS framework, since it improves the population diversity by only using information extracted from the decision space, rather than trading off the objective function value and the constraint satisfaction like most advanced constrained optimization methods.

The parameters in the DM method are set as follows: $T = 5 \times \lambda$, $\nu = 10$ and $\varepsilon = 0.1$ assuming a $[-10, 10]$ search region (we can make this assumption come true by scaling). More details are in Section 7.

5 Surrogate Modeling

This section discusses the implementation of Step 4 of GPEEC for the surrogate modeling and prescreening.

5.1 *Surrogate Modeling Based on Improved SMAS*

The idea of SMAS [20] is used in GPEEC. Different from other SAEA frameworks, SMAS concentrates both its search and its surrogate modeling on the current promising region and gradually moves the promising region for exploration. It uses the λ best candidate solutions to form its working population, and only evaluates the exact function value of the estimated best candidate solution.

Training data points with a higher quality can be generated using the SMAS framework than using an SAEA framework which relies on a standard EA. The solutions in P (Step 1) are not necessarily far away from one another since they are the current best candidate solutions. Also, because at most a single new solution enters P in each iteration, the selected estimated best solutions (which are generated from P in step 3 and which will serve as training data points) in several consecutive iterations will not be far away from one another. As a consequence, most training data points are also in or near the current promising region. Therefore, a high-quality surrogate model for this region can be constructed for prescreening newly generated solutions. That is why SMAS is a surrogate model-aware search mechanism. In contrast, new solutions often spread in different regions in standard EAs and thus often no sufficient number of training data points are around candidate solutions to be prescreened, affecting the surrogate model quality negatively.

To select training data points, the median of the λ new solutions for each decision variable is computed to construct a vector $\mathbf{m}_v$. The τ available training data points that are nearest to $\mathbf{m}_v$ are selected to construct the surrogate model. τ should be set between $5 \times d$ and $7 \times d$. $\tau = 6 \times d$ is used in all the experiments.

Note that the LCB prescreening is used only for the objective function, while for constraints the predicted value is used (i.e., $\omega = 0$). The reason is to prevent that many near-feasible solutions are selected, since in SMAS only a single candidate solution is selected and evaluated in each iteration.

5.2 *The Adaptive Surrogate Model Updating Method*

The ASU method in GPEEC adaptively decides whether the surrogate model will be updated or not, with the goal of reducing the computational overhead of surrogate modeling. This is important for GPEEC, because the computational overhead of surrogate model construction for constrained expensive optimization problems can be tremendous (see Section 1).

Since the goal is to minimize $f(\mathbf{x})$, the quality of the surrogate model for $f(\mathbf{x})$ does matter. For this reason, the surrogate model for $f(\mathbf{x})$ is updated at every iteration. As to the surrogate model for each $g_i(\mathbf{x})$, there are different considerations for the different search stages. In the early stage, the purpose of most surrogate models for $g_i(\mathbf{x})$ is to estimate $max\{0, g_i(\mathbf{x})\}$. Therefore, one needs high-quality models at this stage in order to estimate the total amount of constraint violations reliably. In contrast, when the search has almost entered the feasible region (the late stage), the surrogate models for $g_i(\mathbf{x})$ serve the purpose as long as the models can distinguish feasible solutions from infeasible ones. Therefore, it is not necessary to update the models of $g_i(\mathbf{x})$ at every iteration in the late stage. For those iterations where the surrogate model updating is not conducted, previous models are used for estimating the constraint function values. Therefore, the ASU method is applied when the total number of feasible solutions in the current database is larger than the parameter T which was introduced in Section 4.

At each iteration t at the late stage ($t \geq T$), the surrogate model is updated for $g_i(\mathbf{x})$ if

- $remaining(t, t_c) = 0$; or
- one of the η most recently evaluated points does not satisfy $g_i(\mathbf{x}) \leq 0$

where t_c and η are control parameters.

Several remarks can be made:

- The update of the surrogate models for constraint functions is conducted at every iteration in the early stage and after every t_c iterations in the late stage to reinforce the reliability of the surrogate models.
- At the late stage, since many candidate solutions in P are deep inside the feasible region, it is unlikely that all of the λ new solutions are infeasible. Thus, there is a high probability that some infeasible solution (considering $g_i(\mathbf{x})$) among the λ new solutions should be predicted as feasible and ranked as the best. In this case, it is possible that the search region is near the boundary of $g_i(\mathbf{x}) = 0$ or that the previous GP model cannot work. Hence, the surrogate model is updated for $g_i(\mathbf{x})$ in this case. For $j \in \{1, \ldots, m\}$ and $j \neq i$, the surrogate model of $g_j(\mathbf{x})$ is then not updated.

$t_c = 10$ and $\eta = 5$ are used in all the experiments. Their settings are discussed in Section 7.

6 Implementation Details

Some implementation details included in Step 1, Step 3 and Step 5 are as follows. Note that various alternative methods can be investigated and applied in the general GPEEC framework:

- In Step 1, the selected experimental design method is Latin hypercube sampling (LHS) [29]. LHS is widely used for the initial database generation in SAEA research. α (the number of initial samples) is often relatively small and the empirical rule for setting α for the SMAS framework is in [20], which is also applicable to GPEEC.
- The selected EA operators in Step 3 are the DE/best/1 mutation operator and the DE crossover operator. Section 2 has provided more details.
- In Step 5, sometimes a modified form of the estimated best candidate solution ($\mathbf{x}^b$) is used. The purpose is to avoid the repeated expensive evaluation of the same candidate solution and to perform local search. A Gaussian distributed random number with zero mean and with 5 % of the search range of each decision variable as variance is added to modify $\mathbf{x}^b$ if $\mathbf{x}^b$ has been evaluated before.

7 Experimental Studies

7.1 Test Problems and Parameter Settings

The GPEEC algorithm is tested with ten problems, which are shown in Table 1. First, 8 hard benchmark test problems for constrained optimization are used (G1 to G10 from the CEC 2006 special session on constrained real-parameter optimization [15], except G3 and G5 which use equality constraints: GN1–GN8 correspond to G1, G2, G4, G6, G7, G8, G9, G10 in [15], respectively), involving many constraints, disconnected feasible regions and active constraints. In addition, to test the ability of GPEEC on handling complex objective and constraint functions, a 20-dimensional test function is constructed with the Ackley and Griewank test functions [30] (please see the Appendix). Finally, a real-world problem from the mm-wave integrated circuit (IC) design field is used to demonstrate the capabilities of GPEEC.

Table 1 Test problems used in the experimental studies

Problem	Opt.	d	ρ (%)	a	m	λ	N_{eval}
GN1 (G1 in [15])	−15	13	0.011	6	9	40	1000
GN2 (G2 in [15])	−0.8036	20	99.99	1	2	40	2000
GN3 (G4 in [15])	−30665.54	5	52.12	2	6	30	800
GN4 (G6 in [15])	−6961.81	2	0.006	2	2	30	1000
GN5 (G7 in [15])	24.31	10	0.00	6	8	40	1000
GN6 (G8 in [15])	−0.0958	2	0.86	0	2	30	800
GN7 (G9 in [15])	680.63	7	0.52	2	4	30	800
GN8 (G10 in [15])	7049.33	8	0.00	3	6	40	1000
GN9	0	20	0.00	0	2	40	1500

Opt. is the globally optimal objective function value of each problem

In Table 1, ρ is an estimate of the ratio of the feasible space to the entire search space. a is the number of active constraints and m is the total number of constraints. All the parameter setting rules of GPEEC have been described above. For GN1–GN9, $\alpha = 50$ is used for problems with 10–20 variables, and $\alpha = 40$ for problems with less than ten variables. The population size λ and the number of exact function evaluations N_{eval} are shown in Table 1. Note that we assume that the calculations of $f(\mathbf{x})$ and the different $g_i(\mathbf{x})$ can be done in a single simulation or can be done in parallel, which is common to many expensive optimization problems nowadays. In all the experiments, DE/best/1 is used and the crossover rate CR is set to 0.8 according to the suggestions in [23]. Note that a relatively large scaling factor F (e.g., $F \in [0.75, 0.95]$) is necessary to promote exploration for the SMAS framework, in order to avoid getting stuck in local optima. F is set to 0.8 in the experiments. ω in LCB prescreening is set to 2 according to [5].

The experiments are carried out on a 2.66 GHz computer with 7.8 Gb RAM in the MATLAB environment on the Linux system. The ooDACE toolbox [8] is used for GP modeling.

7.2 *The GPEEC Performance and Analysis*

7.2.1 Reference Results

To evaluate the solution quality of GPEEC, a state-of-the-art constrained optimization method (without surrogate modeling) is used to provide a reference result. The method is the self-adaptive penalty function (SAPF) method from [31]. In [31], the real-coded genetic algorithm is used as the search engine. For fair comparison, the same DE as in GPEEC is used with $\lambda \times N_{eval}$ evaluations (the average is about 40,000 evaluations). Twenty runs are performed for each case. The results are shown in Table 2.

Table 2 Statistics of the best function values obtained by the first reference method SAPF for GN1–GN9 over 20 runs ($\lambda \times N_{eval}$ function evaluations)

Problem	Best	Worst	Mean	Median	Std	R_{inf}
GN1	−15	−12.01	−14.51	−15	1.06	0
GN2	−0.75	−0.53	−0.61	−0.60	0.06	0
GN3	−30665.53	−30664.04	−30665.07	−30665.03	0.16	0
GN4	−6961.22	−6898.73	−6935.50	−6937.11	18.56	0
GN5	25.37	29.57	26.60	26.02	1.49	0
GN6	−0.0958	−0.0958	−0.0958	−0.0958	1.9e−17	0
GN7	680.64	680.65	680.64	680.64	0.0034	0
GN8	8075.23	15419.02	9545.32	8681.38	2.30e+3	0
GN9	0.61	0.83	0.74	0.75	0.0761	0

R_{inf} refers to the percentage of runs providing an infeasible final result

Table 3 Statistics of the best function values obtained by the second reference method SF for GN1–GN9 over 20 runs ($\lambda \times N_{eval}$ function evaluations)

Problem	Best	Worst	Mean	Median	Std	R_{inf}
GN1	−15	−11	−13.44	−13	1.34	5 %
GN2	−0.77	−0.42	−0.62	−0.62	0.11	0
GN3	−30665.53	−30664.30	−30664.05	−30665.03	0.41	0
GN4	−6961.81	−6944.68	−6959.42	−6961.81	6.48	30 %
GN5	24.31	512.25	73.65	24.31	154.12	0
GN6	−0.0958	−0.0958	−0.0958	−0.0958	1.03e−17	0
GN7	680.63	680.63	680.63	680.63	5.73e−4	0
GN8	7050.92	20101.53	9880.37	7222.31	4295.13	55 %
GN9	0	0.020	0.011	0.015	0.0095	0

R_{inf} refers to the percentage of runs providing an infeasible final result.

Compared to [31], the results of SAPF for GN1, GN2, GN4, and GN8 are worse than the published results and the results of GN3, GN5, GN6, GN7 are better than the published results. This is because different search engines with different parameter settings are used inside SAPF.

Then, as a second reference, the SF method [3] is used with DE. Also, $\lambda \times N_{eval}$ evaluations are used. The results are shown in Table 3. Note that the results of some runs are infeasible, so the statistics only considers the runs that provide feasible final results.

It can be seen that some SF results for GN1, GN4, and GN8 are infeasible. For GN1 and GN5, it can be observed that premature convergence largely harms the performance in some runs. The SAPF method shows clear advantages on these problems. On the other hand, for GN7 and GN9, the SF method performs better than the SAPF method. This indicates that in some cases when the additional diversity enhancement does not help much, the SF method shows its advantage of fast convergence (being more efficient). Tables 2 and 3 will be used for comparison with GPEEC.

7.2.2 The Effect of the DM Method

The method to simulate the SMAS framework is shown in [20]. The same method is used here for the GPEEC framework. To simulate it, the GP modeling and prescreening are removed. Instead, exact function evaluations are conducted on all the λ child solutions in each iteration, and randomly select one from the top β solutions. $\beta = 5$ is used. By this simulation, the search (optimization) ability of the GPEEC framework can be analyzed and then it can be seen whether the surrogate modeling and prescreening/prediction work as expected or not. We first simulate the GPEEC framework without the DM method (without Step 2 in Section 3). Similar results are obtained compared to Table 3, the SF method. Then, the DM method is added to simulate the GPEEC framework. These results are in Table 4.

Table 4 Statistics of the best function values obtained by the simulated GPEEC framework for GN1–GN9 over 20 runs

Problem	Best	Worst	Mean	Median	Std	R_{inf}
GN1	−15	−13	−14.51	−14.92	0.71	0
GN2	−0.73	−0.50	−0.61	−0.63	0.09	0
GN3	−30665.53	−30664.17	−30664.81	−30665.04	0.42	0
GN4	−6961.81	−6928.37	−6957.16	−6961.74	10.41	0
GN5	24.47	25.08	24.73	24.62	0.25	0
GN6	−0.0958	−0.0958	−0.0958	−0.0958	1.02e−17	0
GN7	680.63	680.66	680.64	680.64	0.0095	0
GN8	7273.61	8331.37	7457.52	7335.19	318.99	0
GN9	0	0.096	0.026	0.017	0.028	0

R_{inf} refers to the percentage of runs providing an infeasible final result

Compared to Table 2 (the SAPF method), it can be seen that, except for GN1, GN2, GN3, and GN6, where the results are comparable to that of the SAPF method, the results are better than the SAPF method for the other five test problems. For GN5, GN8, and GN9 considerably better results are observed. Compared to Table 3 (the SF method), it can be seen that no infeasible solution has been provided by GPEEC. For GN1, GN4, GN5, and GN8, which are difficult to solve by the SF method, the DM method is seen to be a very effective solution method. Therefore, it can be concluded that the simulated GPEEC framework incorporates both a high constraint handling ability and a fast convergence, which is effective for problems with complex constraints. Moreover, experiments show that the DM method can be triggered more than once and to more than one variable for problems with complex constraints, while it often is not triggered for problems that can be solved well by the SF method.

7.2.3 Integrating the GP Modeling

GPEEC with surrogate modeling and prescreening but without the ASU method is then tested. N_{eval} evaluations are used (see Table 1). The results are in Table 5.

It can be seen from Table 5 that: (1) All the final solutions are feasible. (2) Except for GN2, the results are comparable to the results provided by the SAPF method. For GN2 with 20 variables, experiments have shown that GP modeling is not very fit for the properties of this function. For GN2 with fewer variables, on the other hand, the performance of GP modeling is good. It is no surprise that a single surrogate modeling method has difficulty to work well on all kinds of problems [13, 16], and the GPEEC method is compatible with hybrid surrogate models. (3) Thanks to the improved SMAS framework, the surrogate modeling works as expected. There is a little degradation compared to the simulation results in Table 4 for some problems, but GN5, GN7, and GN8 show better results than the simulation results. The advantages of using surrogate models in terms of solution quality have been

Table 5 Statistics of the best function values obtained by the GPEEC framework without the ASU method for GN1–GN9 over 20 runs (N_{eval} function evaluations)

Problem	Best	Worst	Mean	Median	Std	R_{inf}
GN1	−15.00	−12.93	−14.29	−14.85	0.94	0
GN2	−0.77	−0.38	−0.53	−0.53	0.12	0
GN3	−30665.53	−30664.01	−30664.33	−30664.21	0.43	0
GN4	−6956.70	−6769.34	−6902.68	−6915.19	58.81	0
GN5	24.31	25.01	24.39	24.32	0.22	0
GN6	−0.0958	−0.0932	−0.0954	−0.0958	8.66e−4	0
GN7	680.63	680.63	680.63	680.63	4.56e−4	0
GN8	7056.87	7583.43	7290.59	7247.08	192.75	0
GN9	1.42e−10	0.014	0.006	0.007	0.006	0

R_{inf} refers to the percentage of runs providing an infeasible final result

Table 6 Statistics of the best function values obtained by GPEEC for GN1–GN9 over 20 runs (N_{eval} function evaluations)

Problem	Best	Worst	Mean	Median	Std	R_{inf}	M/S
GN1	−14.99	−12.88	−14.36	−14.96	0.98	0	1610/17 %
GN2	−0.75	−0.40	−0.51	−0.48	0.12	0	1623/27 %
GN3	−30665.42	−30661.14	−30663.79	−30664.21	1.21	0	1869/36 %
GN4	−6961.54	−6764.85	−6886.90	−6898.43	75.31	0	0/0
GN5	24.31	25.03	24.60	24.32	0.36	0	4816/56 %
GN6	−0.0958	−0.0918	−0.0952	−0.0957	0.0012	0	666/29 %
GN7	680.63	680.63	680.63	680.63	9.27e−4	0	2216/58 %
GN8	7050.20	7310.41	7119.88	7097.64	89.20	0	3041/46 %
GN9	1.25e−9	0.72	0.08	2.39e−9	0.24	0	2038/47 %

R_{inf} refers to the percentage of runs providing an infeasible final result. M/S indicates the number of surrogate model constructions using the ASU method, and the percentage compared to updating every surrogate model in each iteration

discussed in [5, 16]. (4) Note that a fixed number of N_{eval} function evaluations are used and the convergence is earlier than that for most problems, which will be illustrated later on.

7.2.4 The GPEEC Performance and the Effect of the ASU Method

At last, when also integrating the ASU method, the full GPEEC algorithm is tested. N_{eval} function evaluations are used. The results are in Table 6.

In terms of optimality, it can be observed that the results are comparable to the results from Table 5, as well as with the SAPF results from Table 2. In terms of the necessary number of surrogate modeling, less than half or about half the surrogate modeling runs are used in most cases compared to not using ASU. For GN1 (13-dimensional, 9 constraints), GN2 (20-dimensional), and GN9 (20-dimensional), when not using ASU, the surrogate modeling time often costs more than 15 h. When using the ASU method, the surrogate modeling only costs a few hours. The ASU method is especially useful when the exact evaluation is not very expensive but with many constraints, or when the number of decision variables is quite large.

7.2.5 Comparisons

The SF method is widely used in constrained optimization; its main advantages are its simplicity and efficiency [3]. To observe the speed enhancement of GPEEC, GPEEC (Table 6) is compared with SF (Table 3). Note that only the objective function values of the feasible runs of the SF method are used for comparison. In the experiments, N_{eval} evaluations are used for GPEEC, but most convergence happens earlier than that. To mark the convergence, a threshold, δ, is used, which means that after N_{ec} evaluations, the current best solution is feasible and the improvement to the objective function is less than δ after that. Thus, it can be considered that GPEEC converges at N_{ec} evaluations. Because the objective function values for different test problems are in different scales (e.g., GN6 between -0.1 and 0, GN8 about 10^4), the selected δ are shown in Table 7.

To make the comparisons, the following information for GPEEC and SF are reported.

- $G_{N_{ec}}$: the median of the best function values obtained using N_{ec} exact function evaluations by GPEEC;
- $S_{N_{ec}}$: the median of the best function values obtained using N_{ec} exact function evaluations by SF;
- $H_{N_{ec}}$: the number of exact function evaluations needed for SF to achieve $G_{N_{ec}}$. If the final results of the SF method after $\lambda \times N_{eval}$ evaluations are worse than the GPEEC result with N_{ec} exact function evaluations, we denote this as N.A.

The comparison results are shown in Table 8. It can be seen that 6 out of 9 of the problems, tens to even more than a hundred times less exact function evaluations are needed by GPEEC.

Table 7 δ for GN1–GN9

Problem	GN1	GN2	GN3	GN4	GN5
δ	0.1	0.01	1	1	0.1
Problem	GN6	GN7	GN8	GN9	
δ	0.001	0.1	1	0.1	

Table 8 Comparisons between GPEEC and SF for GN1–GN9 over 20 runs

Problem	N_{ec}	$G_{N_{ec}}$	$S_{N_{ec}}$	$H_{N_{ec}}$	Speedup
GN1	268	−14.9	−7.0	N.A.	>150
GN2	1959	−0.47	−0.33	4880	2.5
GN3	209	−30663.7	−29961.1	840	4
GN4	134	−6898.4	−4520.9	1280	10
GN5	920	24.3	1008.4	24640	27
GN6	535	−0.0957	−0.0916	840	1.6
GN7	523	680.4	1738.5	8040	15
GN8	992	7097.6	14615.3	N.A.	>40
GN9	724	0.094	148.29	12800	18

7.2.6 Discussions on Parameter Setting

The parameters of GPEEC can be classified into three categories: (1) parameters for the improved SMAS framework, (2) parameters for the DM method, and (3) parameters for the ASU method.

The parameter setting for the SMAS framework has followed the practice in [17, 20] where the setting has been discussed in detail.

As to parameter setting in the DM method, ε is the threshold for $D_k(P)$ to measure the diversity of the current population on x_k. A small number is needed depending on the range of each decision variable. $\varepsilon = 0.1$ is used assuming a $[-10, 10]$ search region (we can make this assumption come true by scaling). T is used to define the division of the two search stages. In the late stage, the search is conducted mainly in the feasible region and a substantial effort is made to optimize the objective function. Although every solution in the population P will be feasible after λ feasible solutions have been generated due to the ranking rules used in Section 4.1, experiments show that many child solutions produced from P are still infeasible until after $3 \times \lambda$ to $4 \times \lambda$ feasible solutions have been generated. Based on this observation, T is set to be $5 \times \lambda$. ν is used to determine whether a variable is trapped or not. A big ν value may not be necessary. It is suggested to be from 5 to 10. In the following, GN5 and GN7 are used as examples to test the impact of T and ν. It can be seen from Table 9 that if ν and T are selected in the suggested ranges, the DM method performs well and is robust.

As to parameter setting in the ASU method, note that this method is only used when the number of generated feasible solutions is larger than T and the surrogate models for constraint functions are used to differentiate feasible solutions and infeasible ones. η is used to judge if the search is near the boundary of the feasible region. A small η value may lead to a wrong judgement, while a large η value may cause unnecessary model updating and thus may waste computational efforts. It is suggested to be set around 5. t_c defines how often the surrogate model for each constraint function must be updated regularly. Considering the help of η and the

Table 9 Statistics of the best function values obtained by the GPEEC framework without the ASU method for GN5 and GN7 over 20 runs (N_{eval} function evaluations) with different DM parameters

Problem	Best	Worst	Mean	Median	Std	R_{inf}
GN5, $\nu = 5, T = 5 \times \lambda$	24.31	25.21	24.61	24.32	0.35	0
GN5, $\nu = 10, T = 4 \times \lambda$	24.31	25.05	24.57	24.34	0.34	0
GN5, $\nu = 10, T = 6 \times \lambda$	24.31	25.01	24.47	24.32	0.30	0
GN7, $\nu = 5, T = 5 \times \lambda$	680.63	680.63	680.63	680.63	0.0014	0
GN7, $\nu = 10, T = 4 \times \lambda$	680.63	680.63	680.63	680.63	2.7e−4	0
GN7, $\nu = 10, T = 6 \times \lambda$	680.63	680.63	680.63	680.63	4.4e−4	0

R_{inf} refers to the percentage of runs providing an infeasible final result

Table 10 Statistics of the best function values obtained by GPEEC for GN5 and GN7 over 20 runs (N_{eval} function evaluations) with different ASU parameters

Problem	Best	Worst	Mean	Median	Std	R_{inf}
GN5, $t_c = 20, W = 5$	24.31	25.07	24.45	24.32	0.28	0
GN5, $t_c = 10, W = 3$	24.31	25.01	24.59	24.36	0.35	0
GN5, $t_c = 10, W = 7$	24.31	25.01	24.47	24.33	0.27	0
GN7, $t_c = 20, W = 5$	680.63	680.63	680.63	680.63	5.3e−4	0
GN7, $t_c = 10, W = 3$	680.63	680.63	680.63	680.63	8.8e−4	0
GN7, $t_c = 10, W = 7$	680.63	680.64	680.63	680.63	0.0027	0

R_{inf} refers to the percentage of runs providing an infeasible final result

main goal of the late stage, the regular updating does not need to be very frequent. t_c is suggested to be set between 10 and 20. In the following, GN5 and GN7 are used as examples to test if the suggested setting is robust. It can be observed from Table 10 that GPEEC performs well when η and t_c are not very far from the suggested values. When using η in the suggested range, t_c is not sensitive.

7.3 mm-Wave IC Design Optimization Example

This section provides a real-world engineering application of GPEEC: the design optimization of a 60 GHz power amplifier in a 65 nm CMOS technology. At mm-wave frequencies, the simple equivalent circuit models typically used for passive components at low frequencies are no longer accurate, and the way left to the designers is "trial and error." Therefore, the global optimization of mm-wave ICs is very important. However, electromagnetic (EM) simulation is needed in the evaluation of candidate designs, which is computationally expensive. In power amplifier design, the 1 dB compression point ($P_{1\,\mathrm{dB}}$), the power added efficiency ($PAE@P_{1\,\mathrm{dB}}$) and the power gain (G_p) are key performances. In practical design, the goal is often to maximize $P_{1\,\mathrm{dB}}$ or $PAE@P_{1\,\mathrm{dB}}$, with constraints on the other two specifications.

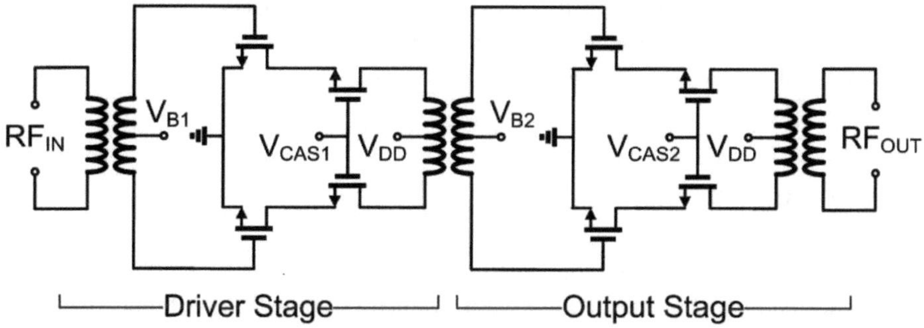

Fig. 2 Schematic of the 60 GHz power amplifier [33]

The problem is defined in (11). The circuit configuration is shown in Figure 2. The design parameters (in total 18) include the inner diameters and metal width of the primary and secondary inductors of every transformer (in total three transformers; each transformer has two inductors), 5 biasing voltages and the number of fingers of the driver stage. The inner diameter has a range from 20 to 100 μm. The metal width has a range from 3 to 10μm. The 5 biasing voltages have ranges from 0.5 to 2 V. The number of fingers can be 2/3/4. This is a simulation-based (black-box) optimization problem, so no explicit analytical expression is available. A Xeon 2.66 GHz computer is used for the synthesis (design optimization). GPEEC is programmed in MATLAB and the simulation is carried out in Cadence and ADS-Momentum (IC and electromagnetic simulation software). All the programs are run on the Linux system. The evaluation of a candidate design of this power amplifier needs 10–13 min using the simulation software to obtain the values of $P_{1\,\mathrm{dB}}$, $PAE@P_{1\,\mathrm{dB}}$, and G_p. The total computational time is restricted to about 2 days to reach the practical requirement of a design automation software tool acceptable in industry.

$$\begin{aligned} &\text{maximize} \quad PAE@P_{1\,\mathrm{dB}} \\ &\text{subject to} \quad P_{1\,\mathrm{dB}} \geq 13\,\mathrm{dBm} \\ &\qquad\qquad\quad G_p \geq 10\,\mathrm{dB} \end{aligned} \tag{11}$$

The initial number of samples α is set to 70. All the other settings are the same as those used in the benchmark problem tests. After 200 exact function evaluations, GPEEC gets the optimized result: $P_{1\,\mathrm{dB}}$ is 14.34 dBm, $PAE@P_{1\,\mathrm{dB}}$ is 9.52 % and G_p is 10.47 dB. The time cost is 41.6 h (wall clock time).

The high quality of this optimized result by GPEEC can be verified by comparing to a manual design [33] using the same circuit structure in the 65 nm technology. The reference result is as follows: $P_{1\,\mathrm{dB}}$ is 10.8 dBm, $PAE@P_{1\,\mathrm{dB}}$ is 4.5 % and G_p is 10.2 dB. It can be seen that the result of GPEEC fully dominates the experience-based manual design result on all the performances. To verify the efficiency of GPEEC, the SF method with the same DE optimizer is used and the optimization time is set to 10 days. The result is $P_{1\,\mathrm{dB}} = 9.44\,\mathrm{dBm}$, $PAE@P_{1\,\mathrm{dB}} = 7.95\,\%$, and $G_p = 12.60\,\mathrm{dB}$. It can be seen that the $P_{1\,\mathrm{dB}}$ constraint is not satisfied and the *PAE* value is much worse than that obtained by GPEEC.

8 Conclusions

This chapter has presented the GPEEC algorithm for dealing efficiently with computationally expensive inequality constrained optimization problems, which is of great importance for the industry. GPEEC has the ability to handle complex constraints in an efficient manner. Thanks to the improved SMAS framework and the ranking method, an efficient SAEA for constrained expensive optimization has been constructed. Thanks to the DM method, complex constraints can be handled effectively under the improved SMAS framework. The ASU method saves more than half the computational effort on surrogate modeling for most test problems compared to updating the surrogate models in each iteration, which is especially useful for problems with several tens of variables or/and with many constraints. In addition, although the ideas behind the key components of GPEEC are not easy, their implementation is straightforward, showing GPEEC potential usage in industrial applications. Experimental studies on a set of widely used test problems have shown that comparable results in terms of optimality can be obtained when compared to a SAPF method (without surrogate model), and that several tens to more than one hundred times less exact function evaluations are needed compared to an efficient SF method. GPEEC is also applied to a mm-wave IC design optimization problem and have obtained a high-performance result with an affordable amount of computational effort.

Appendix

Benchmark test problems:

GN9

$$
\begin{aligned}
&\text{minimize } f(\mathbf{x}) = 1 + \textstyle\sum_{i=1}^{d} \frac{(100*x_i)^2}{4000} - \prod_{i=1}^{d} cos(\frac{(100*x_i)}{\sqrt{i}}) \\
&\text{subject to } g_1(\mathbf{x}) = -20e^{-0.2\sqrt{\frac{1}{d}\sum_{i=1}^{d} x_i^2}} - e^{\frac{1}{d}\sum_{i=1}^{d} cos(2\pi x_i)} \\
&\qquad -5 \leq 0 \\
&\qquad g_2(\mathbf{x}) = -\textstyle\sum_{i=1}^{20} x_i - 10 \leq 0 \\
&\qquad x_i \in [-6, 6], i = 1, \ldots, 20 \\
&\qquad minimum: f(\mathbf{x}^*) = 0
\end{aligned}
\tag{12}
$$

References

1. Alexandrov, N.M., Lewis, R.M., Gumbert, C.R., Green, L.L., Newman, P.A.: Approximation and model management in aerodynamic optimization with variable-fidelity models. J. Aircr. **38**(6), 1093–1101 (2001)
2. Basudhar, A., Dribusch, C., Lacaze, S., Missoum, S.: Constrained efficient global optimization with support vector machines. Struct. Multidiscip. Optim. **46**, 1–21 (2012)
3. Deb, K.: An efficient constraint handling method for genetic algorithms. Comput. Methods Appl. Mech. Eng. **186**(2), 311–338 (2000)
4. Dennis, J., Torczon, V.: Managing approximation models in optimization. In: Multidisciplinary Design Optimization: State-of-the-Art, SIAM pp. 330–347 (1997)
5. Emmerich, M., Giannakoglou, K., Naujoks, B.: Single-and multiobjective evolutionary optimization assisted by gaussian random field metamodels. IEEE Trans. Evol. Comput. **10**(4), 421–439 (2006)
6. Forrester, A., Sóbester, A., Keane, A.: Engineering Design via Surrogate Modelling: A Practical Guide. Wiley, New York (2008)
7. Goh, C., Lim, D., Ma, L., Ong, Y., Dutta, P.: A surrogate-assisted memetic co-evolutionary algorithm for expensive constrained optimization problems. In: 2011 IEEE Congress on Evolutionary Computation (CEC), pp. 744–749. IEEE, New York (2011)
8. Gorissen, D., Couckuyt, I., Demeester, P., Dhaene, T., Crombecq, K.: A surrogate modeling and adaptive sampling toolbox for computer based design. J. Mach. Learn. Res. **11**, 2051–2055 (2010)
9. Jones, D.: A taxonomy of global optimization methods based on response surfaces. J. Glob. Optim. **21**(4), 345–383 (2001)
10. Jones, D., Schonlau, M., Welch, W.: Efficient global optimization of expensive black-box functions. J. Glob. Optim. **13**(4), 455–492 (1998)
11. Kleijnen, J.P., Beers, W.V., Nieuwenhuyse, I.V.: Constrained optimization in expensive simulation: Novel approach. Eur. J. Oper. Res. **202**(1), 164–174 (2010)
12. Koziel, S., Leifsson, L.: Surrogate-Based Modeling and Optimization. Applications in Engineering, Springer (2013)
13. Le, M.N., Ong, Y.S., Menzel, S., Jin, Y., Sendhoff, B.: Evolution by adapting surrogates. Evol. Comput. **21**(2), 313–340 (2013)
14. Lee, C., Ahn, J., Bae, H., Kwon, J.: Efficient global optimization incorporating feasibility criterion for the design of a supersonic inlet. Proc. Inst. Mech. Eng. G J. Aerosp. Eng. **226**(11), 1362–1372 (2011)
15. Liang, J., Runarsson, T.P., Mezura-Montes, E., Clerc, M., Suganthan, P., Coello, C.A.C., Deb, K.: Problem Definitions and Evaluation Criteria for the CEC 2006 Special Session on Constrained Real-Parameter Optimization, IEEE (2006)
16. Lim, D., Jin, Y., Ong, Y., Sendhoff, B.: Generalizing surrogate-assisted evolutionary computation. IEEE Trans. Evol. Comput. **14**(3), 329–355 (2010)
17. Liu, B., Aliakbarian, H., Ma, Z., Vandenbosch, G., Gielen, G., Excell, P.: An efficient method for antenna design optimization based on evolutionary computation and machine learning techniques. IEEE Trans. Antennas Propag. **62**(1), 7–18 (2014)
18. Liu, B., Deferm, N., Zhao, D., Reynaert, P., Gielen, G.: An efficient high-frequency linear RF amplifier synthesis method based on evolutionary computation and machine learning techniques. IEEE Trans. Comput. Aided Des. Integr. Circuits Syst. **31**(7), 981–993 (2012)
19. Liu, B., Gielen, G., Fernández, F.: Automated design of analog and high-frequency circuits. Stud. Comput. Intell. **501** (2014)
20. Liu, B., Zhang, Q., Gielen, G.: A Gaussian process surrogate model assisted evolutionary algorithm for medium scale expensive optimization problems. IEEE Trans. Evol. Comput. **18**(2), 180–192 (2014)
21. Mallipeddi, R., Suganthan, P.: Ensemble of constraint handling techniques. IEEE Trans. Evol. Comput. **14**(4), 561–579 (2010)

22. Ong, Y., Lum, K.Y., Nair, P.: Hybrid evolutionary algorithm with hermite radial basis function interpolants for computationally expensive adjoint solvers. Comput. Optim. Appl. **39**(1), 97–119 (2008)
23. Price, K., Storn, R., Lampinen, J.: Differential evolution: a practical approach to global optimization. Springer, New York (2005)
24. Rasmussen, C.: Gaussian Processes in Machine Learning. Advanced Lectures on Machine Learning, Springer pp. 63–71 (2004)
25. Runarsson, T., Yao, X.: Stochastic ranking for constrained evolutionary optimization. IEEE Trans. Evol. Comput. **4**(3), 284–294 (2000)
26. Runarsson, T.P.: Constrained evolutionary optimization by approximate ranking and surrogate models. In: Parallel Problem Solving from Nature-PPSN VIII, pp. 401–410. Springer, Berlin (2004)
27. Sacks, J., Welch, W., Mitchell, T., Wynn, H.: Design and analysis of computer experiments. Stat. Sci. **4**(4), 409–423 (1989)
28. Sasena, M.J., Papalambros, P., Goovaerts, P.: Exploration of metamodeling sampling criteria for constrained global optimization. Eng. Optim. **34**(3), 263–278 (2002)
29. Stein, M.: Large sample properties of simulations using Latin hypercube sampling. Technometrics **29**, 143–151 (1987)
30. Suganthan, P., Hansen, N., Liang, J., Deb, K., Chen, Y., Auger, A., Tiwari, S.: Problem definitions and evaluation criteria for the CEC 2005 special session on real-parameter optimization. Nanyang Technological University, Singapore, Technical Report **2005005** (2005)
31. Tessema, B., Yen, G.: A self adaptive penalty function based algorithm for constrained optimization. In: IEEE Congress on Evolutionary Computation (CEC 2006), pp. 246–253. IEEE, New York (2006)
32. Venkatraman, S., Yen, G.: A generic framework for constrained optimization using genetic algorithms. IEEE Trans. Evol. Comput. **9**(4), 424–435 (2005)
33. Zhao, D., He, Y., Li, L., Joos, D., Philibert, W., Reynaert, P.: A 60 GHz 14 dBm power amplifier with a transformer-based power combiner in 65 nm CMOS. Int. J. Microwave Wireless Technolog. **3**(02), 99–105 (2011)
34. Zhou, Z., Ong, Y., Nair, P., Keane, A., Lum, K.: Combining global and local surrogate models to accelerate evolutionary optimization. IEEE Trans. Syst. Man Cybern. C Appl. Rev. **37**(1), 66–76 (2007)

Sobol Indices for Dimension Adaptivity in Sparse Grids

Richard P. Dwight, Stijn G.L. Desmedt, and Pejman Shoeibi Omrani

Abstract Propagation of random variables through computer codes of many inputs is primarily limited by computational expense. The use of sparse grids mitigates these costs somewhat; here we show how Sobol indices can be used to perform dimension adaptivity to mitigate them further. The method is compared to conventional adaptation schemes on sparse grids (Gerstner and Griebel, Computing **71**(1), 65–87, 2003), and seen to perform comparably, without requiring the expense associated with a look-ahead error estimate. It is demonstrated for an expensive computer model of contaminant flow over a barrier.

Keywords Uncertainity Quantification • Surrogate Modelling • Navier-Stokes • CFD

MSC code: 62K20

1 Introduction

Uncertainty quantification (UQ) in computational modelling consists of taking into account lack-of-knowledge (epistemic uncertainty) and physical randomness (aleatory uncertainty) when making predictions with simulation codes [16]. One sub-problem of UQ is uncertainty *propagation* (UP): given inputs of a code with known uncertainty, determine uncertainty on the code output. This can be regarded as a kind of sensitivity analysis. However, when the uncertainty is specified via random variables, this is also a necessary component in more general statistical modelling, such as Bayesian calibration [13, 27].

R.P. Dwight (✉) • S.G.L. Desmedt
Faculty of Aerospace, TU Delft, Delft, Netherlands
e-mail: r.p.dwight@tudelft.nl

P.S. Omrani
TNO, Delft, Netherlands
e-mail: pejman.shoeibiomrani@tno.nl

S. Koziel et al. (eds.), *Simulation-Driven Modeling and Optimization*,
Springer Proceedings in Mathematics & Statistics 153,
DOI 10.1007/978-3-319-27517-8_15

Problem statement 1 (Uncertainty propagation). *Given a standard multivariate random variable*

$$\mathbf{X} : \Omega = [0,1]^d \to \mathbb{R}^d,$$

with probability density function (pdf) $\rho_X(\mathbf{x})$, and a computer code regarded as a function of d-inputs and a scalar output

$$f : \mathbb{R}^d \to \mathbb{R},$$

approximate statistical moments of the random variable

$$Y = f(\mathbf{X}),$$

which are defined by ($m \in \mathbb{N}^+$):

$$\mathbb{E}_X Y^m := \int_{\mathbb{R}^d} [f(\mathbf{x})]^m \rho_X(\mathbf{x}) \, \mathrm{d}\mathbf{x}. \tag{1}$$

We may also be interested in the pdf of Y, denoted $\rho_Y(y)$.

The primary difficulty in solving this problem is the computational expense. If $f(\cdot)$ is a simulation of a complex system or fluid, it is likely computationally costly – with many hidden equations and degrees-of-freedom. If the dimension of the input parameter space d is large, then the integral in (1) will require a large number of evaluations of $f(\cdot)$ at different input values. In practice, there is often a computational budget of no more than 50–100 evaluations of f allowed to perform this calculation [22].

What makes this a realistic proposition is that requirements on the accuracy of the moment approximation are also low. This is due to the fact that, typically: (a) input uncertainties $\mathbf{X}$ are not specified to high accuracy, and (b) the resulting uncertainty on Y is not the primary output, but serves more the role of an error estimate [6].

The goal is to get a rough approximation of $\mathbb{E}Y$ and $\mathrm{Var}(Y)$ with as few evaluations of f as possible. As a result we are less than usually interested in the convergence behaviour in the limit of small error, and more interested in coarsely approximating the large-scale structure of the model response in the parameter space.

1.1 Solution methods

At a high level of abstraction there are only two classes of numerical approach to the UP problem:

1. Monte-Carlo [18] and related techniques, notably: quasi-Monte-Carlo [4], which gives improved convergence for small d using low-discrepancy sequences [24];

multi-level Monte-Carlo, which blends samples from fine and coarse versions of f [11]; importance sampling, or adaptive Monte-Carlo, which can improve convergence if an approximation of f is known [5], etc.

2. Surrogate modelling methods, notably: polynomial chaos [10], stochastic collocation [3], and probabilistic collocation [15], all of which use truncated polynomial series to approximate f, either with collocation or Galerkin projection; Gaussian-process regression [1, 2, 13], which approximates f with a random process (also known as Kriging); moment methods [19] which approximate f as a Taylor-series, usually about $\mathbb{E}X$, etc.

There are many more methods in both classes not mentioned here. Any surrogate model can be applied to evaluate integral (1), and methods have been developed that re-purpose reconstruction techniques developed in other areas to the UP problem. A good example is Simplex-Stochastic Collocation [7, 28], which is WENO reconstruction [14] developed for simulating compressible fluids (with shocks), applied to the parameter space in order to deal with discontinuities there.

Surrogate models can be applied in the parameter space $\mathbb{R}^d$ directly, or the transformed space $[0, 1]^d$, where the cumulative distribution of X_i, denoted $F_{X,i} : \mathbb{R} \to [0, 1]$ provides the mapping for each component of X separately, assuming they are independent.

Sparse grids, being a form of quadrature/interpolation designed to scale well as d increases, are a natural fit for the UP problem. They have been used by a number of authors, Matthies & Keese (2005) [17], Xiu & Hesthaven (2005) [29], and Ganapathysubramanian & Zabaras (2007) [8], and Nobile et al. (2008) [20, 21] are a few examples.

1.2 Outline

Sparse grids are introduced in Section 2, together with a brief description of their application to uncertainty propagation. In Section 3 Sobol indices are introduced, their interpretation as a form of global sensitivity analysis, and their computation using sparse grids. Section 4 introduces a novel technique of using Sobol indices to drive dimension adaptivity in sparse grids. It is compared to a reference adaptive method from Gerstner & Griebel [9]. Finally in Section 5 the adaptive method is applied to a case of practical interest: the release of an explosive heavy gas from an industrial facility or tanker accident. Three uncertain parameters control specifics of the release, and the ambient wind conditions; the scalar quantity-of-interest is the distance from the release point at which the concentration of the gas has dropped to a safe level.

2 Sparse-Grid Surrogate Models

Sparse grids are a class of numerical quadrature/interpolation methods, designed to scale well to high-dimensional problems. They were first introduced by Smolyak (1963) [23]. Sparse grids uses tensor products of hierarchical 1-dimensional quadrature rules to construct an d-dimensional grid on a hypercube, which can be used for multivariate integration on that volume. The major difference from standard tensor-product rules – as used in uncertainty quantification by polynomial chaos, etc. – is that only carefully chosen subsets of all possible products of the 1d-rules are used. The hierarchical nature of the 1d rules ensures that many sample points of these product rules coincide – leading to an efficient sampling plan.

We follow the notation of Gerstner and Griebel [9] to form the numerical quadrature rule for smooth functions $f^{(d)}(x) : \mathbb{R}^d \to \mathbb{R}$ over the d-dimensional hypercube $\Omega = [-1, 1]^d$. The rule should approximate the integral:

$$If^{(d)} := \int_{\Omega} f^{(d)}(x)\,\mathrm{d}x \tag{2}$$

by a sum:

$$If^{(d)} \simeq Qf^{(d)} := \sum_{i=1}^{N} w_i f^{(d)}(\mathbf{x}_i),$$

with some weights w and sample-locations $\mathbf{x}_i$.

The construction begins with a set of hierarchical 1-dimensional quadrature formulas for a univariate function $f^{(1)}$:

$$Q_l f^{(1)} := \sum_{i=1}^{N_l} w_i^l f^{(1)}(x_i^l)$$

where l indicates the *level* of the rule in the hierarchy, and superscripts of w and x indicate level. So for example if Clenshaw-Curtis quadrature [12, 26] is chosen for the 1d rule, Q_1 is the level 1 rule with a single quadrature point ($N_1 = 1$), Q_2 is the level 2 rule with three points ($N_2 = 3$), and Q_3 has five points ($N_3 = 5$). The rule is hierarchical, so that, e.g., $\{x_1^2, \ldots, x_{N_2}^2\} \subset \{x_1^3, \ldots, x_{N_3}^3\}$, i.e. all points at the lower levels are included in the higher-level rules.

"Difference" formulas are then defined in 1d by:

$$\Delta_l f^{(1)} := (Q_l - Q_{l-1}) f^{(1)} \text{with}$$
$$Q_0 f^{(1)} := 0.$$

Note that, since Q_l are hierarchical, Δ_l has the same support as Q_l; only the weights are different.

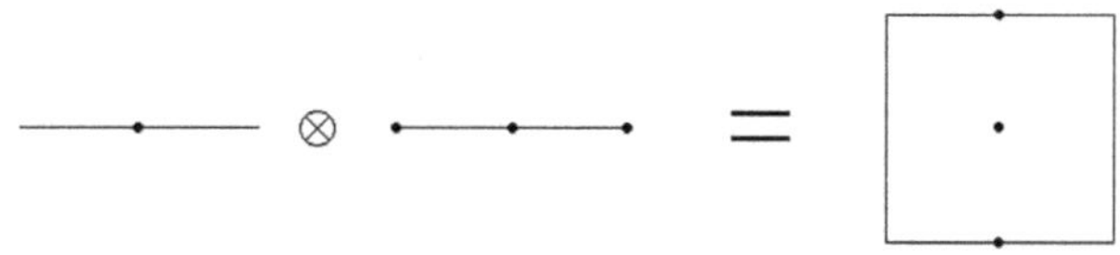

Fig. 1 Tensor product of a 1-point rule and a 3-point rule, resulting in a 2D grid containing 3 points

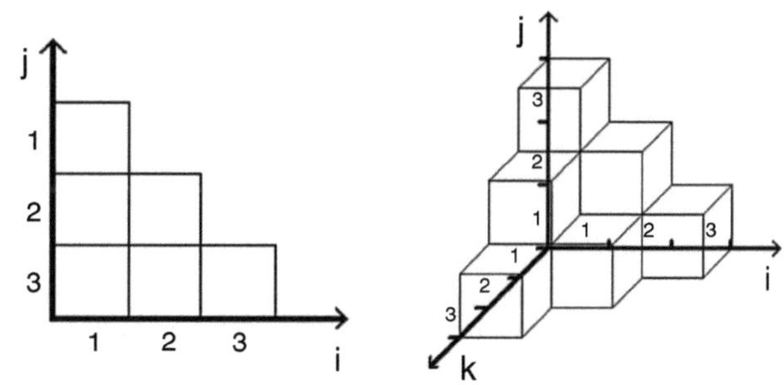

Fig. 2 A level 3 sparse grid simplex in 2 (left) and 3 (right) dimensions

Now define a *multi-index* $\mathbf{k} \in \mathbb{N}^d$ which, for each dimension of the parameter space, specifies the level of rule to use in that dimension. Together with the hierarchy Q_l, $\mathbf{k}$ specifies a tensor-product integration rule in d-dimensions, see Figure 1. Finally define the *simplex index-set* $\mathscr{K}_{S,L} := \{\mathbf{k} : |\mathbf{k}|_1 \leq L + d - 1\}$, then a sparse quadrature rule is given by Smolyak's formula:

$$Q_L^{\mathrm{sp}} f^{(d)} := \sum_{\mathbf{k} \in \mathscr{K}_{S,L}} (\Delta_{k_1} \otimes \cdots \otimes \Delta_{k_d}) f^{(d)}. \tag{3}$$

This is called a level L sparse-grid quadrature rule.

The index-set $\mathscr{K}$ as defined above is visualized in Figure 2 for $L = 3$. Each multi-index in $\mathscr{K}$ is represented as a block. The constraint $|k|_1 \leq L + d - 1$ leads to a simplex-form of the multi-indices contributing to the sum in (3), with the highest-level rule L being achieved in one dimension only if the level in all other dimensions is 1.

Limiting the level in each direction in this way clearly has an impact on the approximation fidelity of the grid. This can be understood in terms of polynomial representation. Consider for example the 2d sparse grid shown on the right of Figure 3. In 1d the level-3 CC rule Q_3 can support polynomials of degree 4 (x^4) as it has 5-points. A full tensor product $Q_3 \otimes Q_3$ can therefore support all polynomials up to x^4y^4. The sparse-grid rule, on the other hand, can support $x^4 \cdot 1$ (via $Q_3 \otimes Q_1$), $1 \cdot y^4$ (via $Q_1 \otimes Q_3$), and x^2y^2 (via $Q_2 \otimes Q_2$), but **not** x^4y^4. In other words, high-order interactions of multiple variables are not captured. It is an empirical fact that, in functions of multiple variables found in nature and engineering, higher-order interactions tend to be dominated by single-variable and low-order interaction effects. If this is the case, then the sparse-grid approximation is appropriate.

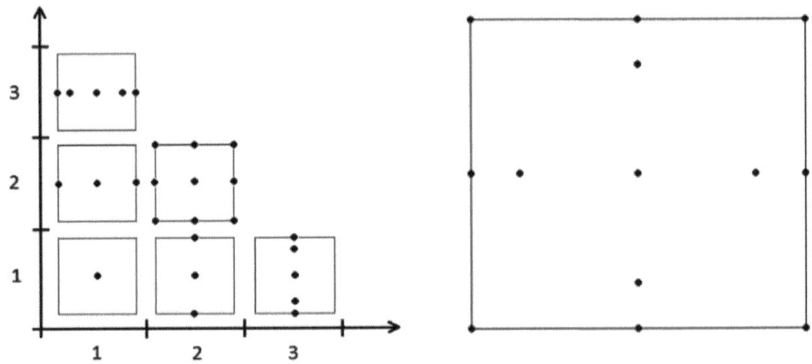

Fig. 3 Sparse-grid components in the 2-dimensional level 3 simplex (left) and the resulting sparse grid (right). In this figure the grid is constructed using nested quadrature rules

The choice of $\mathscr{K}$ used above is standard; however, the construction (3) is a valid quadrature rule whenever the index-set satisfies the basic requirement (*admissibility*):

$$\mathbf{k} \in \mathscr{K},\ k_i > 1 \implies \mathbf{k} - \mathbf{1}_i \in \mathscr{K}, \quad \forall i \in \{1, \dots, d\},$$

where $\mathbf{1}_i \in \mathbb{N}_0^d$ is the unit vector in the i-th direction $\mathbf{1}_i = (0, \dots, 0, 1, 0, \dots, 0)$. This condition demands only that all rules of lower level in every dimension are included in the sum (3). The condition does not therefore require adding any more points to the sparse grid.

This flexibility in the choice of $\mathscr{K}$ immediately provides a framework for dimension-adaptivity: if dimension i requires more resolution, we can adapt the index-set in that dimension, see, for example, the index-sets in Figure 9. Gerstner & Griebel [9] use a traditional look-ahead error estimate to decide which multi-indices to include at each adaptation step. Later we offer an alternative approach using Sobol indices to assess the importance of each dimension, see Section 4. First, however, we introduce Sobol indices.

3 Global Sensitivity of Functions: Sobol Indices

Consider a multivariate real function $f : \mathbb{R}^d \to \mathbb{R}$. Sobol indices are one measure of the *global* sensitivity of the function with respect to its arguments $\mathbf{x}$. Usually sensitivity measures are *local*, that is, they concern themselves only with the linearized behaviour of f function at a reference point $\mathbf{x}_0 \in \mathbb{R}^d$. For instance local sensitivities might be defined as:

$$S_i := \sigma_i \left.\frac{\partial f}{\partial x_i}\right|_{\mathbf{x}_0} \qquad i \in \{1, \dots, d\}, \tag{4}$$

where σ_i is a measure of how much variable x_i is expected to vary. This definition ignores the behaviour of f away from $\mathbf{x}_0$. For example $f(x) = 10^{10}x^2$ at $x_0 = 0$ is considered to be "insensitive" to x, and $f(x) = H(x)$, (with $H(\cdot)$ the Heaviside function) is "insensitive" to x everywhere (except possibly at the origin). Furthermore S provides no information on the effect of interactions of multiple variables on f.

Global sensitivity measures attempt to address these limitations. The first step is to specify what is meant by "global". In the case of variance-based sensitivity indices (of which Sobol indices are an example) this is achieved by defining a probability density function (pdf) for each input variable, specifying the range of that variable which is of interest:

$$\rho(x_1), \ldots, \rho(x_d),$$

with the corresponding random variables denoted $\mathbf{X} = (X_1, \ldots, X_d)$. These are comparable in purpose to σ_i in the local case. To continue the derivation of Sobol indices, the analysis of variance (ANOVA) decomposition must be introduced.

3.1 ANOVA Decomposition

Assume that $f(\mathbf{x})$ is square-integrable[1] with respect to the metric generated by $\mathbf{X}$. Furthermore let $X_1, \ldots, X_d \sim \mathscr{U}(0, 1)$ be independently uniformly distributed on $[0, 1]$. Any input space can be transformed onto this unit hypercube, so there is no loss of generality. Then $f(\mathbf{X})$ is a random variable with finite variance, which we represent in the form:

$$f(\mathbf{X}) = f_\emptyset + \sum_{s=1}^{d} \sum_{1 \le i_1 < \cdots < i_s \le d} f_{i_1 \ldots i_s}(X_{i_1}, \ldots, X_{i_s}). \tag{5}$$

Or in long-hand:

$$\begin{aligned} f(\mathbf{X}) &= f_\emptyset \\ &+ f_1(X_1) + \cdots + f_d(X_d) \\ &+ f_{12}(X_1, X_2) + f_{13}(X_1, X_3) + \cdots + f_{d-1d}(X_{d-1}, X_d) \\ &+ f_{123}(X_1, X_2, X_3) + \ldots \\ &+ \ldots \\ &+ f_{1\ldots d}(X_1, \ldots, X_d) \end{aligned}$$

[1] Incidentally, a much weaker condition on f than that required by (4).

The most convenient form is the third form:

$$f(\mathbf{X}) = \sum_{u \subseteq \mathscr{U}} f_u(X_u) \tag{6}$$

where the indexing set u is here called a *dimension-index*, $\mathscr{U} = \{1, 2, \ldots, d\}$, and the sum is over all subsets of $\mathscr{U}$. Now X_u is the set of random variables whose indices lie in u, and f_u is the component function only dependent on X_u. If it is true that – in physical and engineering models – low-order interactions of variables have the main effect on the output, and if this is captured by the decomposition above, then we should be able to truncate the sum without substantial loss of fidelity. Compare this to sparse grids, in which high polynomial order interactions of multiple variables are also preferentially eliminated.

This formula is called an ANOVA decomposition if and only if:

$$\int f_u(x_{u_1}, \ldots, x_{u_s}) \mathrm{d}\rho(x_i) = 0 \quad \text{for} \quad i \in u. \tag{7}$$

This implies:

$$\mathbb{E} f_u := \int f_u(x_u) \mathrm{d}\rho(x_u) = 0 \quad \text{for} \quad u \neq \emptyset, \tag{8}$$

i.e., all f_u have zero mean, with the exception of $f_\emptyset$, and

$$\mathrm{cov}(f_u, f_v) = \int f_u(x_u) f_v(x_v) \mathrm{d}\rho(x_u \cup x_v) = 0 \quad \text{for} \quad u \neq v, \tag{9}$$

i.e., f_u, f_v are orthogonal. Let u' be the complement of u, so that $\{u \cup u'\} = \mathscr{U}$ and $\{u \cap u'\} = \emptyset$. These properties are satisfied when the component functions f_u are defined as:

$$\begin{aligned} f_\emptyset &= \int f(\mathbf{x}) \mathrm{d}\rho(\mathbf{x}), \\ f_u &= \int f(\mathbf{x}) \mathrm{d}\rho(x_{u'}) - \sum_{w \subset u} f_w(x_w) \quad \text{for} \quad u \neq \emptyset, \end{aligned}$$

which can be rewritten in terms of conditional expectations:

$$\begin{aligned} f_\emptyset &= \mathbb{E} f, \\ f_i &= \mathbb{E}(f|X_i) - f_0, \\ f_{ij} &= \mathbb{E}(f|X_i, X_j) - f_i - f_j - f_0, \\ &\ldots \end{aligned}$$

at which point the terms can be interpreted. For example, it is evident that f_i captures the effect of varying X_i alone, with all other variables integrated out. And f_{ij} captures the effect of varying X_i and X_j simultaneously, *minus* the effect of their individual variations, and so on.

3.2 Sobol Variances

The variances of these terms are therefore our desired sensitivity measures:

$$D_u := \mathrm{var}(f_u) = \int f_u^2 \mathrm{d}\rho(x_u), \tag{10}$$

which implies that all Sobol variances are non-negative. It can be shown that D_u simplifies to:

$$\begin{aligned} D_u &:= \int \left(\int f(x) \mathrm{d}\rho(x_{u'}) \right)^2 \mathrm{d}\rho(x_u) - \sum_{w \subset u} \int (f_w(x_w))^2 \, \mathrm{d}\rho(x_u), \\ &= \int \left(\int f(x) \mathrm{d}\rho(x_{u'}) \right)^2 \mathrm{d}\rho(x_u) - \sum_{w \subset u} D_w, \end{aligned} \tag{11}$$

which is a readily computable expression for D_u, and allows computation in order of increasing order, first D_i, then D_{ij}, then D_{ijk}, etc. As well as $D_u \geq 0$ we have

$$D := \mathrm{var}(f) = \sum_{u \subseteq \mathscr{U}} D_u,$$

i.e., the variance of f has been decomposed into the effects due to individual combinations of variables. This property suggests the definition of Sobol *indices*, which are just normalized Sobol variances:

$$S_u := \frac{D_u}{D}.$$

We say that S_u is a Sobol index of *order p*, if $|u| = p$, where $|u|$ is cardinality of u. For example, given $u = \{0, 1\}$ we speak of the 2nd-order interaction between variables x_0 and x_1.

Sobol indices provide a quantitative means to verify the assertion we made in Section 2, that "higher-order interactions tend to be dominated by single-variable and low-order interaction effects", and thereby justify the use of sparse grids.

3.3 Computation of Sobol Indices on Sparse Grids

Computation of Sobol indices boils down to evaluating the integrals in (11). Sparse grids are ideally suited to this task [25], with some important caveats:

1. Sparse grids deliberately limit interaction between variables. Consider a standard level-2 sparse grid in 2d: i.e., with $\mathscr{K} = \{\mathbf{k} : k_1 + k_2 \leq 3\}$. Tensor-product rules $Q_1 \otimes Q_1$, $Q_1 \otimes Q_2$, and $Q_2 \otimes Q_1$ are included, but the 1-point Q_1 rule approximates a variable's influence as constant. Therefore the sparse-grid approximation will model no interaction between the two variables[2]. Hence, when evaluated on the sparse grid, $S_{12} \equiv 0$, regardless of the level of interaction in the true underlying function. The same holds for higher orders and higher dimensions: on level-3 sparse grids third-degree interactions are necessarily zero: $S_{123} \equiv 0$, and so on.
2. In order to evaluate the double integral in (11), it is necessary to: first (a) perform sparse-grid integration of f over a subset of the variables u', then (b) reconstruct the result of the integration *squared* in the remaining variables u, then (c) integrate over this reconstruction. Consider again a polynomial representation of the sparse-grid model $\phi(x)$ – this time in 1-variable: if our 1d rule has $M + 1$ points, it supports polynomials ϕ up to degree M. Then $(\int \phi \, dx)^2$ is a polynomial of degree $2M+2$. The $M+1$-point rule is not sufficient to integrate this accurately, which can lead to gross errors in the computed indices.

Item 2. is solved by interpolating the sparse-grid approximation of f to a globally refined sparse grid before computing the integrals (no additional evaluations of the underlying function are required). In the case of polynomial approximations the exact Sobol indices of the sparse-grid approximation can be obtained by using two levels of global refinement.

Item 1. is not so easy to address – and the user must be aware of this limitation when considering high-degree Sobol indices. For example this is critical to the Sobol-based adaptive procedure, which we discuss next.

4 Dimension Adaptive Sparse-Grid Refinement

This section introduces two adaptive grid refinement methods based on sparse grids. The first, Gerstner and Griebel (G&G), is a (now) standard look-ahead approach [9]. The second is our proposed Sobol-based adaptive approach. Both require the following definitions:

[2]This can be seen by considering what space of polynomials can be reconstructed on the sparse grid, namely $\phi(x, y) = a_0 + a_1x + a_2x^2 + a_3y + a_4y^2$, which include no interaction (xy) terms.

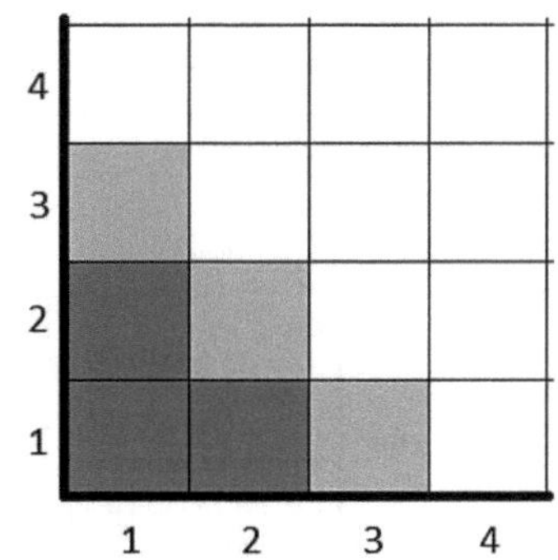

Fig. 4 Diagram showing the forward neighbours of a level 2, 2-dimensional sparse grid. The multi-indices of the sparse grid are shown in dark-grey, while the three forward neighbours are light-grey

Definition 1. The *forward neighbours* of a multi-index $\mathbf{k}$ are defined as the d multi-indices $\{\mathbf{k} + \mathbf{1}_j, 1 \le j \le d\}$. Generalizing, the forward neighbours of an index-set $\mathscr{K}$ is the set of all forward neighbours of elements of $\mathscr{K}$, which are not in $\mathscr{K}$ themselves.

As an example consider a level-2, 2-d sparse grid. The index-set is: $\{(1,1),(2,1),(1,2)\}$. The set of forward neighbours for this index-set is: $\{(3,1),(2,2),(1,3)\}$, see Figure 4.

Definition 2. The *backward neighbours* of a multi-index are the set of multi-indices:

$$\{\mathbf{k} - \mathbf{1}_j : k_j > 1, 1 \le j \le d\}.$$

Compare this definition to the requirement on admissibility of an index-set in Section 2. We have: an index-set $\mathscr{K}$ is admissible if all backward neighbours of all $\mathbf{k} \in \mathscr{K}$ are in $\mathscr{K}$.

4.1 Gerstner and Griebel Adaptivity

The Gerstner and Griebel adaptive approach is explained in detail in [9] together with implementation optimizations; we briefly develop the main idea here. Start with an initial index-set $\mathscr{K}$, e.g., $\{(1,\ldots,1)\}$. Let $\mathscr{A}$ be the set of forward neighbours of $\mathscr{K}$ with the property that for all $\mathbf{a} \in \mathscr{A}$ the index-set $\mathscr{K}(\mathbf{a}) := \mathscr{K} \cup \{\mathbf{a}\}$ is admissible. The set $\mathscr{A}$ is called the *active* set. For each element $\mathbf{a}$ of $\mathscr{A}$ compute the error measure:

$$g_{\mathbf{a}} := \left| Q^{\mathrm{sp}}_{\mathscr{K}} f - Q^{\mathrm{sp}}_{\mathscr{K}(\mathbf{a})} f \right|,$$

where $Q^{\mathrm{sp}}_{\mathscr{K}}$ is the sparse-grid quadrature of (3) using the index-set $\mathscr{K}$. This error measure is with respect to the function of interest f, and therefore requires evaluation

Algorithm 4.1: GERSTNER&GRIEBEL($\eta_{\min}$)

$\mathscr{K} := \{(1,\ldots,1)\}$
$r := Q^{sp}_{\mathscr{K}} f$
while $(\eta > \eta_{\min})$
do
 $\mathscr{A} := \{\mathbf{a} : \mathbf{a} \notin \mathscr{K}, \mathscr{K} \cup \{\mathbf{a}\}$ is admissible$\}$
 for a in $\mathscr{A}$
  do $\begin{cases} \mathscr{K}^+ := \mathscr{K} \cup \{\mathbf{a}\} \\ g_{\mathbf{a}} := |Q^{sp}_{\mathscr{K}^+} f - r| \end{cases}$
 $\mathbf{a}^+ := \arg\max_{a \in \mathscr{A}} g_{\mathbf{a}}$
 $\mathscr{K} := \mathscr{K} \cup \{\mathbf{a}^+\}$
 $r := Q^{sp}_{\mathscr{K}} f$
 $\eta := \sum_{a \in \mathscr{A} \setminus \mathbf{a}^+} g_{\mathbf{a}}$
return (r)

Fig. 5 Pseudo-code for Gerstner and Griebel sparse-grid adaptation

of f on the grid implied by $\mathscr{K} \cup \mathscr{A}$ – i.e., a global refinement of the sparse grid. A global error estimate η is then

$$\eta = \sum_{a \in \mathscr{A}} g_{\mathbf{a}}. \tag{12}$$

The adaptive iteration proceeds by selecting $\mathscr{K}(\mathbf{a}^+)$ as the new index-set where

$$\mathbf{a}^+ := \arg\max_{a \in \mathscr{A}} g_{\mathbf{a}},$$

a new active set is found and the iteration continues. The iteration finally terminates when η becomes less than some user-specified threshold $\eta_{\min}$ (Figure 5).

This algorithm has the benefit of an accurate error estimate in η, but in order to compute it, it must test all members of the active set. This constitutes "global refinement" in a sparse-grid sense, and requires many evaluations of the underlying computational code. For example, consider the case $d = 3$, with a Clenshaw-Curtis rule, and a simplex index-set of level 3:

$$\mathscr{K}_{S,3} := \{\mathbf{k} : |\mathbf{k}|_1 \leq 3 + d - 1\}.$$

Computing the integral approximation requires 25 code evaluations. Computing η requires an *additional* 44. For $d = 10$, the integral costs 221 evaluations, the error estimate an additional 1360 evaluations. What is more, η itself is only reliable when the error approaches zero, yet this is unlikely ever to be achieved with a code-eval budget of maximally 100. The benefit is a refinement scheme guaranteed to converge to the correct solution.

4.2 Sobol-Adaptive Refinement Algorithm

Rather than use a conventional error estimate to drive dimensional refinement, we propose to use Sobol indices to indicate which variables to adapt. The idea is that, in practice, only a small subset of variables and interactions are likely to be responsible for the vast majority of the variance. These variables and interactions should be approximated finely, the remainder coarsely. Since the Sobol indices are not an error estimate, they do not provide a termination criteria. Our method is as follows:

1. **Initialization:** Begin with $\mathscr{K} = \mathscr{K}_{S,2}$ – i.e., a one-factor-at-a-time sample plan. Note: on this grid all Sobol indices of 2nd-degree and higher are zero (see Section 3.3).
2. **Compute Sobol:** Compute all Sobol indices representable on the sparse grid.
3. **Dimension-index selection:** Sort all dimension-indices according to their Sobol indices from largest to smallest: $u_1, u_2, \ldots$ with $S_{u_1} \geq S_{u_2} \geq \ldots$. Find the smallest m such that

$$\sum_{i=1}^{m} S_{u_i} \geq \bar{S}$$

where the threshold $\bar{S} = 0.95, 0.98$, or 0.99 according to the accuracy required. The set of dimension-indices to adapt is:

$$U := \{u_i | 1 \leq i \leq m\}.$$

4. **Dimension-index augmentation:** If an interaction v is not represented on the current sparse grid, and if $u \in U$ for all $u \subset v$, then augment U with v. That is, flag a dimension-index v for adaptation if all its "parents" are flagged, and its own Sobol index is 0 on the sparse grid. This step is necessary to overcome the catch-22: $S_v = 0$, so v will not be adapted, so $S_v = 0$. It is the only means by which new higher-order interactions are introduced.
5. **Index-set extension:** We now have U, the set of dimension-indices we wish to adapt, and we need a corresponding adapted index-set. Let $\mathscr{A}$ be the active set of $\mathscr{K}$, as in the previous section. Let $F(\mathbf{a}) = \{i \mid a_i > 1\}$, then for each $\mathbf{a} \in \mathscr{A}$ augment $\mathscr{K}$ with $\mathbf{a}$ if $F(\mathbf{a}) \in U$. That is, use all members of the active set that represent interactions in U. This may be more than one multi-index per dimension-index in U.
6. **Termination criteria:** If the maximum number of samples N_{tol} is exceeded, stop.
7. Goto 2.

To clarify the importance of Step 4, consider the 2d case at the first iteration, so $\mathscr{K} = \{(1,1), (1,2), (2,1)\}$. Assume $S_0 = S_1 = \frac{1}{2}$, i.e., both variables are significant, so both $(3,1)$ and $(3,1)$ should be added to $\mathscr{K}$. However no interaction

Algorithm 4.2: SOBOLDIMENSIONADAPTIVE($\bar{S}$,N_{tol})

$\mathscr{K} := \mathscr{K}_{S,2} = \{\mathbf{k} : |\mathbf{k}|_1 \leq d+1\}$
$N := 2d+1$
while $(N < N_{\text{tol}})$
do
- **for** $u \subseteq \{1,\ldots,d\}$
  - **do** Compute S_u using $Q^{\text{sp}}_{\mathscr{K}}$
- Sort $u \subseteq \{1,\ldots,d\}$ as $u_1, u_2, \ldots$ such that $S_{u_1} \geq S_{u_2} \geq \ldots$
- Find minimum m such that $\sum_{i=1}^{m} S_{u_i} \geq \bar{S}$
- $U := \{u_i \mid 1 \leq i \leq m\}$
- **if** $v \notin$ Interactions$(\mathscr{K})$ **and** $\forall u \subset v, u \in U$
  - **then** $U := U \cup \{v\}$
- $\mathscr{A} := \{\mathbf{a} : \mathbf{a} \notin \mathscr{K}, \mathscr{K} \cup \{\mathbf{a}\}$ is admissible$\}$
- **for a** in $\mathscr{A}$
  - **do** **if** $F(\mathbf{a}) = \{i \mid a_i > 1\} \in U$
    - **then** $\mathscr{K} := \mathscr{K} \cup \{\mathbf{a}\}$
- $r := Q^{\text{sp}}_{\mathscr{K}} f$
- $N :=$ [number of nodes of $Q^{\text{sp}}_{\mathscr{K}}$]

return (r)

Fig. 6 Pseudo-code for Sobol-index sparse-grid adaptation

is approximated on $\mathscr{K}$, so $S_{01} = 0$ on this grid, and $(2,2)$ is not added in Step 3. The criteria of Step 4 are satisfied, however, so $(2,2)$ is added, and S_{01} is potentially non-zero at the next iteration.

This method is sensitive to the choice of threshold $\bar{S}$, and the criteria for introducing higher-order interactions in Step 4. It will not in general converge as $N_{\text{tol}} \to \infty$, therefore should only be used with small N_{tol}. The exception to this if for $\bar{S} = 1$ so that all interactions are included, in which case the algorithm simply produces a standard sparse grid. It does not use look-ahead, in the sense that no new samples of the computer code are needed to determine the next adaptation step – unlike G&G.

In the following section the two adaptive methods are compared for an industrial case (Figure 6).

5 Industrial Application – Heavy Gas Release

This section introduces an industrial case, against which the performance of the standard sparse grid, the Gerstner and Griebel adaptive sparse grid, and the new Sobol-adaptive sparse grid are compared. Section 5.1 will present the specifics of the case. Section 5.2 presents the results for the three different methods. Finally, Section 5.3 will compare and discuss the differences between the methods.

5.1 Test-case Definition

The case that will be studied is a release of a heavy gas - in this case propane - over a barrier located downstream of the release point. The quantity of interest is the effect distance: the distance from the release point where the molar concentration propane gas drops below 1%, which is roughly half of the lower explosion limit (LEL). The effect distance is measured at a height of 1 m above the ground.

The scenario being modelled is a truck transporting a heavy gas becoming involved in an accident, causing an uncontrolled release. The effect distance determines the area which must be potentially evacuated, and within which special precautions must be taken by emergency services. Since the release, environment, and potential obstacles are uncontrolled, it is necessary to consider a range of scenarios. Uncertainty is specified on a subset of the unknowns, using knowledge of possible and likely conditions.

Figure 7 details the geometry. Parameters are:

- Barrier height (fixed at 4 m),
- Gas release location (60 m upstream of the barrier),
- Gas release height of (1 m),
- Release gap diameter (1.243 m),
- Wind direction (0 degrees, i.e., perpendicular to the barrier),
- Atmospheric boundary-layer velocity U_{ABL} (range of 3 to 7 m/s),
- Gas release velocity U_{rel} (range of 18 to 22 m/s),
- Gas release temperature T_{rel} (range of 270 to 310 K).

A preliminary one-factor-at-a-time parameter study identified three input parameters influential on the effect distance: U_{ABL}, U_{ref}, and T_{rel}. Two cases are considered: uniform distributions on each parameter, and truncated-normal distributions on each parameter. Variability in other parameters is neglected to save computational effort. In particular, the geometry is fixed.

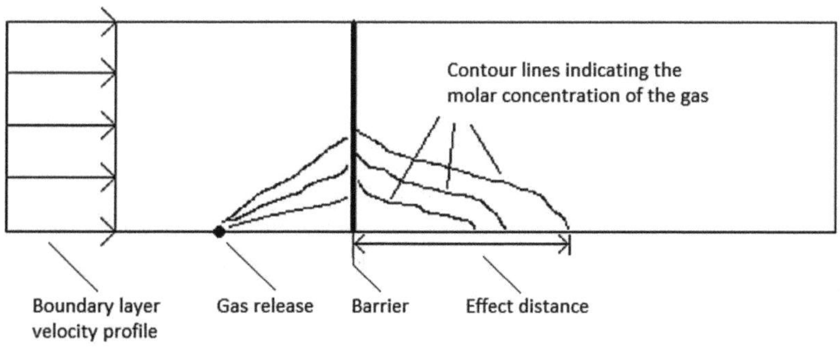

Fig. 7 Top view diagram of the release of propane gas flowing over a barrier

The problem has reflectional symmetry, so only half of the domain is modelled. The ground is a no-slip boundary, the other four boundaries are permeable to both air and propane. The problem is modelled using steady-state RANS equations.

5.2 *Sparse-grid results*

A total of 69 simulations were performed, corresponding to a standard sparse grid of level $L = 4$ in 3-dimensions (index-set $\mathscr{K}_{S,4}$). All simulations were run until the solver residual converged to 10^{-6}. A single computation took between 12 and 24 hours on one core. This data-set is used to test both adaptive methods described in the Section 4.

5.2.1 Standard sparse grid

The sparse grid sampling plan for levels $L = 3$ and $L = 4$ for the heavy gas release is shown in Figure 8. Sobol variances are computed in the two cases: uniform and truncated-normal distributions for the three uncertain inputs. Their values are given for levels $L \in \{2, 3, 4\}$ in Table 1. The level 1 grid is omitted since it approximates the total variance as zero. These results demonstrate the fact that on a level p sparse grid, Sobol variances of order p and higher are zero. We observe roughly the same mean and Sobol indices in the uniform and normal cases, but a much greater variance in the uniform case.

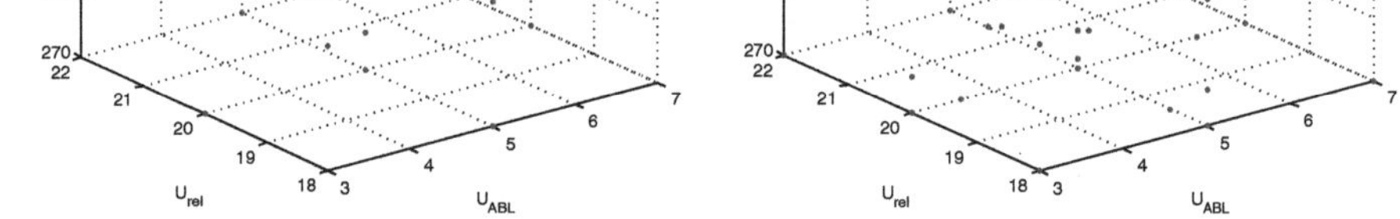

Fig. 8 Full sparse grids for the heavy gas release. Left: level $L = 3$ (index-set $\mathscr{K}_{S,3}$); right: $L = 4$ (index-set $\mathscr{K}_{S,3}$)

Table 1 Sobol variances calculated on a standard sparse grid. Indices 0, 1, and 2 correspond to the atmospheric boundary-layer velocity U_{ABL}, the propane release velocity U_{rel} and the propane release temperature T_{rel}, respectively.

Uniform input parameters			
Variable	Level 2	Level 3	Level 4
Grid points	7	25	69
Mean	184.7	183.1	182.8
Variance	446.2	363.9	346.6
D_0	375.9	261.3	253.0
D_1	60.18	73.59	75.73
D_2	10.15	0.5088	0.9390
$D_{0,1}$	0	3.826	2.417
$D_{0,2}$	0	24.86	13.24
$D_{1,2}$	0	0.002394	0.1490
$D_{0,1,2}$	0	0	1.096
Truncated-normal input parameters			
Variable	Level 2	Level 3	Level 4
Grid points	7	25	69
Mean	182.8	181.3	180.9
Variance	264.4	166.3	158.6
D_0	222.7	114.6	111.0
D_1	35.66	40.30	40.56
D_2	6.014	1.391	2.560
$D_{0,1}$	0	1.343	0.8433
$D_{0,2}$	0	8.651	3.536
$D_{1,2}$	0	0.0008375	0.03190
$D_{0,1,2}$	0	0	0.09199

5.2.2 Gerstner and Griebel adaptive sparse-grid results

Iterations of the Gerstner and Griebel are given in Table 2, showing the current estimate of the mean, and the multi-index to be included in the index-set on the subsequent iteration.

The algorithm is initialized with the index-set $\mathcal{K} = \{(1,1,1)\}$, and begins by refining only in the direction of the first parameter, until it reaches the multi-index $(4,1,1)$, i.e., a 9-point rule in the direction U_{ABL}. The forward neighbour $(5,1,1)$ is not part of the pre-computed data-set, so further refinement is no longer possible in this variable. We manually remove $(5,1,1)$ from the active set and continue with the adaptation. We can justify this choice as follows: it is a general principle when performing polynomial interpolation of potentially noisy data-sets, to limit the maximum degree of the approximation. With $(5,1,1)$ the rule would have 17-points, or polynomial degree 16 in U_{ABL}. A tiny amount of noise in the code output would lead to oscillation in the response, and likely to poor error estimates.

The iteration continues by refining T_{rel} once, followed by interactions of U_{ABL} and T_{rel}. The index-set after 6 iterations is shown in Figure 9 (right).

Table 2 Iterations of Gerstner and Griebel adaptivity

Uniform input parameters			
Iteration	Grid points (+active)	Mean	Added multi-index
0	1 (7)	180.04	(2, 1, 1)
1	3 (9)	184.65	(3, 1, 1)
2	5 (13)	182.44	(4, 1, 1)
3	9 (21)	182.22	(1, 1, 2)
4	11 (27)	182.41	(2, 1, 2)
5	15 (31)	183.00	(3, 1, 2)
6	19 (31)	182.81	–
Truncated-normal input parameter			
Iteration	Grid points (+active)	Mean	Added multi-index
0	1 (7)	180.04	(2, 1, 1)
1	3 (9)	182.77	(3, 1, 1)
2	5 (13)	181.02	(4, 1, 1)
3	9 (21)	180.68	(1, 1, 2)
4	11 (27)	180.79	(2, 1, 2)
5	15 (31)	180.99	(3, 1, 2)
6	19 (31)	180.91	–

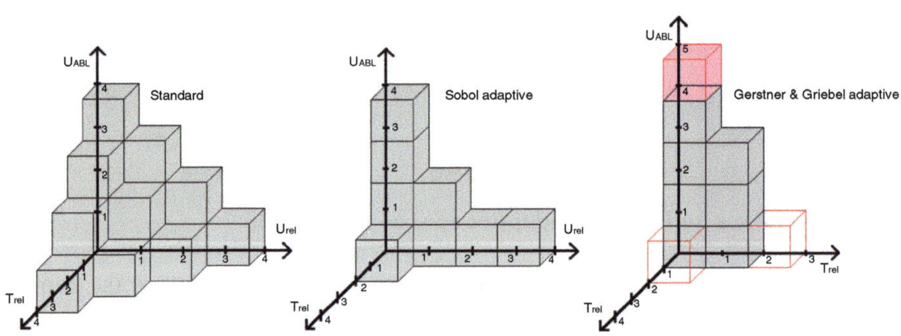

Fig. 9 Index-sets of the final sparse grids for the heavy gas release case. From left to right: simplex, Sobol, and Gerstner and Griebel sparse grids. In the latter, the red outlines represent multi-indices included for the error estimate computation, the red filled cube represents a limit on the adaptation in that direction

The progress of the adaptation is identical for the uniform and truncated-normal input parameter distributions. In both cases the error in the mean obtained (using 31 grid points) is the approximately that of the level 4 simplex sparse grid (requiring 69 points).

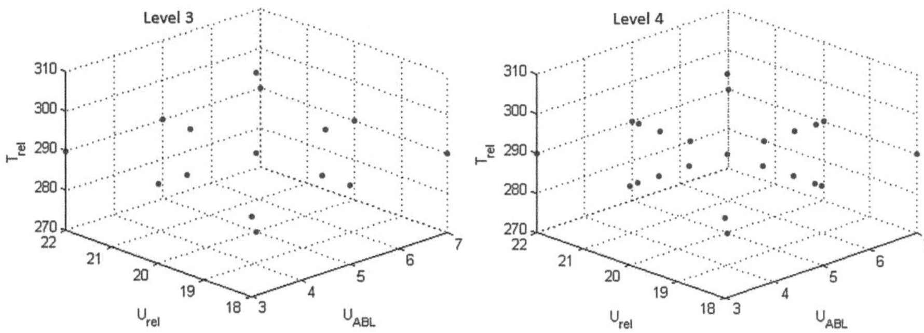

Fig. 10 Evolution of the Sobol-adapted sparse grid for the heavy gas release. Left $l = 3$, right $l = 4$

5.2.3 Sobol-adaptive sparse-grid algorithm

Unlike G & G, Sobol adaptivity can add multiple multi-indices at each adaptation iteration. As a consequence only two iterations are needed to reach an accuracy comparable to the level 4 simplex sparse grid. The sample points after 1- and 2-iterations are shown in Figure 10, denoted "Level 3 and 4", respectively. In this case too the adapted grids were identical for the uniform and truncated-normal distributed cases.

The Sobol variances on the adaptation iterations are given in Table 3. In this table, the indices 0, 1, and 2 correspond to U_{ABL}, U_{rel}, and T_{rel}, respectively. It can be seen that the variances change somewhat from one iteration to the next, but not substantially. This suggests that coarse-grid indices are accurate, and therefore reliable as adaptation indicators. The exception is those indices that are not resolved on the coarse sparse grid, and therefore zero.

Already on the initial grid, U_{ABL} and U_{rel} dominate the total variance. The cutoff value has been set to $\bar{S} = 0.95$, given which T_{rel} is not adapted, and refinement only occurs in U_{ABL} and U_{rel}, including their interaction (multi-index $(2, 2, 1)$). The interaction is seen to be negligible ($S_{0,1} < 1 - \bar{S} = 0.05$), and not adapted further. Variances $D_{0,2}$, $D_{1,2}$, and $D_{0,1,2}$ are necessarily approximated as 0. The second refinement iteration only adds points in the direction of U_{ABL} and U_{rel}. The final index-set is plotted in Figure 9 (centre).

5.2.4 Convergence comparison

The mean effect-distance from Tables 1, 2, and 3 is plotted in Figure 11 for uniformly distributed inputs. Truncated-normal results are substantially identical. Both adaptive methods lead to a good approximation of the mean with a significant reduction in the sample points. In the G & G case we plot the total number of samples, including those needed for the error estimate. On the final iteration, we incorporate all samples into the approximation for the mean – hence two different

Table 3 Evolution of Sobol variances on the Sobol-adapted sparse grid

Uniform input parameters			
Variable	Iter 0	Iter 1	Iter 2
Grid points	7	15	23
Mean	184.7	182.5	182.4
Variance	446.2	309.6	312.5
D_0	375.9	222.6	225.6
D_1	60.18	72.71	72.80
D_2	10.15	10.15	10.15
$D_{0,1}$	0	3.826	4.013
$D_{0,2}$	0	0	0
$D_{1,2}$	0	0	0
$D_{0,1,2}$	0	0	0
Truncated-normal input parameters			
Variable	Iter 0	Iter 1	Iter 2
Grid points	7	15	23
Mean	182.8	181.1	180.8
Variance	264.4	150.3	151.1
D_0	222.7	102.9	103.4
D_1	35.66	40.01	40.08
D_2	6.014	6.014	6.014
$D_{0,1}$	0	1.343	1.546
$D_{0,2}$	0	0	0
$D_{1,2}$	0	0	0
$D_{0,1,2}$	0	0	0

approximations for 31 grid points for G & G. Sobol adaptivity needs no such "look-ahead", and therefore at every iteration, all samples are used in the estimate of the mean.

Ultimately the two methods perform similarly, also in terms of the index-set they select (see Figure 9). This suggests Sobol adaptivity is performing almost optimally, without the benefit of "look-ahead".

5.3 *Analysis of parametric variability*

To judge the performance of the surrogate model, it is informative to look at the response surface. Sobol indices indicate that U_{ABL} and U_{rel} are the two influential parameters, therefore to visualize the 3D response we fix $T_{\mathrm{rel}} = 270,\ 290,\ 310\,\mathrm{K}$, and plot surfaces of the other two variables, see Figure 12. The level 4 simplex grid is used – i.e., all available data. The following trends are visible:

- The effect distance increases with increasing U_{rel} and with *decreasing* U_{ABL} and T_{rel}. U_{ABL} is the most important parameter followed by U_{rel}.

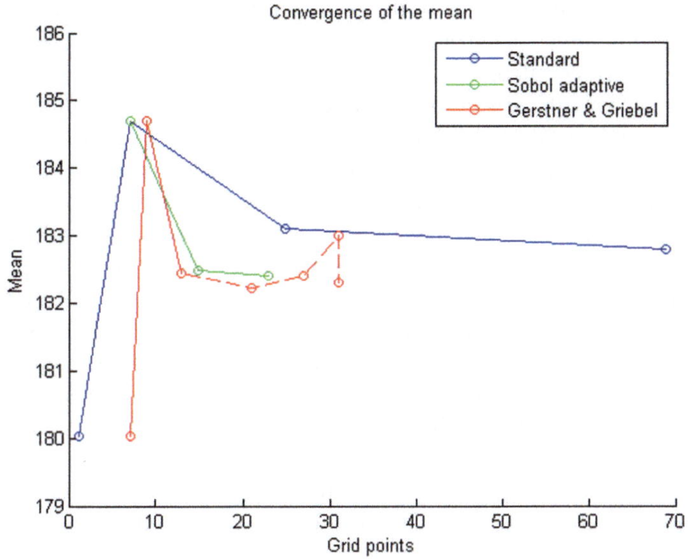

Fig. 11 Convergence of the mean for the standard sparse grid, Sobol-adaptive sparse grid and the Gerstner and Griebel adaptive sparse grid, assuming a uniform input parameter distribution

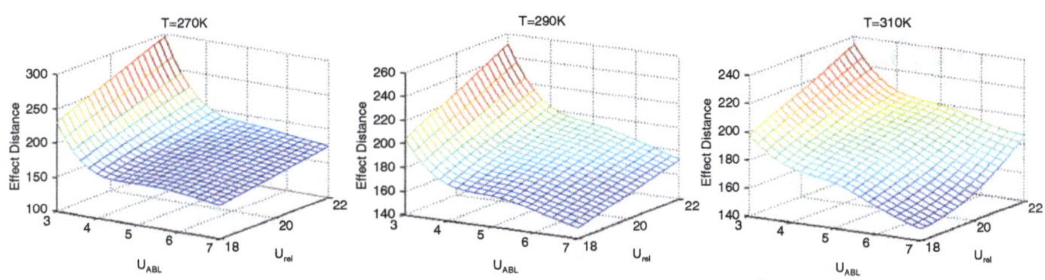

Fig. 12 The effect distance as a function of U_{ABL} and U_{rel} for $T_{\text{rel}} = 270K$, $290K$, and $310K$

- T_{rel} does not have a large effect on its own, but there is an interaction between U_{ABL} and T_{rel}. When both values are low, the effect distance increases substantially more than in the separate contributions. Neither of the adaptive schemes catch this interaction.
- The variation of effect distance with U_{ABL} comes mainly from values between 3 and 4 m/s, where the gradient becomes much steeper. This effect would not have been visible in a linear sensitivity analysis at the nominal conditions.
- The maximum effect distance occurs for $U_{\text{ABL}} = 3m/s$, $U_{\text{rel}} = 22m/s$, and $T_{\text{rel}} = 270K$. The minimum effect distance occurs for $U_{\text{ABL}} = 7m/s$, $U_{\text{rel}} = 18m/s$, and $T_{\text{rel}} = 310K$. The maximum distance therefore is a clear function of the width of the uncertain parameter values. If only maximum distance was of interest, an optimization could have been performed instead.

We turn to study the physics of the problem: in Figure 13 molar concentration of propane at 1 m above the ground is shown, for $U_{\mathrm{ABL}} = 3,\ 5,\ 7\,\mathrm{m/s}$. These fields come without further calculation, from the existing samples. Of course we choose to look at changes with respect to U_{ABL} due to its large Sobol variance.

These plots suggest the behaviour of the gas just upstream of the barrier is critical to the effect distance (ED). The time the propane needs to reach the barrier depends on wind to a large extent (U_{ABL}), and the release velocity of the gas to a lesser extent (U_{rel}). The propane gas has a higher density than air and will settle to the ground – but the speed of settling depends on the concentration of propane. As the air and propane gas mix, this effect will becomes weaker. A lower U_{ABL} gives more time and space for the propane to mix, so that by the time the barrier is reached the mixture has concentration low enough to overcome it. This can be seen in Figure 13 (top), where penetration occurs not at high- or low-concentrations, but off-centre, at intermediate values. Hence, the counter-intuitive result that less wind results in larger ED.

The release temperature plays a role in the gas density too, but as the sensitivity analysis shows, in this case, the variation in density is very small, so that the effect of the release temperature is insignificant. The molar concentrations at the maximum and minimum values of ED are plotted in Figure 14.

6 Conclusions

Sparse grids are an effective way of reducing the effort required for uncertainty propagation through expensive computer simulations. They also allow for simple computation of Sobol indices – a measure of global sensitivity. Isotropic sparse grids are rarely optimal, however, as the influence of different parameters on the output quantity-of-interest can vary by orders of magnitude. Dimension adaptivity should automatically select those variables for which a higher resolution is needed. We have compared two dimension adaptive approaches: that of Gerstner & Griebel (2003) [9], based on a standard error estimator, and our own based on Sobol indices computed on the sparse grid. For the twin requirements of: low number of samples (e.g. $\leq$ 100), and relatively low accuracy (e.g. $\simeq$ 1%), which are common in industrial problems, our method performs better thanks to not requiring a look-ahead error estimate. This is demonstrated for a computationally expensive industrial problem.

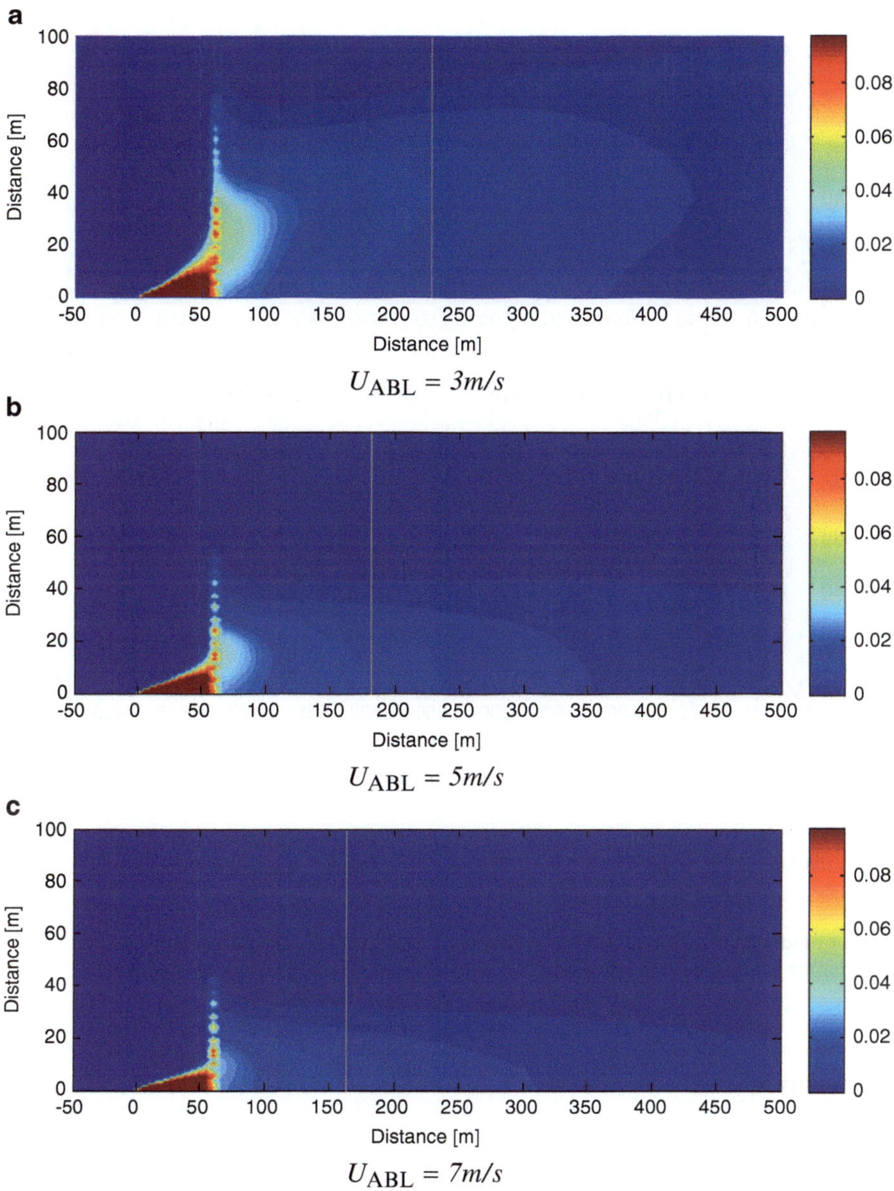

Fig. 13 Molar concentration of propane gas at a height of $1m$ for 3 values of U_{ABL}. Here $T_{\text{rel}} = 290\,\text{K}$ and $U_{\text{rel}} = 20\,\text{m/s}$. The thin gray lines indicate effect distance

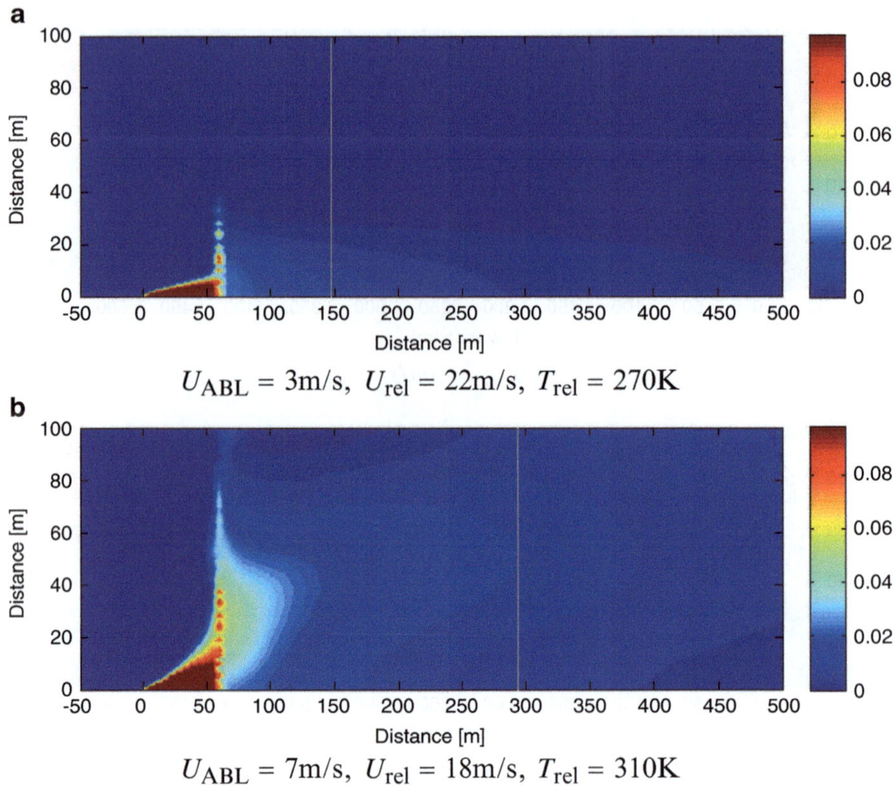

Fig. 14 Molar concentration of propane gas at a height of $1m$ for (a) minimum and (b) maximum effect distance

Acknowledgements Special thanks to Andreas Mack for providing modelling and CFD simulation for the heavy gas release test-case.

References

1. de Baar, J., Dwight, R., Bijl, H.: Improvements to gradient-enhanced Kriging using a Bayesian interpretation. Int. J. Uncertain. Quantif. **4**(3), 205–223 (2013)
2. de Baar, J., Scholcz, T., Dwight, R., Bijl, H.: Exploiting adjoint derivatives in high-dimensional metamodels: can we observe the expected speedup? AIAA J. **53**(5), 1391–1395 (2015)
3. Babuska, I., Nobile, F., Tempone, R.: A stochastic collocation method for elliptic partial differential equations with random input data. SIAM J. Numer. Anal. **45**(3), 1005–1034 (2007)
4. Caflisch, R.E.: Monte carlo and quasi-monte carlo methods. Acta Numer. **7**, 1–49 (1998)
5. Denny, M.: Introduction to importance sampling in rare-event simulations. Eur. J. Phys. **22**(4), 403 (2001)
6. Dwight, R., Marques, S., Badcock, K.: Reducing uncertainty in aeroelastic flutter boundaries with experimental data. International Forum on Aeroelasticity and Structural dynamics, IFASD-2011-71 (2011)

7. Edeling, W., Dwight, R., Cinnella, P.: Improved simplex-stochastic collocation method for higher dimensional uncertainty quantification problems. J. Comput. Phys. (2015)
8. Ganapathysubramanian, B., Zabaras, N.: Sparse grid collocation schemes for stochastic natural convection problems. J. Comput. Phys. **225**, 652–685 (2007)
9. Gerstner, T., Griebel, M.: Dimension–adaptive tensor–product quadrature. Computing **71**(1), 65–87 (2003)
10. Ghanem, R.G., Spanos, P.D.: Stochastic Finite Elements: A Spectral Approach. Dover, New York (1991)
11. Giles, M.B.: Multilevel Monte Carlo path simulation. Oper. Res. **56**(3), 607–617 (2008). doi:10.1287/opre.1070.0496
12. Johnson, L.W., Riess, R.D.: Numerical Analysis. Addison-Wesley, Reading (1982)
13. Kennedy, M., O'Hagan, A.: Bayesian calibration of computer models (with discussion). J. R. Stat. Soc. B. **63**, 425–464 (2001)
14. Liu, X.D., Osher, S., Chan, T.: Weighted essentially non-oscillatory schemes. J. Comput. Phys. **115**(1), 200–212 (1994)
15. Loeven, G.: Efficient uncertainty quantification in computational fluid dynamics. Ph.D. Thesis, TU Delft, Department of Aerodynamics (2010)
16. Lucor, D., Xiu, D., Su, C., Karniadakis, G.: Predictability and uncertainty in CFD. Int. J. Numer. Methods Fluids **43**, 483–505 (2003)
17. Matthies, H.G., Keese, A.: Galerkin methods for linear and nonlinear elliptic stochastic partial differential equations. Comput. Methods Appl. Mech. Eng. **194**(12), 1295–1331 (2005)
18. Metropolis, N., Ulam, S.: The Monte Carlo method. J. Am. Stat. Assoc. **44**(247), 335–341 (1949)
19. Najm, H.N.: Uncertainty quantification and polynomial chaos techniques in computational fluid dynamics. Ann. Rev. Fluid Mech. **41**, 35–52 (2009)
20. Nobile, F., Tempone, R., Webster, C.G.: An anisotropic sparse grid stochastic collocation method for partial differential equations with random input data. SIAM J. Numer. Anal. **46**(5), 2411–2442 (2008)
21. Nobile, F., Tempone, R., Webster, C.G.: A sparse grid stochastic collocation method for partial differential equations with random input data. SIAM J. Numer. Anal. **46**(5), 2309–2345 (2008)
22. Smith, R.C.: Uncertainty Quantification: Theory, Implementation, and Applications, vol. 12. SIAM, Philadelphia (2013)
23. Smolyak, S.: Quadrature and interpolation formulas for tensor products of certain classes of functions. Dokl. Akad. Nauk SSSR **4**, 240–243 (1963)
24. Sobol, I.M.: On quasi-Monte Carlo integrations. Math. Comput. Simul. **47**(2), 103–112 (1998)
25. TANG, G., Eldred, M., SWILER, L.P.: Global sensitivity analysis for stochastic collocation expansion. In: CSRI Summer Proceedings 2009, p. 100. Citeseer (2010)
26. Trefethen, L.N.: Is gauss quadrature better than Clenshaw-Curtis? SIAM Rev. **50**(1), 67–87 (2008)
27. Wikle, C., Berliner, L.: A Bayesian tutorial for data assimilation. Phys. D **230**, 1–16 (2007)
28. Witteveen, J.A., Iaccarino, G.: Simplex stochastic collocation with random sampling and extrapolation for nonhypercube probability spaces. SIAM J. Sci. Comput. **34**(2), A814–A838 (2012)
29. Xiu, D., Hesthaven, J.S.: High-order collocation methods for differential equations with random inputs. SIAM J. Sci. Comput. **27**(3), 1118–1139 (2005)

Index

S. Koziel et al. (eds.), *Simulation-Driven Modeling and Optimization*,
Springer Proceedings in Mathematics & Statistics 153,
DOI 10.1007/978-3-319-27517-8

F

G

H

MIX
Papier aus verantwortungsvollen Quellen
Paper from responsible sources
FSC® C105338

If you have any concerns about our products,
you can contact us on
ProductSafety@springernature.com

In case Publisher is established outside the EU,
the EU authorized representative is:
Springer Nature Customer Service Center GmbH
Europaplatz 3, 69115 Heidelberg, Germany

Printed by Libri Plureos GmbH
in Hamburg, Germany